Vertebrates, Phylogeny, and Philosophy

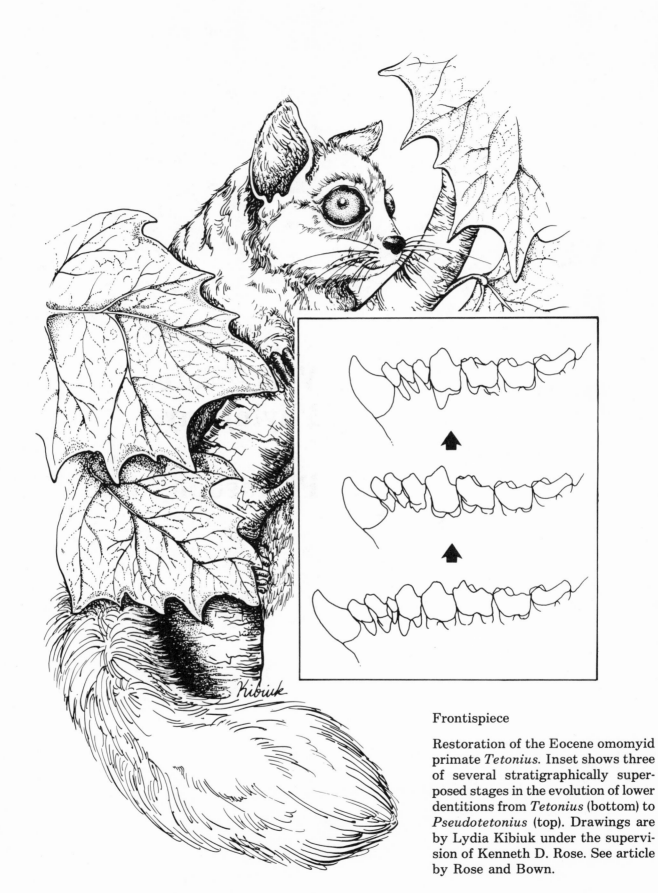

Frontispiece

Restoration of the Eocene omomyid primate *Tetonius*. Inset shows three of several stratigraphically superposed stages in the evolution of lower dentitions from *Tetonius* (bottom) to *Pseudotetonius* (top). Drawings are by Lydia Kibiuk under the supervision of Kenneth D. Rose. See article by Rose and Bown.

Special Paper 3

Vertebrates, Phylogeny, and Philosophy

Edited by
Kathryn M. Flanagan
Jason A. Lillegraven

Contributions to Geology
The University of Wyoming

The following is an example of the preferred mode of citation of papers:

> Rose, K. D., and Bown, T. M., 1986, Gradual evolution and species discrimination in the fossil record, *in* Flanagan, K. M., and Lillegraven, J. A., eds., Vertebrates, phylogeny, and philosophy: Contributions to Geology, University of Wyoming, Special Paper 3, p. 119-130.

Department of Geology and Geophysics
The University of Wyoming
Laramie, Wyoming, U.S.A. 82071-3006

Copyright © 1986
The University of Wyoming
ISBN 0-941570-02-9
Library of Congress Catalog Card Number: 86-50857
Printed in the United States of America

CONTENTS

Contributors vii

Editors' preface ix

Introduction 1
    Laurence M. Gould

George Gaylord Simpson: empirical theoretician 3
    Philip D. Gingerich

The Jurassic "bird" *Laopteryx priscus* re-examined 11
    John H. Ostrom

Brain evolution in Mesozoic mammals 21
    Zofia Kielan-Jaworowska

Origin and transformation of the mammalian stapes 35
    Michael J. Novacek and André Wyss

New Late Cretaceous, North American advanced therian mammals
that fit neither the marsupial nor eutherian molds 55
    William A. Clemens and Jason A. Lillegraven

Paraphyly in *Catopsalis* (Mammalia: Multituberculata) and its biogeographic implications 87
    Nancy B. Simmons and Miao Desui

Competitive exclusion and taxonomic displacement in the fossil record:
the case of rodents and multituberculates in North America 95
    David W. Krause

Gradual evolution and species discrimination in the fossil record 119
    Kenneth D. Rose and Thomas M. Bown

Nycticeboides simpsoni and the morphology, adaptations,
and relationships of Miocene Siwalik Lorisidae 131
    R. D. E. MacPhee and Louis L. Jacobs

The Paleogene record of the rodents: fact and interpretation 163
    Robert W. Wilson

Machaeroides simpsoni, new species, oldest known sabertooth creodont (Mammalia),
of the Lost Cabin Eocene 177
    Mary R. Dawson, Richard K. Stucky, Leonard Krishtalka, and Craig C. Black

Early Eocene artiodactyls from the San Juan Basin, New Mexico,
and the Piceance Basin, Colorado 183
    Leonard Krishtalka and Richard K. Stucky

Early Eocene rodents from the San Jose Formation, San Juan Basin, New Mexico 197
   Kathryn M. Flanagan

Fossil vertebrates from the latest Eocene, Skyline channels, Trans-Pecos Texas 221
   John A. Wilson and Margaret S. Stevens

Systematics and evolution of *Pseudhipparion* (Mammalia, Equidae) from the late Neogene of the
   Gulf Coastal Plain and the Great Plains 237
   S. David Webb and Richard C. Hulbert, Jr.

Species longevity, stasis, and stairsteps in rhizomyid rodents 273
   Lawrence J. Flynn

The late Miocene radiation of Neotropical sigmodontine rodents in North America 287
   Jon Alan Baskin

Very hypsodont antelopes from the Beglia Formation (central Tunisia),
   with a discussion of the Rupicaprinae 305
   Peter Robinson

Faunal provinces and the Simpson Coefficient 317
   John J. Flynn

Evolutionary epicycles 339
   Jon Marks

The evolutionary synthesis today: an essay on paleontology and molecular biology 351
   Everett C. Olson and Clifford F. Brunk

CONTRIBUTORS

Jon Alan Baskin
    Department of Geosciences, Texas A&I University, Kingsville, Texas 78363

Craig C. Black
    Office of the Director, Natural History Museum of Los Angeles County, Los Angeles, California 90007

Thomas M. Bown
    Paleontology and Stratigraphy Branch, U.S. Geological Survey, Denver, Colorado 80225

Clifford F. Brunk
    Department of Biology, University of California, Los Angeles, California 90024

William A. Clemens
    Department of Paleontology, University of California, Berkeley, California 94720

Mary R. Dawson
    Section of Vertebrate Fossils, Carnegie Museum of Natural History, Pittsburgh, Pennsylvania 15213

Kathryn M. Flanagan
    Department of Geology and Geophysics, The University of Wyoming, Laramie, Wyoming 82071

John J. Flynn
    Department of Geological Sciences, Rutgers University, New Brunswick, New Jersey 08903

Lawrence J. Flynn
    Department of Anthropology, Peabody Museum, Harvard University, Cambridge, Massachusetts 02138

Philip D. Gingerich
    Museum of Paleontology, The University of Michigan, Ann Arbor, Michigan 48109

Laurence M. Gould
    Department of Geosciences, The University of Arizona, Tucson, Arizona 85721

Richard C. Hulbert, Jr.
    Department of Zoology, The University of Florida, Gainesville, Florida 32611

Louis L. Jacobs
    Department of Geological Sciences, Southern Methodist University, Dallas, Texas 75275

David W. Krause
    Department of Anatomical Sciences, Health Sciences Center, State University of New York, Stony Brook, New York 11794

Zofia Kielan-Jaworowska
    Zaklad Paleobiologii, Polska Akademia Nauk, Warszawa, Poland

Leonard Krishtalka
    Section of Vertebrate Fossils, Carnegie Museum of Natural History, Pittsburgh, Pennsylvania 15213

Jason A. Lillegraven
    Departments of Geology/Geophysics and Zoology/Physiology, The University of Wyoming, Laramie, Wyoming 82071

R. D. E. MacPhee
    Department of Anatomy, Duke University Medical Center, Durham, North Carolina 27710

Jon Marks
    Department of Genetics, University of California, Davis, California 95616

Miao Desui
    Department of Geology and Geophysics, The University of Wyoming, Laramie, Wyoming 82071

Michael J. Novacek
    Department of Vertebrate Paleontology, The American Museum of Natural History, New York, New York 10024

Everett C. Olson
    Department of Biology, University of California, Los Angeles, California 90024

John H. Ostrom
    Department of Geology and Geophysics, Peabody Museum of Natural History, Yale University, New Haven, Connecticut 06511

Peter Robinson
    Museum, University of Colorado, Boulder, Colorado 80309

Kenneth D. Rose
    Department of Cell Biology and Anatomy, Johns Hopkins University School of Medicine, Baltimore, Maryland 21205

Nancy B. Simmons
    Department of Paleontology, University of California, Berkeley, California 94720

Margaret S. Stevens
    Department of Geology, Lamar University of Orange County, Orange, Texas 77710

Richard K. Stucky
    Section of Vertebrate Fossils, Carnegie Museum of Natural History, Pittsburgh, Pennsylvania 15213

S. David Webb
    Florida State Museum, The University of Florida, Gainesville, Florida 32611

John A. Wilson
    Department of Geological Sciences, The University of Texas at Austin, Austin, Texas 78758

Robert W. Wilson
    Museum of Natural History and Department of Systematics and Ecology, The University of Kansas, Lawrence, Kansas 66045

André Wyss
    Department of Geological Sciences, Columbia University, New York, New York 10027

EDITORS' PREFACE

Dr. George Gaylord Simpson, one of the most important evolutionary biologists of the Twentieth Century, was born on June 16, 1902 and died on October 6, 1984. His contributions to science include not only a modern synthesis of evolutionary thought, but original research on anthropology, mammalogy, paleontology, general biology, and statistics. His prolific writings were intended for scientific and nonscientific communities alike. He helped and encouraged many who now work in the fields of paleontology and evolutionary biology. Contributors to this book dedicate their efforts as tribute to his memory.

Included authors are colleagues, former students, and friends of Dr. Simpson's. They represent but a few of the people he would have included in these categories. The book is intended to suggest only a sampling of the diversity of George Gaylord Simpson's impact on present vertebrate paleontology, from its most senior to its very junior participants.

Ms. Flanagan's letter of invitation entreated the following from potential authors: "In the spirit of Dr. Simpson's own writings, we encourage imaginative contributions that would be just a little different from items expected in a regular scientific journal." The title of the volume (*Vertebrates, Phylogeny, and Philosophy*) reflects that request. Though individual articles deal almost exclusively with fossil mammals, emphases cross the spectrum of evolutionary biology, including systematic paleontology, considerations of adaptation, ontogeny, analyses of evolutionary tempo and mode, biogeographic procedure, and paleogeography. Philip Gingerich's contribution stresses the crucial importance of solid empirical research to the foundations upon which theoretical/philosophical writings should be based. Mesozoic and Cenozoic taxa are considered, and two articles discuss the modern union of molecular biology, genetics, and paleontology. Most articles benefited directly from the pioneering writings of George Simpson, yet the breadth of concerns of this volume covers only a small fraction of the interests exhibited in his lifetime of evolutionary research.

Kathryn Flanagan served as principal correspondent with authors and reviewers. Jason Lillegraven had principal responsibility for manuscript editing and considerations of production.

We take this opportunity to thank the thirty-two authors for their contributions. Similarly, more than fifty individuals served as unpaid reviewers, and we give our most sincere thanks for their generosity of time and effort. Also, we thank Linda E. Lillegraven for creating the cover design.

Laramie, Wyoming
July 7, 1986

Jason A. Lillegraven
Kathryn M. Flanagan

George and Anne Roe Simpson
Spring of 1984

Introduction

LAURENCE M. GOULD *Department of Geosciences, University of Arizona, Tucson, Arizona 85721*

I find a proper introduction of George Gaylord Simpson in the citation for the honorary degree of Doctor of Science which the University of Arizona awarded him in 1982.

> "Simpson is not only a paleonologist rare amongst his colleagues, he has made himself master of all the disciplines involved in the synthetic theory of evolution and particularly of taxonomy which makes him a great biologist. He is not only a biologist but a man of science with the widest horizon and experience."

There is not adequate space to even outline Simpson's magnificent scientific achievements. He did within 50 years travel to every continent and every state, usually accompanied with his partner-wife, Anne Roe Simpson, who not only made significant discoveries of vertebrate fossils but read and critiqued his writing. Surely a heroic task, for Simpson authored some 800 books and articles.

An event destined to have great influence on the life of Simpson occurred in 1831. In that year the 22 year old Charles Darwin enlisted on H.M.S. Beagle to participate in a voyage that took four years to encircle the globe. Early on that voyage a stop was made in lower South America which enabled Darwin to explore parts of Patagonia and bring back a significant collection of fossils. One century and one year later the youthful Simpson followed in Darwin's Patagonian steps and beyond them, bringing back the most important collection of vertebrate fossils yet found there. "Attending Marvels" is the delightful and informative record of that trip and remains one of his most popular books.

To a large extent Simpson's life and achievements have paralleled those of Darwin who became his idol. Indeed during the years of his teaching at the University of Arizona Simpson held annual parties for his students to celebrate Darwin's birthday, February 12, 1809—the same as that of Abraham Lincoln.

I doubt that anyone else ever knew as much about Darwin as Simpson did. He once told me that he had read everything Darwin had ever written. A mammoth achievement indeed and perhaps a unique one. I like to think of George Gaylord Simpson as a twentieth century Darwin.

To learn more about this remarkable man I recommend his autobiography which is one of the most interesting biographies I have ever read. The following lines from it suggest the quality of George Gaylord Simpson.

> "There is currently much discussion about the motivation of scientists and other professionals. The motivations are numerous and usually complex even for any one individual. Those of us so oriented ponder our own motivations. In the main, mine is surely that I was born with or somehow very early acquired an uncontrollable drive to know and understand the world in which I live. I have not aspired to promotion and office, on the contrary I have refused opportunities for them. I have received many tokens of recognition such as honorary degrees, medals and other awards, honorary memberships in learned societies, and other such things. I could not quarrel with anyone who felt that I have received more recognition than I have deserved, and I cannot deny that I enjoyed it. But I have never done anything motivated by desire for recognition and I have never sought it. I have suffered about as much over mistakes in my pursuit of knowledge and understanding as I have rejoiced in the apparent successes.
>
> Outside of my profession I have had a great deal of pain and emotional sorrow. I do not regret that, but rather value it as a part of the complete fullness of life. It is part of my good fortune that the pain and sorrow are more than overbalanced by pleasure and delight. Here in the closing years of my life I have a loving and fascinating wife and family and many good friends. I am enjoying more than ever before, the company of students and colleagues at the University with which I am still connected."

Those of us who knew George Gaylord Simpson as one of his colleagues or one of his students at the University of Arizona will always cherish these words.

George Gaylord Simpson: empirical theoretician

PHILIP D. GINGERICH *Museum of Paleontology, The University of Michigan, Ann Arbor, Michigan 48109*

ABSTRACT

George Gaylord Simpson published some 21 books and monographs, 79 notes, and 271 research articles from 1925 through 1971. This primary literature totals 371 titles and 12,656 pages; 4,451 pages (35%) are devoted to mammals, and 2,363 pages (19%) are devoted to evolution. Simpson published primarily on Mesozoic and Paleocene mammals, but he also contributed significantly to the study of Eocene and Pleistocene mammals as well. Early work was concentrated on North American faunas, but interest later shifted to South America. Simpson published some 224 titles and 5,785 pages of empirical work, much of it during the first 20 years of his career. He published 109 titles and 6,675 pages of theoretical work. Research collections and museum support were important throughout Simpson's working life. The concentration of empirical research early in Simpson's career, with later emphasis on theoretical questions, affirms that observation and experience are important in generating ideas of lasting value.

INTRODUCTION

Physics represents what most people think science should be. When I was an undergraduate in the late nineteen-sixties, many geology departments in universities across the country changed their names, adding geophysics to their titles to reflect expansion in this direction. My department went even farther, changing its name to "Department of Geological and Geophysical Sciences" —lest anyone think that geology, even geophysics, might not be science. The extended name was a nuisance from the beginning for anyone trying to fit it on an envelope, and I suspect that many still wonder at the department's vision of itself every time they address a letter.

A little more thought about old Alma Mater is appropriate here, because adding "geophysics" and "science" to the department's name proved not to be enough. The professor of paleobotany was allowed to retire without replacement. Then a half-endowed professorship in vertebrate paleontology was allowed to lapse. Finally, last year, the department decided to give away its paleontology research collections. The department was founded by a vertebrate paleontologist and maintained an outstanding national and international reputation, indeed distinguished tradition, in vertebrate paleontology for most of a century. In discussing "deaccession" of the vertebrate collections, I and other members of an assembled advisory committee protested that the department was now making a precipitous decision precluding later participation in an important scientific endeavor—it is impossible to rebuild, once lost, the libraries and research collections necessary for quality research in vertebrate paleontology. The department responded, "Oh, we may be giving away the research collections, but we're not necessarily giving up paleontology. If we were to hire a professor of vertebrate paleontology, we would undoubtedly hire an outstanding theoretician, one who wouldn't require research collections—someone like George Gaylord Simpson."

Simpson, architect of the modern evolutionary synthesis (with Dobzhansky, Mayr, and others), is undoubtedly one of the leading paleontologists of the twentieth century in terms of his contribution to important issues in biogeography, evolution, systematics, and other fields. But it is, I think, a misrepresentation to promote George Gaylord Simpson as a theoretical paleontologist whose genius developed independently of collection-oriented empirical research. Simpson made many collecting expeditions to Argentina, Brazil, Florida, Montana, and New Mexico, doing original field work as long as his health permitted. He wrote more than two hundred papers describing new fossils collected on these and other expeditions. Analysis of Simpson's career is warranted to better understand the role empirical research plays in generating paleontological and evolutionary theory.

SIMPSON'S RESEARCH PUBLICATIONS

Publications provide one source of information about a person's professional career. Simpson was a proficient writer, publishing some 667 titles and 19,934 pages during the years 1925 through 1971 (Table 1). All were not written alone. Simpson's highly successful biology textbook *Life: An Introduction to Biology*, for example, involved collaboration with C. S. Pittendrigh and L. H. Tiffany on the first edition, and W. S. Beck on the second. Some titles and many pages included in Simpson's bibliography are reprints and translations of work published previously. However, Simpson was the sole author of the vast majority of titles, and the principal author on virtually all the rest. His publications undoubtedly reflect both his interests and his prodigious productivity.

Simpson's professional writing was accomplished first as a graduate student in the Peabody Museum at Yale

TABLE 1. PUBLICATIONS BY GEORGE GAYLORD SIMPSON FROM 1925 TO 1971.

Genre	Titles	Pages
Abstracts	14	28
Bibliography (G.G.S.)	1	6
Books and monographs	21	7,149
Books (edited)	1	557
Books (popular)	1	295
Book reviews	148	276
Comments	6	16
Forewords	4	24
Letters	15	27
Notes	79	218
Obituaries	14	113
Popular articles	38	207
Reprinted books and articles	39	5,562
Research articles	271	5,289
Secretarial contributions	14	164
Sound recording	1	3
Totals:	667	19,934

University, then as a professional paleontologist at the American Museum of Natural History, the Museum of Comparative Zoology at Harvard, and finally at the University of Arizona. Simpson continued to write long beyond 1971, of course, but I have limited analysis of his publications to the years 1925-1971—the span of his regular working career. Simpson's bibliography through 1971, the source of information for this analysis, was published by Hecht and others (1972).

Published titles and pages do not measure the importance of a writer's work. Some subjects are more easily serialized. Some journal pages are shorter than others. Some authors have access to the large blocks of time necessary to write books and monographs, while others suffer constant interruption. It would be difficult to compare the volume of Simpson's writing to that of other paleontologists without making corrections for subject, style, and the demands of other responsibilities. The value of this exercise lies not in comparing Simpson's productivity with that of other authors, nor even in comparing early Simpson with later Simpson. My goal is rather to characterize Simpson's changing interests, and to examine the relationship of his empirical and theoretical work.

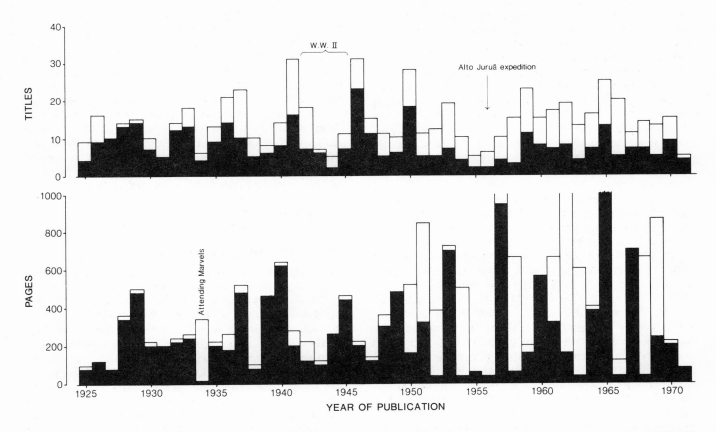

Figure 1. Titles and pages published by George Gaylord Simpson from 1925 through 1971. Open histograms represent all 667 titles and 19,934 pages, including reprints and translations, listed in Hecht and others (1972). Solid histograms are 371 titles and 12,656 pages considered here to represent primary scientific literature. Simpson served in U.S. Army during World War II, from December 3, 1942 until October 7, 1944. He was seriously injured by a falling tree, and had to be evacuated from a remote part of the Amazon Basin of Brazil during 1956 Alto Juruá expedition. *Attending Marvels* is a popular book based on Simpson's 1930-1931 expedition to Patagonia in Argentina. Karen Klitz drew this and following figures.

PRIMARY AND SECONDARY LITERATURE

The pattern of Simpson's productivity over time is shown graphically in Figure 1. The two highest peaks in published titles lie on either side of his two-year term of military service (1942-1944) during World War II. Military service itself is reflected by a low valley in published titles for these years. A second low valley occurs in 1955-1956, presumably reflecting much of a sabbatical year (1954-1955) spent in Brazil and Argentina, work on the textbook *Life*, several months of field work in Brazil in 1956, and, only incidentally, Simpson's accident on the Alto Juruá at the end of the 1956 expedition.

The most important publications listed in Table 1 are the 21 original scientific books and monographs (including substantially revised second editions), 79 notes, and 271 research articles. Also original but of lesser importance are the abstracts, a personal bibliography, an edited book, a popular book (*Attending Marvels*), numerous book reviews, several comments, four forewords in books by others, letters to editors, obituaries, popular articles (principally in *Natural History*), secretarial contributions (for the Explorer's Club and Society of Vertebrate Paleontology), and a sound recording. Reprinted books and articles include those originally published in English and translated with little modification into other languages. It is sometimes difficult to draw a line between comments, letters, and notes—shorter contributions of substantial original or summary value are here identified as notes and regarded as primary literature. Notes also include a number of brief reviews contributed to encyclopedias; these did not, in the end, fit comfortably into established analytical categories, and they might better have been listed separately from the start.

Eliminating abstracts and other secondary publications, the remaining data base includes 371 scientific books and monographs, research articles, and notes, totaling 12,656 printed pages (Table 2 and solid histograms in Fig. 1). These can be divided by subject, by geological age, by geographic location, and by theoretical content, yielding patterns offering insight into the course of Simpson's research career.

MAMMALS AND EVOLUTION

Simpson is most widely known for his work on evolution. As shown in Table 2, 47 of his 371 primary publications (13%), and 2,363 of 12,656 printed pages (19%) were devoted to evolution. For comparison, 198 primary publications (53%) and 4,451 printed pages (35%) were devoted to the morphology and systematics of living and fossil mammals, roughly twice the number of titles and pages devoted to evolution. The large number of pages published on general biology reflects two editions of the textbook *Life*. In terms of published pages, biogeography, quantitative methods, and systematics appear to have been next most important to Simpson.

Figure 2 illustrates the temporal pattern of Simpson's work on mammals (solid histograms) and evolution (shaded areas). Work on the morphology and systematics

TABLE 2. PRIMARY PUBLICATIONS BY GEORGE GAYLORD SIMPSON, 1925-1971, ORGANIZED BY SUBJECT.

Subject	Titles	Pages
Amphibians	1	2
Biogeography	17	621
Biology (general)	8	1,812
Biostratigraphy	13	137
Birds (penguins)	8	201
Ethnography	2	477
Evolution	47	2,363
Exploration	1	2
Faunas	1	11
Fishes	1	6
General (misc.)	9	80
Geology	3	82
History of paleontology	7	122
Mammals	198	4,451
Paleontology (general)	4	219
Philosophy	11	104
Quantitative methods	7	900
Reptiles	14	181
Systematics (general)	15	823
Techniques	1	6
Variation	3	56
Totals:	371	12,656

of living and fossil mammals dominated the first fifteen years of his professional career. Simpson's book *Quantitative Zoology,* with Anne Roe, was published in 1939, and his book-length ethnographic monograph, *Los Indios Kamarakotos,* was published in 1940. Publication on evolution began with a 1941 article on the role of the individual, followed by Simpson's classic *Tempo and Mode in Evolution,* completed in 1942 and published in 1944. *Principles of Classification and a Classification of Mammals* was also completed in 1942, but publication was delayed until 1945. This work is certainly evolutionary in perspective, but more strictly systematic than evolutionary in subject; hence it is not shaded in Figure 2. Other publications on evolution include *The Meaning of Evolution* (1949), *The Major Features of Evolution* (1953), *This View of Life* (1964), a revised edition of *The Meaning of Evolution* (1967), and *Biology and Man* (1969). *This View of Life* and *Biology and Man* are (largely) collections of previously published essays on evolution, and arguably they could have been classified as secondary reprints.

MESOZOIC AND CENOZOIC MAMMALS

Simpson is best known for his work on Mesozoic and early Cenozoic mammals. He published 37 titles and 902 pages on Mesozoic mammals, 38 titles and 1,384 pages on Paleocene mammals, and 32 titles and 561 pages on Eocene mammals (Table 3). Simpson published extensive-

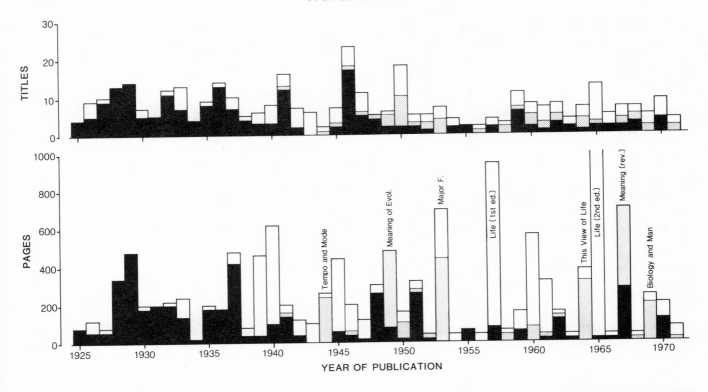

Figure 2. Titles and pages on mammals (solid histograms) and evolution (shaded areas) published by George Gaylord Simpson from 1925 through 1971, outlined against a silhouette of all primary literature published during this interval. Titles of books on evolution and two editions of textbook *Life: An Introduction to Biology* also are shown.

TABLE 3. PRIMARY PUBLICATIONS ON MAMMALS BY GEORGE GAYLORD SIMPSON, 1925-1971, ORGANIZED BY AGE.

Age	Titles	Pages
Recent	3	32
Pleistocene	19	583
Pliocene	1	4
Miocene	5	137
Oligocene	14	200
Eocene	32	561
Paleocene	38	1,384
Mesozoic	37	902
Totals:	149	3,803

ly on Pleistocene mammals, but relatively little on those of the Oligocene, Miocene, and Pliocene.

Simpson's doctoral dissertation on American Mesozoic mammals was completed in 1926, and a revised version was published in 1929. Simpson was given access to Mesozoic mammals collected by the American Museum's Central Asiatic Expeditions (published in part with W. K. Gregory in 1926), and he spent the academic year 1926-1927 in London studying European and African specimens in the British Museum of Natural History, publishing a monograph on British Museum specimens in 1928. By the end of 1928 Simpson could claim to have studied all of the Mesozoic mammals known at the time. This early emphasis shows up clearly in Figure 3, where Mesozoic mammals dominate Simpson's first five years of publication.

Simpson began work at the American Museum in 1927 and initiated work on Paleocene mammals immediately, describing Barnum Brown's small Paskapoo fauna in 1927, and the Bear Creek fauna in 1928 and 1929. Simpson's major paper of 1929 was a revision of Paleocene multituberculates published with Walter Granger. He initiated field work in the early Paleocene of New Mexico in 1929. Simultaneously, Simpson worked on late Tertiary and Pleistocene mammals of Florida. Field work in Patagonia (Argentina) began in 1930 to clarify the beginning of the age of mammals in South America. Two monographs on South American Paleocene (and Eocene) faunas were eventually published in 1948 and 1967.

In 1932 Simpson was asked to write a monograph on the U.S. National Museum collection of Paleocene mammals from the Crazy Mountain Field in Montana, and the summers of 1932 and 1935 were spent in the field in Montana. In connection with this work, Simpson published a three-part revision of the type Tiffanian land-mammal age (late Paleocene) fauna based on American Museum collections from Colorado.

Simpson's Eocene research was never so clearly focused. He published an 87-page study on the Eocene

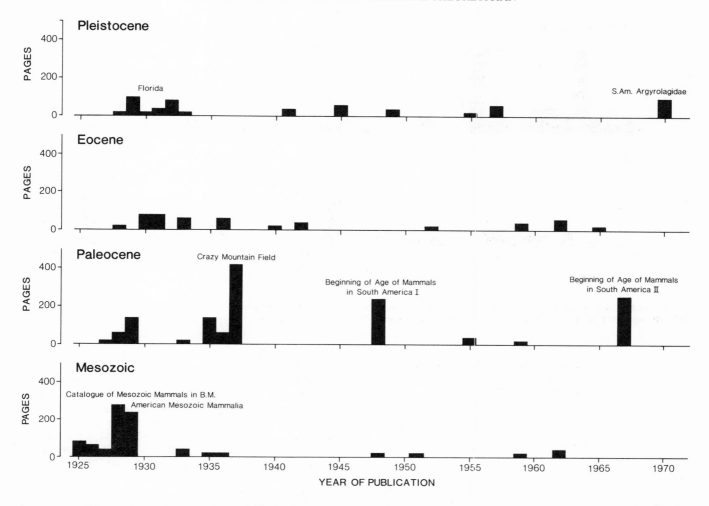

Figure 3. Titles and pages on Pleistocene, Eocene, Paleocene, and Mesozoic mammals (solid histograms) published by George Gaylord Simpson from 1925 through 1971. Note progression from early work on Mesozoic mammals, including two important monographs, to work on Paleocene mammals (three important monographs), and Eocene mammals (monographs on *Beginning of Age of Mammals in South America* also include Eocene faunas). Simpson's early research on Pleistocene mammals was concentrated in Florida. Later work on Pleistocene mammals included South American Argyrolagidae.

edentate *Metacheiromys* in 1931, and the monographs on the beginning of the age of mammals in South America include Eocene faunas, but the rest of Simpson's Eocene work, like that on the Oligocene, Miocene, and Pliocene, consists of short articles and notes.

Pleistocene research began, as noted above, in Florida, and virtually all of Simpson's Pleistocene work from 1928 through 1933 was concentrated in Florida. Jaguar bones and footprints took him to Tennessee in 1941, and peccaries to Missouri in 1946. Other major studies, on tapirs, mastodonts, and argyrolagids were all based on South American fossils.

NORTH AMERICAN AND SOUTH AMERICAN MAMMALS

Most of Simpson's research on fossil mammals concerned those of the Western Hemisphere. He published some 87 titles and 2,014 pages on the mammals of North America, and 56 titles and 1,521 pages on the mammals of South America (Table 4). As outlined above, and illustrated in Figure 4, intensive study of North American faunas preceded a similar effort in South America. Simpson's first paper on South American mammals, published in 1928, concerned the affinities of Polydolopidae—marsupials with a convergently multituberculate-like dentition. This study and an early review of post-Mesozoic marsupials undoubtedly motivated further interest in Paleocene and Eocene mammals of South America; field work began in 1930, yielding a steady stream of research articles and the 1948 and 1967 monographs on the beginning of the age of mammals in South America.

Simpson's principal study of African mammals was a review of Tertiary lorisiform primates published in 1967. He published one article in 1940 on Antarctica as a faunal migration route (Antarctic mammals were, of course, unknown at the time—when found, interestingly, the first were polydolopids like those that drew Simpson to South America). Most of Simpson's papers on

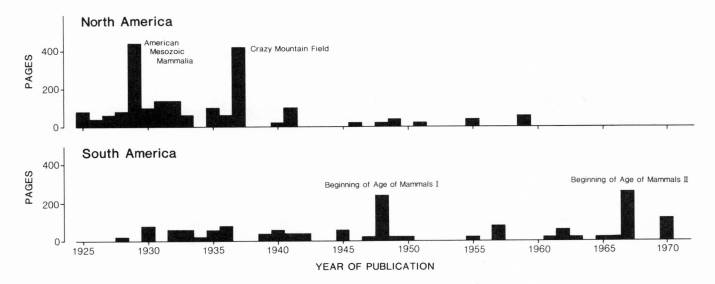

Figure 4. Titles and pages on North American and South American mammals (solid histograms) published by George Gaylord Simpson from 1925 through 1971. Note shift in emphasis from North American to South American faunas.

TABLE 4. PRIMARY PUBLICATIONS ON MAMMALS BY GEORGE GAYLORD SIMPSON, 1925–1971, ORGANIZED GEOGRAPHICALLY.

Continent	Titles	Pages
Africa	4	33
Asia	9	122
Australia	2	30
Europe	8	283
North America	87	2,014
South America	56	1,521
Totals:	166	4,003

Asian mammals concerned Cretaceous remains from Mongolia.

Two papers on extant monotremes (one on the dentition of the platypus, and one on the ear region of the platypus and echidna), and one on historical zoogeography constitute Simpson's entire publication record on Australian mammals (the latter included here under biogeography; several other papers dealt with penguins). Most of Simpson's papers on European fossils concerned Mesozoic mammals, but he reviewed Paleocene and early Eocene mammals in 1929 and 1936, and added short articles on French *"Menatotherium" (Plesiadapis)* and British *Hyracotherium* in 1948 and 1952.

SIMPSON AND THEORY IN PALEONTOLOGY

Simpson published some 224 titles and 5,785 pages of empirical work, most descriptive and systematic, although a few papers were functional and more interpretive. He published 109 titles and 6,675 pages of theoretical and synthetic work, beginning with his first classification of mammals in 1931 and including two editions of *Quantitative Zoology* in 1939 and 1960, *Tempo and Mode in Evolution* in 1944, *The Principles of Classification and a Classification of Mammals* in 1945, two editions of *The Meaning of Evolution* in 1949 and 1967, *The Major Features of Evolution* in 1953, two editions of *Life: An Introduction to Biology* in 1957 and 1965, *Principles of Animal Taxonomy* in 1961, *This View of Life* in 1964, *The Geography of Evolution* in 1965, and *Biology and Man* in 1969.

The pattern and proportion of empirical and theoretical work in Simpson's career is shown in Figure 5. Fifty-three titles and 1,295 pages of primary empirical work were published before his first theoretical contribution. One hundred twenty-one titles and 2,806 pages of primary empirical work were published before *Quantitative Zoology* in 1939, and 142 titles and 3,560 pages were published before *Tempo and Mode in Evolution* in 1944.

Two stages are evident in Simpson's professional career. During the first empircal stage, lasting some twenty years from 1925 through about 1944, Simpson averaged seven titles and 178 pages of primary empirical work per year. During the second and largely theoretical stage, from 1945 through 1971, Simpson still averaged three titles and 82 pages of empirical work per year. The choice of 1944 as a division between the two stages is, of course, arbitrary, reflecting both the end of Simpson's active military service and publication of *Tempo and Mode*. It is noteworthy that for 47 years, from 1925 through 1971, Simpson published at least one primary empirical study every year except 1958 (many book reviews were published in 1957 and 1958, following the Alto Juruá accident).

It is fair to say that the comparative work required during the first, predominantly empirical, stage of Simp-

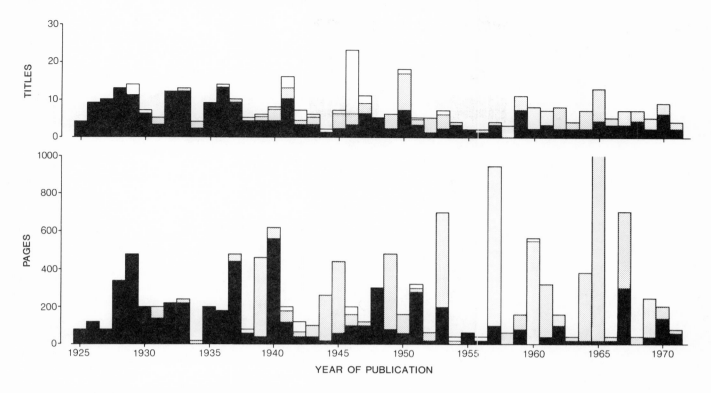

Figure 5. Titles and pages of empirical (solid histograms) and theoretical work (shaded areas) published by George Gaylord Simpson from 1925 through 1971, outlined against a silhouette of all primary literature published during this interval. Note initiation of theoretical work in 1931, and shift in emphasis from predominantly empirical to predominantly theoretical work in early 1940s.

son's career could not have been carried out without access to museum collections for comparative purposes, and conservation of reference material would not have been possible without museum support. The second, predominantly theoretical, stage of Simpson's career also made use of collections and libraries in research museums (first at the American Museum, and later the Museum of Comparative Zoology). Museums are a resource essential for research in natural science, and research is their most important function.

It is popular in some paleontological circles to argue that theory develops independently of observation, that science is deductive rather than inductive, and that compilation of empirical catalogues in hopes of generalization is effort both misdirected and nonscientific. Simpson's research career stands as evidence to the contrary. His career began with many years of empirical collecting and cataloging. From this grew integration, synthesis, and theory. Simpson's citation on presentation of the Geological Society's Penrose Medal credits "a foundation of hard work and a dedication to the principle that big thoughts grow from many little facts" (Jepsen, 1953).

No one would accuse Simpson of counting pebbles and describing colors in a gravel pit, and yet much of his work involved both counting and describing in vacua so large (Mesozoic and early Cenozoic mammals in general, and those of South America in particular) that the task cannot have been approached with particular views or preconceived theories in mind. The work was undertaken to provide basic information necessary for integration and synthesis, making theory possible. If South American mammals are interesting to paleontologists outside South America, it is largely because Darwin, Ameghino, Simpson, and others, have made them so.

Returning to Alma Mater, wouldn't it be nice to hire a theoretician like Simpson, . . . someone who wouldn't require museum resources or research collections.

REFERENCES CITED

Hecht, M. K., Schaeffer, B., Patterson, B., Frank, R. van, and Wood, F. D., 1972, George Gaylord Simpson: his life and works to the present, in Dobzhansky, T., Hecht, M. K., and Steere, W. C., eds., Evolutionary biology: New York, Appleton-Century-Crofts, Volume 6, p. 1-29.

Jepsen, G. L., 1953, Presentation of Penrose Medal to George Gaylord Simpson: Geological Society of America, Annual Report for 1952, Proceedings Volume, p. 51-53.

MANUSCRIPT RECEIVED JUNE 6, 1986
MANUSCRIPT ACCEPTED JUNE 6, 1986

The Jurassic "bird" *Laopteryx priscus* re-examined

JOHN H. OSTROM *Department of Geology and Geophysics, Peabody Museum of Natural History, Yale University, New Haven, Connecticut 06511-8161*

ABSTRACT

The holotype specimen of *Laopteryx prisca* (YPM 1800), described by Marsh in 1881 as the cranium of a North American Jurassic bird, has been cited several times by subsequent authors as of unlikely avian identity or a possible pterosaur. Careful re-examination and illustration of the infrequently examined type specimen, and a review of the characters noted by Marsh, refute the avian identification and confirm later suspicions. A single tooth originally associated with the cranium, long thought to be lost, is also re-studied and illustrated, and is judged to be unrelated to *Laopteryx*.

INTRODUCTION

George Simpson's Ph.D. dissertation, revised and published (1929) under the title "American Mesozoic Mammalia," became a classic almost the day it was published and it still stands today as one of his most important works. Much of the material described there is from Yale's Quarry Nine—the famed mammal quarry in the Jurassic Morrison Formation at Como Bluff, Wyoming—collections made for O.C. Marsh in the 1870s and 80s. Today, a full century later, that collection from Quarry Nine remains the largest and most important assemblage of Jurassic mammals in the world.

Among the hundreds of specimens recovered from that site there was one that attracted George's attention that had been described by Marsh (1881) as avian: *Laopteryx priscus* (YPM 1800; Figs. 1-4). It attracted special attention because when originally described it was only the third known specimen of a Jurassic bird (following discovery of the London [1861] and Berlin [1877] specimens of *Archaeopteryx*). That was still the situation in the 1920s when George examined YPM 1800, and *Laopteryx* remains today as one of less than a dozen purported avian specimens of Jurassic age. The specimen consists of a partial, crushed braincase that displays very few certifiable structures and has remained of uncertain identity (Simpson, 1926a; Lambrecht, 1933; Brodkorb, 1971, 1978). Although Marsh (1881) was unequivocal in his announcement of its avian identity, George was less than certain in his (1926a) Quarry Nine paper.

> "The status of *Laopteryx* is less certain and further study is necessary to affirm or deny its avian nature in a definitive manner."
> Simpson, 1926a, p. 4.

During a conversation with George many years ago, he admitted to me in a passing remark that he had never been able to satisfy himself that YPM 1800 was in fact a bird.

The name *Laopteryx*, perhaps not widely known, is well-known to paleo-ornithologists as that of a doubtful Jurassic bird nearly contemporaneous with *Archaeopteryx*. As such, it is of great importance, and it is surprising that so little attention has been given to this specimen. Also surprising is the fact that, despite having been cited a number of times, it has not been described further since Marsh's (1881) original (and now inadequate) description, nor has it been well illustrated. These deficiencies are corrected here—in memory of George.

STRATIGRAPHIC AND LOCALITY DATA

YPM 1800 was collected during August of 1880, possibly by Marsh's collector William Reed, and received in the Yale College Museum (precursor of Yale's present Peabody Museum) on September 18th, 1880. The locality was recorded as "Quarry 9" (which is situated in SW ¼ sec. 12, T. 22 N., R. 77 W.), Albany County, Wyoming. The Quarry is still recognizable low on the north-facing ridge slope at the east end of Como Bluff just west of the Fish Hatchery Road that extends north from U.S. Highway 30. The stratigraphic level occurs in a lenticular grey sandy claystone sequence near the middle of the Morrison Formation (Kimmeridgian/Portlandian) of Late Jurassic age (see Loomis, 1901; Simpson, 1926b; Ostrom and McIntosh, 1966). Immediately overlying the bone producing level is a persistent, one meter resistant sandstone layer (No. 24 of Loomis' [1901] section). At the Quarry 9 site, this is approximately 25m (80 ft) below the ridge capping Cloverly Formation (Dakota Sandstone of earlier literature).

Quarry 9 was re-opened in 1968-69 by a joint Yale-American Museum effort in an attempt to recover additional mammal specimens from the Jurassic. The mammal-bearing claystone was re-exposed, and over two field seasons produced several dozen additional specimens. As exposed then, the claystone varied from 0 to

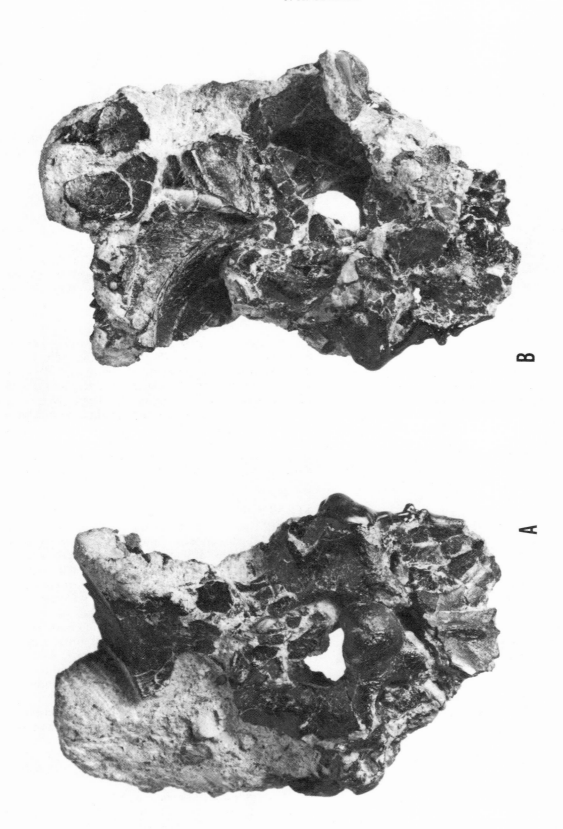

Figure 1. Holotype specimen of *Laopteryx priscus* (YPM 1800). *A*, Posterior aspect of crushed braincase; and *B*, anterior aspect of the same. See Figure 2 for explanation. Scale units equal 1.0 mm.

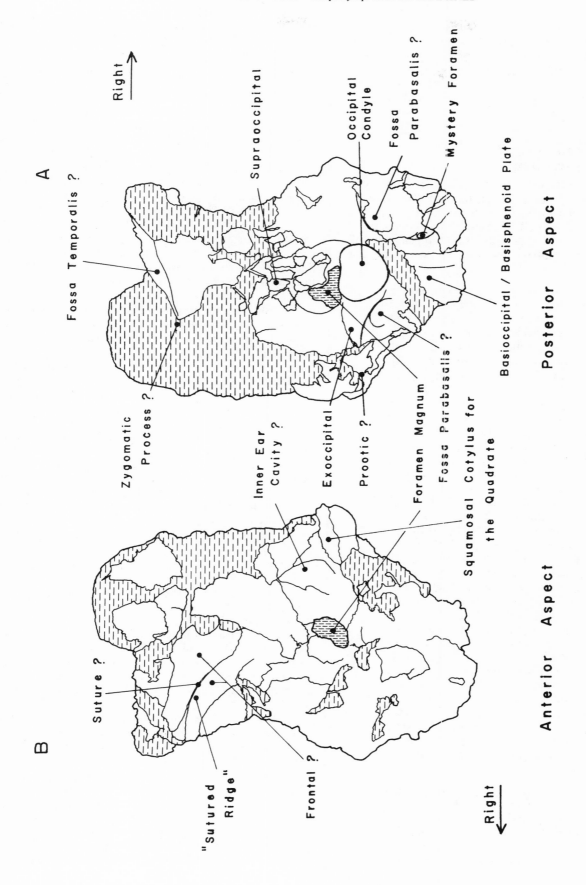

Figure 2. Line drawing interpretation of photographs of Figure 1. Horizontal hatching indicates rock matrix. Same scale as Figure 1.

approximately 30 cm in thickness where it occurred closer to 28m (90 ft) below the base of the lowest massive sandstone of the Cloverly Formation. The clay seam is sparsely fossiliferous, but it did yield a variety of rhynchocephalian and mammalian jaws, isolated fish vertebrae, crocodilian teeth, turtle fragments and an occasional theropod tooth.

ASSESSMENT OF MARSH'S ORIGINAL DESCRIPTION

Marsh's (1881) brief description is the only one published to date (repeated in part by Lambrecht [1933]), despite the several subsequent references and the obvious potential significance of YPM 1800. For evaluation and comparative purposes, that description is repeated in full here. Bracketed numbers designate specific items for discussions that follow.

> "The type specimen of the present species is the posterior portion of the skull, which indicates a bird rather larger than a Blue Heron *(Ardea herodias)*. The braincase is so broken that its inner surface is disclosed, and in other respects the skull is distorted, but it shows characteristic features. [1] The bones of the skull are pneumatic. [2] The occipital condyle is sessile, hemispherical in form, flattened and slightly grooved above. [3] There is no trace of a posterior groove. [4] The foramen magnum is nearly circular, and small in proportion to the condyle. [5] Its plane coincides with that of the occiput, [6] which is slightly inclined forward. The bones around the foramen are firmly coössified, but the supra-occipital has separated somewhat from the squamosals and parietals. [7] Other sutures are more or less open. [8] On each side of the condyle, and somewhat below its lower margin, there is a deep rounded cavity, perforated by a pneumatic foramen. [9] The cavity for the reception of the head of the quadrate is oval in outline, and its longer axis, if continued backward, would touch the outer margin of the occipital condyle. [10] This cavity indicates that the quadrate had an undivided head. [11] The braincase was comparatively small, but the hemispheres were well developed. They were separated above by a sharp mesial crest of bone. [12] A low ridge divided the hemispheres from the optic lobes, which were prominent." Marsh, 1881, p. 341.

Before discussing the several items marked in the above quotation, it should be noted that Marsh's description failed to do justice to the subject considering its taxonomic importance. We should not fault Marsh though; the specimen is so badly fragmented and crushed that it is not immediately recognizable as skull remnants except, for presence of the occipital condyle. Nevertheless, that he assigned it without question to the avian class is a puzzle. Illustrations presented here demonstrate the sad state of preservation of YPM 1800. Beginning with the annotated (above) features described by Marsh, my objectives are to comment on those features, redescribe and illustrate the specimen, and attempt to establish its identity. Ideally, the following comments should include comparisons with the best known Jurassic avian specimens, *Archaeopteryx*. Unfortunately that cannot be done here for the simple reason that not one of the 12 items annotated above is available for comparison in any of the present specimens of *Archaeopteryx*.

1. Although described as pneumatic by Marsh, only some of the cranial bones of *Laopteryx* are in fact "hollow" (or more accurately, cancellous), as opposed to being demonstrably pneumatic. Marsh did refer to pneumatic foramina in the specimen, but new preparation (undertaken by Charles Schaff at the author's direction) show that these are direct canals into the endocranial cavity and are not pneumatic foramina. Presence of hollow or cancellous bone is not an exclusively avian condition, having been reported in most archosaurs (thecodonts, crocodilians, theropods, prosauropods, sauropods, ornithopods, and pterosaurs). Accordingly, this feature is of no use in assigning *Laopteryx* taxonomically.

2. The occipital condyle is "sessile" (without a constriction or neck) and hemispherical, and somewhat flattened and slightly grooved on its dorsal aspect. The description by Marsh is accurate, but it is not uniquely avian—nor is it true of most birds. The condyle *is* necked or constricted at its base in some birds. But more important is the fact that this description applies equally well to most adequately preserved pterosaurian condyles. Thus, while possibly true for some birds, the nature of the condyle in *Laopteryx* is not definitive.

3. Again, the description by Marsh is accurate. There is no trace of a posterior groove (which I interpret to refer to a slight vertical sulcus that traverses the posterior surface of the occipital condyle, such as in some crocodilians). Such a groove is absent in birds and apparently in pterosaurs as well, although the pterosaur record is almost non-existent.

4. The foramen magnum is small, and appears to have been nearly circular. However, the foramen is *not* small relative to the condyle, but rather is slightly larger than the condyle. The foramen can only be approximated as nearly circular because of the fractured and distorted condition of the occiput. Neither the inferred relative size nor shape of the foramen magnum is distinctly avian. In fact, the restored size of the foramen magnum relative to the well-preserved occipital condyle strongly indicates a non-avian assignment for *Laopteryx*.

5. That the plane of the foramen magnum "coincides" with that of the occiput is hardly noteworthy in view of the fact that so little of the *Laopteryx* occiput is uncrushed. Only those bones immediately adjacent to (and defining) the foramen on the left side are preserved in the "plane of the foramen." The others are fragmented, separated, and displaced. It should be noted that

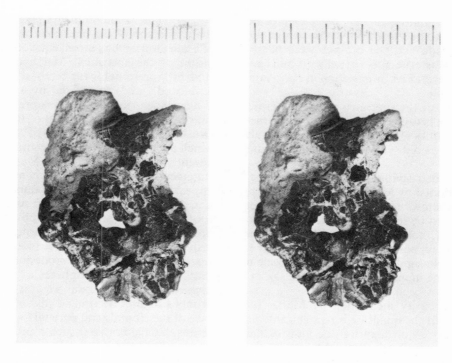

Figure 3. Stereophotographs of *Laopteryx priscus,* holotype specimen (YPM 1800), posterior aspect. Scale units = 1.0 mm.

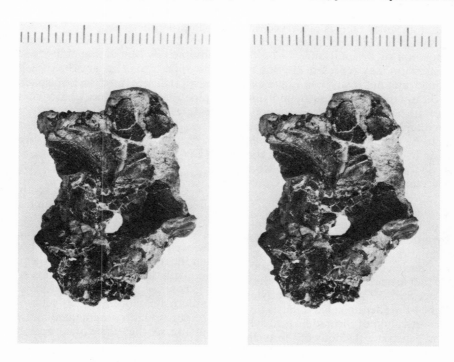

Figure 4. Stereophotographs of *Laopteryx priscus,* holotype specimen (YPM 1800), anterior aspect. Scale units = 1.0 mm.

these bones on the left are oriented at approximately 100° to the longitudinal axis of the condyle. But that probably has little significance since the "plane of the occiput" in the vicinity of the foramen is always approximately perpendicular to the axis of the emerging spinal nerve cord, which for obvious reasons must nearly parallel the axis of the occipital condyle. This feature appears to have no significance for identification purposes.

6. Marsh described the occiput of *Laopteryx* as being inclined forward. The preserved state of the specimen does not permit verification of that, unless one arbitrarily assumes a horizontal axis for the occipital condyle. But

the orientation of the condyle in *Laopteryx* cannot be established. All that can be shown is that the axis of the condyle and the "plane" of the occiput were nearly perpendicular, which is true for virtually (if not) all vertebrates (see '5' above). This information in *Laopteryx* has no systematic value.

7. The condition of the specimen does not permit unequivocal distinction between "open sutures" and fractures, nor can the identities of the squamosals or parietals be established. These data are useless for identification purposes.

8. The deep rounded cavities on each side of the occipital condyle mentioned by Marsh are real. However, that on the left side shows not one, but three distinct foramina. All three can be seen to perforate through to the endocranial cavity, and thus cannot be pneumatopores, but must represent passageways for cranial nerves or vascular elements. No similar foramen-bearing occiput fossae are known to me in any modern bird, but an exhaustive search has not been made.

9. The articular cavity mentioned by Marsh for reception of the head of the quadrate is recognizable on the left side. This depression may well be an articular cotylus for the quadrate. It is oriented as Marsh described, not markedly different from the lateral cotylus of modern birds; but the crushed and fragmented condition of this specimen leaves the matter very much in question. Unfortunately, no similar articular facet is preserved on the right side of the specimen to confirm this identification, or to verify its position or orientation.

10. The above depression may represent the squamosal cotylus for the quadrate, but if so, then I take exception to Marsh's conclusion that it indicates an undivided condition of the quadrate head. As preserved, the concavity is incomplete medially where indeterminate bone chips of what appears to be highly vascularized bone is all that remains. This is in the probable region of the prootic-opisthotic, but no identification is possible. Most important is that no undamaged bony surface is preserved, so presence or absence of a prootic cotylus for a medial head of the quadrate cannot be established.

This last point is of key importance. Birds are uniquely characterized by a double headed quadrate—two separate proximal articulations: an external condyle articulating in a deeply concave cotylus in the squamosal, and an internal condyle that articulates with a less distinct cotylus on the prootic (see Saif, 1974). The two condyles are separated by a slight to moderate sulcus that is occupied by an anterior diverticulum of the middle ear cavity (a second diverticulum penetrates the body of the avian quadrate; Jollie, 1957). Perhaps at the Jurassic stages of *Laopteryx* and *Archaeopteryx*, the "primitive" single-headed quadrate condition prevailed. If so, it is a condition not found in other birds. Whether the quadrate is single- or double-headed in *Archaeopteryx* is under debate: see Whetstone, 1983; Walker, 1985; and Bühler, 1985. It is presently under investigation by Haubitz, Wellnhofer, and me. Unfortunately, the condition of the quadrate in the holotype of *Laopteryx* will never be known.

11. The final comments by Marsh concerning the shape and size of the braincase and evidence suggestive of brain morphology are all equivocal, in view of the condition of the specimen. Marsh described the "hemispheres" (presumably the cerebral hemispheres) as "well developed" and "separated by a sharp mesial crest." While some evidence of this is preserved, it is of doubtful value. The anterior aspect of the specimen preserves two large concavities separated by a prominent ridge which appears to be a mid-line sutural junction. This apparently is the feature that led Marsh to make the above statements. The ridge and the concavities on either side appear to be symmetrical, and thus may be equated with a mid-line structure or a major part of the endocranial roof. But what Marsh failed to observe is that on the opposite surface of the bone(s) forming this ridge is a corresponding groove or trench; no similar osteological feature is known to me in modern birds. Despite the fact that the cerebral hemispheres of all modern birds are divided by a conspicuous sagittal fissure, there usually is only a slight reflection (if any) of this on the inner surface of the frontals and parietals—the crista frontalis internus. I have not seen such a prominent internal midline crest in any bird skull, nor in any tetrapod, for that matter, except in the skull of a very young (3-6 months?) alligator. Consequently, I consider it unlikely that this feature represents an internal mid-line feature, either avian or otherwise.

12. Possibly the above feature is the "low ridge" observed by Marsh that "divided the hemispheres from the optic lobes." If so, then I failed to see his "mesial crest" that separated the hemispheres. I am forced to conclude that Marsh and I have interpreted several of the features of YPM 1800 differently, which is not surprising.

NEW DESCRIPTION

Posterior Aspect

Orienting the specimen so that the observer looks directly at the foramen magnum and the occipital condyle in their normal positions as key reference points, several details are apparent. First, only the left side of the occiput is relatively undisturbed. That left side shows that the foramen magnum was centered in a slight depression with the exoccipitals flaring slightly backward. This is in contrast to the *convex* backward flaring of the typical modern avian occiput. The supraoccipital portion of the occiput is highly fragmented and sufficient portions are missing so that the dorsal occipital surface is beyond meaningful description and the upper margin of the foramen magnum cannot be delineated. The right exoccipital is more or less uncrushed except for its superior and external portions, but it has been broken away from the basioccipital and condyle so that the occiput appears distinctly asymmetrical.

The occipital condyle is prominent, sessile (not stalked), and nearly spherical except for its dorsal surface. Here it is distinctly flattened and slightly grooved in the mid-line. There is no sign whatsoever of a posterior axial indentation or groove. Because of the slightly flattened dorsal surface, the height of the condyle is less than its width (4.2 mm x 4.9 mm). The condyle protrudes approximately 3.1 mm beyond the exoccipitals. The exact shape and size of the foramen magnum are indeterminate, but from the left side of the occiput it appears to have been nearly circular and very close to 6.5 mm wide by 5.0 mm high in original dimensions.

Extending laterally on each side from the condyle the exoccipitals form stout ridges or buttresses delimiting the occipital surfaces described above from well-defined and prominent deep depressions below. Each of these fossae lies just lateral to the lateral limits of the foramen magnum, and are slightly less in diameter than the occipital condyle. That on the right is crushed and poorly preserved, but the left depression appears to be intact. It is slightly oval and contains three well-preserved foramina—two large oval canals laterally, and a much smaller nearly round opening medially. There is a doubtful, smaller fourth opening. All three main foramina communicate directly into what must have been the postero-ventral region of the endocranial cavity. The opposite (anterior) aspect of the specimen is such a jumble of fragments that precise original locations of the internal exits of these canals are uncertain, and identifications of the enclosing bones are not possible.

An exhaustive search for comparable occipital fossae or a similar array of foramina in modern birds was not attempted, but a cursory search was unsuccessful. On the other hand, a similar triangular arrangement of three prominent foramina in the same general area (but not within a distinct fossa) of the modern avian occiput is widespread. In modern birds, these are the foramina for exits of the vagus nerve (most medial of the three) and branches of the cerebral carotid and external ophthalmic arteries. Topologically, the foramina in *Laopteryx* appear best equated with these three modern avian features, in which case the prominent depression housing them might be termed the fossa parabasalis.

Directly ventral to the occipital condyle, but broken away from it, what remains of the basioccipital/basisphenoid plate is marked by a prominent mid-line ridge. On the right side this molds smoothly into what is now an incomplete but natural bony edge that appears to have defined the medial margin of another large foramen. The left side is incomplete, and does not show such an opening. No comparable opening has been observed in this region in modern bird crania. The foramen obviously is too small, too close to the mid-line, and too posteriorly located to be part of the middle ear chamber complex. Its identity is unknown.

Above the foramen magnum, and clearly separated from the occiput components, is a large bony element preserving a prominent and sharply defined ridge that separates two large, distinctly concave but asymmetrical surfaces. Both concavities lead to natural edges, one of which is quite robust. That surface preserves a series of subtle, non-parallel grooves and faint ridges that diverge away from the ridge terminus. Although not exactly like that of modern birds familiar to me, if this ridge-bearing element was derived from the skull, this ridge terminus might represent the zygomatic process of the squamosal (if *Laopteryx* is a bird). That would make the striated concavity a portion of the fossa temporalis, the site or origin of the M. temporalis. The unstriated concave surface then must represent that lateral portion of the squamosal just dorsal and lateral to the cotylus for the quadrate. This is only hypothetical, but might be confirmed by further study.

Other areas exposed in the posterior aspect of YPM 1800 are too fractured and dis-arranged to permit clear identification, and warrant no further description here.

Anterior Aspect

Unfortunately, the anterior surfaces defy interpretation and probable indentification, so badly fractured and disarrayed are the bony remnants. Aside from the "quadrate cotylus" and "sutured ridge" mentioned earlier in the remarks on Marsh's description, only one other distinctive feature stands out: the apparently uncrushed cavern-like sinus in the left prootic-opisthotic complex. The chamber preserved there probably is part of the inner ear cavity. It has been explored as far as possible by Charles Schaff's (now at Harvard's Museum of Comparative Zoology) preparation skills, but confirmatory details cannot be seen. The precise position of this chamber in the undeformed skull is no longer determinable, but in the crushed specimen it is situated lateral (left) to the occipital condyle and above and medial to the "quadrate cotylus" observed by Marsh. Only by sectioning this region of the specimen might the diagnostic foramina be exposed to confirm this tentative interpretation.

The Associated Tooth

Marsh (1881, p. 342) noted that:

> "In the matrix attached to this skull, a single tooth was found, which most resembles the teeth of birds, especially those of *Ichthyornis*. It is probable that *Laopteryx* possessed teeth, and also biconcave vertebrae."

That last statement by Marsh is a perfect example of 'seeing what we want to see,' and the all too prevalent tendency of extrapolating beyond available evidence. The tooth in hand is *not* like those of *Ichthyornis,* and no vertebrae are associated with the *Laopteryx* specimen.

The single tooth ascribed by Marsh to *Laopteryx* has long been thought to be missing (Lambrecht, 1933; Molnar, 1985). One tooth, however, together with several small bone fragments contained in glass vials labeled "1800," are present in the Yale Peabody collections together with the type skull (Fig. 5). The tooth was separated from the skull matrix many years ago and, although there is no way to certify that this is the same tooth mentioned by Marsh, there is no obvious reason to doubt that association. Three views of this tooth are included here.

This tooth has not been described since Marsh mentioned it, and obviously it has had no influence on subse-

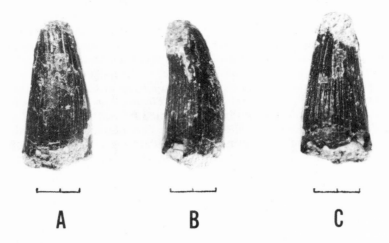

Figure 5. Solitary tooth associated with YPM 1800, mentioned by O. C. Marsh (1881), shown in external *(A)*, axial *(B)*, and medial *(C)* views. Scale units = 1.0 mm.

quent deliberations on the possible systematic assignment of *Laopteryx*. It probably should have no influence here either. But its prior association with the type specimen makes it mandatory that I include it in this report, despite the fact that its size seems *much* too large for the fragmentary skull of *Laopteryx*. The preserved length is greater than any dimension of the occipital condyle!

The tooth is conical in form, tapering gradually and uniformly toward the missing apex. The preserved length is 6.6 mm, and the greatest crown width is 3.0 mm. I estimate that approximately 1.5 mm has been abraded from the apex, giving an original crown length of approximately 8.0 mm. The crown curves slightly medially (?). All crown surfaces are markedly fluted with fine grooves and ridges that are most conspicuous on the concave inner (?) surface. The crown is only slightly oval to subcircular in section, in sharp contrast to the highly sectorial and transversly compressed teeth of *Ichthyornis*. In my opinion, this tooth is not similar to those of *Ichthyornis*.

The tooth in question, currently catalogued under YPM 1800, is indistinguishable from several other isolated teeth (mostly larger) in the Yale Peabody collections from Quarry 9 and other localities in the Morrison Formation. Usually these have been considered crocodilian and attributed to *"Goniopholis."*

DISCUSSION

Only one set of features of YPM 1800 provides incontrovertible evidence and permits necessary comparative details relevant to the systematic assignment of *Laopteryx prisca* (gender corrected by Brodkorb, 1971): the occipital condyle and the foramen magnum. As preserved, the condyle provides accurate dimensions and morphology (reported above). The foramen magnum is preserved in part, especially on the left side, and permits excellent reconstruction of the shape and dimensions (also reported above). Comparison of the two leaves no question in my mind that they establish a non-avian identification of YPM 1800. *Laopteryx prisca* is not a bird.

As registered in my notes '1' through '12' above, none of the features described by Marsh is (uniquely or otherwise) avian. A random sample of relative dimensions of the occipital condyle vs. foramen magnum in Recent birds versus reptiles established some interesting data. Of two dozen Recent species both of carinates and ratites, the avian foramen magnum width ranged from a high of 8 times the width of the occipital condyle to a low of 1.4 times. Most were in the 2+ range. That of a dozen Recent reptile species ranged from 1.2 to 0.8 times, with most falling below 1.0. The data for reptiles reflect the large diameter of the medulla/spinal cord emergence from the braincase in lower vertebrates relative to condyle size. In YPM 1800, the foramen magnum is close to 1.3 times the condyle width, or above the extreme upper limit found in the reptilian grade (or well outside the extreme lower limit of the avian grade).

A possible alernative identification of *Laopteryx* is that of pterosaur, as was suggested by Brodkorb (1971, 1978), and as I hope to show here. Unfortunately, uncrushed pterosaur cranial remains are rare, and none is known to the author. Very few specimens exist to provide observation of parameters mentioned above in their pristine states. In the Yale collections (see Eaton, 1910), the type of *Pteranodon longiceps* (YPM 1177) bears an occipital condyle estimated to have had an uncrushed width of approximately 8.0 mm and a height of about 7.0 mm. The condyle of YPM 1179, the type of *Pteranodon occidentalis,* has a width of 7.5 mm, but the height is indeterminate. The original widths of the two foramen magna in these specimens are indeterminate, both specimens having been crushed laterally. But in both, the foramen appears to have been less than that of the condyle. *Pteranodon occidentalis* (YPM 1179) is reconstructed with a foramen width of less than 5.0 mm, or about 60 percent that of the condyle. In the University of Kansas collection, an excellent but distorted skull (KU 976) of *Pteranodon ingens* preserves the following occipital dimensions: condyle width, 8.7 mm; condyle height, 7.6 mm; estimated foramen magnum width, 5.1 mm; and estimated foramen magnum height, 6.0 mm.

In short, evidence from the above *Pteranodon* specimens, limited though the evidence may be, is suggestive of a pterosaurian identity of *Laopteryx*, rather than avian. That identification is consistent with other evidence from Quarry 9 (Marsh, 1878; Galton, 1981), in which unmistakable pterosaurian remains *(Dermodactylus = Pterodactylus montanus* Marsh, [1878] and *Comodactylus ostromi* [Galton, 1981] have been reported. Accordingly, *Laopteryx* is here assigned to the class Reptilia, order Pterosauria.

CONCLUSIONS

Laopteryx prisca (YPM 1800) is judged to be pterosaurian, not avian, on the basis of: 1, non-avian relative sizes of the foramen magnum and the occipital condyle; 2, hemispherical form of the condyle, a characteristic of pterosaurs as well as birds (also ceratopsians); 3, the fact that all cranial features noted above in YPM 1800 are not solely avian—these occur elsewhere; and 4, the associated tooth is *not* like those of *Ichthyornis*, but is more probably referable to the order Crocodylia.

Except for its small size (relatively), there is no reason to conclude that *Laopteryx* was a bird.

REFERENCES CITED

Brodkorb, P., 1971, Origin and evolution of birds, *in* Farner, D.S., King, J.R., and Parkes, K.C., eds., Avian biology, v. 1: New York Academic Press, p. 19-55.

―――― 1978, Catalogue of fossil birds: Part 5 (Passeriformes): Florida State Museum, Bulletin, v. 23, p. 139-228.

Bühler, P., 1985, On the morphology of the skull of *Archaeopteryx*, *in* Hecht, M. K., Ostrom, J.H., Viohl, G., and Wellnhofer, P., eds., The beginnings of birds: Eichstätt, West Germany, Proceedings of the International *Archaeopteryx* Conference, p. 135-140.

Eaton, G.F., 1910, Osteology of *Pteranodon*: Connecticut Academy of Arts and Sciences, Memoir, v. 2, p. 1-38.

Galton, P.M., 1981, A rhamphorhynchoid pterosaur from the Upper Jurassic of North America: Journal of Paleontology, v. 55, p. 1117-1122.

Jollie, M.T., 1957, The head skeleton of the chicken and remarks on the anatomy of this region in other birds: Journal of Morphology, v. 100, p. 389-436.

Lambrecht, K., 1933, Handbuch der Palaeornithologie: Berlin, Borntraeger, 1024 p.

Loomis, F.B., 1901, On Jurassic stratigraphy in southeastern Wyoming: American Museum of Natural History, Bulletin, v. 14, p. 189-197.

Marsh, O.C., 1878, New pterodactyl from the Jurassic of the Rocky Mountains: American Journal of Science, v. 16, p. 233-234.

―――― 1881, Discovery of a fossil bird in the Jurassic of Wyoming: ibid., v. 21, p. 341-342.

Molnar, R. E., 1985, alternatives to *Archaeopteryx:* a survey of proposed early or ancestral birds, *in* Hecht, M. K., Ostrom, J. H., Viohl, G., and Wellnhofer, P., eds., The beginnings of birds: Eichstätt, West Germany, Proceedings of the International *Archaeopteryx* Conference, p. 209-217.

Ostrom, J. H., and McIntosh, J. S., 1966, Marsh's dinosaurs: the collections from Como Bluff: Yale University Press, New Haven, Connecticut, 388 p.

Saif, E. I., 1974, The middle ear of the skull of birds: The Procellariiformes: Zoological Journal of the Linnean Society, v. 54, p. 213-240.

Simpson, G. G., 1926a, The fauna of quarry 9: American Journal of Science, v. 12, p. 1-11.

―――― 1926b, The age of the Morrison formation: ibid., v. 12, p. 196-216.

―――― 1929, American Mesozoic Mammalia: New Haven, Connecticut, Peabody Museum of Natural History Memoir, v. 3, Yale University, 235 p.

Walker, A. D., 1985, The braincase of *Archaeopteryx*, *in* Hecht, M. K., Ostrom, J. H., Viohl, G., and Wellnhofer, P., eds., The beginnings of birds: Eichstätt, West Germany, Proceedings of the International *Archaeopteryx* Conference, p. 123-134.

Whetstone, K. N., 1983, Braincase of Mesozoic birds: I. New preparation of the "London" *Archaeopteryx*: Journal of Vertebrate Paleontology, v. 2, p. 439-452.

ADDENDUM

Just hours before this manuscript was to be returned to the editors, a startling coincidence occurred. Mary Ann Turner, Peabody Collection Manager, handed me a small box with Marsh's handwriting *"Laopteryx* lower jaw." Uncovered during our collection reorganization project, the box contained two lumps of walnut-size Quarry Nine matrix with small fragments of what appears to be a lower jaw containing at least one tooth. The tooth appears to be a simple, slightly curved, and unfluted cone. The specimen will require much careful preparation before further information is available.

These fragments were accessioned, but never catalogued. The box bears accession number 1401, and our records indicate it was collected by S. W. Williston and received in New Haven on October 16, 1880. The type specimen of *Laopteryx* YPM 1800 was accessioned under number 1394, and received on September 18, 1880, a month earlier. The Como Bluff Quarry Nine provenance of both specimens is beyond question, but associations in the quarry are unknown. Marsh's questionable reference of these fragments to *Laopteryx* is beyond verification, but results of preparation will be forthcoming.

MANUSCRIPT RECEIVED JANUARY 13, 1986
REVISED MANUSCRIPT RECEIVED MAY 19, 1986
MANUSCRIPT ACCEPTED JUNE 18, 1986

Brain evolution in Mesozoic mammals

ZOFIA KIELAN-JAWOROWSKA *Zaklad Paleobiologii, Polska Akademia Nauk, al. Zwirki i Wigury 93, P1 02-089 Warszawa, Poland*

ABSTRACT

Endocranial casts of Mesozoic mammals and of some cynodonts are reviewed. New tentative reconstructions of brains of *Probainognathus* and *Therioherpeton* are given. It is claimed that the endocast of *Amblotherium* is an artefact. Brains of Mesozoic mammals were lissencephalic, with no flexure, had very large olfactory bulbs, relatively extensive cerebral hemispheres diverging posteriorly, and large paraflocculi. Within this pattern two types are designated: the cryptomesencephalic type (large vermis, no dorsal midbrain exposure, and no cerebellar hemispheres) which occurs in Triconodonta and Multituberculata; and the eumesencephalic type (wide cerebellum, cerebellar hemispheres, and large dorsal midbrain exposure) which occurs in Cretaceous Tribosphenida. Overlap of the midbrain took place in individual lines of the Tribosphenida at different times during the Tertiary. If advanced cynodonts *(e.g., Probainognathus* and *Therioherpeton)* had narrow cerebellum and exposed midbrain, then both types could develop from them: the cryptomesencephalic by overlap of the midbrain by an enlarged vermis, and the eumesencephalic by acquisition of enlarged cerebellar hemispheres. If, however, the midbrain was overlapped in advanced cynodonts, then they belong to the cryptomesencephalic type. If so, the eumesencephalic type would have developed from cryptomesencephalic by secondary exposure of the midbrain and acquisition of enlarged cerebellar hemispheres. This latter is less likely, as it would involve the reduction of an already expanded vermis. The expansion of cerebral hemispheres suggests that neocortex was possibly present in all Mesozoic mammals and in some cynodonts.

INTRODUCTION

The first scientific love of George Gaylord Simpson was Mesozoic mammals. When he published his two great monographs ("A Catalogue of the Mesozoic Mammalia in the Geological Department of the British Museum," 1928; and "American Mesozoic Mammalia," 1929) he became, at the age of twenty-seven, a leading authority in the field. In addition to these monographs, prolific as he was all his life, he published 25 other papers on Mesozoic mammals between 1926 and 1929. Later on, his scientific interests covered an exceptionally broad spectrum, but he "revisited" early mammals from time to time throughout his life (e. g., Simpson, 1959, 1961, 1971a,b, Clemens and others, 1979).

During the first period of his work on Mesozoic mammals, Simpson (1927) described the first known endocranial cast of a Jurassic triconodont *(Triconodon)* and compared it with that of a cynodont *(Nythosaurus)*. Subsequently (1928), he described and refigured the triconodont specimen. For more than a half century this endocast was the only mammalian endocranial cast known from the first 140 million years of mammalian evolution. The knowledge of origin and evolution during the first two thirds of mammalian history increased considerably since Simpson's papers. The problem of the reptilian-mammalian transition and early mammalian evolution has been widely discussed recently, mostly from the viewpoints of the evolution of the masticatory system, braincase, postcranial skeleton, and homeothermy, but with little contribution to structure of endocasts (see Lillegraven and others, 1979 and Kemp, 1982 and 1983 for summaries). Jerison's book (1973) and the paper of Quiroga (1980a) are notable exceptions.

Although only six years have passed since the thorough review by Quiroga, the new material of endocasts from the late Cretaceous of the Gobi Desert (Kielan-Jaworowska, 1983, 1984, Kielan-Jaworowska and Trofimov, 1980, in press, Kielan-Jaworowska, Presley, and Poplin, in press, allows one to interpret some parts of endocasts of early mammals, especially the cerebellum and midbrain, differently than previous authors did.

The aim of my paper is to review existing knowledge on endocasts at the reptilian-mammalian transition and in Mesozoic mammals, and to consider its implications to relationships of early mammals.

ABBREVIATIONS

BM - British Museum (Natural History), London.
PIN - Paleontological Institute, USSR Academy of Sciences, Moscow.
PVL - Section of Vertebrate Paleontology, Miguel Lillo Foundation, Tucuman, Argentina.
ZPAL - Institute of Paleobiology, Polish Academy of Sciences (Zaklad Paleobiologii, Polska Akademia Nauk), Warsaw.

AVAILABLE SAMPLE

Endocranial Casts of Cynodonts

Endocranial casts or endocranial cavities have been described in the following genera of cynodonts: *Diademodon* and *Nythosaurus* (see Watson, 1913); cynodont B (Olson, 1944); *Oligokyphus* (see Kühne, 1956); *Bienotherium* (see Hopson, 1964); *Procynosuchus* (see Kemp, 1979, 1982); *Trirachodon* (see Hopson, 1979); *Massetognathus* and cf. *Probelesodon* (see Quiroga, 1979); *Probainognathus* (see Quiroga, 1980a, b); and *Therioherpeton* (see Quiroga, 1984). Of these forms, the early Triassic galesaurid *Nythosaurus larvatus* (possibly synonymous with *Thrinaxodon liorhinus*—see Hopson, 1979) and the middle Triassic *Probainognathus jenseni* and *Therioherpeton cargnini* stay relatively close to animals that gave rise to mammals.

Nythosaurus larvatus, BM R 1713 (Figs 1*A*, *B*), described originally by Watson (1913), was redescribed by Simpson (1927) and Hopson (1969, 1979). In addition to the cast of the nasal cavity, the long and tubular endocast has been preserved with relatively large olfactory bulbs. The cerebral hemispheres are narrow, roughly parallel-sided, slightly constricted at the end, in front of a much broader cerebellar region. Simpson (1927) reconstructed anterior (optic) colliculi at the level of this constriction, although the boundary between the cerebral hemispheres and the colliculi is not seen on the cast. He based his reconstruction upon the fact that the posterior end of the hemispheres may be located at the level (or a little behind the level) of the parietal foramen, which has been preserved. The cerebellum is relatively broad, with prominent paraflocculi. Hopson (1979) drew the exists of cranial nerves IX and X, and a double exit of the hypoglossal nerve as seen on the endocast. If Simpson's reconstruction is correct, *Nythosaurus* had smooth cerebral hemispheres, the midbrain (anterior colliculi) was exposed on the dorsal surface, relatively wide cerebellum, no cerebellar hemispheres, and large paraflocculi.

The endocast of *Probainognathus jenseni*, PVL 4169, described by Quiroga (1980a, b), appears more complete than those of other cynodonts. Quiroga reconstructed it (see Fig. 1*C* in present paper) as having: large olfactory bulbs with long olfactory peduncles; very long cerebral hemispheres diverging and slightly widening posteriorly; the boundary between the neocortex and paleocortex (referred to by him as the "slope") present in the caudal portion of the hemispheres; anterior colliculi exposed on the dorsal side; pineal region inserted between the caudal parts of the hemispheres; relatively wide and roughly rectangular cerebellum with median sulcus; and large paraflocculi.

As the endocast of *Probainognathus* is somewhat distorted in its posterior parts, it may be reconstructed in several ways. I present here (Figs. 1*D-E*) two alternative reconstructions. Quiroga (1980a, Figs. 2-4) designated the structure to which he gave the notation *S* as the "slope of the dorsal surface of the cerebral hemispheres." This "slope" corresponds (Quiroga, 1980a, p. 328) to the "caudal boundary of the neocortex," where the rhinal fissure should be expected. However, the rhinal fissure, if preserved on the endocasts, is not marked as a ridge, but as a furrow. According to Quiroga's interpretation, this boundary is seen on the dorsal side of the endocast as a wide ridge. I suspect that this ridge might rather be a cast of a relatively wide vessel, similar to the cast of the transverse sinus present immediately behind the posterior boundary of cerebral hemispheres on the endocranial casts of primitive mammals. On the lateral wall of the endocast this boundary in Quiroga's interpretation is marked as a cast of a minute vessel which continues on the olfactory bulb, very different from the one on the dorsal side (Quiroga, 1980b, Pl. II*B*).

In both of my reconstructions (Figs. 1*D-E*), I place the posterior boundary of cerebral hemispheres in the same place as Quiroga did. However, I regard the "slope" as a cast of a vessel, not a rhinal fissure, although I believe that the neocortex may have been developed in *Probainognathus*. I also regard the colliculus in Quiroga's sense as a cast of the transverse sinus. In my first interpretation (Fig. 1*D*), I recognize the part of the cast inserted between caudal ends of the hemispheres as colliculi. I reconstruct the cerebellum relatively narrower than Quiroga did, possibly consisting only of the vermis and paraflocculi. Drawbacks of this interpretation are that: (1) the longitudinal furrow which separates the colliculi is irregular in shape, and may be an artefact; and (2) the boundary between the apparent colliculi and the cerebellum is hardly seen on the endocast.

In my alternative interpretation (Fig. 1*E*), I accept that the longitudinal furrow in the posterior part of the endocast is an artefact and, consequently, reconstruct the cerebellum as consisting of a large vermis and paraflocculi. The vermis is pointed anteriorly, and inserted between caudal ends of the hemispheres. It cannot be excluded that the structure which I recognize as the cast of the transverse sinus is, in fact, a laterally placed colliculus, as reconstructed by Quiroga.

It is possible to reconstruct the brain of *Probainognathus* differently, accepting that the "slope" (as called by Quiroga) is the cast of the tranverse sinus. If so, the caudal quarter of the hemispheres in Quiroga's reconstruction would correspond to the anterior colliculi. I do not provide a drawing of such a reconstruction, as it would have the cerebral hemispheres barely separated from the colliculi, which is unlikely.

The brain cast of *Therioherpeton cargnini*, PVL - no number, was reconstructed by Quiroga (1984, Fig. 3) as having the midbrain widely exposed on the dorsal side of a relatively short and wide cerebellum, similar to those in the Theria (although Quiroga did not mention this similarity). The endocranial cast figured by Quiroga (1984) in his Figure 1 looks different from that illustrated by him in Figure 2, although both drawings derived from the same specimen. The right and left colliculi as recognized by Quiroga in his Figure 2 are separated by a ridge. However, the colliculi on mammalian endocasts are always separated by a furrow, not by a ridge. A cast of the natural endocast of *Therioherpeton*, which Dr. Quiroga kindly sent me, is similar to that in his Figure 1, but not as in his Figure 2. As the posterior part of the endocast is not well preserved, I present here two tentative reconstructions of its brain. In the first (Fig. 1*G*), I accept (as did Quiroga, 1984), that the midbrain is exposed on the dorsal side, even though a longitudinal furrow separating the colliculi is not recognizable. A cast

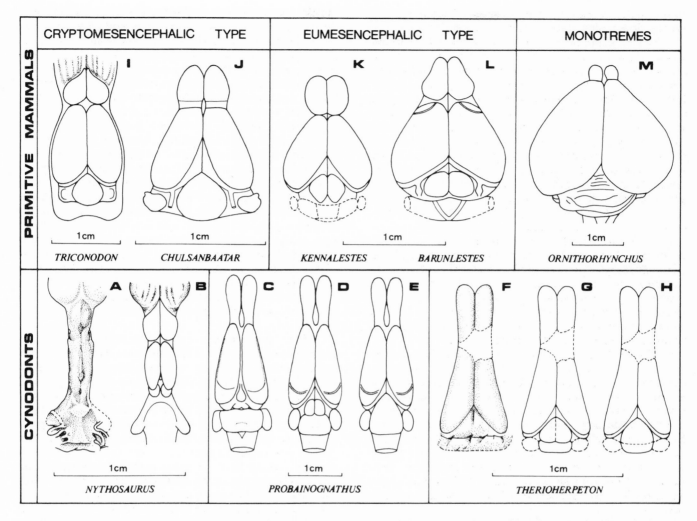

Figure 1. Brain constructions in some cynodonts and primitive mammals, dorsal view. *A, B, Nythosaurus larvatus, A*, redrawn from specimen by Hopson (1979), *B*, reconstructed by Simpson (1927); *C, D, E, Probainognathus jenseni, C*, reconstructed by Quiroga (1980a), *D, E*, present paper reconstructions; *F, G, H, Therioherpeton cargnini, F*, reconstructed by Quiroga (1984), *G, H*, present paper reconstructions; *I, Triconodon mordax*, based on Simpson's reconstructions (1927), emended; *J, Chulsanbaatar vulgaris* (from Kielan-Jaworowska, 1983, emended); *K, Kennalestes gobiensis* (from Kielan-Jaworowska, 1984, simplified); *L, Barunlestes butleri* (from Kielan-Jaworowska and Trofimov, *in press*); *M, Ornithorhynchus anatinus* (from Hines, 1929).

of the transverse sinus (poorly seen) is preserved between the caudal ends of the cerebral hemispheres and the colliculi. I interpret the cerebellum as being small, consisting only of a short and narrow vermis and possibly the paraflocculi.

In an alternative reconstruction (Fig. 1*H*), I accept that the midbrain is not exposed on the dorsal side and that the vermis is deeply inserted between caudal ends of the hemispheres. The transverse ridge on the vermis might represent a folded (due to depression) boundary between the dorsal and occipital surfaces of the vermis.

The remaining endocasts of cynodonts do not provide much information. All have relatively large olfactory bulbs, narrow and tubular cerebrum, no cerebellar hemispheres, and large paraflocculi. The caudal border of the hemispheres is not discernible in any of them, but it seems probable that the anterior colliculi are exposed on the dorsal surface. In most genera, the cerebellum appears relatively narrow, but Kemp (1979) reconstructed a wide cerebellum in *Procynosuchus*. Hopson (1979) tentatively reconstructed incipient cerebellar hemispheres in *Trirachodon*, situated in front of flocculi (paraflocculi). Rather, I interpret these "cerebellar hemispheres" as colliculi exposed on the dorsal side.

Endocranial Casts of Triconodonts

The oldest known mammalian endocranial casts belong to triconodonts. Patterson and Olson (1961; see also Edinger, 1964) figured a fragmentary endocast of the Liassic triconodont *Sinoconodon rignei*. It represents the dorsal aspect of the caudal end of the hemispheres and the anterior part of the cerebellum. Quiroga (1984) recently commented on this endocast, and argued that in the posterior part it is the midbrain which is exposed on the dorsal side, rather than the cerebellum.

Kermack and others (1981) reconstructed the lateral aspect of the brain of the Liassic triconodont *Morganucodon*. More complete is an endocast of the Purbeckian *Triconodon mordax*, reconstructed by Simpson (1927, 1928) on the basis of three dorsal aspects, most complete of which is BM 47763. According to Simpson's reconstruction, in addition to the endocast, a cast of the nasal cavity has been preserved. The olfactory bulbs are relatively large, roughly pyriform, and the cerebral hemispheres are smooth and oval. At the caudal end of the hemispheres, Simpson mentioned presence of two parallel grooves, and marked this region as the midbrain. The midbrain, as demonstrated by Edinger (1964), however, is not exposed on this endocast.

As demonstrated by Bauchot and Stephan (1967), endocasts of the midbrain in primitive mammals, if exposed dorsally, often are partly obscured by the cast of the transverse sinus, and appear smaller than on the brain. However, in *Triconodon* there is only a small area between the hemispheres and the vermis (filled by the cast of the transverse sinus), and there is no room for the midbrain to be exposed on the dorsal surface.

BM 47763 shows that the vermis in *Triconodon* possibly was larger than as reconstructed by Simpson. Structure of the petrosal in triconodonts (Kermack, 1963, Kermack and others, 1981) shows that the subarcuate fossa was relatively large, and consequently the parafloccули had to be large. Edinger (1964, p. 14) stated: "since Simpson's labelling L-lateral lobe of cerebellum further preparation revealed, to the right and left of the vermis cast, a transverse pocket in the brain case. This fossa subarcuata lodged a large flocculus." In Figure 1I, I give an emended reconstruction of the endocast of *Triconodon*.

The lateral aspect of triconodont brain, as reconstructed by Kermack and others (1981), is similar to that of multituberculates (Kielan-Jaworowska, 1983) in having a shallow telencephalon, deep rhombencephalon, and large parafloccули.

Endocranial Casts of Multituberculates

Of all known Mesozoic mammal endocasts, those of the multituberculates are most complete. The oldest known multituberculate skulls derive from the Kimmeridgian of Portugal. On the basis of skull structure and a comparison with Simpson's (1937) reconstruction of an endocast of the Paleocene multituberculate *Ptilodus*, Hahn (1969) reconstructed the dorsal aspect of the endocast of the Kimmeridgian *Paulchoffatia delgadoi*. Best preserved multituberculate endocasts derive from the Late Cretaceous Barun Goyot Formation (? middle Campanian) of Mongolia (Figs. 1J, 2, and 3). I described (Kielan-Jaworowska, 1983; see also Kielan-Jaworowska, Presley, and Poplin, *in press*) the complete endocranial cast of *Chulsanbaatar vulgaris*, ZPAL MgM-I/88, prepared by removing the braincase and cutting off most of the cast of the nasal cavity. More information was obtained (Kielan-Jaworowska, Presley, and Poplin, *in press*) by serially sectioning two multituberculate skulls (*Nemegtbaatar gobiensis*, ZPAL MgM-I/76, and *Chulsanbaatar vulgaris*, ZPAL MgM-I/84) on a Jung microtome, at thickness 25 μm and 20 μm, respectively. On the basis of sections, we made a wax model of the cast of the endocranial cavity of *Nemegtbaatar*, including casts of certain nerves and blood vessels (see also Kielan-Jaworowska and others, 1984).

Multituberculate endocasts have large olfactory bulbs, which taper anteriorly. Simpson (1937) reconstructed the olfactory bulbs in *Ptilodus* as widening anteriorly. Examination of the material on which he based his reconstruction, however, shows that this is not the case (Kielan-Jaworowska and Krause, *in preparation*). In *Chulsanbaatar* there is a transverse furrow at the posterior part of the bulb, and a single fusiform structure along the median line between the bulbs and the cerebral hemispheres. In both *Chulsanbaatar* and *Nemegtbaatar* the olfactory bulb, seen from the side, is tall at the rear, and slopes off sharply toward the front. In *Chulsanbaatar* the olfactory tract has been preserved, extending posteriorly from the inflated part of the olfactory bulbs.

The vermis is large, deeply inserted between the cerebral hemispheres in *Chulsanbaatar* and *Ptilodus*, but possibly relatively shorter in *Nemegtbaatar*. There are no cerebellar hemispheres. The parafloccули are large, oval in *Chulsanbaatar* and spherical in *Nemegtbaatar*. In both genera on the parafloccular cast there is a large cuspule, which is a cast of the structure designated the posttemporal recess (Kielan-Jaworowska, Presley, and Poplin, *in press*), apparently housing vessels of the posttemporal recess. Casts of two branches of the mandibular nerve, connected with two skull openings (foramen ovale inferius and foramen masticatorium), have been preserved in sections of *Nemegtbaatar*; the maxillary and ophthalmic nerves have been reconstructed. Anteriorly in *Nemegtbaatar*, a cast of the semilunar ganglion merges with the cast of the front part of the cavum epiptericum, which communicates with the cranial cavity through the optic foramen. In *Nemegtbaatar* a cast of one more nerve in this region has been preserved; possibly this is a deep temporal nerve, which runs dorsally from the anterior part of the semilunar ganglion and passes through the dorsal foramen in the anterior lamina of the petrosal.

The pons is not prominent, situated posteriorly to exits of the trigeminal nerve. The rhombencephalon is deep in comparison with the relatively shallow telencephalon.

Endocranial Casts of Mesozoic Theria

The only known Jurassic therian endocast is that of *Amblotherium nanum*, BM 47758, reconstructed by Edinger (1964, Fig. 7B). The specimen originally was described by Owen (1871, p. 50, Pl. II, Fig. 18) as *Stylodon posillus*, and then redescribed by Simpson (1927, p. 262, and 1928, p. 137, Pl. X, Fig. 3). Despite detailed examination of the specimen, neither Owen nor Simpson noticed the presence of a natural endocast. Simpson (1927, p. 262) described "The brain of *Amblo-

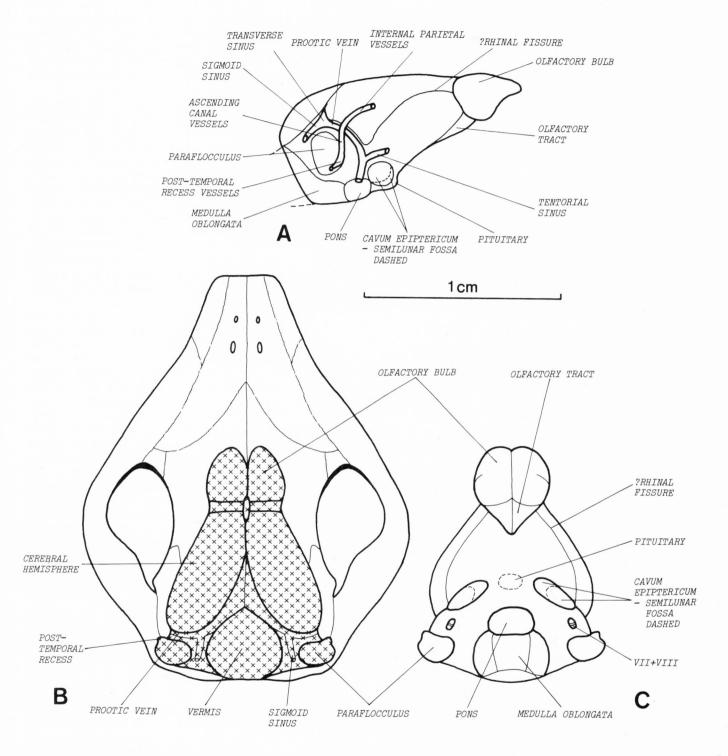

Figure 2. *Chulsanbaatar vulgaris.* Reconstructions. *A,* endocast, lateral view; *B,* skull and endocast (hatched), dorsal view; *C,* endocast, ventral view (from Kielan-Jaworowska, 1983, emended).

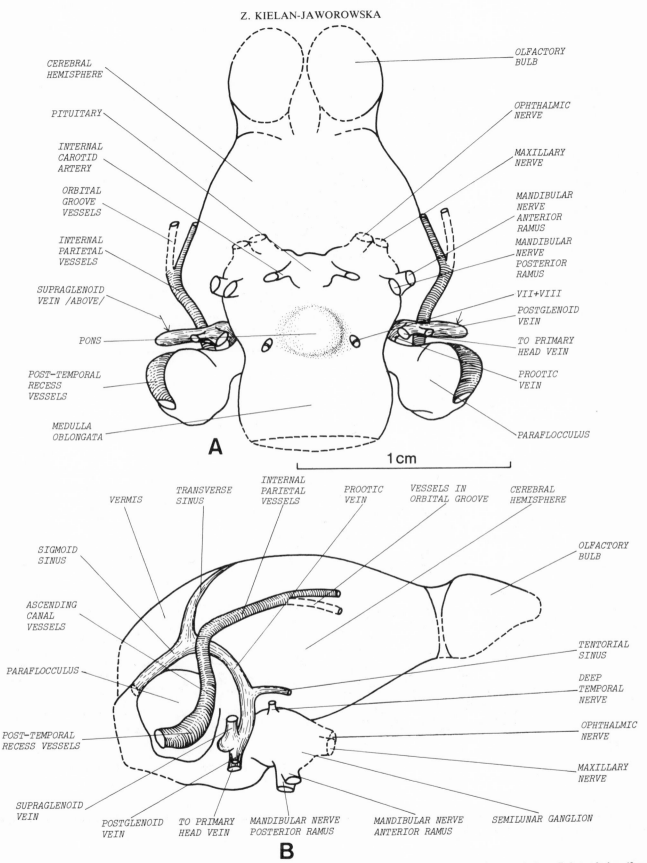

Figure 3. *Nemegtbaatar gobiensis*. Reconstructions of endocast and vascular system based on sections. *A*, ventral view; *B*, lateral view (from Kielan-Jaworowska, Presley, and Poplin, *in press*, simplified and emended).

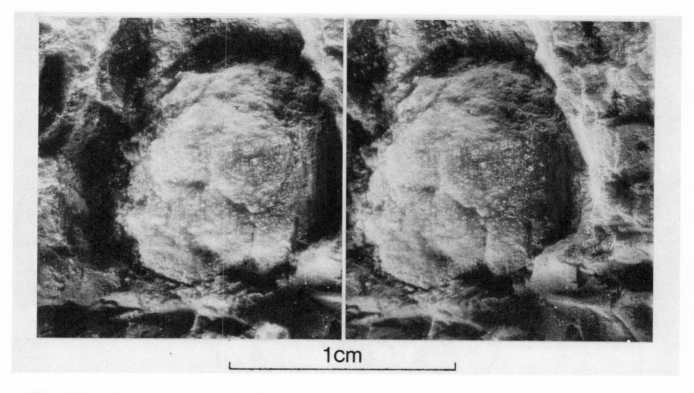

Figure 4. Stereo-photograph of latex cast, covered with ammonium chloride, of apparent endocast of *Amblotherium nanum*, BM 47758, which might be an artefact.

therium" on basis of: "a badly crushed skull of *Amblotherium nanum* (British Museum 47758) in which a large part of the internal aspect of the shattered cranial roof may be made out." On the basis of internal aspect, he concluded that the olfactory bulbs apparently are: "... similar to but relatively slightly smaller than those of *Triconodon*. The cerebral hemispheres are much like those of *Triconodon* in general character, but somewhat more expanded posteriorly. The dorsal exposure of the cerebellum has about the same relative extent."

Subsequently, Edinger (1964, p. 12) stated that Simpson's hesitant deductions from the broken bones "were found to be correct when a convexity laying alongside the concave cranial remains was recognized as the endocranial cast. This must have been mud already hardened when the skull was flattened and splintered. Fallen out of the cranium and turned over, it presents the dorsal surface of cerebrum and cerebellum."

The mode of endocast preservation described by Edinger appears improbable. Unfortunately the original of the apparent endocast is not to be seen anymore. I have been informed by Dr. J. J. Hooker of the Department of Palaeontology of the British Museum (Natural History) that "the slab containing the crushed skull of *Amblotherium nanum* has been prepared in acid, thus isolating the various elements and allowing them to be seen from both sides" (letter of July 25, 1985). I examined the cast of BM 47758, and the apparent endocast preserved on the slab is figured in the present paper as Figure 4. In my opinion it is an artefact.

Therefore, the oldest known therian endocranial casts are those of Late Cretaceous eutherians from Mongolia. We described (Kielan-Jaworowska and Trofimov, 1980 and *in press*) endocasts of *Barunlestes butleri*, PIN 3142-701 and 3142-702, and I described (Kielan-Jaworowska, 1984) endocasts of *Kennalestes gobiensis*, ZPAL MgM-I/14 and MgM-I/16, *Asioryctes nemegetensis*, ZPAL MgM-I/56, and *Zalambdalestes lechei*, ZPAL MgM-I/16. *Kennalestes* and *Zalambdalestes* are from the Djadokhta Formation (? late Santonian and/or ? early Campanian), while *Asioryctes* and *Barunlestes* are from the Barun Goyot Formation (? middle Campanian), or its stratigraphic equivalent.

Endocasts of these four genera (Figs. 1*K-L* and 5-6), though differing in proportions, are of the same general pattern. They all have: large olfactory bulbs; strong posterior divergence of cerebral hemispheres; an extensive midbrain exposure on the dorsal side; and a comparatively short and wide cerebellum consisting of vermis, cerebellar hemispheres, and a large paraflocculi. An accessory olfactory bulb tentatively has been recognized in *Asioryctes*. In all known endocasts of Cretaceous eutherians, the midbrain consists of one pair of large colliculi.

In *Barunlestes butleri*, PIN 3142-702 (Kielan-Jaworowska and Trofimov, *in press*, and Fig. 1 in the present paper), there is a small cuspule, lateral to the colliculus, which might be confused with the posterior colliculus. As, however, it apparently lies near the point of divergence of the transverse and sigmoid sinuses, and

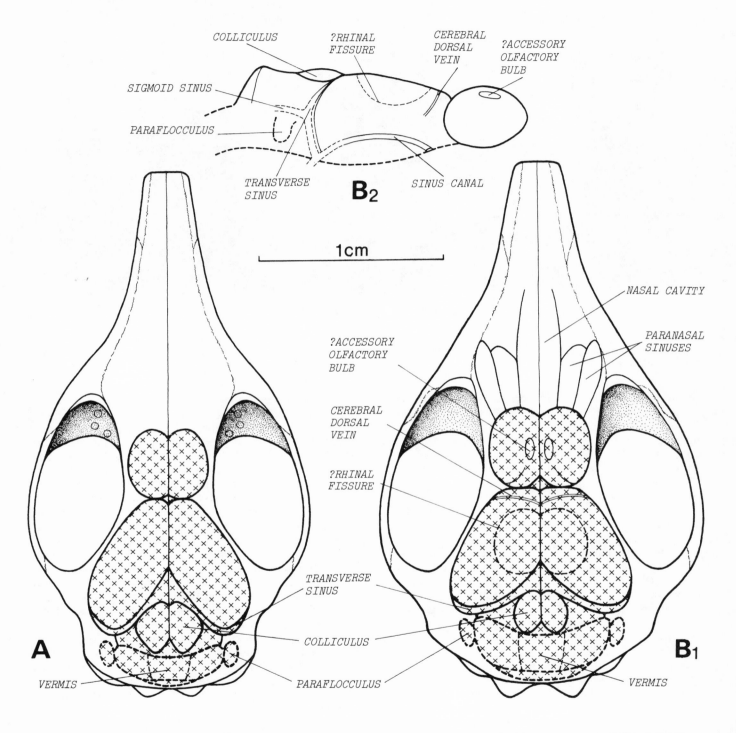

Figure 5. Reconstructions. *A, Kennalestes gobiensis,* skull and endocast (hatched), dorsal view; *B, Asioryctes nemegetensis, B₁,* skull and endocast, dorsal view; *B₂,* endocast, lateral view (from Kielan-Jaworowska, 1984).

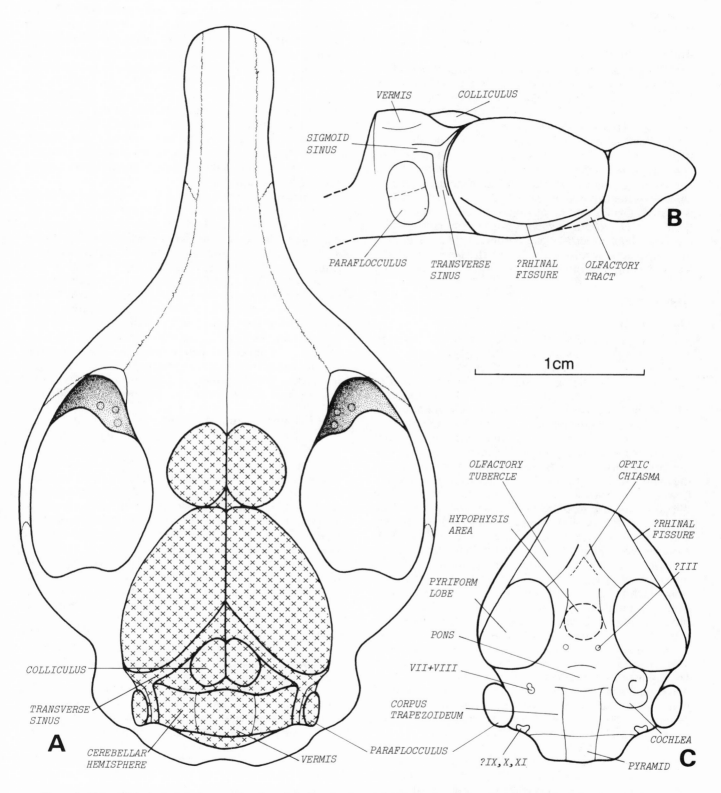

Figure 6. *Zalambdalestes lechei*. *A*, reconstruction of skull and endocast (hatched), dorsal view; *B*, reconstruction of endocast, lateral view; *C*, diagrammatic drawing of ZPAL MgM-I/16 endocast, ventral view (from Kielan-Jaworowska, 1984).

appears continuous with these vessels, we tentatively regarded it as belonging to the cast of part of the sigmoid sinus.

Brain of Monotremes

The gross anatomy of monotreme brain was studied by numerous authors, more recently among others by Hines (1929) and Abbie (1934); see also Griffiths (1968, 1978) for summaries. The brains both of *Ornithorhynchus* (Fig. 1M) and *Tachyglossus* are relatively short, with small olfactory bulbs and enormous cerebral hemispheres with extensive neocortex, lissencephalic in *Ornithorhynchus* and gyrencephalic in *Tachyglossus*. The midbrain, consisting of large anterior colliculi and smaller posterior colliculi, is not exposed on the dorsal side. The cerebellum is relatively wide, with no obvious cerebellar hemispheres (at least in *Ornithorhynchus*). Paraflocculi have flocculi placed at their antero-lateral corners. The pons is prominent, placed wholly posterior to the insertion of the Vth (trigeminal), cranial nerve.

DISCUSSION

The Neocortex

Quiroga (1980a) discussed the problem of development of the neocortex at the reptilian-mammalian transition, from the dual viewpoints of neurophysiology and paleontology. At the same time, the book edited by Ebbesson (1980) was published, showing how complicated and unclear are problems of homology of different parts of the telencephalon between reptiles and mammals.

From the beginning of the twentieth century until the late sixties, neurophysiologists generally accepted that the dorsal cortex of reptiles represents a homologue of the mammalian neocortex. Investigations of recent years, however, demonstrated that the functions of the reptilian dorsal cortex are insufficiently understood. As put by Peterson (1980, p. 381): "Superficially, at least, the existing behavioral data suggest that the dorsal cortex is more closely analogous to the mammalian limbic system, specifically to the hippocampus and septum, than to its motor or sensory neocortex." It is obvious that the question which parts of the reptilian cortex are homologues of the mammalian neocortex may be answered by neurophysiologists, not paleontologists. Neurophysiologists generally accept homology of the neocortex in all extant mammals (*e. g.*, Diamond and Hall, 1969, Lende, 1969, Welker and Lende, 1980 and personal information from Dr. John I. Johnson, Michigan State University). If so, one should expect presence of a tissue homologous to the mammalian neocortex in earliest mammals, and even in cynodonts. However, evidence of existence of such a tissue could hardly be preserved in endocasts.

Quiroga (1980a) claimed that the neocortical plate was present in *Probainognathus*. I argued above that the apparent rhinal fissure of *Probainognathus* might rather be the cast of a vessel. Nevertheless, the shape of the expanded cerebral hemispheres both in Quiroga's and my reconstructions (Figs. 1C-E), as well as in *Therioherpeton* (Figs. 1F-H), suggests presence of neocortex. Similarly, the expanded cerebral hemispheres of triconodonts may suggest presence of neocortex, although a rhinal fissure has not been found. Neither has a rhinal fissure been found in multituberculates, but I claimed (Kielan-Jaworowska, 1983) that the convex dorso-lateral part of the cerebral hemispheres might correspond to the extensive neocortex.

I demonstrated (Kielan-Jaworowska, 1984) that a rhinal fissure has not been preserved in the Cretaceous eutherian *Kennalestes*; it was tentatively identified in *Asioryctes* and *Zalambdalestes*. But in *Barunlestes* (Kielan-Jaworowska and Trofimov, *in press*), which is closely related to *Zalambdalestes,* a rhinal fissure has not been found.

The conclusion on existence of the neocortex based on presence or absence of a rhinal fissure may be misleading. In the extant marsupial mole *(Notoryctes typhlops),* a rhinal fissure is not perceptible macroscopically; microscopic sections, however, clearly show a rhinal fissure pattern (Schneider, 1968). Presence of a rhinal fissure does signify the existence of neocortex, but its absence cannot prove lack of the neocortex. In this situation, it is difficult to demonstrate unequivocally at which level of mammalian or cynodont evolution the neocortex made its appearance. Nevertheless, on the basis of expansion of cerebral hemispheres, one may tentatively conclude that the neocortex possibly was well developed in all Mesozoic mammals, and in some (*e. g., Probainognathus* and *Therioherpeton),* but not all, cyondonts.

The Cerebellum and Midbrain

Ariëns Kappers and others (1960) characterized the mammalian cerebellum as follows: "The cerebellum of mammals, as compared with that of lower vertebrates, is characterized by the greater development of its transverse diameter. In most mammals this increase is to be noted not only in lobus medius, from which the hemispheres are chiefly formed, but also in the floccular portion of the posterior lobe. To the above statement the monotremes are an exception."

Studies on endocranial casts of Mesozoic mammals allow one to emend the above statement; not only monotremes, but also triconodonts and multituberculates provide exceptions in this respect. Among mammals, only marsupials and placentals are characterized by development of the transverse diameter of the cerebellum and an acquisition of cerebellar hemispheres. In all other mammals, the cerebellum consists only of a vermis and large paraflocculi.

In mammals which have not developed an increase of the transverse diameter of the cerebellum, the vermis is large. It is longer and wider than in primitive eutherians of similar endocast length. It is also deeply inserted between caudal ends of the cerebral hemispheres, and pointed anteriorly. Apparently it contributes to overlap of the midbrain, which is not exposed on the dorsal surface.

Midbrain exposure was discussed by Starck (1964) and Edinger (1963, 1964). Starck demonstrated that midbrain overlap in extant Theria depends upon three factors: folding of the midbrain itself; development of the neopallium; and development of the cerebellum. Edinger (1964, p. 18) stated: "In extremely primitive macrosmatic mammals of 180 to 140 million years ago, an expanded but still palaeencephalic cerebrum overlapped a midbrain which was still of the reptilian type as were the hindbrain and other characters in these most ancient mammals.... In early Neozoic eutherian mammals, the midbrain had again emerged to the brain surface. Now it was stretched between a cerebrum that had acquired a distinct neocortex, and a fully mammalian cerebellum."

The problem appears to be more complex. It follows from preceding descriptions that the midbrain was not exposed on the dorsal surface in Liassic or Purbeckian triconodonts, or in Kimmeridgian, Late Cretaceous, or Paleocene multituberculates. It may be presumed that in all non-therian mammals (*i. e.*, triconodonts, multituberculates, and possibly docodonts, which probably originated from triconodonts), the midbrain became overlapped early in the evolution of these groups and never emerged again to the brain surface.

Midbrain exposure in evolution of the Theria is less clear. Endocasts of the earliest Theria, the symmetrodonts, are not known. If the endocast of *Amblotherium* is an artefact, then it is not known whether the midbrain was exposed in eupantotheres. The midbrain was widely exposed in all known Late Cretaceous and possibly in most (or all) Paleocene Eutheria (Kielan-Jaworowska, 1984; Edinger, 1929, 1956; Russell and Sigogneau, 1965). As the midbrain is exposed in various primitive extant marsupials (*e. g.*, in most didelphids), it may be presumed that it might have been exposed in all early Tribosphenida (an infraclass or legion of Theria with tribosphenic molars, embracing Aegialodontia, Pappotherida, Deltatheroida, Marsupialia, and Eutheria; see McKenna, 1975 and Dashzeveg and Kielan-Jaworowska, 1984). Overlap of the midbrain in various eutherian lines occurred at different times during the Tertiary, while in other lines the midbrain remained exposed until today (Edinger, 1929, 1948, 1956; Starck, 1962, 1964; Stephan and Spatz, 1962; Jerison, 1973; Radinsky, 1976, 1977a,b, 1981). The record of brain evolution for marsupials is less complete.

In all known endocasts of Late Cretaceous eutherians, the midbrain roof consists of one pair of large colliculi. Possibly, the same holds for most Paleocene endocats. However, in the late Paleocene miacid described originally as a bat (Edinger, 1964, but see Jepsen, 1966), small anterior and large posterior colliculi are present.

It is not known whether the midbrain roof consisted of but the corpora bigemina (optic colliculi) in brains that produced Cretaceous eutherian endocasts. In some reptiles (snakes and some lizards; see Ariëns Kappers and others, 1960), the midbrain externally shows four protuberances. However, in most reptiles (including cynodonts; see Hopson, 1979; Kemp, 1982), the midbrain roof consists but of the optic lobes. If unpaired colliculi of Cretaceous Eutheria correspond to the corpora bigemina of reptiles, then this would be an extremely primitive feature retained in these brains. However, it cannot be excluded that one pair of colliculi in these brains has been overlapped. Lastly, a third possibility is that the lack of a furrow separating the anterior and posterior colliculi is an artefact due to the state of preservation of these endocasts.

Phylogenetic Implications

Brain endocasts of Mesozoic mammals display no flexure, and were lissencephalic and macrosmatic, with relatively larger olfactory bulbs and less expanded cerebral hemispheres than in those of Tertiary and living mammals. Within this basic pattern, two different types are recognized, designated herein as cryptomesencephalic and eumesencephalic.

Cryptomesencephalic type (Fig. 1*I-J;* Greek *cryptos* —hidden, *mesos*—middle, *engkephalos*—brain) occurs in non-therian mammals (Triconodonta and Multituberculata). It is not known in Docodonta. Its characteristic features are: the cerebellum is deeply inserted between the cerebral hemispheres and consists of a large vermis and paraflocculi; the cerebellar hemispheres are not developed; the midbrain is not exposed on the dorsal side. In multituberculates the insertion of Vth (trigeminal) nerve is placed wholly anterior to the pons. This structure is unknown in Triconodonta, but it occurs in Monotremata, which are discussed below.

Eumesencephalic type (Fig. 1*K-L;* Greek *eu*-well, *mesencephalon*-as above) occurs in Cretaceous (and possibly in most or all Paleocene) Tribosphenida. In this type, in contrast to the former, the transverse diameter of the cerebellum is greatly increased, and cerebellar hemispheres are developed. The midbrain is widely exposed on the dorsal surface, and its roof consists (at least in Cretaceous Eutheria) of one pair of colliculi. Insertion of the Vth cranial nerve is placed posterior to the pons.

Questions arise how these two types of brains could have evolved from those of cynodonts, and what the relationships between them are. Before answering, I should stress that endocranial casts of Mesozoic mammals reflect relatively well the gross anatomy of the brains. However, this is not the case with the few known endocranial casts of cynodonts, which may be interpreted in a variety of ways. This is because cynodont brains possibly did not fill the endocranial cavity as tightly as brains of mammals do. Another reason is that known cynodont endocasts are represented by single specimens, which often are distorted. Perhaps if endocasts of *Probainognathus* and *Therioherpeton* were known from several specimens, reconstructions of their brains would be less ambiguous. Thus because of highly disparate interpretations that can be drawn from endocranial casts of *Probainognathus* and *Therioherpeton,* discussion that follows on brain evolution at the reptilian-mammalian transition must be regarded as speculative.

If the reconstructions of cynodont brains made by Quiroga (Fig. 1*C* and 1*F* in the present paper) are correct, it would be difficult to visualize evolution of cryptomesencephalic (triconodont and multituberculate) brains from them. In both *Probainognathus* and *Therioherpeton,* Quiroga reconstructed a wide cerebellum, similar to that in Theria (in particular so in *Therioherpeton*), possibly with at least incipient cerebellar hemispheres, although he did not mention this similarity. In late Jurassic triconodonts and in all known multituberculates, the cerebellum consists only of an expanded vermis and parafloculi. Such a cerebellum could not develop from the laterally expanded one of *Therioherpeton* as reconstructed by Quiroga (Fig. 1*F*); this would involve a secondary reduction of already formed cerebellar hemispheres.

In the first of my alternative reconstructions of brains of *Probainognathus* and *Therioherpeton* (Fig. 1*D* and 1*G*, respectively), I presume that the midbrain was exposed on the dorsal side in both these genera; but I reconstruct the cerebellum as relatively narrow, consisting only of the vermis and parafloculi. Both the cryptomesencephalic and eumesencephalic brains could be developed from such type of brain. Development of the cryptomesencephalic type would involve enlargement of the vermis, which would overlap the midbrain. If the midbrain really was exposed on the dorsal surface in the Liassic triconodont *Sinoconodon* (see Patterson and Olson, 1961; Quiroga, 1984), one may presume that overlap of the midbrain in Triconodonta occurred during the Jurassic. In all known multituberculate endocasts, the midbrain is completely overlapped, and it is not known when this overlap took place. In evolution both of triconodonts and multituberculates, the midbrain did not emerge again on the dorsal surface, and cerebellar hemispheres did not develop. The similarity of known endocasts of triconodonts and multituberculates (at least in dorsal aspect) suggests that these groups may be more closely related than hitherto accepted.

The eumesencephalic type of brain developed differently. Here, the midbrain remained exposed on the dorsal side, possibly during the whole of the Mesozoic (as it is exposed in all known late Cretaceous and most Paleocene Tribosphenida), and is exposed in some conservative lines even now. In most eutherian and marsupial lines, it became overlapped at various times during the Tertiary.

It was demonstrated recently (Archer and others, 1985, Kielan-Jaworowska, Crompton, and Jenkins, *in press*), that monotremes belong to Theria. According to the latter interpretation, monotremes diverged from the main therian line leading to the Tribosphenida possibly during the Jurassic. Endocranial casts of the Early Cretaceous monotreme *Steropodon* and the Miocene monotreme *Obdurodon* (see Woodburne and Tedford, 1975) are known. Brains of modern monotremes are highly specialized. Both in *Ornithorhynchus* (Fig. 1*M*) and *Tachyglossus* there are large anterior and posterior colliculi, entirely obscured by the cerebellum and by large cerebral hemispheres. There are no obvious cerebellar hemispheres in *Ornithorhynchus,* but incipient ones are present in *Tachyglossus*. In both these genera the cerebellum is relatively wider than in triconodonts or multituberculates, but narrower than in Cretaceous and modern Theria.

One may speculate that the monotreme brain originated from the pre-eumesencephalic type (with exposed midbrain and only incipient or no cerebellar hemispheres), and that overlap of the midbrain in recent monotremes is secondary, parallel to that in various therian lines. A characteristic feature of the eumesencephalic brain is development of cerebellar hemispheres. As these are only incipiently developed in modern monotremes, one may presume that they were not developed as yet, or were only incipient in common ancestors of the Monotremata and Tribosphenida; they started to develop in the line leading to the Tribosphenida at about the time of the Monotremata-Tribosphenida divergence, during the Jurassic.

However, as I demonstrated above, endocasts of *Probainognathus* and *Therioherpeton* are somewhat distorted and may be interpreted in an alternative way (Fig. 1*E* and 1*H*, respectively). If these reconstructions are correct, then some cynodonts would acquire cryptomesencephalic brains. If so, one would presume that all of the earliest mammals, both Theria and non-Theria, had cryptomesencephalic brains (as accepted by Edinger, 1964 and Quiroga, 1984). Then the eumesencephalic type would have originated from the cryptomesencephalic type by secondary exposure of the midbrain and the acquisition of cerebellar hemispheres. The drawback of this hypothesis is the difficulty in visualizing reduction of the strongly expanded vermis of the cryptomesencephalic type to obtain the non-expanded one of the eumesencephalic type. Theoretically, the former hypothesis presented above appears the more probable. Also, it is more parsimonious; it does not involve reduction of earlier-developed brain structures, such as the size of the vermis or cerebellar hemispheres, in evolution of the mammalian brain.

ACKNOWLEDGMENTS

I thank Dr. Jeremy J. Hooker (British Museum, Natural History, London) for sending casts of British cynodont and Mesozoic mammal natural cranial endocasts, and Dr. Juan C. Quiroga (Museo Universidad National, La Plata) for stereophotographs and casts of brain endocasts of *Probainognathus* and *Therioherpeton*. I am also grateful to Dr. John I. Johnson (Department of Anatomy, Michigan State University, East Lansing), Dr. Jon H. Kaas (Department of Psychology, Vanderbilt University, Nashville), and Dr. Mervyn Griffiths (Deakin, Canberra) for valuable comments. The following persons from the technical staff of the Institute of Paleobiology in Warsaw helped me in preparation of this paper: Ms. E. Mulawa took the photograph, and Ms. Elzbieta Gutkowska and Ms. Danuta Slawik made the drawings.

REFERENCES CITED

Abbie, A. A., 1934, The brain stem and cerebellum in *Echidna aculeata*: Royal Society of London, Philosophical Transactions, ser. B, v. 224, p. 1-74.

Archer, M., Flannery, T. F., Ritchie, A., and Molnar, R. E., 1985, First Mesozoic mammals from Australia—an early monotreme: Nature, v. 318, p. 363-366.

Ariëns Kappers, C. U., Huber, C. C., and Crosby, E. C., 1960, The comparative anatomy of the nervous system of vertebrates, including man: New York, Hafner Publishing Company, 3 vols., 1845 p.

Bauchot, R., and Stephan, H., 1967, Encéphales et moulages endocraniens de quelques insectivores et primates actuels: Problèmes actuels de paléontologie (Evolution des Vertébrés), Colloques Internationaux du CNRS, no. 163; Paris, Editions du CNRS, p. 575-586.

Clemens, W. A., Lillegraven, J. A., Lindsay, E. H., and Simpson, G. G., 1979, Where, when, and what—a survey of known Mesozoic mammal distribution, in Lillegraven, J. A., Kielan-Jaworowska, Z., and Clemens, W. A., eds., Mesozoic mammals: the first two-thirds of mammalian history: Berkeley, University of California Press, p. 7-58.

Dashzeveg, D., and Kielan-Jaworowska, Z., 1984, The lower jaw of an aegialodontid mammal from the Early Cretaceous of Mongolia: Linnean Society of London, Zoological Journal, v. 82, p. 217-227.

Diamond, I. T., and Hall, W. C., 1969, Evolution of neocortex: Science, v. 164, p. 251-262.

Ebbesson, S. O. E., ed., 1980, Comparative neurology of the telencephalon: New York, Plenum Press, *xxi* + 506 p.

Edinger, T., 1929, Die fossilen Gehirne: Zeitschrift für die gesamte Anatomie, Abt. III, v. 28, p. 1-249.

———— 1948, Evolution of the horse brain: Geological Society of America, Memoir 25, 177 p.

———— 1956, Objets et résultats de la paléoneurologie: Annales de Paléontologie, v. 42, p. 95-116.

———— 1963, Meanings of the midbrain exposure, past and present: XVI International Congress of Zoology, Proceedings, v. 3, Specialized Symposia, Washington, D. C., p. 225-228.

———— 1964, Midbrain exposure and overlap in mammals: American Zoologist, v. 4, p. 5-19.

Griffiths, M., 1968, Echidnas: Oxford, Pergamon Press, *ix* + 282 p.

———— 1978, The biology of the monotremes: London, Academic Press, *vii* + 367 p.

Hahn, G., 1969, Beiträge zur Fauna der Grube Guimarota Nr. 3. Die Multituberculata: Palaeontographica, Abt. A, v. 133, 100 p.

Hines, M., 1929, The brain of *Ornithorhynchus anatinus*: Royal Society of London, Philosophical Transactions, ser. B, v. 217, p. 155-287.

Hopson, J. A., 1964, The braincase of the advanced mammal-like reptile *Bienotherium*: Postilla, no. 87, 30 p.

———— 1969, The origin and adaptive radiation of mammal-like reptiles and non-therian mammals: New York Academy of Sciences, Annals, v. 167, p. 199-216.

———— 1979, Paleoneurology, in Gans, C., Northcutt, R. G., and Ulinski, P., eds., Biology of the Reptilia, v. 9, Neurology A: London, Academic Press, p. 39-146.

Jerison, H. J., 1973, Evolution of the brain and intelligence: New York, Academic Press, *xiv* + 482 p.

Jepsen, G. L., 1966, Early Eocene bat from Wyoming: Science, v. 154, p. 1333-1339

Kemp, T. S., 1979, The primitive cynodont *Procynosuchus:* functional anatomy of the skull and relationships: Royal Society of London, Philosophical Transactions, ser. B, v. 285, p. 73-122.

———— 1982, Mammal-like reptiles and the origin of mammals: London, Academic Press, 363 p.

———— 1983, The relationships of mammals: Linnean Society of London, Zoological Journal, v. 77, p. 353-384.

Kermack, K. A., 1963, The cranial structure of the triconodonts: Royal Society of London, Philosophical Transactions, ser. B, v. 246, p. 83-103.

Kermack, K. A., Mussett, F., and Rigney, H. W., 1981, The skull of *Morganucodon:* Linnean Society of London, Zoological Journal, v. 71, p. 1-158.

Kielan-Jaworowska, Z., 1983, Multituberculate endocranial casts: Palaeovertebrata, v. 13, p. 1-12.

———— 1984, Evolution of the therian mammals in the Late Cretaceous of Asia. Part VI. Endocranial casts of eutherian mammals, in Kielan-Jaworowska, Z., ed., Results of the Polish-Mongolian Palaeontological Expeditions, Part X: Palaeontologia Polonica, no. 46, p. 157-171.

Kielan-Jaworowska, Z., Crompton, A. W., and Jenkins, F. A., Jr., in press, Monotreme relatives: Nature.

Kielan-Jaworowska, Z., and Krause, D. M., *in preparation,* Endocranial cast of *Ptilodus* (Multituberculata).

Kielan-Jaworowska, Z., Poplin, C., Presley, R., and de Ricqlés, A., 1984, Preliminary note on multituberculate cranial anatomy studied by serial sections, in Reif, W. E., and Westphal, F., eds., Third symposium on Mesozoic terrestrial ecosystems, short papers: Tübingen, Attempto Verlag, p. 123-128.

Kielan-Jaworowska, Z., Presley, R., and Poplin, C., *in press,* The cranial vascular system in taeniolabidoid multituberculate mammals: Royal Society of London, Philosophical Transactions.

Kielan-Jaworowska, Z., and Trofimov, B., 1980, Cranial morphology of Cretaceous eutherian mammal *Barunlestes:* Acta Palaeontologica Polonica, v. 26, p. 167-185.

———— *in press,* Endocranial cast of Cretaceous eutherian mammal *Barunlestes:* ibid.

Kühne, W. G., 1956, The Liassic therapsid *Oligokyphus:* London, British Museum (Natural History), 149 p.

Lende, R. A., 1969, a comparative approach to the neocortex: localization in monotremes, marsupials and insectivores: New York Academy of Sciences, Annals, v. 167, p. 262-276.

Lillegraven, J. A., Kielan-Jaworowska, Z., and Clemens, W. A., eds., 1979, Mesozoic mammals: the first two-thirds of mammalian history: Berkeley, University of California Press, 311 p.

McKenna, M. C., 1975, Toward a phylogenetic classification of the Mammalia, in Luckett, W. P., and Szalay, F. S., eds., Phylogeny of the primates: New York, Plenum Press, p. 21-46.

Olson, E. C., 1944, Origin of mammals based on the cranial morphology of the therapsid suborders: Geological Society of America, Special Paper no. 55, p. 1-136.

Owen, R., 1871, Monograph of the fossil Mammalia of the Mesozoic formations: Palaeontographical Society, Monographs, no. 24, 115 p.

Patterson, B., and Olson, E. C., 1961, A triconodontid mammal from Triassic of Yunnan, in Vandebroek, G., ed., International colloquium on the evolution of lower and non specialized mammals: Brussels, Koninklijke Vlasmse Academie voor Watenschappen, Letteren en Schone Kunsten van Belgie, pt. 1, p. 129-191.

Peterson, E., 1980, Behavioral studies of telencephalic function in reptiles, in Ebbesson, S. O. E., ed., Comparative neurology of the telencephalon: New York, Plenum Press, p. 343-388.

Quiroga, J. C., 1979, The brain of two mammal-like reptiles (Cynodontia-Therapsida): Journal für Hirnforschung, v. 20, p. 341-350.

―――― 1980a, The brain of the mammal-like reptile *Probainognathus jenseni* (Therapsida, Cynodontia). A correlative paleoneonneurological approach to the neocortex at the reptile-mammal transition: ibid, v. 21, p. 299-226.

―――― 1980b, Sobre un molde endocraneano del cinodonte *Probainognathus jenseni* Romer, 1970 (Reptilia, Therapsida) de la formacion Ischichuca (Triasicio medio), La Rioja, Argentina: Ameghiniana, v. 17, p. 181-190.

―――― 1980, The endocranial cast of the mammal-like reptile *Therioherpeton cargnini* (Therapsida, Cynodontia) from the Middle Triassic of Brazil: Journal für Hirnforschung, v. 25, p. 285-290.

Radinsky, L., 1976, Oldest horse brains: more advanced than previously realized: Science, v. 194, p. 626-627.

―――― 1977a, Brains of early carnivores: Paleobiology, v. 3, p. 333-349.

―――― 1977b, Early primitive brains: facts and fiction: Journal of Human Evolution, v. 6, p. 79-86.

―――― 1981, Brain evolution in extinct South American ungulates: Brain, Behavior and Evolution, v. 18, p. 169-187.

Russell, D. E., and Sigogneau, D., 1965, Etude de moulages endocraniens de Mammifères Paléocènes: Muséum National d'Histoire Naturelle, Mémoires, v. 16, p. 1-36.

Schneider, C., 1968, Beitrag zur Kenntnis des Gehirnes von *Noryctes typhlops*: Anatomischer Anzeiger, v. 123, p. 1-24.

Simpson, G. G., 1927, Mesozoic Mammalia. IX. The brain of Jurassic mammals: American Journal of Science, v. 214, p. 259-268.

―――― 1928, A catalogue of the Mesozoic Mammalia in the Geological Department of the British Museum: London, Oxford University Press, x + 215 p.

―――― 1929, American Mesozoic Mammalia: Peabody Museum (Yale University), Memoires, v. 3, New Haven, Yale University Press, xv + 171 p.

―――― 1937, Skull structure of the Multituberculata: American Museum of Natural History, Bulletin, v. 73, p. 727-763.

―――― 1959, Mesozoic mammals and the polyphyletic origin of mammals: Evolution, v. 13, p. 405-414.

―――― 1961, Evolution of Mesozoic mammals, in Vandebroek, G., ed., International colloquium on the evolution of lower and non specialized mammals: Brussels, Koninklijke Vlaamse Academie voor Wetenschappen, Letteren en Schone Kunsten van Belgie, pt. 1, p. 57-95.

―――― 1971a, Recent literature on Mesozoic mammals: Journal of Paleontology, v. 45, p. 862-868.

―――― 1971b, Concluding remarks: Mesozoic mammals revisited, in Kermack, D. M., and Kermack, K. A., eds., Early mammals: Linnean Society of London, Zoological Journal, v. 50, Suppl. no. 1, p. 181-198.

Starck, D., 1962, Die Evolution des Säugetier-Gehirns: Wissenschaftliche Geselschaft an der Johann Wolfgang Goethe-Universität Frankfurt am Main, Sitzungsberichte, v. 1, p. 1-60.

―――― 1964, "Freiliegendes Tectum mesencephali" ein Kennzeichen des primitiven Säugetiergehirns?: Zoologischer Azeiger, v. 171, p. 350-359.

Stephan, H., and Spatz, H., 1962, Vergleichend-antomische Untersuchungen an Insektivorengehirnen. IV. Gehirne afrikanischer Insektivoren. Versuch einer Zuordnung von Hirnbau und Lebensweise: Morphologisches Jahrbuch, v. 103, p. 108-174.

Watson, D. M. S., 1913, Further notes on the skull, brain and organs of special sense of *Diademodon*: Annals and Magazine of Natural History, ser. 8, v. 12, p. 213-228.

Welker, W., and Lende, R. A., 1980, Thalamocortical relationships in Echidna *Tachyglossus aculeatus*, in Ebbesson, S. O. E., ed., Comparative neurology of the telencephalon: New York, Plenum Press, p. 449-481.

Woodburne, M. O., and Tedford, R. H., 1975, The first Tertiary monotreme from Australia: American Museum Novitates, no. 2588, 11 p.

MANUSCRIPT RECEIVED OCTOBER 10, 1985
REVISED MANUSCRIPT RECEIVED MAY 9, 1986
MANUSCRIPT ACCEPTED JUNE 4, 1986

Origin and transformation of the mammalian stapes

MICHAEL J. NOVACEK *Department of Vertebrate Paleontology, The American Museum of Natural History, New York, New York 10024*

ANDRÉ WYSS *Department of Geological Sciences, Columbia University, New York, New York 10027*

ABSTRACT

The mammalian stapes is the subject of considerable investigation, but ambiguity remains with respect to the primitive condition in higher-level mammal clades and the pattern of subsequent modifications. We question the widely held belief that the strongly bicrurate, stirrup-like stapes represents the ancestral mammalian and therian state (Goodrich, 1930). The primitive stapes in the common ancestor of mammals and therapsids was probably columelliform and had a stapedial foramen for the passage of the stapedial artery. However, additional modifications are required to produce the bicrurate structure characteristic of many eutherians. Moreover, distributional evidence does not rule out the possibility that the columelliform-imperforate stapes seen in adult monotremes was ancestral for therians or a group comprising therians and monotremes.

INTRODUCTION

Perhaps more has been written about the auditory ossicles than any other aspect of mammalian skeletal anatomy. This is of course partly due to the unique nature of two ossicles, the incus and malleus, whose appearance provides an important characterization for the mammals. During the 19th century, one body of workers held that, in addition to the stapes, the malleus and incus were derived from the reptilian columella auris (Dollo, 1883; Cope, 1888; Gadow, 1901). This view was contradicted by studies of ontogeny of the ossicles, which demonstrated that the malleus was derived from the reptilian articular, and the incus was derived from the quadrate (Reichert, 1837). It was a modern version of the latter theory that found broad acceptance (Gaupp, 1913; Klaauw, 1923; Goodrich, 1930). Under this view, only the mammalian stapes is a derivative of the columella auris.

We do not question the essential homology between the reptilian columella and the mammalian stapes. However, we find that an overly simplified view of this problem has fostered ambiguities concerning the origin and transformation of the stapes in mammals. The majority of mammals has a bicrurate stapes, with a large opening (stapedial foramen) for passage of the stapedial artery (Figs. 1A, 1D). However, most marsupials, monotremes, some edentates, and pangolins have a stapes more reminiscent of that in many other tetrapod groups. In such cases, the stapes is columelliform and is either pierced near its base by a very small stapedial foramen or slit-like depression, or is lacking this foramen altogether (Fig. 1B, C, E). Another type of mammalian stapes, a massive, inflated element of variable form, is also either imperforate or has a small opening (Fig. 1F). This type, which is found primarily in marine mammals, is recognized as a highly derived condition (Doran, 1878; Fleischer, 1973; Novacek, 1982; remarks below).

In his comprehensive survey of the mammalian ossicles, Doran (1878) characterized the columelliform stapes as a "low type" that represented a basic "sauropsid" character. Following Doran, authorities on mammalian classification (e. g., Gregory, 1910) and specialists on auditory anatomy (e. g., Segall, 1970) concluded that a columelliform-imperforate stapes represents the primitive mammalian condition. Novacek (1982) argued

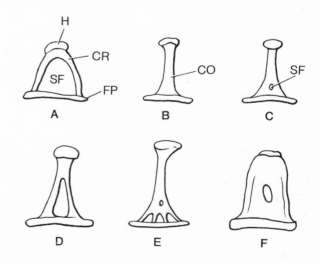

Figure 1. Variation of stapes: *A*, bicrurate (stirrup-like); *B*, columelliform-imperforate; *C*, columelliform-"microperforate"; *D*, bicrurate; *E*, columelliform-"microperforate" (Aves type); and *F*, massive-microperforate. See text for discussion. Symbols are: CO, columella; CR, stapedial crura; FP, stapedial footplate; H, stapedial head; and SF, stapedial foramen. Not to scale.

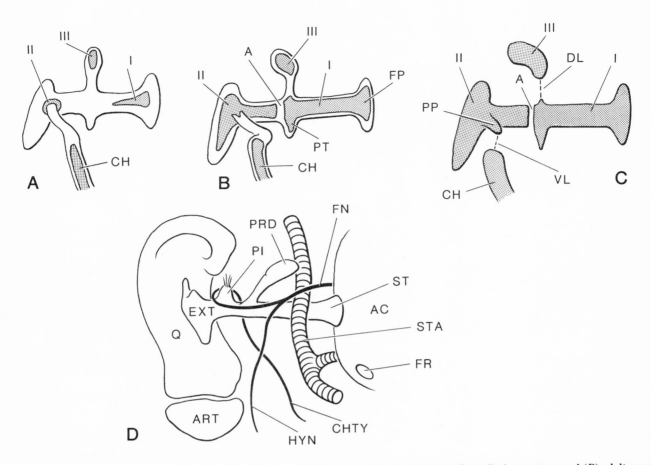

Figure 2. Development of stapes in *Lacerta* (after Versluys, 1903) during earlier (*A, B*), later (*C*) cartilaginous stages and (*D*) adult osseous stage. Blastema outlined, cartilage indicated by stippled area. I, otostapedial cartilage; II, hyostapedial; III, pars dorsalis of dorsal process. Symbols are: A, temporary limit between I and II; AC, auditory capsule; ART, articular; CH, cornu hyale; CHTY, chorda tympani; DL, ligamentous vestige of base of processus dorsalis; EXT, extra-stapedial; FN, hyomandibular branch of facial nerve; FP, stapedial footplate; FR, fenestra rotunda; HYN, hyoid nerve; PI, processus internus; PP, processus posterior; PRD, dorsal process; Q, quadrate; ST, stapes; STA, stapedial artery; and VL, vestige of pars interhyalis.

that a columelliform stapes either lacking or having a small stapedial foramen is primitive. These conclusions run against the views of Kuhn (1971), Henson (1974), Fleischer (1978), and other workers who have accepted Goodrich's (1930) contention that a bicrurate, stirrup-like stapes is primitive for mammals.

At first inspection, Goodrich's (1930) evidence for his argument is compelling. He noted (*ibid.*, 1930, p. 457): "In many marsupials (Dasyuridae, Phalangistidae, Peramelidae) and in monotremes the stapes is imperforate and columelliform, but this condition appears to be secondary, since, in *Ornithorhynchus* and *Trichosurus* at all events, it surrounds the stapedial artery in the blastematous stage." This influential statement highlights a basic problem in discussions of evolution of the stapes. Various conditions often refer to at least two traits—the shape of the stapes, and the absence or presence of the stapedial foramen for passage of the stapedial artery. These traits are not in perfect correspondence; some columelliform stapes are without a foramen, and some with (Fig. 1*B, C, D, E*). Moreover, some groups show a broad-

ly open, bicrurate stapes where there is no stapedial artery passing through it.

Hence, a theory for transformation of the stapes should account for the mosaic of relevant traits present in embryonic and adult stages of various groups. Here we examine the components of the stapes in somewhat more detail, and review arguments for their modifications at different levels within Mammalia. Our review is, of course, largely dependent on information provided in the remarkable documents of Doran (1878), Gaupp (1913), and Goodrich (1930) among others. We can hardly claim more than a reinterpretation of aspects of these studies. Even within the current milieu of comparative biology, these works have a thoroughly modern stamp. Indeed, these studies display a forthright and comprehensive treatment of comparative data that seems so peculiarly absent in many recent publications.

We are also honored to present this essay as part of an issue commemorating the life and works of George Gaylord Simpson. The great breadth of Simpson's scientific achievements included a focus on the problems of

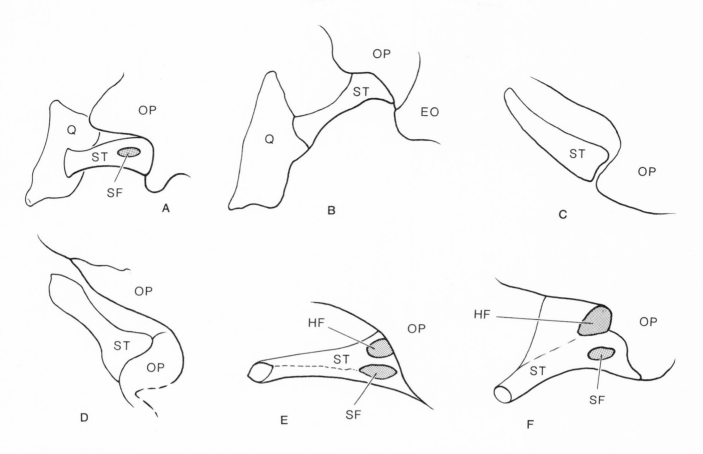

Figure 3. Left stapes in various fossil reptiles: *A, Millerina; B, Nyctiphruretus; C, Seymouria; D, Kotlassia; E, Captorhinus;* and *F, Edaphosaurus* (all views are posterior left stapes). Symbols are: EO, exoccipital; HF, hyomandibular foramen; OP, opisthotic; Q, quadrate; SF, stapedial foramen; and ST, stapes (after Romer, 1956, from various sources). Not to scale.

mammalian origins and the evolution of early mammalian traits. Studies on embryology and homology of the ossicle system were relevant to these concerns. We know he would have inspected our arguments with a keen and interested eye.

STAPES CONDITIONS

From the foregoing it is apparent that several components come into play in the transformation of the mammalian stapes. The cast of characters includes: the form or shape of the stapes (columelliform, stirrup-like, or some other variation), the presence or absence of the stapedial foramen, the relative size of the stapedial foramen, the presence, absence, development, and location of the stapedial artery, the nature of the contact between the stapes and the auditory capsule (stapedial footplate round, oval, cartilaginous, or bony, *etc.*), the articulation of the stapes with its adjacent ossicle, the incus (or, in the case of sauropsids, the relationship between the proximal stapes and the extra-stapedial elements that contact the tympanic membrane), and the development of the attachment for the stapedius muscle. This account emphasizes aspects relating to stapes form, development of the stapedial artery and foramen, and their functional implications. For purposes of comparison we recognize four different stapedial conditions:

A. Columelliform-perforate—This condition is generally distributed among sauropsids and other non-mammalian vertebrates. The stapes has a rod-like or columellar outline. The stapedial foramen (presumably for passage of the stapedial artery in extinct forms as well) is moderate to large in size, but not to the extent that it radically alters the columellar structure of the stapes. The foramen is located either centrally or basally (*i. e.,* near the contact of the stapes with the auditory capsule), but the stapes retains much ossification both proximal and distal to the foramen (Figs. 3*A, E-F,* 4*A, E-H*). There is less bone anterior and posterior to the foramen, but these sections usually do not approach the delicate form of the stapedial crura in condition *B.*

B. Bicrurate—The bicrurate stapes is essentially dominated by the large stapedial foramen. The ossified stapes is represented only by slender crura anterior and posterior to the foramen, a stapedial head (or capitulum) for articulation with the incus and insertion of the stapedius muscle, and a stapedial footplate that contacts the fenestra ovalis. The classic form of the bicrurate

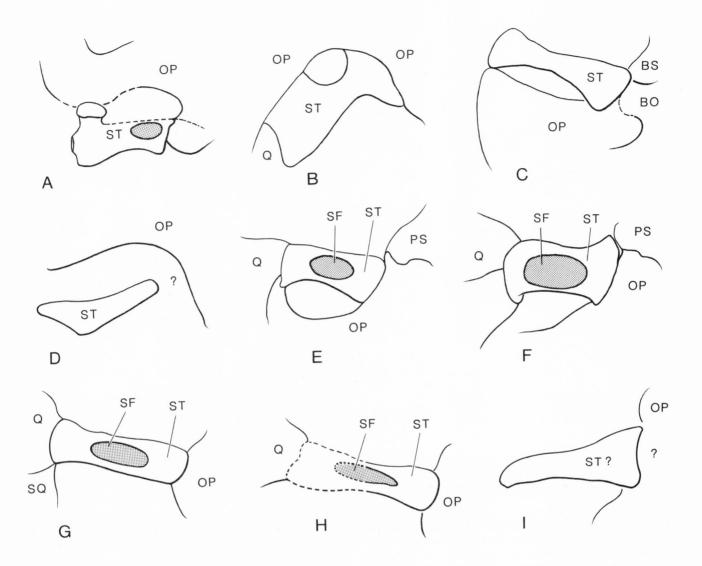

Figure 4. Stapes in therapsid reptiles: *A, Scylacops; B, Ulemosaurus; C, Bauria; D, Lycedops; E, Procynosuchus; F, Thrinaxodon; G, Cynognathus; H, Probainognathus;* and *I, Bienotherium. A, B,* and *D* are posterior views of left stapes. *C, E, F, G, H,* and *I* are ventral views of right stapes. Symbols are: BO, basioccipital; BS, basisphenoid; OP, opisthotic; PS, presphenoid; Q, quadrate; SF, stapedial foramen; SQ, squamosal; and ST, stapes (after: Young, 1947; Romer, 1956; Estes, 1961; Allin, 1975; and Parington, 1979). Not to scale. Dashed line indicates reconstruction of damaged or missing areas.

stapes is that of a "horse-shoe" or "stirrup" (Figs. 1*A*, 6*A, G-I, K-L*), where the expansive foramen is flanked by very delicate and strongly curved crura. In some marsupials and eutherians the stapedial foramen is moderate to large in size, but the crura diverge only toward the region of the footplate; and usually there is some ossification between the capitulum and the foramen (Fig. 1*D,* 5*E-F,* 6*F*). This state is difficult to categorize. While the stapes in such cases is bicrurate it does not show the essential form of a "stirrup," a distinction noted at least as far back as Doran's (1878) review. This problem has some bearing on the transformation analyses discussed below (see Fig. 7) and, in certain instances, it is useful to distinguish the simple bicrurate condition (Fig. 1*D*), from the bicrurate, stirrup-shaped condition (Fig. 1*A*).

C. Columelliform-imperforate (or microperforate) —Here the stapes is columellar in appearance, and although it may have a well-defined capitulum and stapedial footplate, typically there is no stapedial foramen. The stapes in *Tachyglossus* (Figs. 1*B,* 5*A*) is highly characteristic of the columelliform-imperforate type, and this condition is essentially duplicated in (the adult) stapes of *Ornithorhynchus* (Fig. 5*B*), numerous marsupial groups (Fig. 5*C, I-J*), pholidotans (Fig. 6*B*), and the anteater *Cyclopes* (Fig. 6*E*). There is, in addition, some variation accommodated by this condition. The imperforate stapes in many sauropsid groups retain a basic columellar form but otherwise show little resem-

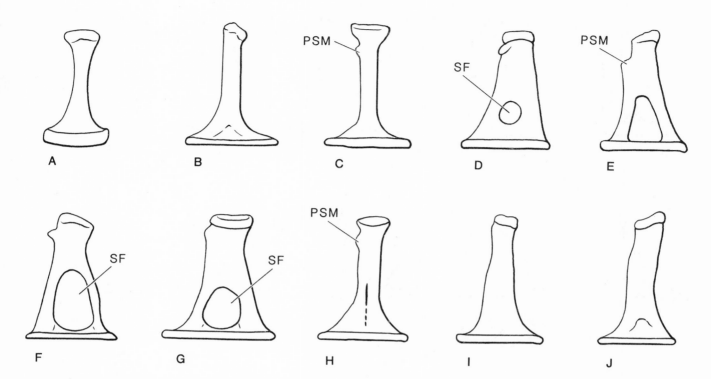

Figure 5. Stapes in monotremes (*A, B*) and marsupials (*C-J*). *A, Tachyglossus; B, Ornithorhynchus; C, Notoryctes; D, Petaurus; E, Thylogale; F, Philander (= Metachirops); G, Pseudochirus; H, Thylacinus; I, Perameles;* and *J, Dasyurus*. Symbols are: PSM, process for insertion of stapedius muscle; and SF, stapedial foramen (after Doran, 1878; Segall, 1970; 1973; Fleischer, 1973; and personal observation). Not to scale.

blance to each other and to the mammalian condition (Figs. 3*B-D*; 4*B-D*). As noted below, it is probable that these sauropsid types represent separate derivations from the columellar-perforate condition described under *A*, and they have little bearing on the problem of stapes transformation within mammals. More interesting is the variation seen in certain eutherians. In some species of *Manis* the columelliform-imperforate stapes is very low, rather than elongate, and has an extremely broad footplate (Fig. 6*C*). This is likely a secondary trend within pholidotans (see Segall, 1973). Bradypodid edentates have an imperforate stapes, but this element has a more fusiform or triangular outline than a columellar one (Fig. 6*D*). Nevertheless, we recognize this stapes as a variant of condition *C*. Finally, there are stapes of strong resemblance to those noted here except that they have a tiny opening or slit, usually denoted as a stapedial foramen. While we recognize that one might categorize these as representatives of condition *A*, these stapes, usually of relatively very small size with a well-defined capitulum and footplate, most strongly resemble condition *C* (Figs. 1*C, E*, 5*D, H, J*). The "stapedial foramen" is so small in these stapes that the appellation "microperforate" seems more fitting than "perforate." The tiny foramen or slit has been regarded as a vestige of a functional stapedial foramen (Doran, 1878), but studies determining what the opening conveys in embryos and adults are lacking. We regard the columelliform-microperforate stapes as a variant of condition *C*, although we recognize that in some cases within mammals (*e. g.*, the rodent *Heliophobius*, see Fig. 6*J*) it is obviously a secondary derivation of the bicrurate condition.

D. Massive-microperforate—An interesting modification of the stapes occurs in certain groups of marine mammals. The stapes is relatively large, with marked ossification that greatly reduces or nearly obliterates the stapedial foramen (Figs. 1*F*, 6*M-O*). This condition, as noted above, is obviously a specialized one, and has little relevance to the transformations considered here. It probably was derived from the bicrurate stapes seen in carnivores and many other eutherians, although transition to its form has not been considered adequately.

It is clear from the above that some stapes do not strictly conform to categories we recognize. Is, for example, the stapes of didelphid marsupials (Fig. 5*F*) better recognized as "columellar-perforate" or "bicrurate"? The answer is not clear-cut, as in any case in which we are forced to summarize a range of variation within a subset of conditions. In sections below we attempt to show where one categorization or another, as in the case of marsupial stapes, is critical to character analysis.

HOMOLOGY OF THE STAPES IN MAMMALS AND OTHER VERTEBRATES

As a premise for any discussion of transformation of the mammalian stapes, homologies of this bone with elements in other vertebrates must be established. Unlike

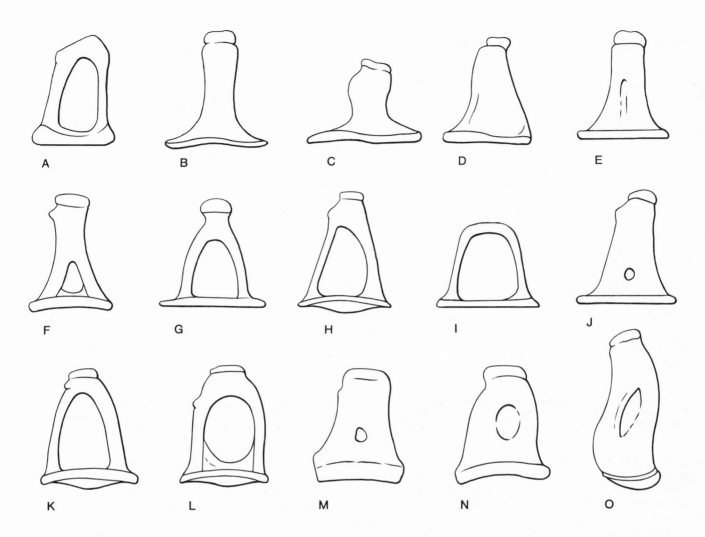

Figure 6. Stapes in eutherians: *A*, Cretaceous form (see Archibald, 1979); *B*, *Manis*; *C*, *Manis*; *D*, *Bradypus*; *E*, *Cyclopes*; *F*, *Tolypeutes*; *G*, *Chlamyphorus*; *H*, *Myrmecophaga*; *I*, *Sorex*; *J*, *Heliophobius*; *K*, *Sciurus*; *L*, *Orycteropus*; *M*, *Balaenoptera*; *N*, *Phoca*; and *O*, *Manatus* (after: Doran, 1878; Segall, 1970, 1973; Fleischer, 1973; and personal observation). Not to scale.

other mammalian ossicles, the stapes is a conservative structure. Its essential form and ontogeny are traceable not only through tetrapods, but also through the fishes. The mammalian stapes, the sauropsid columella, and the fish hyomandibula are all derivatives of the dorsal hyoid arch, and they show similar location relative to certain cranial nerves and vessels (Goodrich, 1930; Lombard and Bolt, 1979). There remains some doubt that the hyomandibula is completely homologous in all fish groups, but this doubt seems largely dispelled by the ontogeny of living forms and by fossil evidence (Goodrich, 1930). Another qualification should be stated from the outset. Several experimental studies (*e. g.*, Noden, 1978) of avian neural crest development would imply that mammalian ossicles are of neural crest origin (Van De Water and others, 1980). However, this argument is inconclusive, due to virtual lack of relevant morphogenetic evidence in mammalian embryos. For the present, we accept the general scheme of homology as derived from abundant work in comparative anatomy and as summarized below.

The hyomandibula in fishes has been broadly homologized with the columella auris in lower tetrapods (Goodrich, 1930, p. 420). Building on this theory, Westoll (1943) and, later, Parrington (1979) identified fine details of structure and function in the hyomandibula of crossopterygian fishes that seemed to be shared by cynodont therapsids and other tetrapods. Although aspects of Parrington's (1967) comparisons have been questioned seriously by Allin (1975), the homology of these elements at a more general level is widely accepted.

Following the revised version of Reichert's (1837) theory, the mammalian stapes is homologous to the reptilian columella auris. However, as Versluys (1903), Gaupp (1913), Goodrich (1930), and others have explained, the columella auris, which extends from the fenestra ovalis of the auditory capsule to the tympanic membrane, is a composite element, not all of which can be equated with the stapes in mammals. The reptilian columella usually forms from three different centers of chondrification, one adjacent to the auditory capsule, another in a process dorsal to the main column, and a third in the distal region near the point of contact with the top of the hyoid arch (Fig. 2A; Versluys, 1903). The proximal cartilage spreads outward, and joins the dorsal process to form a region called the otostapes (Fig. 2B). The outer "extrastapedial" cartilage grows inward to meet the otostapes, forms a small interhyal process, and develops a small processus inferior which contacts the tympanum (Fig. 2B). This outer region is referred to as the hyostapes. The procartilaginous connection between the hyoid cornu and the interhyal process degenerates to a ligament. At later stages, the hyostapes fuses with the otostapes (Fig. 2C). The latter ossifies as a slender rod (stapes), and a joint may form between its distal end and the hyostapes (Fig. 2D). These events are summarized in Goodrich (1930).

The importance of this pattern is demonstrated by its common occurrence. Development of the columella is essentially the same in birds, squamates, crocodilians, and chelonians. Amphibians exhibit a more complex pattern, involving development of a cartilaginous or ossified operculum and a rod-like columella or plectrum, as well as an "extra stapedial" process. The operculum and plectrum may develop directly within the membrane enclosing the fenestra ovalis, while the "extra-stapedial" has a separate source (Gaupp, 1913). The columella shows diverse patterns of modification or degeneration in modern groups of amphibians. However, the middle ear of early tetrapods (embolomeres, rhachitomes, and stegocephalians) suggests that the columella auris of modern amphibians may have been derived from a columella similar to that of primitive reptiles (Goodrich, 1930). Thus, the ancestral tetrapod pattern would essentially conform to development of the columella described above.

Variations on this theme, in addition to modifications seen in amphibians, are poor development of the hyostapes (chelonians and snakes, and probably, cotylosaurian reptiles), and elaboration of the dorsal process (*Sphenodon*) and extra-stapedial element (crocodilians and birds). Of importance here is the observation that the stapedial artery (often denoted the arteria stapedialis or a. facialis in sauropsids), which always arises from dorsal eminence of the hyoid arterial arch in tetrapods, can vary in its relationship with the stapes. In certain geckonids and birds it pierces the stapes via the stapedial foramen, but in certain other geckonids and in *Sphenodon* it runs ventrally; in the majority of modern "reptiles" it passes dorsally to the stapes (Fig. 2D). Despite this variation, Goodrich (1930) and others are emphatic in arguing that the intra-stapedial course of this vessel is the primitive condition, noting that this passage is evident in early fossil sauropsids and amphibians.

In mammals, the adult stapes is derived from a continuous blastema from the dorsal hyoid arch, which is bent at a strong angle so its most proximal edge is continuous with the membrane for the fenestra ovalis. Although some have held that much of the stapes in mammals and the otostapes in the reptile groups is an outgrowth of the auditory capsule rather than the hyoid (Doran, 1878), it has been demonstrated clearly that the capsule, at most, contributes to the base of the stapes in these forms (Goodrich, 1930). As noted above, the stapes often develops in blastematous stages as a ring surrounding the stapedial artery and chondrifies separately from the ear capsule. More distal elements, said to correspond to the reptilian hyostapes as well as the dorsal process, are poorly formed and often ephemeral. Two diminutive cartilages present in certain mammals—one in the tendon of the stapedius muscle ("the element of Paauw") and the other above the chorda tympani between the stapes and the hyoid cornu ("the element of Spence")—have been identified as phylogenetic "remnants" of the reptilian extra-stapedial region (Klaauw, 1923, 1931; Westoll, 1944). Shute (1956), however, disputed these claims because he observed: (1) that the cartilages of Spence and Paauw appeared very late in ontogeny; and (2) that there is no continuity between Spence's cartilage and the stapes during development.

Several features of development are then essential information to analysis of the mammalian stapes. First, the mammalian stapes is homologous strictly with the "reptilian" and bird otostapes, and the latter is a columella-like bone (Fig. 2D), with or without a moderate sized opening for passage of the stapedial artery. Second, the columellar appearance of the otostapes is not lost with loss of the extra-stapedial. In fact, the otostapes of certain birds and certain modern groups of squamates bears a striking resemblance to the stapes seen in monotremes and many marsupials (Figs. 1, 5). Third, the presence of a stapedial foramen seems to be a primitive character for tetrapods, even though the foramen is absent in many tetrapod groups where the stapedial artery is either lost or passes around the stapes.

THE STAPES IN "MAMMAL-LIKE REPTILES"

It is now important to consider the form and variation of the stapes in synapsids and, more specifically, in the therapsid groups thought to be close mammal relatives. Even if Goodrich's (1930) theory for the primitive tetrapod plan of the stapes is accepted, it is conceivable that this plan may have been altered in the common ancestor of mammals and their nearest sister taxon.

In some primitive sauropsids and synapsids the stapes is typically massive (Fig. 3E-F), with a stout, proximal-dorsal process articulating with the paroccipital process (Romer, 1956; Allin, 1975). In certain forms (*e.g., Ophiacodon, Dimetrodon, Captorhinus*) the stapedial shaft shows evidence for attachment of a ligament from the hyoid cornu (Romer and Price, 1940; Parrington, 1946). Thus, these taxa display the primitive persistence of a hyo-stapedial connection, as in living *Sphenodon*. Such a connection exists in early ontogenetic stages of all amniotes (Goodrich, 1930; Allin, 1975; remarks above).

Despite its massive appearance, the stapes in these groups is typically elongate and rod-like, with its distal and proximal ends flared (Fig. 3). The presence of a stapedial foramen is variable. In some, the foramen is a relatively small, elliptical or circular hole near the proximal end of the stapes (Fig. 3A). In other primitive sauropsids this opening is absent (Figs. 3B-D). In *Captorhinus* and *Edaphosaurus,* the complex stapes shows two openings in its proximal moiety (Fig. 3E-F). The more ventral of these is probably homologous with the stapedial foramen. The larger, dorsal opening probably allowed passage of the hyomandibular branch of the facial nerve, as it is simply the gap between the stapes and the opisthotic that is closed laterally by the expanded dorsal process. Hence, this foramen corresponds precisely in position with the gap for the hyomandibular nerve reconstructed for primitive tetrapods (Goodrich, 1930, Fig. 449).

In therapsids, the stapes is considerably smaller in mass, but it retains an end-on-end contact with the quadrate either as a cartilaginous joint or a bony articulation (Allin, 1975). There is, in some cases, an ossified dorsal process situated in the more extreme distal region of the stapes (Fig. 4B). Although there is evidence for retention of the hyo-stapedial contact in certain dicynodonts, this connection apparently is lost in most therapsids. Instead, a contact is formed between the hyoid cornu and the paroccipital process, as in the great majority of living "reptiles" and mammals (Allin, 1975).

The stapes varies in form in therapsids, although it is, as in pelycosaurs, a short rod-like structure showing some expansion at either or both its extremities (Fig. 4C). In some gorgonopsians the stapes has distinct bony processes and a small elliptical foramen near its proximal end (Fig. 4A). This foramen is absent in certain dinocephalians and in therocephalians (Fig. 4B-D). The stapedial foramen is present, however, in various cynodonts (Fig. 4E-H), and these taxa clearly conform in basic construction of the stapes. Although the cynodont stapes retains its rod-like appearance, the stapedial foramen is a large, elongate slit or ellipse located in the central region of the shaft.

Development of the stapedial foramen in cynodonts lends to the stapes a somewhat bicrurate appearance, but one quite distinct from the bicrurate stapes in mammals. In the latter, the stapedial foramen nearly reaches the proximal stapedial extremity. The foramen is bounded by narrow bony crura, a basal stapedial footplate, and a small head for articulation of the incus (Fig. 1A). In contrast, cynodonts, while having a large stapedial foramen, still retain a columellar stapes with well-developed proximal and distal moieties (Figs. 4E-H). With exception of the stapedial foramen, it is difficult to observe any other features of special resemblance between the stapes in cynodonts and most mammals.

It is finally noteworthy that *Bienotherium*, a tritylodontid (a group which has been identified by some as closer relatives of mammals than other cynodont families, see Kemp, 1983; Sues, 1985), has been shown by Young (1947) to have a rod-like, imperforate stapes. However, it is possible that this structure is instead a piece of ossified hyoid arch (J. Wible and J. A. Hopson, *personal communication*) and that the actual form of the stapes is unknown in tritylodonts. Nevertheless, if Young's (1947) identification is correct, then one would have to reason that the nearest mammalian sister group lost the stapedial foramen.

THE STAPES IN MAMMALS

Clues to the primitive condition of the mammalian stapes are found in characters both of the nearest outgroups to mammals and the higher-level mammalian clades. In the former case, we have noted some degree of variation, although it seems plausible that the common ancestor of cynodonts and mammals had a columelliform-perforate stapes described above as condition *A*.

When we turn to major clades of mammals for information, we confront some serious gaps in the evidence. Despite all the discussions and speculations on the evolution of ossicles in Mesozoic mammals, these elements in triconodonts and other Mesozoic mammals are virtually unknown (Crompton and Sun, 1985), except for some bits and pieces of isolated stapes associated with *Morganucodon* (see Kermack and others, 1981) and a stapes of a Late Cretaceous eutherian (Archibald, 1979). In taxa which otherwise retain many primitive cranial features (Crompton and Jenkins, 1979; Kemp, 1983), we can only speculate on what might have been present in the middle ear. For example, Crompton and Sun (1985) identified in a Triassic triconodont a small facet for the incus which closely resembles this feature in *Ornithorhynchus*. In the latter taxon, this facet sets the outside limit of the incus and stapes. Crompton and Sun's (1985, p. 113) reconstruction holds that such a facet in triconodonts would, of necessity, bring the malleus far away from contact with the dentary, an idea suggested by absence of any groove in the dentary for the post-dentary bones. From this indirect evidence, they concluded that triconodonts had a three-ossicle system as in living mammals.

Twelve stapedial footplates with portions of the crura were described for the Triassic triconodont *Morganucodon* by Kermack and others (1981, p. 103-104, Figs. 84-85). These elements are very small compared with the stapes of many cynodonts but, as Kermack and others (*ibid.*, p. 103) argued, they "may be more usefully compared with that of therapsids than that of modern mammals." The stapes have a large stapedial foramen flanked anteriorly and posteriorly by slender crura. However, there is appreciable ossification between the foramen and the footplate, and the size of the stapedial process of the quadrate indicates that the stapes were distally flared, as in the case of *Thrinaxodon* (Fig. 4F). Hence, a reconstruction of the complete stapes would plausibly yield a columelliform-perforate structure (condition *A*), close to that in cynodonts, and not one that was as strongly bicrurate (condition *B*) as in eutherian mammals.

Critical to this study would be information of the stapes in diverse lineages of other Mesozoic mammals, including multituberculates, dryolestids, and early therians. For example, was a three-ossicle system unambiguously present in these forms? Did the stapes more resemble that of triconodonts and therapsids than that of monotremes or eutherians? Any evidence of stapes structure in such lineages would have strong effect on outgroup analysis of character transformations for monotremes and therians. Unfortunately no such evidence is as yet available.

Given the poor information on the stapes in Mesozoic mammals, the monotreme condition offers a unique reference point. It provides the only evidence for a putative intermediate stage between therian mammals and triconodonts. It is understandable then, why Goodrich's (1915, 1930) account of stapedial ontogeny in *Ornithorhynchus* was so crucial to acceptance of his theory.

The stapes in adult monotremes is more like the imperforate, proximal columella of many birds and squamates than the stapes of cynodont therapsids or triconodonts (*cf.* Figs. 1, 4, 5). However, most workers have accepted Goodrich's (1930) conclusion that the monotreme condition is a secondary one for mammals. The reasons for this argument are the presence of the stapedial foramen in many tetrapod groups and the appearance, in early stages of the ontogeny of *Ornithorhynchus,* of a stapedial opening in the blastematous stapes for passage of the stapedial artery (Goodrich, 1915, p. 149).

The matter of ontogeny of the monotreme stapes presents an intriguing problem. The stapedial foramen forms early and disappears later in the ontogeny of *Ornithorhynchus,* where a large stapedial artery remains without passing through the stapes (Tandler, 1899; Wible, 1984). There is absolutely no evidence of this foramen at any stage during the ontogeny of the stapes in *Tachyglossus* (see Doran, 1878; Gaupp, 1908; Griffiths, 1968; Kuhn, 1971), where the stapedial artery involutes before the stapes blastema appears (Gaupp, 1908). It has long been established that, even in procartilaginous stages, the stapes of *Tachyglossus* takes on the form of an imperforate columella, an ontogenetic pattern very similar to that in several metatherians (especially *Vombatus* and *Phascolarctos*) and the eutherian *Manis* (see Doran, 1878; Gregory, 1947; Griffiths, 1968, p. 60, 62, 72). Despite their different ontogenies, the adult stapes of *Tachyglossus* and *Ornithorhynchus* are closely similar in structure (Fig. 5*A-B*). Which of the two ontogenies is the more basic (primitive) for monotremes or, for that matter, monotremes and therians? The distribution of known ontogenies, as reviewed below, does not decisively settle this issue, but does offer some insight on transformation of the stapes.

When we look to other groups of Recent mammals, we find that the columelliform stapes is present, though its structure may differ in detail from that in adult monotremes. Variation among marsupials is most interesting in this regard. Goodrich's (1930) note that the stapedial foramen is present in ontogeny of *Trichosurus* (Phalangeridae) is not surprising, as this genus and certain other marsupials retain the foramen as adults. However, the pattern of stapes architecture in marsupials is variable. Fleischer (1973), in his excellent survey of the mammalian middle ear, stated (p. 143) that, in marsupials, "der Stapes kann zwei Crura aufweisen, oder auch Columella-artig geformt sein." His survey was limited to four marsupials, which variously show: (1) a stapes of remarkable similarity to that in adult monotremes (*Notoryctes,* Fig. 5*C*); (2) a somewhat triangular stapes with a very small perforation (*Petaurus,* Fig. 5*D*); or (3) a stapes with a moderately large stapedial foramen (*Thylogale,* Fig. 5*E*; *Philander* [*Metachirops*], Fig. 5*F*). The broader sampling of Doran (1878) is more illuminating. Therein it is shown that the majority of marsupial groups has either an imperforate, elongate stapes, or one with a very small opening. The essentially imperforate condition is found in caenolestids, dasyurids, thylacinids, myrmecobiids, phascolarctids, peramelids, vombatids, and notoryctids (see Figs. 5, 9). For this reason, Doran (1878, p. 486-487) proposed that most marsupials retained a primitive, columelliform condition, best exemplified by peramelids, and that only macropodids, didelphids, and some phalangerids (also *Dromiciops,* see Segall, 1969) attained a more advanced, bicrurate stapes. However, even in these latter forms, the stapes does not clearly match the strongly bicrurate, stirrup-like bone of most eutherians (*cf.* Figs. 5 and 6). Instead, the stapedial foramen is more elliptical or triangular in outline, and the stapedial crura converge some distance below the capitulum so that the stapes maintains, at least distally, a columella-like outline. This comparison prompted Segall (1971, p. 55) to remark, "the marsupial stapes has in all forms, generalized and specialized alike, a columella-like shape, similar to that of the monotreme stapes."

Taking note of these differences, Archibald (1979) suggested that the primitive therian stapes probably was more like that in *Didelphis* than the strongly bicrurate form of most eutherians. We agree with Archibald, but emphasize that the primitive therian condition also may have more resembled that of marsupials other than didelphids. The broad distribution of columelliform stapes in marsupials, as well as in monotremes, is compelling in this regard (see discussion below). Perhaps this is another case in which didelphids do not fulfill their conventionally-visualized role as "ancestral" marsupials.

Moving to a consideration of eutherians, we find that a strongly bicrurate stapes may have been a very ancient feature of this group. The earliest putative eutherian stapes, as described by Archibald (1979), was of this form (Fig. 6*A*). However, the distribution of this type is not universal within Eutheria. Noted exceptions are the columelliform-imperforate (or microperforate) structure in pholidotans (Fig. 6*B, C*), bradypodids (Fig. 6*D*), and the myrmecophagid *Cyclopes* (Fig. 6*E*), the somewhat didelphid-like stapes of the dasypodid *Tolypeutes* (Fig. 6*F*), and the massive-microperforate stapes of various marine mammals (Fig. 1*F*, 6*M-O*).

There seems little doubt that the stapes of marine mammals is highly derived (Doran, 1878). The ear ossicles of these forms are characteristically massive, with densely compact bone. This trait was related by Fleischer (1978) to special hearing adaptations of marine mammals. He noted (*ibid.*, p. 41) that, like the bicrurate stapes and unlike the typical columella, force applied to head of the stapes is directed to the perimeter of the footplate, and not to its center. Also, the stapes is, particularly in cetaceans, anchored by an annular ligament of highly specialized microstructure. Complementing these functional arguments is phyletic distributional evidence. The nearest relatives, respectively, of pinnipeds, sirenians, and cetaceans (see, McKenna, 1975; Novacek, 1982) all have stapes of the bicrurate type. Thus, one might infer that, in the massive stapes of marine mammals, the crura have become so robust that the intercrural (stapedial) foramen is nearly or entirely closed (see Fleischer, 1978). Clearly needed is modern ontogenetic study of derivation of this type of stapes.

Of greater interest to the question of transformation in the mammalian stapes is the columellar type exemplified by manids and some edentates (Fig. 6B-E). Parker (1886) observed that the stapes of manids maintained an imperforate columella both in cartilaginous and bony stages of ontogeny (see also Starck, 1941). Likewise, Parker (1886) noted that the manid stapes resembled the columella-like stapes of bradypodids and the anteater *Cyclopes* (Fig. 6D-E), and these structures, taken together, represented a "metatherian level" (ibid., p. 78). These views followed Doran (1878), who emphasized that the stapes of *Manis* (ibid., p. 476) is "more absolutely Sauropsidan in every respect than any other placental mammal." (It is noteworthy, however, that the stapes in *Bradypus* shows an obliteration of the small foramen during ontogeny in a manner similar to that in *Ornithorhynchus;* see Schneider, 1955.) Doran (1878, p. 476) also argued that only certain dasypodids and myrmecophagids among edentates showed a more advanced, bicrurate stapes, whereas the stapes of the dasypodid *Tolypeutes* (Fig. 6F) "resembles that of the Kangaroos and other Marsupials, or even some Birds, but not the complete columelliform type of *Manis,* and most Aves." More recent studies of the stapes in manids (Segall, 1973) show some variation in this structure. Notably, the stapes in certain Asian terrestrial pangolins (Fig. 6C) is anomalous in its compressed form. However, all members of this group retain a columelliform-imperforate structure. The general condition for the order is likely represented by *Manis crassicaudata* (Fig. 6B).

With the above exceptions noted, there apparently is little significant variation remaining in the form of the eutherian stapes. Virtually all orders of this group show a strongly bicrurate stapes (Fig. 6A, G-I, K-L), with a large intracrural (stapedial) foramen (Doran, 1878; Fleischer, 1973). Variations on this plan include modifications in relative size of the crura, shape of the foramen, and stapedial footplate, and the presence or absence of a process for insertion of the stapedius muscle. Although certain rodents, mustelid carnivores, and ungulates show slight or marked reduction of the stapedial foramen (Fig. 6J), doubtless this is a secondary condition that arose within these groups (Doran, 1878).

ALTERNATIVE TRANSFORMATIONS

Introduction

Before we consider the possible pathways of modification of the mammalian stapes it is worthwhile to review a few observations and inferences.

First, the stapes in diverse living and fossil tetrapods is an elongate, rod-like structure with a large elliptical foramen for passage of the stapedial artery. It is probable that this is the primitive condition for tetrapods.

Second, the mammalian stapes is homologous with the proximal columella of sauropsids and other tetrapods, but not with the extra-stapedial elements (distal columella) of these forms. Plausibly the latter elements were functionally analogous, at least in some cases, to the malleus and incus of mammals, as they form the osseous connection between the proximal columella and the tympanic membrane.

Third, the stapes in therapsids and other synapsids is variably constructed. However, most forms have a large elliptical foramen, probably representing the general condition for these groups.

Fourth, even where the stapedial foramen is present, the therapsid stapes does not resemble the typical bicrurate stapes of eutherian mammals. The former is a large columelliform structure with a centrally located foramen. The latter is a very small horseshoe-shaped element with narrow crura that converge directly at the head of the stapes—there is no columellar form maintained with this structure. Moreover, the small, columelliform-imperforate or microperforate stapes in monotremes, most marsupials, and some eutherians is more reminiscent of the proximal columella of modern birds and lizards, than the stapes of therapsids.

Fifth, fragmentary stapes of *Morganucodon* show that this element had a stapedial foramen. However, the stapes in other clades of Mesozoic mammals is as yet unknown.

Sixth, monotremes show two different ontogenetic patterns, one in which the stapedial foramen is present briefly in procartilaginous stages (*Ornithorhynchus*) and one in which the stapedial foramen is absent from the ontogeny (*Tachyglossus*). Both patterns yield essentially the same adult structure—a columelliform-imperforate stapes.

Seventh, what ontogenetic data are available suggest that the imperforate, columellar stapes in therians develops either in a manner similar to that in *Tachyglossus (e. g., Vombatus, Phascolarctos, Manis)* or that in *Ornithorhynchus (e. g., Bradypus)*.

Eighth, the majority of marsupial clades has a stapes that is columellar, and either imperforate or microperforate. A moderately large stapedial foramen is characteristic only of phalangerids, petaurids, didelphids, microbiotheriids, and macropodids. However, in the latter taxa the stapes is usually columelliform distal to the convergence of the crura—the element is not bicrurate and stirrup-shaped.

Ninth, the great majority of eutherians has a bicrurate stapes. However, a columellar-imperforate

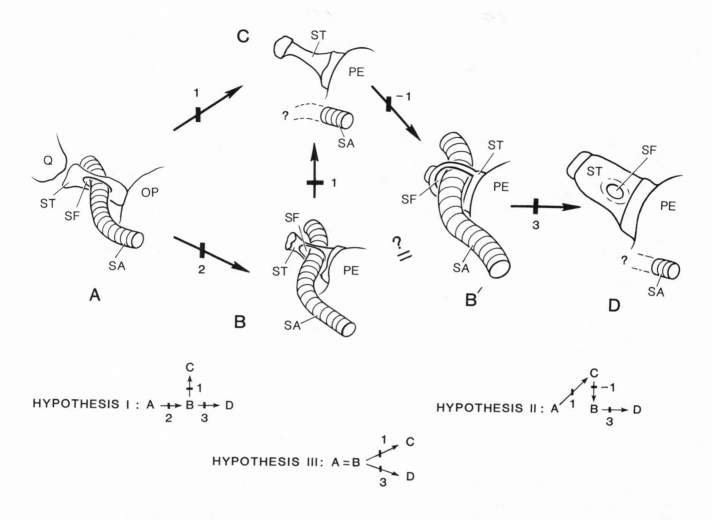

Figure 7. Hypotheses for transformation of mammalian stapes: A, outgroup and primitive mammal condition, (columelliform-perforate); B, bicrurate condition; B', bicrurate stirrup-shaped condition; C, columelliform-imperforate (or microperforate) condition; and D, massive-microperforate condition. Events are: (1) loss of stapedial foramen; (2) enlargement of stapedial artery and foramen, bicrurate development; (3) further ossification, enlargement of stapes, reduction of stapedial foramen; and (-1) regain of stapedial foramen. PE, petrosal. For other symbols, see Figures 1-6. Hypotheses I, II, and III show relative economies of different transformations (see text).

stapes is present in *Manis* and some edentates. Certain other edentates have a microperforate stapes or a columelliform stapes with a larger foramen, resembling this element in didelphids. In some eutherians, notably marine mammals, the stapes is massive, somewhat rod-like, and micro-perforate. It is likely that this represents a secondary derivation from the bicrurate stapes.

Tenth, the only eutherian conditions for the stapes also found in monotremes and marsupials are the types noted for pholidotans and some edentates.

From these points it is evident that the matter of what is primitive for the stapes at various levels within Mammalia is more complicated than usually assumed. At the very least, the taxonomic distributional evidence suggests that the stapedial foramen disappeared or reappeared several times in the phylogeny of mammals (Figs. 8, 9, 10). It is thus worthwhile to examine the alternative pathways in their simplest form, and then assess their relative strengths in light of assumed patterns of relationships and comparative ontogeny.

Figure 7 is a basic scheme for transformation of the stapes in mammals. Condition A, the presence of a columelliform-perforate stapes, is accepted here as the ancestral condition in mammals and therapsids. Enlargement of this stapedial foramen coupled with reduction in mass of the stapes (event 2) gives rise to a bicrurate stapes. For purposes of simplification, the bicrurate, somewhat columelliform stapes (B) is equated with the bicrurate, stirrup-shaped stapes (B'). As noted above, however, there may be reason to distinguish the two conditions. Alternative pathways yield the columelliform imperforate stapes (C) through loss of the stapedial foramen (event 1) either as a derivation from A or B. Finally, the massive-microperforate stapes (D) is derived through increased density and ossification and great reduction or obliteration of the stapedial foramen (event 3).

45

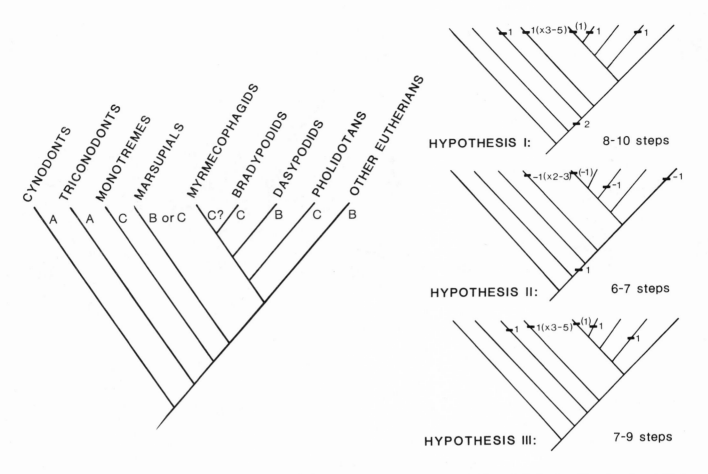

Figure 8. Alternative distributions for stapedial modifications. 1, -1, and 2 are events described in Figure 7. Derivation of massive-microperforate stapes (condition *D*, Fig. 7) not shown. Left, distributions of traits superposed on a framework of higher-level relationships suggested elsewhere (modified from Novacek, 1982). *A, B,* and *C,* are conditions diagrammed in Figure 7. ? indicates uncertainty in morphotypical condition for terminal taxon. Right, Hypotheses I, II, and III (of Fig. 7) superimposed on branching framework of taxa where alternative primitive conditions are suggested for the monotreme-therian node. Events for marsupials are multiplied by a range of possible events given in Figure 9. Parentheses indicate event occurs *within* terminal taxon (see Fig. 10). Total number of events (steps) are given for each hypothesis. For further explanation, see text.

Below the schematic are flow charts for three hypotheses. Hypothesis I argues that the columelliform-imperforate stapes is a derivative of *B*. Hypothesis II argues that *B* can be derived from *C* through a secondary gain of the stapedial foramen (-1). Hypothesis III is a further simplification; conditions *A* and *B* are equated, and *C* is a derivative of the more general condition (*A, B*).

Phylogenetic Distributional Evidence

Alternatives as displayed in Figure 7 clearly show hypotheses I and III to be the most economic ones. However, the schematics may not reflect the actual number of events required, given the distribution of stapes traits in various clades of mammals and their outgroups. Figure 8, left is a general cladogram of mammalian groups (and the cynodont outgroups) with probable morphotypical conditions (or their alternatives) for each terminal taxon. Figure 8, right shows events necessary under each of the three hypotheses given in Figure 7.

Distribution of stapedial conditions in marsupials and edentates is complicated, and alternative solutions for stapedial transformation within these groups are presented in Figures 9 and 10. Note that interpretations of marsupial phylogeny (Kirsch, 1977, Figs. 23, 24; Archer 1982, Fig. 8) are more forgiving to the hypothesis that the columelliform-imperforate stapes (condition *C*) is primitive for the group. Interestingly, the more economic pattern favoring the bicrurate stapes (*B*) as the primitive marsupial condition requires reversal events, namely, the gain of the stapedial foramen that was lost at earlier stages in the phylogeny. The case for *C* as morphotypical, based on ingroup analysis, is strengthened if monotremes, which have condition *C*, are taken as an outgroup to therians (Fig. 8).

The solution for eutherians is more ambiguous. Following Engelmann's (1985) study of edentate relationship, we find that the columelliform-imperforate stapes (*C*) is present both in bradypodids and in *Cyclopes* (outgroup to the two other living anteater genera). Dasypodids are characterized by the bicrurate stapes (although note the

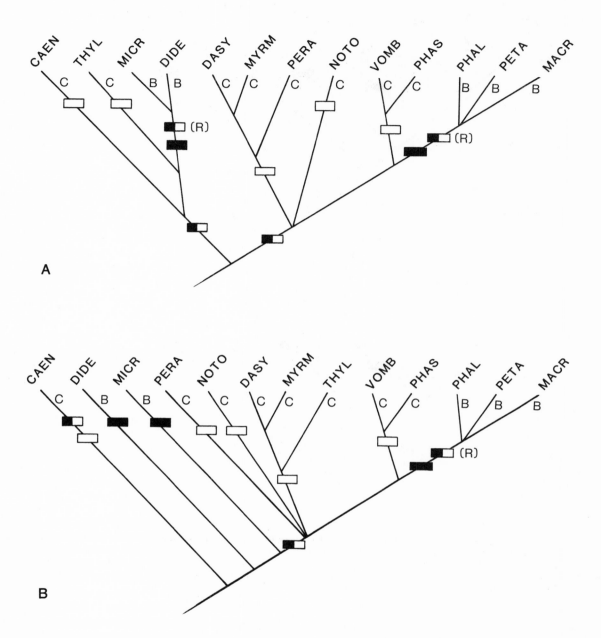

Figure 9. Possible stapes modifications given suggested phylogenies for marsupials: *A*, phylogeny after Kirsch (1977, Figs. 23, 24), and *B*, phylogeny after Archer (1982, Fig. 8). Conditions *B*, *C* for terminal taxa are shown in Figure 7. Solid boxes are changes required if the columelliform-imperforate (or microperforate) stapes (condition *C*) is primitive for marsupials. Open boxes, changes required if the bicrurate stapes (condition *B*) is primitive and no reversals are allowed. Semi-solid boxes, changes required if condition *B* is primitive and reversals (R) are allowed. Abbreviations for taxa are: CAEN, caenolestids; DASY, dasyurids; DIDE, didelphids; MACR, macropodids, MICR, microbiotheriids; MYRM, myrmecobiids; NOTO, notoryctids; PERA, peramelids; PETA, petaurids; PHAL, phalangerids; PHAS, phascolarctids; THYL, thylacinids; and VOMB, vombatids.

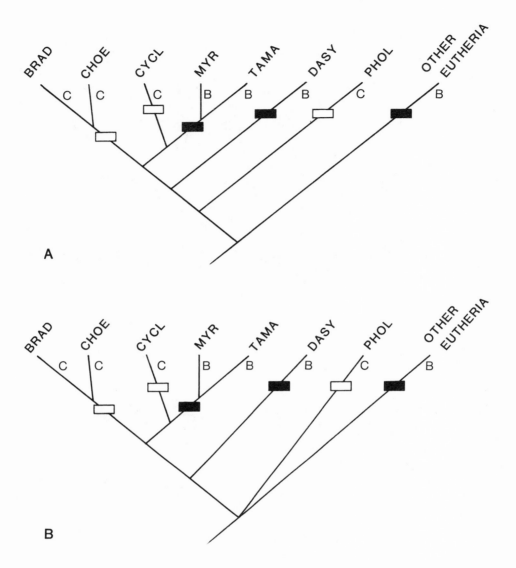

Figure 10. Possible stapes modifications given two alternative theories of higher eutherian relationships: *A*, relationships after Novacek and Wyss *(in press)*; and *B*, relationship after Novacek (1982), Goodman and others (1985) with modifications. *B, C* are conditions for terminal taxa shown in Figure 7. Solid boxes, changes required if *C* is primitive for eutherians. Open boxes, changes required if *B* is primitive for eutherians. Reversals yield equally or less parsimonious solutions, and are not shown. Abbreviations for taxa are: **BRAD**, *Bradypus*; **CHOE**, *Choelepus*, **CYCL**, *Cyclopes*; **DASY**, *Dasyurus*; **MYR**, *Myrmecophaga*; **PHOL**, pholidotans; and **TAMA**, *Tamandua*.

stapes condition in *Tolypeutes,* Fig. 6F). The morphotypical condition for the eutherian stapes is not affected by a choice between the two patterns shown in Figure 10. It makes no difference whether pangolins are made a sister-group of edentates (Fig. 10A, following Novacek and Wyss, 1986) or are simply left unresolved in a trichotomy with edentates and other Eutheria (Fig. 10B). The minimum number of steps required for positing B or C as the primitive eutherian condition is always three.

We acknowledge that if eutherians are simply left as an unresolved polytomy, the hypothesis that the columelliform-imperforate stapes was primitive for the group would seem highly unlikely; it would necessitate independent evolution of the bicrurate stapes in many eutherian lineages. However, an unresolved polytomy for eutherians cannot be taken as a higher level pattern that provides insight on the transformation of this, or any other character. For the present, we follow recent studies in comparative anatomy and molecular biology (McKenna, 1975; Goodman and others, 1985; Engelmann, 1985) which suggest that edentates are a remote eutherian clade.

Ingroup solutions to eutherians and, especially, marsupials are critical to comparing the three hypotheses in Figure 8. In hypotheses I it is argued that the bicrurate stapes (requiring event 2) is primitive at the level of therians + monotremes. This scheme requires 8-9 steps, the minimum number of events within marsupials being between 3 and 5, depending on the marsupial phylogeny chosen from Figure 9. Hypothesis II, in which the columelliform-imperforate (or microperforate) stapes is primitive at the level of therians + monotremes, requires between 6 & 7 steps. Finally, if the bicrurate condition is simply equated with the condition in therapsids, as shown in hypothesis III, the transformation requires between 7 and 9 steps.

Note that this analysis is designed to give maximum support to the hypothesis I, namely that the bicrurate stapes is the primitive condition for therians plus monotremes. For example, the bicrurate stapes in marsupials is equated with that in eutherians, despite the above noted distinctions. Moreover, the dasypodid stapes is characterized as bicrurate, even though this element in *Tolypeutes* is closer in structure to the perforate stapes of marsupials. Despite these provisions, hypothesis II is the most economic one, unless it is argued that there are no events necessary to convert the columelliform-perforate stapes into the bicrurate stapes typical of many eutherians. Even if this unlikely argument is accepted, the resultant solution is no more parsimonious than hypothesis II.

In making these comparisons, we do not dispute the idea that the columelliform-imperforate stapes might be a derivative of a more primitive condition in which the stapedial foramen was present. We merely question the level of universality at which such a derivation occurred. We do not blindly accept the theory that this stapes type was derived independently in many lineages *within* major mammalian clades. Of course, more definitive solutions will come when it is determined whether other therian outgroups (dryolestids, multituberculates, *etc.*) have a columelliform-imperforate or a bicrurate stapes.

Ontogenetic Evidence

It is possible to examine relative strengths of hypotheses I, II, and III in Figure 7 from an ontogenetic perspective. Unfortunately, ontogenetic information is sketchy, as only a few of the relevant taxa have been well studied. However, a general pattern can be suggested. The bicrurate (or columelliform-perforate stapes) always develops from a blastemous stage in which the foramen is present. One can envision, then, the stapes developing around the vascular embryonic tissue for the stapedial artery. Appearance of the foramen is a very early developmental event. By contrast, the columelliform-imperforate stapes is known to form in two different ways. The stapedial foramen either is never present in ontogeny of this element (*Tachyglossus, Manis,* some birds and lacertilians, *Vombatus,* and *Phascolarctos*) or it appears very early, but is subsequently reduced or obliterated (*Ornithorhynchus, Bradypus*). Recourse to some form of the biogenetic law might lead one to accept the ontogenetic pattern in *Ornithorhynchus* as more primitive, because the more widely distributed embryonic stage (*i. e.,* in two out of three documented ontogenies) is one wherein the stapedial foramen appears. (This follows, despite the fact that *Ornithorhynchus* shows an anomalous combination in the adult—it has an imperforate stapes, but is the only non-eutherian with a well developed intra-tympanic stapedial artery). Under this view, the ontogeny of *Tachyglossus* simply represents the early occurrence of developmental events that are terminal in other ontogenies.

In any case, it is difficult to decisively choose whether the ontogeny of *Ornithorhynchus* or that of *Tachyglossus* is the more primitive for monotremes. However, it is important to emphasize that if the ontogeny in *Ornithorhynchus* were accepted as more generalized, this would not preclude the hypothesis that the columellar-imperforate stapes is the primitive *adult* condition for monotremes and therians. The ontogeny of *Tachyglossus* would simply be regarded as a more derived ontogeny that also gives rise to the columelliform-imperforate stapes.

These considerations still leave open a critical question. If the columelliform-imperforate stapes were a primitive condition at some higher level within mammals, this would require a reversal event (Fig. 8, hypothesis II), a reacquisition of the stapedial foramen (and concomitant penetration of stapes by the stapedial artery). At face value, this modification seems implausible. No known ontogeny shows initial lack of the foramen and its appearance at later developmental stages. Nevertheless, we submit that absence of this ontogeny is not really relevant to the problem. Where the stapedial foramen is present, it always forms at the earliest stages of development, because the stapes must be molded around the stapedial artery. Hence, "reversals" necessary for hypothesis II could be explained either by regaining the proximal stapedial artery or, more simply, its shift in position to the area of the stapes. Such modifications, especially the latter, do not seem improbable given the variation both in embryos and adults with regard to the vertebrate cephalic arterial system (Bugge, 1974). This scheme would even accommodate cases in which the stapedial foramen is present but the stapedial artery is absent in adults (*e. g.,* some marsupials, dermopterans), if that vessel existed during earlier embryonic stages.

We thus maintain that interplay between development of the stapes and the stapedial artery is more plastic than one might intuitively assume. Taxonomic distributional evidence is highly suggestive in this regard. Even if the bicrurate (or perforate) stapes is taken as the primitive condition for marsupials, the most efficient transformation sequence would require at least one event in which the stapedial foramen re-appears (Fig. 9). Rejecting the possibility of reacquisition would require the unlikely scenario (Fig. 9) that the stapedial foramen was lost independently in at least five different marsupial clades. Thus, distribution of stapes types and suggested higher-level marsupial relationships conflicts with the notion that stapes can lose, but never regain, the stapedial foramen in phylogeny.

Functional Questions

To this point, we have not considered functional aspects of these alternative transformations. The stapedial foramen is present and obvious where the stapedial artery takes its course through the stapes, although the bicrurate stapes in some mammals is retained even when the stapedial artery has been lost. Accordingly, Fleischer

(1978) postulated that the stirrup-like stapes might be maintained for reasons other than those relating to presence of the stapedial artery. He suggested that this design confers a mechanical advantage over the typical columelliform stapes, because the driving force from the incus to the head is transferred directly through the thin crura to the periphery of the stapedial footplate. The annular ligament usually is much thicker where the crura join the footplate than it is in other areas (Fleischer, 1978, Fig. 19). Hence, the force is applied directly where stapedial attachment shows greatest elasticity. By contrast, the driving force on the columelliform stapes is applied directly down the shaft to the center of the footplate. This causes more tensional stress on the footplate, because the reaction force of the annular ligament does not directly oppose the driving force applied in the center of the attachment.

This common sense model is appealing but problematic, as it is not experimentally documented. Moreover, Fleischer's (1978, p. 40-41) adaptational argument for evolution of the stapes is unsatisfying when he deals with exceptions to the basic pattern. He proposed (*ibid.*, p. 40-41) that the columella-like stapes in *Notoryctes, Manis, Bradypus,* and African mole-rats *(Heliophobius)* are secondary reversions to the primitive condition because "All these forms are slow and/or subterranean and thus do not need highly developed sensory organs. Apart from the monotremes, *Manis* and *Bradypus* may even be called living fossils. All this shows that advanced and successful mammals have a stirrup-like stapes and that only some forms in small ecologic niches may secondarily return to a columella-like stapes."

Fleischer required this assessment because he assumed (1978, p. 40) that the bicrurate stapes was the primitive condition for mammals (but note contradiction with the above quoted statement). He acknowledged that the columelliform stapes was also present in monotremes, but that these (p. 40) "may not be true mammals." However, his mechanical hypothesis can be salvaged without this complicated eco-adaptive scenario. It only requires that the columelliform condition be a retained primitive feature in some mammals. This, of course, does not rule out the possibility of a secondary development of the columelliform stapes in some forms (*e. g., Heliophobius*), but it frees one from making a broad and unconvincing selection argument for the "independent reversion" to the columelliform condition in several different lineages.

In addition to these mechanical factors, interplay of the stapes and proximal stapedial artery, as noted above, is of prime interest. A simple theme in this relationship is evident. In most non-mammalian tetrapods the proximal stapedial artery is the major or sole supplier of blood to the supraorbital, infraorbital, and mandibular regions (Hafferl, 1933). In therapsids and other tetrapods of moderate to large size, the relatively large stapes could accommodate passage of a well-developed stapedial artery. More variation is apparent in *Tachyglossus* and marsupials, in which the stapedial artery in adults is lost and is replaced in function by other vessels. In *Ornithorhynchus,* the stapedial artery circumvents the small, columellar stapes rather than burrowing through it.

Although the stapes is unknown in most lineages of Mesozoic mammals, these forms were very small, and it is possible that the same "packing problems" applied for the course of the stapedial artery through the middle ear. If the stapedial artery were retained, at least two solutions to this packing problem in small mammals seem feasible. One solution would be detour of the artery away from the stapes (as in *Ornithorhynchus*). The other solution, as is well shown in shrews (Fig. 6*I*), rodents, and other eutherians, would be marked enlargement of the stapedial foramen and concomitant reduction of the bony stapes to a stirrip-like structure. Because either solution seems plausible, functional arguments do not contribute to identity of the primitive condition for a group including therians and monotremes.

The bicrurate stapes is an obvious accommodation to this vessel's passage through the middle ear cavity. Hence, a stapes of this form must be tied to presence of a relatively large stapedial artery in the confined spaces of the tympanic region. It cannot simply be correlated with presence of a stapedial artery. In larger mammals, the stapedial artery no longer seems adequate as the sole supplier of its target regions, and its function is usurped by other vessels (Bugge, 1974). Accordingly, the stapedial artery often degenerates in these mammals, although the bicrurate structure of the stapes is retained. Such a retention does not necessarily require Fleischer's (1978) mechanical hypotheses. As noted above, the bicrurate structure might simply be a holdover from an embryonic condition in which an ephemeral stapedial artery was present. This possibility certainly warrants thorough investigation.

CONCLUSIONS

1. Common occurrence of the rod-like, perforated stapes in cynodonts, other therapsids, and other tetrapods suggests that this is the most likely ground plan for derivation of the mammalian stapes (Fig. 7*A*). This condition probably persisted in triconodonts.

2. It is uncertain whether the stapes had a small foramen for passage of the stapedial artery at the level of ancestry for monotremes plus therians, therians, or eutherians. The general scheme of the transformation would suggest that retention of the stapedial foramen is a more economical pattern (Fig. 7, hypotheses I and III) than is the loss of the foramen at these levels (Fig. 7, hypothesis II). However, when likely topologies for actual mammalian clades are considered (Fig. 8), hypothesis II is equally or more efficient. Regardless of the hypothesis (Fig. 7) accepted, ontogeny of the stapes in *Ornithorhynchus* could be inferred as the more primitive for monotremes, because at least a vestige of the stapedial foramen is retained in early ontogeny. However, this would not preclude possibility that the columelliform-perforate stapes was a primitive *adult* condition for therians and monotremes.

3. Other events that probably accompanied derivation of the modern mammalian stapes were marked decrease in size and modifications for articulation with the incus in the ossicle chain (Fig. 7). These factors may partly account for strong differences in the stapes form between therapsids and living mammalian groups with columelliform or bicrurate stapes. (The fragmentary stapes in *Morganucodon* is very small, but according to Kermack and others (1981) is more comparable to that in therapsids than in more modern groups of mammals). It is probable that such modifications parallel those occurring in small "reptiles," where the stapes is also reduced and displays a complex extra-stapes that attaches to the tympanic membrane.

4. The strongly bicrurate, stirrup-shaped stapes is a secondary feature in mammals. Its essential morphology is not found in most marsupials, monotremes, or non-mammalian tetrapods. Moreover, the bicrurate, stirrup-shaped stapes is not common to all eutherians, as the more columelliform stapes is seen in bradypodids, some myrmecophagids, and pholidotans. We maintain that the bicrurate stapes (Fig. 7B) is a departure from the columelliform-perforate stapes (Fig. 7A) of many tetrapods. Moreover, we note that bicrurate stapes of didelphid, macropodid, and phalangerid marsupials often have a columelliform outline rather than a stirrup-form characteristic of most eutherians.

5. The probable factors in play for derivation of the bicrurate stapes included marked enlargement of the stapedial artery, which required maximal space for its passage through a small stapes. In addition, there may have been a mechanical advantage to a stapes of bicrurate design as an impedance matching device (Fleischer, 1978; and remarks above). If ancestral mammals had a perforate stapes (Fig. 7A, hypothesis I), the bicrurate stapes in some marsupials and most eutherians is a logical extension of this trend. If ancestral therians and eutherians had an imperforate stapes (Fig. 7C), an extra step (regaining the stapedial foramen; Fig. 7, modification -1) was required for derivation of the bicrurate stapes (Fig. 7, hypothesis II).

6. The massive-microperforate stapes of cetaceans, sirenians, and marine carnivorans (Fig. 7D) clearly is a derivative of the bicrurate stapes, and was acquired independently in these groups.

We have, in presenting this series of arguments, attempted to give fair attention to their likely alternatives. We cannot decisively resolve the question of the condition of the stapes for the ancestral therian. Was this element with or without a stapedial foramen? Information from paleontology and embryology is not sufficient at this time to settle the issue. However, we emphasize that assumptions of higher level branching sequences (Figs. 8-10) do not clearly favor the standard view that the strongly bicrurate stapes was necessarily primitive for therians. Thus, we question the widely accepted theory offered by Goodrich (1930) and other comparative anatomists.

ACKNOWLEDGMENTS

Thank you Malcolm C. McKenna, Lawrence J. Flynn, John Wible, Miao Desui, Mark Norell, Bobb Schaeffer, J. A. Hopson, Jacques Gauthier, Michael Diamond, and Tim Rowe for discussing with us many of the issues treated in this paper. In addition, John Wible provided incisive criticisms, corrected some embarrassing errors, and probably remains unconvinced by some of our arguments. We are also grateful to the editors of this volume, Katherine M. Flanagan and Jason A. Lillegraven for their invitation of our contribution to honor the memory of George Gaylord Simpson. Lisa Lomauro prepared final versions of the figures. Support for the paper was provided by the Frick Laboratory Endowment Fund in Vertebrate Paleontology, and a Columbia University Faculty Fellowship.

REFERENCES CITED

Allin, E. F., 1975, Evolution of the mammalian middle ear: Journal of Morphology, v. 147, p. 403-438.

Archer, M., 1982, A review of Miocene thylacinids (Thylacinidae, Marsupialia), the phylogenetic position of the Thylacinidae and the problem of apriorisms in character analysis, *in* Archer M., ed., Carnivorous marsupials: Mosman, Australia, Royal Zoological Society of New South Wales, v. 2, p. 445-476.

Archibald, J. D., 1979, Oldest known eutherian stapes and a marsupial petrosal bone from the Late Cretaceous of North America: Nature, v. 281, p. 669-670.

Bugge, J., 1974, The cephalic arterial system in insectivores, primates, rodents and lagomorphs with special reference to systematic classification: Acta Anatomica, v. 87, suppl. 62, p. 1-160.

Cope, E. D., 1888, Hyoid and otic elements of the skeleton in the Batrachia: Journal of Morphology, v. 2, p. 297-310.

Crompton, A. W., and Jenkins, F. A., 1979, Origin of mammals, *in* Lillegraven, J. A., Kielan-Jaworowska, Z., and Clemens, W. A., eds., Mesozoic mammals: the first two-thirds of mammalian history: Berkeley, University of California Press, p. 59-73.

Crompton, A. W., and Sun, A. L., 1985, Cranial structure and relationship of the Liassic mammal *Sinoconodon**: Linnean Society, Zoological Journal, v. 85, p. 99-119.

Dollo, J., 1883, On the malleus of Lacertilia and the malar and quadrate bones of Mammalia: Quarterly Journal of Microscopal Sciences, v. 23, p. 579-596.

Doran, A., 1878, Morphology of mammalian ossicula auditûs: Linnean Society, London, Transactions, 2nd series, Zoology, v. 1 (1879), part VII, p. 371-497.

Engelmann, G. F., 1985, The phylogeny of the Xenarthra, *in* Montgomery, G., ed., The evolution and ecology of armadillos, sloths, and vermilinguas: Washington, Smithsonian Institution Press, p. 51-64.

Estes, R., 1961, Cranial anatomy of the cynodont reptile *Thrinaxodon liorhinus*: Museum of Comparative Zoology, Bulletin, v. 6, p. 165-180.

Fleischer, G., 1973, Studien am Skelett des Gehörorganes der Säugetiere, einschliesslich des Menschen: Säugetierkundliche Mitteilungen, v. 21, p. 131-239.

_____ 1978, Evolutionary principles of the mammalian middle ear: Advances in Anatomy, Embryology and Cell Biology, v. 55, 70 p.

Gadow, H., 1901, Evolution of auditory ossicles: Anatomischer Anzeiger, v. 19, p. 396-411.

Gaupp, E., 1908, Zur Entwickelungsgeschichte und vergleichenden Morphologie des Schädels von *Echnida aculeata,* var. *typica:* Semon, Zoologische Forschungsreisen in Australien, v. 6, p. 539-788.

_____ 1913, Die Reichertsche Theorie (Hammer-, Amboss-, und Kieferfrage): Archif für Anatomie und Physiologie, Abteilung Anatomie, Supplement-Band, p. 1-413.

Goodman, M., Czelusniak, J., and Beeber, J. E., 1985, Phylogeny of primates and other eutherian orders: a cladistic analysis using amino acid and nucleotide sequence data: Cladistics, v. 1, p. 171-185.

Goodrich, E. S., 1915, The chorda tympani and middle ear in reptiles, birds, and mammals: Quarterly Journal of Microscopal Sciences, v. 61, p. 137-160.

_____ 1930, Studies on the structure and development of vertebrates: London, Macmillan and Co., 837 p.

Gregory, W. K., 1910, The orders of mammals: American Museum of Natural History, Bulletin, v. 27, p. 1-524.

_____ 1947, The monotremes and the palimpsest theory: *ibid.,* v. 88, p. 1-52.

Griffiths, M., 1968, Echidnas: Oxford, Pergamon Press, 282 p.

Hafferl, A., 1933, Das Arterien System, *in* Bolk, L., Göppert, E., Kallius, E., and Lubosch, W., eds., Handbuch der vergleichenden Anatomie der Wirbeltiere, v. 6, p. 563-684.

Henson, O. W., 1974, Comparative anatomy of the middle ear, *in* Keidel, W. D., and Neff, W. D., eds., Handbook of sensory physiology, vol. 5: Berlin, Springer Verlag, p. 39-110.

Kemp, T. S., 1983, The relationships of mammals: Linnean Society, Zoological Journal, v. 77, p. 353-484.

Kermack, K. A., Musset, F., and Rigney, H. W., 1981, The skull of *Morganucodon: ibid.,* v. 71, p. 1-158.

Kirsch, J. A. W., 1977, The comparative serology of Marsupialia, and a classification of marsupials: Australian Journal of Zoology, Supplementary Series No. 52, p. 1-152.

Klaauw, C. J., van der, 1923, Die Skelettstücke in der Sehne des Musculus stapedius und nahe dem Ursprung der Chorda tympani: Zeitschrift für Anatomie und Entwicklungsgeschichte, v. 69, p. 32-83.

_____ 1931, The auditory bulla in some fossil mammals: American Museum of Natural History, Bulletin, v. 62, p. 1-352.

Kuhn, H.-J., 1971, Die Entwicklung und Morphologie des Schädels von *Tachyglossus aculeatus:* Abhandlungen der Senckenbergischen Naturforschenden Gesellschaft, 528, 192 p.

Lombard, R. E., and Bolt, J. R., 1979, Evolution of the tetrapod ear: an analysis and reinterpretation: Linnean Society, Biological Journal, v. 11, p. 19-76.

McKenna, M. C., 1975, Toward a phylogenetic classification of the Mammalia, *in* Luckett, W. P., and Szalay, F. S., eds., Phylogeny of the Primates: New York, Plenum Press, p. 21-46.

Noden, D. W., 1978, The control of avian cephalic neural crest cytodifferentiation. I. Skeletal and connective tissues: Developmental Biology, v. 67, p. 296-312.

Novacek, M. J., 1982, Information for molecular studies from anatomical and fossil evidence on higher eutherian phylogeny, *in* Goodman, M., ed., Macromolecular sequences in systematic and evolutionary biology: New York, Plenum Publishing Corporation, p. 3-41.

Novacek, M. J., and Wyss, A., *in press,* Higher-level relationships of the Recent eutherian orders: morphological evidence: Cladistics, v. 2.

Parker, W. K., 1886, On the structure and development of the skull in the Mammalia: Part I, Edentata, Part II, Insectivora: Royal Society, London, Philosophical Transactions, v. 176, p. 1-275.

Parrington, F. R., 1946, On the cranial anatomy of cynodonts: Zoological Society of London, Proceedings, v. 116, p. 181-197.

_____ 1967, The origins of mammals: Advancement in Science, v. 24, p. 1-9.

_____ 1979, The evolution of the mammalian middle and outer ears: a personal review: Biological Review, v. 54, p. 369-387.

Reichert, C., 1837, Ueber die Visceralbogen der Wirbelthiere im Allgemeinen und deren Metamorphosen bei den Vögeln und Säugethieren: Archif für Anatomie, Physiologie und wissenschaftliche Medizin (Leipzig), p. 120-222.

Romer, A. S., 1956, Osteology of the reptiles: Chicago, University of Chicago Press, 772 p.

Romer, A. S., and Price, L. I., 1940, Review of the Pelycosauria: Geological Society of America, Special Paper, v. 28, 538 p.

Schneider, R., 1955, Zur Entwicklung des Chondrocraniums der Gattung *Bradypus:* Morphologisches Jahrbuch, v. 95, p. 209-309.

Segall, W., 1969, The middle ear of *Dromiciops:* Acta Anatomica, v. 72, p. 489-501.

_____ 1970, Morphological parallelism of the bulla and auditory ossicles in some insectivores and marsupials: Fieldiana: Zoology, v. 51, p. 169-205.

_____ 1973, Characteristics of the ear, especially the middle ear in fossorial mammals, compared with those in the Manidae: Acta Anatomica, v. 86, p. 96-110.

Shute, C. C. D., 1956, The evolution of the mammalian eardrum and tympanic cavity: Journal of Anatomy, v. 90, p. 261-281.

Starck, D., 1941, Zur Morphologie des Primordialkraniums von *Manis javanica* Desm.: Morphologisches Jahrbuch, Leipzig, v. 86, p. 1-122.

Sues, H. D., 1985, The relationships of the Tritylodontidae (Synapsida): Linnean Society, Zoological Journal, v. 85, p. 205-217.

Tandler, J., 1899. Zur vergleichenden Anatomie der Kopfarterian bei den Mammalia: Denkschriften Akademie der Wissenschaften, Wien, mathematisch-naturwissenschaftliche Klasse, v. 67, p. 677-784.

Van De Water, T. R., Maderson, F. A., and Jaskoll, T. F., 1980, The morphogenesis of the middle and external ear: Birth Defects: Original Article Series, v. 16, p. 147-180.

Versluys, J., 1903, Entwicklung der Columella auris bei den Lacertilien. Ein Beitrag zur Kenntnis der schalleitenden Apparate und des Zungenbeinbogens by den Sauropsiden: Zoologisches Jahrbuch, Abteilung für Anatomie und Ontogenie der Tiere, v. 19, p. 107-188.

Westoll, T. S., 1943, The hyomandibular of *Eusthenopteron* and the tetrapod middle ear: Royal Society, London, Proceedings, B, v. 131, p. 393-414.

―――― 1944, New light on the mammalian ear ossicles: Nature, v. 154, p. 770-771.

Wible, J. R., 1984, The ontogeny and phylogeny of the mammalian cranial arterial pattern [Ph.D. thesis]: Durham, North Carolina, Duke University, 705 p.

Young, C. C., 1947, Mammal-like reptiles from Lufeng, Yunnan, China: Zoological Society, Proceedings, v. 117, p. 537-597.

MANUSCRIPT RECEIVED JANUARY 16, 1986
REVISED MANUSCRIPT RECEIVED JUNE 13, 1986
MANUSCRIPT ACCEPTED JUNE 27, 1986

New Late Cretaceous, North American advanced therian mammals that fit neither the marsupial nor eutherian molds

WILLIAM A. CLEMENS — *Department of Paleontology, University of California, Berkeley, California 94720*

JASON A. LILLEGRAVEN — *Departments of Geology/Geophysics and Zoology/Physiology, The University of Wyoming, Laramie, Wyoming 82071*

ABSTRACT

Upper molars of two new genera and species of mammals (*Falepetrus barwini* and *Bistius bondi*) are described from rocks of Late Cretaceous age of Montana, Wyoming, and New Mexico. Although these teeth are evolutionarily advanced in being fully tribosphenic, they have combinations of characters that preclude identification of the animals that bore them as either marsupials or eutherians. A review of available dental features useful in classification of known, fully-tribosphenic mammals from the Late Cretaceous suggests the presence of four principal groups: (1) Marsupialia; (2) Eutheria; (3) "deltatheridians;" and (4) others. Groups "1" and "2" are recognized as formal taxonomic units, defined by anatomically diverse suites of derived characters. Groups "3" and "4" are recognized informally and, as a grade, dubbed "tribotheres," mammals with tribosphenic dentitions lacking documented specializations characteristic of either marsupials or eutherians. Although group "3" may represent an evolutionary clade equivalent in taxonomic rank to marsupials or eutherians, members of group "4" (including *Falepetrus* and *Bistius*) comprise a heterogeneous conglomeration whose members have uncertain relationships to members of groups "1-3." In addition to the evolutionary radiations of contemporary marsupials and eutherians, the tribotheres provide evidence of at least a third, if not several, broad mammalian radiations during the Cretaceous. However, available dental criteria are inadequate to allow development of a useful, phylogenetically-based classification of the tribotheres.

INTRODUCTION

In his monographic reviews of Mesozoic mammals Professor George Gaylord Simpson (1928; 1929) summarized and reanalyzed data provided by the then available fossil record of these rare but intriguing members of the Class. These contributions still stand as definitive references. Other early studies, for example research on Cretaceous mammals of Asia (note Gregory and Simpson, 1926), added data supporting the then widely accepted Cope-Osborn theory of origin of the "trituberular" (upper) and "trituberular-sectorial" (lower) cheek teeth composing dentitions of early marsupials and eutherians (see Osborn, 1907). In 1936 Simpson presented a thorough review of the Cope-Osborn theory of origin of these types of teeth, which he renamed the tribosphenic type of dentiton. This paper provided ". . . another step toward clarifying and filling in the outline. . ." of the Cope-Osborn theory, but did not present major modifications of its arguments (Simpson, 1936, p. 3).

Twenty years later Simpson's colleague, Bryan Patterson, advanced another hypothesis for origin of the tribosphenic type of molar dentition (Patterson, 1956). Simpson (1961) reviewed this and other interpretations of homologies and evolution of cusps of tribosphenic molars ". . . in a skeptical way . . ." (Simpson, 1971, p. 194). But, a decade later he commented that, "further comparisons, especially of *Kuehneotherium*, leave some but only a little question that Patterson's views . . . were probably correct" (Simpson, 1971, p. 194).

Patterson's (1956) new hypothesis was the outgrowth of his analysis of fragmentary tribophenic teeth found in late Early Cretaceous localities in Texas. These fossils lacked (possibly because of their fragmentary nature) characters diagnostic either of marsupials or eutherians, and Patterson gathered them together in an informal grouping awkwardly dubbed "therians of metatherian-eutherian grade." Subsequently, additional material of tribosphenic mammals from these sites has been collected, described, and several new genera, including *Pappotherium, Holoclemensia,* and *Kermackia,* established (see Slaughter, 1981, and references cited). Some genera founded on fossils from other areas (e. g., *Aegialodon, Kielantherium,* and *Deltatheridium*) have been referred to this group (see Kielan-Jaworowska, Eaton, and Bown, 1979, and references cited).

Attempts to determine phylogenetic relationships of these therians of metatherian-eutherian grade resulted in markedly different suprageneric classifications. Slaughter (1968a, 1968b, and 1981), Fox (1975, 1980), and others maintained that at least *Pappotherium* and *Holoclemensia* could be referred to the Eutheria and Marsupialia, respectively. Turnbull (1971 and, later, Butler 1978a), held contrary views and, arguing that these taxa could not clearly be identified as eutherians or marsupials, established coordinate ordinal or higher level taxa to include many of the genera. Kielan-Jaworowska, Eaton, and Bown (1979) maintained that *Pappotherium* and *Holoclemensia* should be included in an informal unit, therians of metatherian-eutherian grade, until they were better known. Then, if appropriate, they could be referred to ordinal taxa.

We choose to follow Kielan-Jaworowska, Eaton, and Bown (1979) and avoid referring the tribotheres to ordinal-level taxonomic units; previously implied relationships do not appear warranted by data in hand. Our gathering of the genera into an informal group, the tribotheres, implies no particular evolutionary relatedness of its members other than to indicate they are: (1) advanced in possession of fully tribosphenic molars; and (2) lacking in molar specializations unequivocally associated either with marsupials or eutherians. The tribotheres comprise a cladistically meaningless, heterogeneous group of uncertain affinities.

Our report has two purposes. The first is to describe two new genera of Late Cretaceous, therian mammals with tribosphenic dentitions that cannot be clearly assigned to the Marsupialia or Eutheria. The second, arising from our inability to satisfactorily assign these new genera within existing taxonomic hierarchies, is to review the utility of available dental characters to see if improvements can be made in the understanding of evolutionary relationships of Late Cretaceous lineages of advanced therians. Data presented in Table 1 are restricted arbitrarily to those provided by Late Cretaceous tribosphenic mammals. Genera of advanced therians also are known from local faunas of Early Cretaceous age (see Kielan-Jaworowska, Bown, and Lillegraven, 1979, and Slaughter, 1981), and are considered in discussions of character polarities.

SYSTEMATIC PALEONTOLOGY

Class MAMMALIA
Subclass THERIA Parker and Haswell 1897
Falepetrus, new genus

Etymology: fale, from Fales plus Gr. *petro,* rock, in recognition of the area of discovery at Fales Rocks, Wyoming.

Type and only known species: Falepetrus barwini, n. sp.

Diagnosis: As for type and only included species.

Falepetrus barwini, new species
Figures 1 and 2

Etymology: Named in honor of John R. Barwin for his discovery of the fossil-bearing level now known as "Barwin Quarry."

Type specimen: AMNH 86316, an isolated, right, upper molariform tooth, crown only, collected by Malcolm C. McKenna, 1966 (Fig. 1).

Type locality: "Barwin Quarry," area of Rattlesnake Hills anticline, southeastern Wind River Basin, Natrona County, central Wyoming. Detailed locality data are on file in the Department of Vertebrate Paleontology, The American Museum of Natural History (also see Lillegraven and McKenna, 1986).

Formation: Lower part of "unnamed middle member," Mesaverde Formation.

Age: Judithian "age" (see Lillegraven and McKenna, 1986).

Diagnosis: Minuscule stylar cusps in positions B, C, and D, stylar shelf narrow, and both anterior and posterior lingual cingula (pre- and postprotoconular cingula) weak but distinct, thus differing from most metatherians. Paracone only slightly larger than metacone and anterior paracrista absent; derived characters shared by many eutherians and metatherians. Protocone not strongly produced lingually, a primitive character retained by most metatherians.

Referred specimen: UCMP 118602, an isolated, left, upper molariform tooth (Fig. 2), UCMP locality V-77083, Judith River Formation, Montana. Detailed locality data are on file at the Museum of Paleontology, University of California, Berkeley.

Description: AMNH 86316 is a complete tooth except for lacking: (1) roots; (2) anterobuccal corner of the crown (parastylar area); and (3) a small chip of enamel from the posterobuccal corner of the crown. Although somewhat broader posteriorly, the tooth essentially forms an equilateral triangle in occlusal view. Its crown is dominated by the tall paracone, metacone, and protocone; conules are small, stylar shelf is narrow, and lingual cingula are weakly developed both anteriorly and posteriorly.

Dimensions of the crown of the type specimen are: length, est. 3.50 mm; anterior width, 3.04 mm; and posterior width, 3.69 mm. Dimensions of the crown of the referred specimen are: anterior width, 2.71 mm and posterior width, 3.08 mm. All measurements were taken with use of an EPOI Shopscope (see Lillegraven and Bieber, 1986) following conventions established by Lillegraven, 1969, Fig. 5.

The paracone is the tallest cusp, with the metacone a close second. The paracone is nearly perfectly conical, with only slight linguobuccal compression. There is no anterior paracrista and the posterior paracrista, which runs from the cusp's apex to the valley at the base of the metacone, is weak. A short, weak ridge runs anterolinguad from the mesostylar area about one-fourth the distance from the base of the paracone toward its apex. Significant wear is observed on the paracone apex and along the length of the posterior paracrista; only the slightest wear facet is seen along the anterior face of the paracone.

The metacone is shorter anteroposteriorly than the paracone, but projects farther lingually at its widest point to contact the premetaconular crista. The buccal surface of the metacone differs from that of the paracone in being flatter and in possessing several subdued, vertical flutings. The anterior surface of the metacone is only

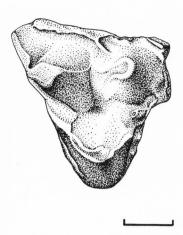

Figure 1. *Falepetrus barwini*, n. gen. and n. sp., holotype, AMNH 86316, right upper molar, occlusal view. Scale bar one millimeter.

moderately convex, and is interrupted by an anterior metacrista that is slightly stronger than the posterior paracrista. The two crests (forming a centrocrista) contact exactly in line in occlusal view between the paraconal and metaconal apices. The posterior edge of the metacone, although worn on AMNH 86316, clearly had a sharply developed postmetacrista that continued posteriorly the straight occlusal line established by the centrocrista. At the posterior edge of the tooth the posterior keel on the metacone is separated by a sharp carnassial-style notch from a strong continuation of the posterior metacrista that angles sharply posterolabiad to terminate at the extreme corner of the crown.

The entire posterolingual surface of the metacone plus continuation of the postmetacrista form a single, slightly convex, essentially vertical plane. The anterolabial surface of the postmetacrista (posterior to the notch) slopes more gently to terminate in a gutter at the base of an ectocingulum. Wear on the metacone is seen on: (1) a narrow but dorsally-expanding surface in the area of the premetacrista; (2) the metaconal apex and along the postmetacristal keel of the cone itself; and (3) the posterolingual edge of the posterolabial continuation of the postmetacrista. Wear is not developed along the broad, vertical, posterolingual surface of the metacone plus its continuing crest.

Except for a narrow but sharply-defined ectocingulum, a stylar shelf is nonexistent lateral to the paracone. The parastylar corner of the crown is broken away, but a sharp-crested cuspule is preserved at the anterolabial base of the paracone, about in the position expected for a stylocone (or cusp "B" position for a generalized marsupial). A mesostylar cusp is represented by an elevation of the ectocingulum having a steep anterior surface and a more gently-sloping posterior surface. As mentioned earlier, a short crest runs anterolinguad from the apex of the mesostyle onto the base of the paracone. Posterior to this crest, the ectocingulum is more strongly developed and is separated from the metacone and its postmetacrista by a narrow but clearly-defined gutter; this is the only part of the ectocingulum that could be considered a stylar shelf. An anteroposteriorly-elongated cuspule is present on the ectocingulum near its posterior end; it is placed directly lateral to the notch that separates the two limbs of the postmetacrista. The nature of the posterior termination of the ectocingulum cannot be determined because the enamel is chipped. The mesostylar area is set just lingual to a straight line drawn between the apices of the anterior- and posterior-most cuspules of the ectocingulum. All three ectocingular cuspules have slight labial expansions, deforming an otherwise straight labial border of the crown; an ectoflexus is lacking. Slight wear is observed along the entirety of the crest of the ectocingulum and is obvious on the posterior-most cuspule.

The conules are small and close to the bases of the paracone and metacone. The paraconule lacks a crestiform postparaconule crista, though a rounded step into the trigon basin does exist. The preparaconule crista runs laterad to the anterolingual base of the paracone. At that point the crest turns sharply dorsad to blend with a vertical wear surface that probably extended laterad as far as the parastylar area; morphology in this area cannot be determined because of a break on the specimen. Wear on the paraconule is restricted to its anterolingual surface and less strongly on the preparaconule crista.

The premetaconule crista is short, terminating at the lingualmost base of the metacone. The postmetaconule crista is convex posteriorly. It extends posterobuccally and dorsally to the posterolingual base of the metacone where the crista continues posterolaterally and more horizontally as a weak but distinct cingulum that terminates at a level just posterior to the notch of the postmetacrista. Wear on the metaconule is restricted to its posterolingual surface and along the postmetaconule crista. Wear on the cingular posterolabial continuation of the postmetaconule crista is insignificant.

The protocone is robust, but not strongly produced transversely. It has strong pre- and postprotocrista that run to the lingual bases of their respective conules. Additionally, a more subdued crest extends labiad from the apex of the protocone into the depth of the trigon basin. The postprotocrista in occlusal view is moderately convex posteriorly while the preprotocrista is somewhat straighter. As in most molariform teeth of therian mammals, the anterior face of the protocone is more vertically set than the posterior face. Significant wear on the protocone is restricted to its apex and postprotocrista (strongly) and its preprotocrista (weakly).

The anterior lingual cingulum (preprotoconular cingulum) is a weakly-defined fold of enamel punctuated by two minuscule cuspules. The ridge is near the enamel base and extends linguobuccally from a point dorsal to the medial end of the paraconule to a point slightly more than half way from the protoconal apex to the lingual edge of the crown. Slight wear is observable on the two cingular cuspules.

The posterior lingual (postprotoconular) cingulum is a slightly more definite structure, lacking distinct cuspules, that extends linguobuccally from the level of

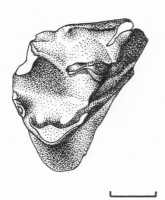

Figure 2. *Falepetrus barwini,* n. gen. and n. sp., referred specimen, UCMP 118602, left upper molar, occlusal view. Scale bar one millimeter.

the center of the metaconule to a point nearly at the lingual end of the protoconal crown. Only the lightest wear is seen along the crest of the posterior lingual cingulum.

Other than in all probability being three-rooted, no information about the roots of AMNH 86316 can be determined; the dorsal part of the tooth shows enamel only, with much of the dentine seemingly dissolved away.

The referred tooth from the Judith River Formation, UCMP 118602, is more extensively damaged than the type: (1) the anterior end of the crown, broken through the paracone, is missing; (2) a small chip is lost from the posteriolabial corner of the crown; (3) the posterior basal part of the protocone is lacking; and (4) the roots have been lost. This tooth is smaller than the type. The longest comparable dimension that can be taken is the distance between the lingual side of the protocone to the external edge of the crown, measured along an axis through the lowest point in the notch formed by the centrocrista. This dimension is 2.9 mm in the tooth from the Judith River Formation and 3.4 mm in the type.

Other differences from the type specimen include much weaker development in the referred specimen of a ridge between the slightly higher mesostylar cusp and the base of the paracone. The conules appear to be slightly smaller than those of the type. Only a weak crenulation is present in the region where the anterior lingual cingulum would be present. Because of breakage, the region in which a posterior lingual cingulum might be developed is missing.

Comments: Only three Mesozoic mammals invite close comparison with *Falepetrus*. One is *Picopsis pattersoni* Fox, 1980. The type upper molar of this genus and species is smaller than that of *Falepetrus*. The referred molars are larger but still not as large as that of the type of *Falepetrus*. *Falepetrus* stands apart in presence of stylar cusp B, smaller size of stylar cusp C, presence of conules, larger protocone, and presence of both pre- and postprotoconular cingula. *Picopsis* lacks conules; they are well developed in *Falepetrus*.

The broad stylar shelf of molars of *Potamotelses,* presence of a large stylar cusp B, and absence of conules as well as pre- and postprotoconular cingula clearly distinguish it from *Falepetrus*.

The last Cretaceous tribosphenic therian requiring close comparison with *Falepetrus* is the unnamed tribosphenic therian from the upper Milk River Formation described by Fox (1982). It differs from *Falepetrus* in its smaller (*c.* two-thirds) size. Its protoconal region is "bulbous," transversely narrow but anteroposteriorly long. The protocone of *Falepetrus* has different proportions (wider and not as long anteroposteriorly) and, unlike the unnamed therian, pre- and postprotoconular cingula are present. Finally, the stylar cusps of the unnamed therian appear to be larger and more distinct. These differences challenge what would be expected as a normal range of intraspecific or intrageneric variation.

Bistius, new genus

Etymology: Named after the former Bisti Trading Post, Hunter's Wash, San Juan County, New Mexico.

Type and only known species: Bistius bondi, n. sp.

Diagnosis: As for type and only included species.

Bistius bondi, new species
Figure 3

Etymology: Named for Mr. and Mrs. Clayton Bond who, over the years, extended hospitality and help to many field parties from The University of Kansas that camped near the Bisti Trading Post.

Type specimen: KU 15589, an isolated, right, upper molar (Fig. 3).

Type locality: KU locality New Mexico 37, Hunter's Wash, San Juan County, New Mexico. Exact locality data are on file at the Museum of Natural History, The University of Kansas.

Formation: Uppermost Fruitland Formation.

Age: Uncertain, either Judithian "age" (*c.* late Campanian) or, more probably, Edmontonian "age" (*c.* early Maastrichtian; see Lillegraven and McKenna, 1986).

Diagnosis: Stylar cusps A, B, C, and D large, thus differing from most known eutherians. Pre- and postprotoconular cingula small but distinct, thus differing from most known metatherians. Paracone and metacone subequal, a derived character shared by many eutherians and metatherians. Anterior paracrista absent, a derived character shared by most eutherians and some metatherians. Width of talon basin about one-third of crown width, a primitive character.

Description: Only one tooth in The University of Kansas collections from the San Juan Basin is referable to *Bistius bondi*. It was the basis of the entry, "eutherian of uncertain ordinal affinities," in an earlier report (Clemens, 1973b).

It is a relatively large tooth by Mesozoic standards (length, 3.98 mm; anterior width, 4.29 mm; posterior width, 4.53 mm — measurements taken using conventions established by Lillegraven, 1969, Fig. 5). Proportions suggest placement near the middle of the molar series. The cusps are low and bulbous; among Cretaceous mammals, they are reminiscent in general appearances of molar cusps of stagodontid marsupials.

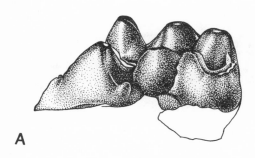

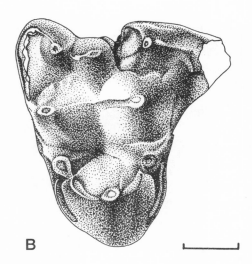

Figure 3. *Bistius bondi*, n. gen. and n. sp., holotype, KU 15589, right upper molar. *A*, labial view; *B*, occlusal view. Scale bar one millimeter.

Stylar cusp A, broken away, apparently was large and salient. Stylar cusp B, separated from the paracone by a deep trench, resembles a cusp on the molars of a multituberculate such as *Meniscoessus*. Its anterior slope is convex, rounded, with a slight crest directed anteriorly. Sharp ridges delimit the concave posterior surface. A small cuspule is present at the posterior end of the labial ridge. The medial ectoflexus is distinct. Stylar cusp C is centrally placed and equivalent in height to stylar cusp D. The latter is linked to the postmetacrista by a distinct, crenulated ridge tracing the rim of the posterolabial margin of the crown.

Paracone and metacone are of approximately equal height and size, although metacone appears slightly more massive. The worn centrocrista appears to have been low. Both conules are large, prominent cusps, and the preparaconule and postmetaconule cristae are distinct. The preparaconule crista extends across the base of the paracone to stylar cusp A; the postmetaconule crista ends posterior to the apex of the metacone. Central conule cristae are not developed. Protocone is a large cusp with pronounced, lingually-convex pre- and postprotocristae extending toward their respective conules and enclosing the talon basin. Short, but distinct pre- and postprotoconular cingula extend lingually to points slightly lingual to the apex of the protocone. The preprotoconular cingulum, beveled by wear, appears to have been linked to the preparaconule crista. A slight undulation in the wear facet suggests presence of a minute cusp near the labial end of this cingulum. The postprotoconular cingulum begins at the lingual edge of the metaconule and forms a gradual curve as it extends lingually. No cusps appear to have been present on this cingulum.

Apices of all preserved stylar cusps, cones, and conules exhibit apical wear. Only on the preparaconule crista, preprotoconular cingulum, and postmetacrista are facets indicative of tooth-on-tooth contact. The bases of the three roots are preserved.

Comments: The upper molar on which *Bistius* is based was compared with molars of a wide variety of Cretaceous and Cenozoic tribosphenic therians. None of the eutherians or other tribotheres has such prominent development of inflated stylar cusps B, C, or D as well as small but distinct pre- and postprotoconular cingula. Detailed comparisons were narrowed to the following marsupials.

Closest approximations to molar morphology of *Bistius* among the species of *Alphadon* were found with *A. rhaister*, particularly its M^1, and *A*. cf. *rhaister* (see Clemens, 1966, Fig. 12). *Bistius* differs from these species in having stylar cusps C and D enlarged, of subequal size, and almost as large as stylar cusp B. As argued below (see Section 12, *Stylar Cusp Development*), we join Fox (1983b) in suggesting that presence of a distinct stylar cusp D probably was primitive for the Marsupialia. Addition of a stylar cusp C occurred, possibly more than once, within the group. Thus *Bistius* is more derived than known species of *Alphadon* in enlargement of stylar cusp C. Also, the curious selene-like structure of the posterior face of stylar cusp B is unmatched in known species of *Alphadon* or other marsupials.

The possibility of close relationship between *Bistius* and *Glasbius* demands attention. To be sure there is a resemblance in their low crowns and inflated cusps, but detailed comparisons suggest that similarities are the result of parallel evolution. Molars of *Glasbius* are more derived in reduction of stylar cusp A and the parastylar area (see Section 12, below). Also, the talon basins of upper molars of both species of *Glasbius* tend to be modified from the primitive condition found in *Bistius* by shifts of the large metaconule to positions posterior to the protocone and reduction or loss of the paraconule. No hints of these derived characters are found in the molar of *Bistius*. In contrast, the molar of *Bistius* is more derived than those of *Glasbius* in the larger size of its stylar cusp C. In *Glasbius,* stylar cusp C either is usually smaller than stylar cusp D *(G. intricatus)* or commonly absent *(G. twitchelli)*. Additionally, although a weak preprotoconular cingulum is present on many molars of *Glasbius,* postprotoconular cingula are rare. When both are present, they differ in their asymmetry from the equally developed cingula of *Bistius*. The latter morphology could be interpreted as more derived, or equally so but

TABLE 1. IMPORTANT LATE CRETACEOUS GENERA FROM NORTH AMERICA, SOUTH AMERICA, ASIA, AND EUROPE. MORPHOLOGICAL FEATURES FROM EXAMINATION OF FOSSILS THEMSELVES, WITHOUT INTERPRETATIONS OR EXTRAPOLATIONS BEYOND THE FOSSILS.

DENTAL FORMULAE

		1	2	3	4
Taxa	Primary Literature	No. of Incisors	No. of Canines	No. of Premolars	No. of Molars
"TRIBOTHERES"					
"Deltatheridians"					
Deltatheridium	Kielan-Jaworowska, 1975b; Butler and Kielan-Jaworowska, 1973	?4/?1	1/1	3/3	3/4
Deltatheroides	Kielan-Jaworowska, 1975b	?/?	1/1	3/3	4/4
Deltatheroides sp.	Lillegraven, 1969; Clemens, 1973a; Fox, 1974	?/?	?/?	?/?	?/?
Hyotheridium	Gregory and Simpson, 1926	?/?	1/1	3/3	3/3
Other "tribotheres"					
Potamotelses	Fox, 1972, 1976a	?/?	?/?	?/?	?/?
Beleutinus	Bashanov, 1972	?/?	?/?	?/?	?/?
Gallolestes	Lillegraven, 1972, 1976; Clemens, 1980	?/?	?/?	?/?	?/?3
Picopsis	Fox, 1980	?/?	?/?	?/?	?/?
unnamed upper molar	Fox, 1982	?/?	?/?	?/?	?/?
Bistius	this paper	?/?	?/?	?/?	?/?
Falepetrus	this paper	?/?	?/?	?/?	?/?
"MARSUPIALS"					
Alphadon	Clemens, 1966; Lillegraven, 1969; Fox, 1971, 1976b, 1979a; Sahni, 1972; Archibald, 1982; Johnston and Fox, 1984	?/4	1/1	3/3	4/4
Peradectes	Archibald, 1982	?/?	?/?	?/?	4/4
Glasbius	Clemens, 1966; Archibald, 1982	?/?	?/1	?/3	4/4
Albertatherium	Fox, 1971	?/?	?/?	?/?	?4/?4
Pediomys	Clemens, 1966; Lillegraven, 1969, 1972; Fox, 1971, 1979b; Sahni, 1972; Archibald, 1982; Johnston and Fox, 1984	?/?	1/1	3/3	4/4
Aquiladelphis	Fox, 1971	?/?	?/?	?/?	?4/?4
Didelphodon	Clemens, 1966, 1968; Lillegraven, 1969; Archibald, 1982	?/?3	?/1	?/3	?/4
Eodelphis	Fox, 1971	?/3	?/1	?/3	4/4
Roberthoffstetteria	Marshall and others, 1983a; Muizon and others, 1984	?/?	?/?	3/?3	4/4

(continued on next page)

THERIANS THAT FIT NEITHER THE MARSUPIAL NOR EUTHERIAN MOLDS

(TABLE 1, CONTINUED)

		DENTAL FORMULAE			
		1	2	3	4
Taxa	Primary Literature	No. of Incisors	No. of Canines	No. of Premolars	No. of Molars
"EUTHERIANS"					
Gypsonictops	Lillegraven, 1969; Sahni, 1972; Clemens, 1973a; Fox, 1977, 1979c; Archibald, 1982	?/2+?	1/1	?4/5	3/3
Kennalestes	Kielan-Jaworowska, 1969, 1981	4/3	1/1	4/4	3/3
Cimolestes	Clemens and Russell, 1965; Lillegraven, 1969; Clemens, 1973a; Johnston, 1980; Archibald, 1982	?2/2+?	1/1	4/4	3/3
Batodon	Lillegraven, 1969; Clemens, 1973a; Archibald, 1982	?/?	?/1	?/4	3/3
Procerberus	Sloan and Van Valen, 1965; Lillegraven, 1969; Archibald, 1982; Johnston and Fox, 1984	?/?	?/?	?/?	3/3
unnamed palaeoryctid	Fox, 1979c	?/?	??	?/?	?/?
Asioryctes	Kielan-Jaworowska, 1975a, 1981	5/4	1/1	4/4	3/3
Telacodon	Simpson, 1929; Clemens, 1973a	?/?3	?/1	?/4	?/?
"Champ-Garimond tooth"	Ledoux and others, 1966	?/?	?/?	?/?	?/?
Paranyctoides	Fox, 1979c, 1984	?/?	?/?	?/?	?/3
Zalambdalestes	Kielan-Jaworowska, 1969, 1975a	?2/3	1/1	4/4	3/3
Barunlestes	Kielan-Jaworowska, 1975a; Kielan-Jaworowska and Trofimov, 1980	?/?	1/1	3/3	3/3
Protungulatum	Sloan and Van Valen, 1965; Van Valen, 1978; Kielan-Jaworowska and others, 1979; Johnston, 1980; Archibald, 1982; Johnston and Fox, 1984	?/?	1/1	4/4	3/3
Perutherium	Grambast and others, 1967; Sige, 1972; Marshall and others, 1983b	?/?	?/?	?/?	?/?
Purgatorius (including specimens from Tullock Fm.)	Van Valen and Sloan, 1965; Clemens, 1974; Kielan-Jaworowska and others, 1979	?/3	?/1	?/4	3/3

(continued on next page)

(TABLE 1, CONTINUED)

UPPER CHEEK TEETH

Taxa	5 General Morphology of Post. Upper Premolars	6 Ratio of Greatest Transverse Width to A-P Length	7 Size and Proportions of Protocone	8 Protoconular Cingula	9 Conular Development	10 Relative Strengths of Paracone and Metacone	11 Strength of Stylar Shelf	12 Stylar Cusp Development
"TRIBOTHERES"								
"Deltatheridians"								
Deltatheridium	Premolariform	1.3:1	Small, low, and anteroposteriorly compressed	Absent	Large, swollen	Paracone slightly taller than metacone	Extremely wide	Moderate stylocone, otherwise weak and virtually absent
Deltatheroides	?Premolariform	1.2:1	Small	?Absent	?	?	Extremely wide	?
Deltatheroides sp.	?	1:1	Small, low, and anteroposteriorly compressed	Absent	Weak	Paracone taller than metacone	Wide	Strong stylocone and metastyle, otherwise weak
Hyotheridium	?Submolariform	?	?	?	?	?	?	?
Other "tribotheres"								
Potamotelses	?	1:1	Small and low	Absent	Absent	Paracone taller than metacone	Wide	Strong stylocone and metastyle, otherwise weak
Beleutinus	?	?	?	?	?	?	?	?
Gallolestes	?	?	?	?	?	?	?	?
Picopsis	?	0.9:1	Small and low	Absent anteriorly, very weak posteriorly	Absent	Paracone taller than metacone	Wide posteriorly, absent anteriorly	Large mesostylar cusp
unnamed upper molar	?	1.2:1 Est.	Transversely narrow but anteroposteriorly long	Absent	Robust	Paracone taller than metacone	Anterior narrow, posterior wide	Series of definite cuspules along margin
Bistius	?	1.1:1	Large	Moderate	Robust	Paracone taller than metacone	Wide	Strong cusps along margin
Falepetrus	?	0.9:1 Est.	Large	Weak	Weak	Paracone taller than metacone	Wide posteriorly, narrower anteriorly	Weak cusps along margin
"MARSUPIALS"								
Alphadon	Premolariform	1.1:1	Large	Weak or, more generally, absent	Strong	Variable	Wide	Usually strong series of cuspules along margin
Peradectes	?	1.3:1	Large	Absent	Moderate	Metacone taller than paracone	Wide	Usually strong cusps along margin
Glasbius	Premolariform	1.1:1	Large and anteroposteriorly long	Strong anteriorly and weak posteriorly or absent	Low but broad	Metacone taller than paracone	Narrowed	Strong cusps along margin
Albertatherium	?	1.3:1	Large and anteroposteriorly short	Absent	Strong	Paracone taller than metacone	Wide	Strong cusps along margin
Pediomys	Premolariform	1.2:1	Large	Usually absent, weak if present	Strong	Paracone and metacone about equal in height	Wide posteriorly, narrow or absent anteriorly	Variously well developed to absent
Aquiladelphis	?	1.1:1 Est.	Large	Absent	Strong	Paracone and metacone about equal in height	Narrowed anteriorly	Strong cusps along margin
Didelphodon	Premolariform, very bulbous	1.3:1	Large	Absent	Strong	Metacone taller than paracone	Wide, especially posteriorly	Very strong cusps
Eodelphis	?	1.1:1	Large	Absent	Strong	Metacone slightly taller than paracone	Wide	Very strong cusps
Roberthoffstetteria	Premolariform	1.3:1	Large	Absent	Strong, merged with protocone to produce distinct crest	Metacone larger than paracone	Wide	Very strong cusps

*Measured on M² where possible

(continued on next page)

THERIANS THAT FIT NEITHER THE MARSUPIAL NOR EUTHERIAN MOLDS

(TABLE 1, CONTINUED)

UPPER CHEEK TEETH

Taxa	5. General Morphology of Post. Upper Premolars	6. Ratio of Greatest Transverse Width to A-P Length	7. Size and Proportions of Protocone	8. Protoconular Cingula	9. Conular Development	10. Relative Strengths of Paracone and Metacone	11. Strength of Stylar Shelf	12. Stylar Cusp Development
"EUTHERIANS"								
Gypsonictops	Molariform	1.7:1	Large and linguo-buccally produced	Very strong	Strong	Paracone taller than metacone	Narrow	Absent
Kennalestes	Submolariform	1.5:1	Large and linguo-buccally produced	Moderate	Moderate	Paracone taller than metacone	Wide	Stylocone, parastyle, and cuspule in metastylar area
Cimolestes	Submolariform	1.6:1	Large and linguo-buccally produced	Absent to strong	Strong	Paracone taller than metacone	Wide to narrowed	Stylocone only, or absent
Batodon	Submolariform	1.5:1	Large and linguo-buccally produced	Strong	Moderate	Paracone taller than metacone	Narrowed	Stylocone only
Procerberus	Submolariform	1.3:1	Large	Weak or absent	Moderate	Paracone taller than metacone	Wide	Absent
unnamed palaeoryctid	?	?	?	?	?	?	?	?
Asioryctes	Submolariform	1.7:1	Large and linguo-buccally produced	Absent	Paraconule strong, metaconule weak	Paracone taller than metacone	Narrowed	Parastyle, stylocone, and metastyle
Telacodon	?	?	?	?	?	?	?	?
"Champ-Garimond tooth"	?	?	?	?	?	?	?	?
Paranyctoides	?	1.2:1	Large and linguo-buccally produced	Strong	Strong	Paracone taller than metacone	Wide posteriorly, narrowed anteriorly	Strong parastyle, small stylocone, plus several marginal cuspules
Zalambdalestes	Submolariform	1.7:1	Large and linguo-buccally produced	Absent	Weak	Paracone taller than metacone	Narrowed	Strong parastyle and metastyle plus several accessory marginal cuspules
Barunlestes	Submolariform	1.5:1	Large	Absent	?	Paracone taller than metacone	Narrowed	Parastyle, stylocone, and metastyle present
Protungulatum	Submolariform	1.3:1	Large and anteroposteriorly long	Very strong	Strong	Paracone taller than metacone	Narrow	Absent
Perutherium	?	?	Large	Moderate	?	?	?	?
Purgatorius (including specimens from Tullock Fm.)	Submolariform	1.5:1	Large	Strong	Strong	Paracone taller than metacone	Narrow	Absent

*Measured on M^2 where possible

(continued on next page)

(TABLE 1, CONTINUED)

LOWER CHEEK TEETH

Taxa	13 General Morphology of Post. Lower Premolars	14 Size and Position of Paraconid Relative to Metaconid	15 Position of Hypoconulid Relative to Entoconid and Hypoconid	16 Strengths of Antero- and Posterolabial Cingulids
"TRIBOTHERES"				
"Deltatheridians"				
Deltatheridium	Premolariform	Taller than metaconid and set at least as far lingually as metaconid	Centrally placed on weakly-dvpt. talonid	Absent
Deltatheroides	Semi-molariform	Taller than metaconid and set at least as far lingually as metaconid	Centrally placed on weakly-dvpt. talonid	Weak
Deltatheroides sp.	?	Taller than metaconid and set at least as far lingually as metaconid	?	Absent
Hyotheridium	?	?	?	?
Other "tribotheres"				
Potamotelses	?	Roughly equal in height to metaconid and set as far lingually	Centrally-placed on moderate talonid	Short anterior cingulid, no posterior cingulid
Beleutinus	?	?	?	?
Gallolestes	?Molariform	Much smaller, lower, and more labially set than metaconid	Slightly closer to entoconid	Short anterior and posterior cingulids
Picopsis	?	May be slightly taller than metaconid, relative position undeterminable	?	Absent anteriorly, ? posteriorly
unnamed upper molar	?	?	?	?
Bistius	?	?	?	?
Falepetrus	?	?	?	?
"MARSUPIALS"				
Alphadon	Premolariform	Nearly as tall and as far lingual as metaconid	Twinned with entoconid	Usually strong
Peradectes	?	Lower than metaconid and set as far lingually	Closer to entoconid, but not closely twinned	Strong anteriorly, variably present or absent posteriorly
Glasbius	Premolariform	Lower than metaconid and set slightly more labially	Twinned with entoconid	Very strong
Albertatherium	?	Lower than metaconid and set slightly more labially	Twinned with entoconid	Strong
Pediomys	Premolariform	Lower than metaconid and set nearly as far lingually	Twinned with entoconid	Strong
Aquiladelphis	?	Lower than metaconid and more labially set	Twinned with entoconid	Strong
Didelphodon	Premolariform, very bulbous	Taller than metaconid and set as far lingually	Twinned with entoconid	Strong anteriorly, weak or absent posteriorly
Eodelphis	Premolariform, enlarged in some species	Taller than metaconid and set as far lingually	Twinned with entoconid	Strong anteriory, weak or absent posteriorly
Roberthoffstetteria	Premolariform	Lower than metaconid and set nearly as far lingually	Reduced in size, closer to entoconid, but not closely twinned	Weak anteriorly and posteriorly

(continued on next page)

(TABLE 1, CONTINUED)

LOWER CHEEK TEETH

Taxa	13 General Morphology of Post. Lower Premolars	14 Size and Position of Paraconid Relative to Metaconid	15 Position of Hypoconulid Relative to Entoconid and Hypoconid	16 Strengths of Antero- and Posterolabial Cingulids
"EUTHERIANS"				
Gypsonictops	Nearly molariform	Lower than metaconid and more labially set	Central to or slightly closer to entoconid	Strong anteriorly, weak or absent posteriorly
Kennalestes	Premolariform	Lower than metaconid and more labially set	Centrally between	Absent
Cimolestes	Submolariform	Lower than metaconid and more labially set	Centrally between	Strong anteriorly, weak or absent posteriorly
Batodon	Submolariform	Lower than metaconid and more labially set	Slightly closer to entoconid	Strong anteriorly, weak or absent posteriorly
Procerberus	Nearly molariform	Lower than metaconid and set only slightly more labially	Centrally between	Weak anteriorly, absent posteriorly
unnamed palaeoryctid	Premolariform	?	?	?
Asioryctes	Premolariform	Lower than metaconid and more labially set	Slightly closer to entoconid	Strong anteriorly, absent posteriorly
Telacodon	Premolariform	?	?	?
"Champ-Garimond tooth"	?	Lower than metaconid and set only slightly more labially	Slightly closer to entoconid	Strong anteriorly, weak posteriorly
Paranyctoides	Premolariform	Much smaller, lower, and more labially set than metaconid	Closer to entoconid, but not twinned	Strong
Zalambdalestes	Submolariform	Lower than metaconid and more labially set	Slightly closer to entoconid	Absent
Barunlestes	Submolariform	Lower than metaconid and more labially set	Slightly closer to entoconid	Absent
Protungulatum	Submolariform	Lower than metaconid and set only slightly more labially	Slightly closer to entoconid	Weak
Perutherium	?	Much smaller than metaconid and set as far lingually	Closer to entoconid, but not twinned	Strong anteriorly, absent posteriorly
Purgatorius (including specimens from Tullock Fm.)	Submolariform	Lower than metaconid, but set as far lingually	Centrally between	Strong anteriorly, absent posteriorly

(end of Table 1)

in a different pattern. *Bistius, Glasbius,* several other Late Cretaceous mammals (for example, *Protungulatum* and *Purgatorius*), and numerous early Cenozoic species share the derived characters of low crowns and inflated cusps. The following review of evolution of therian dentitions demonstrates this cannot be interpreted as retention of a primitive condition, but documents parallel changes in a variety of lineages of tribosphenic therians. Differences in other derived characters of molars of *Bistius* and *Glasbius* indicate they probably represent distinct lineages.

The stagodontids *Didelphodon* and *Eodelphis* are distinctly different from *Bistius.* They share the primitive marsupial morphology in lacking a stylar cusp C or having only a small cuspule in this position. In contrast, they are more derived than *Bistius* in the slightly (*Eodelphis*) or distinctly (*Didelphodon*) smaller size of paracone relative to metacone.

Finally, turning to the South American record (Marshall and others, 1983a, and Muizon and others, 1984), upper molars of the marsupial *Roberthoffstetteria nationalgeographica* show some similarities to the molar of *Bistius.* However, *Bistius* is easily distinguished by the presence of pre- and postprotoconular cingula and by the positions of the conules approximately equidistantly removed and labial to the protocone. Upper molars of *Rob-*

erthoffstetteria lack protoconular cingula, and the conules are situated almost as far lingually as the protocone, to which they are closely approximated.

DISCUSSION

Introduction

In the last decade the number of fossils of Late Cretaceous age representing various tribotheres has greatly increased. This new material was gained primarily through further investigations of Upper Cretaceous strata in Asia and South America, combined with application of underwater screening techniques to collecting fossiliferous deposits in North America. New records of mammals of Early Cretaceous age, for example, discovery of *Arguimus* (see Dashzeveg, 1979) in Mongolia and a variety of mammals in Russia (see Nesov, 1982, 1984, 1985a,b; Nesov and Golovneva, 1983; Nesov and Gureyev, 1981; and Nesov and Trofimov, 1979) have begun to fill in knowledge of mammalian evolution during this interval of earth history. Discoveries in the Middle Jurassic, Forest Marble of England made by Freeman (1976a,b) and associates provide new information concerning stocks closely related to mammals characterized by tribosphenic dentitions (Clemens, 1985). Also, specimens such as the curious "pantothere" molar from the Late Jurassic at Porto Pinheiro, Portugal (Krusat, 1969) and the remarkable new mammal from the Jurassic of China, *Shuotherium* (see Chow and Rich, 1982), point to a greater diversity of early mammals than previously perceived. In spite of these significant discoveries we still are faced with major gaps in the fossil record of therian evolution just prior to origin of the tribosphenic dentition.

Although *Falepetrus* and *Bistius* clearly represent dentally advanced therian mammals new to science, their reference to major groups known from Upper Cretaceous strata has proven frustrating; were they marsupials, eutherians, or perhaps something else? Our inability to make these referrals led us to consider exactly what suites of dental characters are available to help make such taxonomic decisions.

Table 1 specifies the traditionally used dental characters for almost all named genera of Late Cretaceous, advanced therian mammals. Omitted genera include *Oxyprimus, Ragnarok,* and *Mimatuta*, which all are anatomically close to *Protungulatum; Daulestes* (Nesov and Trofimov, 1979), which resembles *Zalambdalestes; Kumsuperus* (Nesov, 1984), which provides little new pertinent information; and genera recently described by Nesov (1985a,b). The table was developed to help recognize dental differences and commonalities within and among: (1) generally accepted marsupials and eutherians (both groups with living representatives and a reasonable Cenozoic record); and (2) the miscellaneous genera here informally referred to as tribotheres. We know of no other generally applicable morphological characters. The data offer hints about directions of evolution of therian dentitions, and are presented in hope that they will provide a useful framework for future analyses as more becomes known of the dental morphology of these mammals. Table 2 summarizes data presented in Table 1, and sets out our view of the "probable primitive conditions" in teeth of the last common ancestor of tribotheres, marsupials, and eutherians.

To set the stage for analysis of dental characters, certain conventions require explanation:

1. Synapsids and mammals

We follow Kemp (1982), Hopson and Barghusen *(in press)*, and other recent students of the amniotes in recognizing a basic dichotomy, with its roots in the Paleozoic, separating reptiles commonly referred to as "synapsid reptiles" or "mammal-like reptiles"). We support the view that early members of the synapsid clade should not be dubbed reptiles, either informally or formally. An appropriate and complete classification of these non-reptilian amniotes has yet to be advanced, although several colleagues are at work trying to fill the gap. For present purposes, we leave the non-reptilian clade of amniotes unnamed, and recognize within it a primitive, paraphyletic group, the synapsids, and a derived group, the mammals.

Mammals provisionally will be diagnosed as members of this non-reptilian clade characterized by evolution of a dentary-squamosal articulation between lower jaw and skull. Students of advanced synapsids (*i. e.,* the cynodontians) and early mammals will recognize that we are begging issues raised by: (1) forms such as *Diarthrognathus* and *Morganucodon,* in which both the dentary and squamosal as well as articular and quadrate functioned jointly as parts of the jaw articulation; and (2) the broader question of monophyly of the Mammalia when defined by evolution of the dentary-squamosal jaw articulation (see Kemp, 1983, Crompton and Sun, 1985). However, given the scope of the present analysis (*i. e.,* an attempt to order Late Cretaceous members of what we take to be a derived monophyletic group of mammals characterized by presence of tribosphenic dentitions), this problem is at a different scale and can be set aside, though not forgotten.

2. Non-therian/therian dichotomy of mammals

In the 1970s students of cynodontians and early mammals marshalled data appearing to support interpretation of a basic dichotomy separating mammals into non-therian (or prototherian) and therian clades (see Kermack and Kermack, 1984). One form of evidence supporting such a view was the mode in which the bony braincase was enlarged. A primary division was based on whether the alisphenoid or an anterior lamina of the petrosal made a major contribution to the lateral wall of the braincase. This interpretation has been challenged effectively by Presley (1981) and Kemp (1983), whom we follow.

Recognition of a non-therian/therian dichotomy also was supported by recognition of two basic patterns of symmetry of molariform cheek teeth. In non-therians, the major cusps of these teeth are aligned in anteroposteriorly oriented rows; in therians, the principal cusps are ordered in triangular patterns. Kemp (1983) argued that the basic pattern of linear cusp orientation is a primitive character, inherited from synapsids, though modified among mammals in a variety of ways (*e. g.*, addition of rows of cusps of similar size in multituberculates or multiplication of complexly interlocking cusps in docodonts). Phylogenetic unity of non-therians no longer can be rigorously defended on the basis of a derived dental morphology. In contrast, evolution of the "reversed triangle" pattern of symmetry, with subsequent modifications in patterns and orientations of wear facets, is thought to have occurred but once, and is accepted as a derived character defining a monophyletic clade (see Prothero, 1981).

3. Non-tribosphenic and tribosphenic therians

Patterson's (1956) hypothesis concerning evolution of the tribosphenic postcanine dentition suggests that new structural units (*i. e.*, the protocone and talonid basin) were added to posterior postcanine teeth of later therians. The hypothesis has been strengthened through tests provided by discovery of new Mesozoic therians and has yet to be falsified.

Three major groups of modern mammals (monotremes, marsupials, and eutherians) can be recognized on the basis of differences in patterns of reproduction involving a large suite of developmental, physiological, and other characters (Lillegraven, 1984, 1985). Until recently, the fossil record was interpreted by most workers (but note Gregory, 1947, and Kemp, 1982) as indicating that the ancestry of monotremes did not include therians with tribosphenic dentitions. This view has been shaken by discovery in Australia of a late Early Cretaceous monotreme-like mammal, *Steropodon,* with cheek teeth of a distinctly tribosphenic-like pattern (Archer and others, 1985). It remains to be determined, however, whether this dental pattern was: (1) characteristic of the common ancestor of all modern mammals; or (2) the Australian fossil documents convergent evolution of a tribosphenic-like dentition.

4. The polychotomy of tribosphenic therians

Modern marsupials and eutherians, when recognized on differences in patterns of reproduction, also can be distinguished on the basis of a number of dental and skeletal traits (*e. g.*, bullar or pelvic structure). Such osteological characters hold true for all known Cenozoic members of the two groups and, when available, form the basis for identification of Cretaceous therians.

Data are accumulating that suggest Late Cretaceous Asian deltatheridians (*i. e., Deltatheridium, Deltatheroides,* and *?Hyotheridium*) plus, possibly, North American forms known only from isolated teeth (Fox, 1974) represent an evolutionary clade equivalent in taxonomic rank to marsupials and eutherians. This possibility was emphasized when in many of the following analyses of particular characters tribotheres were clearly segregated into deltatheridians and others (see Tables 1 and 2).

Finally, many Cretaceous tribosphenic therians are known only from isolated teeth or, at best, fragments of jaws with constellations of characters that argue against reference either to marsupials or eutherians. Some of these species lack dental specializations usually thought definitive of these groups; others illustrate perplexing mixtures of derived dental characters. The records of tribotheres including deltatheridians appear to represent lineages that, with marsupials and eutherians, differentiated in the early evolutionary radiation of tribosphenic mammals.

Dental Embryology and Relevance to Evolutionary Interpretations

Studies of dental embryology of modern mammals should give pause in evaluating abilities to recognize: (1) dental homologies among various tribosphenic mammalian groups (living and extinct); (2) which teeth should be identified as "incisors," "canines," "premolars," or "molars"; and (3) which teeth in the adult are derived from what usually are called the "deciduous" (primary) or "permanent" (secondary) tooth bud series.

The basic processes of embryonic origin and morphogenesis of mammalian teeth are well known on a descriptive level (*e. g.*, see Gaunt and Miles, 1967). Paleontologists, unfortunately, do not have access to such information for most ancient species with which they deal, and must depend upon interpretation of developmentally complete (or nearly complete), strongly mineralized structures.

The currently widely used definitions of basic units of adult mammalian dentitions imply that conservatism exits within extant therians in regard to positions and numbers of incisiforms, caniniforms, and teeth posterior to the caniniform. Key aspects of these definitions include limitation of all upper incisors to the premaxillary, and position of the alveolus of the upper canine either at or just adjacent to the suture joining the premaxillary and maxillary bones. One might predict, therefore, that if taxonomically systematic data were collected on embryological development of tooth germs, dental homologies (even among distant relatives) readily could be determined. But such has not been the case, as witnessed by the large and frequently conflicting literature on interpretations of homologies of pre- and postcanine teeth between marsupials and eutherians (as examples see: Wilson and Hill, 1897; Berkovitz, 1968a,b; McKenna, 1975; Archer, 1978; Osborn, 1978; and Westergaard, 1980).

Confusion and lack of general agreement in recognition of tooth homologies (even among living therians) are understandable when the extent of embryogenetic lability in mammalian tooth development is appreciated. The specimen of the eutherian *Pararyctes* recently analyzed by Fox, 1983a, illustrates that this is not a problem limited to modern species. Discussions of a few general features of this variability drawn from a review by Schwartz (1982) highlight some of the major problems.

Because there is no special linkage between the developing mammalian canine and the premaxillary-maxillary suture, that tooth is variously positioned at the suture, or immediately anterior or posterior to it; similar variability is seen in positions of posterior incisiform teeth. During their development canines and posterior incisors are known to migrate from one bone to another

TABLE 2. SUMMARY OF DATA FROM TABLE 1, GENERALIZED FOR MAJOR GROUPS OF LATE CRETACEOUS MAMMALS, WITH PRESUMED PRIMITIVE STATES SPECIFIED.

DENTAL FORMULAE

Late Cretaceous Groups	1. Number of Incisors	2. Number of Canines	3. Number of Premolars	4. Number of Molars
"TRIBOTHERES"	Inadequately known to be useful; deltatheridians with 4 or more in upper and 1 (or possibly 2) in lower jaw	1/1 where known (Asiatic deltatheridians only)	Known only in Asiatic deltatheridians, with 3/3	Variable where known, at 3-4/3-4
"MARSUPIALS"	Inadequately known to be useful, but at least 4 lower incisors in some species	1/1 where known	Inadequately known, but 3/3 consistently	Appears consistent at 4/4
"EUTHERIANS"	Poorly known, except to show high variability, with 5/4 a maximum	1/1 where known	Some evidence of 4/5, generally 4/4, rarely reduced to 3/3	Appears consistent at 3/3
Probable primitive condition	5 upper and 4 lower, or possibly more	Count of 1/1 (possibly double rooted)	Primitive count uncertain, possibly 5/5	Primitive count uncertain, possibly 4/4

UPPER CHEEK TEETH

Late Cretaceous Groups	5. General Morphology of Posterior Upper Premolars	6. Ratio of Greatest Transverse Width to A-P Length	7. Size and Proportions of Protocone	8. Protoconular Cingula
"TRIBOTHERES"	Known only in Asiatic deltatheridians and are variably pre- and submolariform	Variation of 0.9-1.3:1; highly variable in width-length proportions	Usually relatively small	Usually absent or weakly developed
"MARSUPIALS"	Where known are consistently premolariform	Variation of 1.1-1.3:1; consistently only slightly wider than long	Large	Usually absent, weak if present
"EUTHERIANS"	Where known are submolariform to molariform	Variation of 1.1-1.7:1; consistently markedly wider than long	Large and usually linguo-buccally produced	Variously absent, weak, or strong, but usually present
Probable primitive condition	Primitive condition uncertain, possibly premolariform	Length similar to width	Small, low protocone	Absence

LOWER CHEEK TEETH

Late Cretaceous Groups	13. General Morphology of Posterior Lower Premolars	14. Size and Position of Paraconid Relative to Metaconid	15. Position of Hypoconulid Relative to Entoconid and Hypoconid
"TRIBOTHERES"	Inadequately known, but variously pre- and submolariform	Highly variable, frequently labial to metaconid; equal in height to or smaller than metaconid	Where known, is usually centrally-placed on a small talonid
"MARSUPIALS"	Where known, consistently premolariform	Usually lower than metaconid and set as far lingually	Consistently twinned with entoconid
"EUTHERIANS"	Variously premolariform, submolariform or, less commonly, molariform	Consistently lower than metaconid and usually more labially set	Either centrally-placed on wide talonid or slightly closer to entoconid
Probable primitive condition	Primitive condition uncertain, possibly premolariform	Subequal or smaller than metaconid and more labially set	Central placement on small talonid

THERIANS THAT FIT NEITHER THE MARSUPIAL NOR EUTHERIAN MOLDS

9 Conular Development	10 Relative Strengths of Paracone and Metacone	11 Strength of Stylar Shelf	12 Stylar Cusp Development
Variously absent to strong	Where known, paracone is taller than metacone	Usually wide	Usually with strong stylocone and metastyle, and with other stylar cusps variously absent or present
Usually strong	Generally metacone is taller than or roughly equal to height of paracone	Usually wide	Usually with strong series of cuspules along margin
Usually strong	Where known, paracone is taller than metacone	Variable, but usually narrowed	Stylocone, parastyle, and metastylar cusps commonly present; other cuspules along margin variously absent or weakly developed
Absence	Paracone taller than metacone	Wide	Stylocone, parastyle, and metastylar cusps commonly present; other cuspules along margin variously absent or weakly developed

16 Strengths of Antero- and Posterolabial Cingulids
Where known, are absent or weak
Commonly both are strong
Usually strong anteriorly (but absent in known Asiatic forms), usually absent or weak posteriorly
Anterior probably present, but weak; absent or weak posteriorly

across regions of adult sutures toward their eventual points of eruption. Thus, the mere position of an erupted tooth relative to the premaxillary-maxillary suture cannot be used independently or with full confidence to identify it as an incisor, canine, or premolar; tooth morphology typically has been applied as the deciding factor. Unfortunately, many specialized mammals develop teeth that are either caniniform or incisiform in unexpected regions of the jaw, thus leading to controversy in how the teeth should be designated and what their homologies are to teeth of other taxa.

Schwartz (1982) also pointed to uncertainty about genesis of tooth shape and its embryonic control (*i. e.*, the "field theory" of Butler, 1963 and 1978b, the tooth "clone model" of Osborn, 1978, or other, still more labile mechanisms). Roth (1984) argued that the basis of homology in a broad sense is the sharing of developmental pathways, and she discussed various levels of homology that might be recognized among elements of mammalian dentitions. However, the very fact that the controlling influences on tooth position and morphology are unknown should inspire lack of confidence in identification of homologies among various specialized species of mammals.

As further reviewed by Schwartz (1982), adult teeth of some mammalian species are derived from tooth germs in embryological positions that in most mammals are dedicated to deciduous teeth. In certain species of *Sorex*, for example, adult antemolar teeth develop from buds on the dental lamina (the usual source of deciduous teeth) and the teeth which are shed *in utero* are derived from external dental epithelium (the usual source of permanent teeth) of their predecessor tooth germs. One cannot tell from inspection of adult dental morphology from which set of embryonic tooth germs the adult teeth have been derived. To further complicate interpretations, an excess number (relative to adult teeth) of embryonic tooth buds is common, and different species develop and regress different combinations of buds in forming the adult dentition. Finally, both embryology and paleontology provide evidence that, contrary to usual interpretation, evolutionary losses of premolars and incisors do not always occur in alveolar positions immediately posterior to and anterior from, respectively, the canine.

Therefore, even when paleontologists are fortunate enough to have fossils with complete dental series, only limited confidence can be given to interpretations of interspecific homologies. Such points must be kept in mind throughout the various analyses that follow.

Analysis of Dental Characters

The numbers preceding discussions of the various characters refer to numbers of the columns in Tables 1 and 2. In preparation of this analysis we have benefited from several recent studies of the evolution of mammalian dentitions, in particular those of Butler (1977), Fox (1975, 1984), Novacek (1982, 1984, 1985), Prothero (1981), and Westergaard (1980, 1983) which, in some instances, reach conclusions differing from ours. Unfortunately the papers by L. A. Nesov (1985a, b) containing descriptions of several new Cretaceous (Albian-Coniacian) tribosphenic therians became available to the authors after the manuscript of this study was completed. Although we reserve judgement on some of the references of these mammals to suprageneric taxa, a preliminary review of the descriptions convincingly points to the presence of several new kinds of mammals.

1. Number of incisors

The traditional definition of upper incisors is those teeth implanted in the premaxillary bone anterior to the canine, which is situated at or adjacent to the premaxillary-maxillary suture. Lower incisors are the occluding teeth set in the dentary anterior to the lower canine.

A survey of primitive cynodontians (*sensu* Kemp, 1982) indicates these synapsids had as many as six or seven incisors in any dental quadrant and some had an additional incisiform tooth in the maxillary. Evolution of advanced cynodontians was characterized by reduction in the number of incisors (frequently to four uppers and three lowers) and loss of incisiform teeth with alveoli in the maxillary (see Kemp, 1982).

Morganucodon, here considered one of the earliest mammals, has three upper incisors and, like primitive cynodontians, an incisiform tooth implanted in the maxillary (Kermack and others, 1981). Mills (1971) found evidence of individual variation in number of lower incisors ranging from four to six. Among the other non-therian mammals some data are available on variation in numbers of incisors in multituberculates where as many as three upper but, as far as known, no more than one lower incisor were present. Amphilestid triconodonts had as many as three and, possibly, four lower incisors; the upper incisor count is unknown (Simpson, 1928). *Dinnetherium*, a triconodont of uncertain family affinities had four lower incisors (Jenkins and others, 1983). The full incisor dentition is known only in one Early Cretaceous tricondontid that had two upper and one lower incisor (Crompton and Jenkins, 1979). However, other species possessed at least two lower incisors and the total number probably was greater (Simpson, 1928). Among docodonts as many as three and possibly four lower incisors were present (see Kron, 1979, and Krusat, 1980). The docodont *Haldanodon* had four upper incisors plus an incisiform tooth set in the maxillary (Krusat, 1980).

The fossil record of non-tribosphenic therians provides only a few hints concerning the numbers of lower incisors present; composition of the upper incisor dentition is unknown. Gill (1974) suggested *Kuehneotherium* had at least four lower incisors. A few specimens demonstrate that dentitions of spalacotheriid symmetrodonts included at least three lower incisors (Cassiliano and Clemens, 1979). As many as four lower incisors were present in some species of eupantotheres (Simpson, 1928, 1929).

Composition of the incisor dentition of Cretaceous tribosphenic therians again is known in only a few forms. Deltatheridians had four or more upper incisors, but only one or two lowers (Kielan-Jaworowska, Eaton, and Bown, 1979). No data are available on composition of the incisor dentitions of tribotheres. The largest known number of incisors (five upper and four lower) occurs in the eutherian *Asioryctes* in which the fifth upper incisiform is situated in the premaxillary-maxillary suture. Kielan-Jaworowska (1981) also reported that in *Kennalestes*, which has four upper incisors and three lowers, both the deciduous last incisor (DI^4) and the erupting last permanent incisor (I^4) are situated in the suture between the premaxillary and maxillary. Thus, both *Asioryctes* and *Kennalestes*, whose allocations to the Eutheria are supported by a variety of dental, cranial, and postcranial characters, had more than the three upper and lower incisors usually described as typical of the dentitions of generalized eutherians.

As long recognized, generalized marsupials with little modified tribosphenic dentitions have a battery of as many as five upper and four lower incisors. At least some species of *Alphadon*, the only Cretaceous marsupials known from material that provides a full count of lower incisors, had at least four. Composition of the upper incisor dentition of Cretaceous marsupials is unknown.

Probable primitive condition — These scraps of information provided by a very few Mesozoic mammals (both non-therians and therians) suggest that for therians, and possibly all mammals, presence of at least four lower incisors was the primitive condition. On the grounds of occlusal relationships of upper and lower dentitions and the frequent occurrence of incisor dentitions in which the number of uppers is equal to or more than the number of lowers, it does not appear out of place to suggest that the dentition of the last common ancestor of tribosphenic therians probably contained at least five upper and four lower incisors.

In many cynodontians, the number of incisiform teeth was reduced to four upper and three lower, and incisiforms set in the maxillary were lost. Kemp (1982, p. 304) suggested that the "*Eozostrodon*" level of his cladistic analysis was characterized by a reversal of this trend toward reduction, and an increase in number of lower incisors occurred. Presumably, increase in number of upper incisors and reappearance of an incisiform in the maxillary also were involved. An alternative hypothesis, that mammals were derived from lineages of cynodontians that did not experience loss of these incisiform teeth, has yet to be refuted.

Situation of the last incisiform tooth in the suture between the premaxillary and maxillary in *Asioryctes* and *Kennalestes*, both relatively long-snouted forms, might well be testimony to variation in developmental patterns rather than evidence of inheritance of a trait from early cynodontians. The same could hold true of *Morganucodon* and *Haldanodon*, although in these non-therian mammals the alveolus of the posterior incisiform is well within the maxillary bone.

2. Number and morphology of canines

Typically the upper canine is identified as the tooth situated at or near the premaxillary-maxillary suture. The lower canine is defined as the member of the lower dentition that occludes directly in front of the upper. Primitively, canines are thought to be larger than immediately adjacent teeth, and have heightened crowns designed for piercing.

Pelycosaurian synapsids (*sensu* Kemp, 1982) document evolution of enlarged caniniform teeth in the maxillary and, possibly later, lower caniniform teeth differentiated. In forms such as *Dimetrodon* the upper and lower caniniforms were widely separated when the jaws were brought together. The lowers came to rest near the premaxillary-maxillary suture and were separated from the upper caniniforms by diastemata and simple conical teeth.

Tight interlocking or regular occlusion of upper and lower canines characterized some early therapsids (*e. g.*, dinocephalians). Large upper and lower canines were present in most cynodontians; in a few, wear facets document patterns of regular occlusion. Few data on ontogeny of canines are available, but in *Thrinaxodon* the lower canines are reported to have been polyphyodont (Crompton, 1963).

Mills (1971) noted the large size of the crowns and roots of canines of *Morganucodon*. The root of the lower canine has distinct vertical grooves on both sides, usually has two pulp canals, and, in some specimens, its tip is divided into two separate rootlets.

Many workers have suggested that multituberculates lost the canines early in the evolution of the group. However, on grounds of their position relative to the premaxillary-maxillary suture and support of the crown by only one root, Hahn (see 1977) argued that the anteriormost premolariform teeth of some paulchoffatiid multituberculates might be modified canines (see Clemens and Kielan-Jaworowska, 1979).

Variation among caniniform teeth of triconodonts is great. A specimen of *Triconodon mordax* (see Simpson, 1928) demonstrates that the lower canine was diphyodont. The deciduous canine is double rooted but its replacement is single rooted. The canine alveolus preserved in a skull fragment of *Priacodon* is bilobed, suggesting that the lateral sides of the root were deeply grooved (Simpson, 1925), but the shape of the alveolus for the lower canine in a referred specimen suggests presence of a single rooted tooth. *Trioracodon* has a single rooted lower canine, but the only known upper canine, which might be a deciduous tooth, is strongly double rooted. The lower canine of *Amphilestes* appears to have been double rooted but that of *Phascolotherium* is single rooted. Three Asiatic amphilestids (*Klamelia, Gobiconodon,* and *Guchinodon*) are characterized by foreshortening of the mandible, and reduction of the number of teeth anterior to the molars to six or less. The canines of these genera have not been certainly identified, but, if present, appear to have been single rooted (Chow and Rich, 1984).

Finally among non-therians, the canines of docodonts exhibit a relatively constant morphological pattern. Upper and lower canines of *Haldanodon* and lowers of *Docodon* have been discovered. All are large teeth, supported by two roots.

Turning to early therians, some data are available on morphology of lower and, rarely, upper canines of symmetrodonts and eupantotheres. In symmetrodonts, *Spalacotherium* (including *Peralestes*) has large, double rooted upper and lower canines. The canine of *Tinodon* is relatively smaller, but might have been supported by a double root. Among other non-tribosphenic therians variation ranges from simple, single rooted, lower canines (e. g., *Paurodon*) to canines supported by anteroposteriorly elongated roots with lateral grooves (e. g., *Amphitherium* and, possibly, *Ambolotherium*) to double rooted canines (e. g., *Laolestes* and *Phascolestes*). *Kurtodon* and *Miccylotyrans* demonstrate that at least two eupantotheres have double rooted upper canines. An unnamed eupantothere described by Butler and Krebs (1973) had double rooted deciduous canines.

Little is known of canine morphology of early tribosphenic therians, and usually data are limited to preservation of part of the alveolus for a large lower canine. Marked exceptions are information provided by material of Asian species. Canines of deltatheridians are large and single rooted. Kielan-Jaworowska (1981) reported that in *Asioryctes* permanent upper and lower canines are double rooted. In *Kennalestes* deciduous upper and lower canines are single rooted, but the permanent canines are double rooted. As far as known all other Late Cretaceous eutherians and marsupials had single rooted canines. In the great majority of the Cenozoic members of these groups that retain canines the teeth are single rooted, but there are exceptions, such as the two rooted upper canines of *Ptilocercus* and *Dendrogale* (see Butler, 1980).

Probable primitive condition — The foregoing summary suggests that enlargement of upper and lower caniniform teeth was an early evolutionary trait of synapsids. The pattern of enlargement of a single large lower caniniform tooth occluding directly in front of an enlarged upper was well established in cynodontian lineages and can be taken as a primitive character both of mammals and tribosphenic therians.

Ontogenetic variation documented in modern mammals (see Schwartz, 1982) indicates that use of position of a tooth relative to the premaxillary-maxillary suture as a criterion for identification of homology definitely is suspect. The history of evolution from the widely separated upper and lower caniniforms in pelycosaurs (involving loss of teeth between upper and lower canines, then loss of incisiform maxillary teeth) also challenges the diagnostic utility of position relative to this suture.

Data provided by advanced cynodontians strongly suggest that mammals primitively possessed single rooted canines. However the widespread, but not universal, occurrence of double rooted canines in non-therian and early therian mammals prevents easy resolution of the question of probable primitive root structure of canines in tribosphenic therians. Stressing limitations of available data it appears most likely that double rooted canines evolved several times within various lineages of early mammals. Among known species of symmetrodonts and eupantotheres, double rooted canines are common. Canines with double roots might be primitive for tribosphenic therians. If this was the case, reduction to a single rooted condition occurred early in their Cretaceous evolutionary radiation.

3 & 4. *Number of premolars and molars*
5 & 13. *General morphology of posterior upper and lower premolars*

The usual definition of premolars is those teeth posterior (distal) to the canines as far as, and including, the last diphyodont teeth. Molars are monophyodont teeth posterior to the premolars. Although these ontogenetic patterns can be determined in many extant mammals, they are not always clear. This is vividly illustrated by current discussions of dental ontogeny of marsupials and the debate concerning homologies of the teeth displaced by P3/3. Are they homologous with DP3/3 or M1/1 of eutherians (note Archer, 1978)?

In groups of extinct mammals, particularly those with little modified tribosphenic dentitions, data on dental ontogeny are limited or lacking, and morphological criteria for distinguishing premolars from molars are employed. Discussion of the validity of these conventions was spurred particularly by McKenna's (1975) analysis of cladistic relationships among mammals, which he founded in part on a radically different interpretation of homologies of therian postcanine teeth. As a result, assumptions concerning the probable primitive morphology of posterior upper and lower premolars cannot be divorced from an analysis of primitive numbers of premolars and molars.

As far as is known, none of the advanced synapsids had reduced the primitive polyphyodont dental ontogeny of postcanine teeth to one in which some anterior postcanines were diphyodont and the posterior monophyodont. However, evolutionary experiments involving reduction of frequency of replacement, increase in durability of individual teeth, or origin of complex occlusal patterns characterize many lineages of cynodontians (see Osborn and Crompton, 1973, Kemp, 1982, and references cited).

Morganucodon had a mosaic of cynodontian and mammalian traits in postcanine ontogeny and morphology. The first four or five postcanines are of premolariform morphology. Both Mills (1971) and Parrington (1971) agree that the most posterior of these was diphyodont. A molariform tooth was replaced by a premolariform whose crown was noticeably higher than the following, monophyodont four or five molars. Parrington (1971) argued that all the premolars probably were diphyodont teeth. Mills (1971) did not find evidence of such widespread diphyodonty but noted that, in cynodontian fashion, at least some of the most anterior premolars were lost and not replaced.

Megazostrodon, known from a single skull, had a battery of five premolars and at least four upper and three lower molars. Unlike *Morganucodon* the last premolars are smaller teeth possibly with lower principal cusps than the first molars (Crompton and Jenkins, 1968). *Erythrotherium* had seven postcanine teeth. The dentition in the single available specimen is so poorly preserved that the pattern of differentiation of premolariform and molariform teeth cannot be determined.

Among later triconodontids premolars are morphologically distinct from molars. The main cusp of the last premolar rises to a greater height than the cusps of the first molar. A specimen of *Triconodon* (see Simpson, 1928) demonstrates that the "permanent" lower last premolar replaces a molariform precursor. The morphological transition from premolars to molars is much more gradual in amphilestids. Jenkins (1984) described two individuals of the amphilestid *Gobiconodon* found in the Cloverly Formation that, on the basis of epiphyseal fusion, appear to have been adults at the times of their deaths. Both, ". . . clearly show 'molar' teeth in the process of being replaced (*ibid.*, p. 43)." In the amphilestids, at least, molariformity of crown morphology is not correlated with monophyodonty. Counts of postcanine teeth in triconodontids and amphilestids classified on morphology of their crowns fall in the range of two to four premolars plus three to five molars, with the largest for a species consisting of four premolars and five molars.

Docodonts exhibit a distinct morphological break between premolars and molars; but, unlike the dentitions of some other groups, the principal cusp of the last premolar is lower than cusps of the first molar. Krusat (1980) presented evidence for replacement of premolars in *Haldanodon* and, most probably, *Pericynodon*. Number of premolars varied between three and four; molars reached maxima of, at least, six upper and eight lower. The maximum for a species is four premolars and eight molars.

Morphological differences are employed to differentiate the premolars and molars of multituberculates (Clemens and Kielan-Jaworowska, 1979). Replacement of teeth occurred early in individual development and went unrecognized for many years (Szalay, 1965). Now diphyodonty has been documented for some incisors and anterior upper premolars in a variety of taxa. The only evidence of diphyodonty of lower premolars was found in paulchoffatiids (Hahn, 1978). Lower premolars of more derived multituberculates appear to have been monophyodont. As many as six upper premolariform teeth are present (note Hahn's, 1977, suggestion that the anteriormost premolariforms of *Paulchoffatia* and *Pseudobolodon* are modified canines). The maximum number of lower premolars is five. All known genera have two upper and two lower molars.

Turning to therians, *Kuehneotherium* is thought to have six lower premolars and five or six molars (Gill, 1974, and Mills, 1984). Gill (1974) determined the position of the last premolar on the basis of change in alveolar size in an edentulous dentary. The relatively larger size of the alveoli for this tooth suggests its crown was larger than that of the following molar. Mills' (1984) studies of isolated teeth suggested that five molars usually were present in each quadrant, and some individuals had a sixth molar of variable size and morphology.

An abrupt change in morphology, which has yet to be correlated with a change in ontogeny, is employed to distinguish premolars from molars in symmetrodonts. The teeth taken to be the last premolars of *Spalacotherium* have higher main cusps than those of the following molar; in *Tinodon* these cusps are of about the same size. The dentition of *Spalacotherium* includes three premolars and up to seven molars; *Tinodon* has four premolars and three molars.

Chow and Rich (1982) suggested that the curious therian or therian-like mammal *Shuotherium* has three lower premolars and four lower molars. The "first molar," however, is not fully molariform, lacking a pseudotalonid anterior to the well developed trigonid.

In almost all dryolestids and paurodontids a distinct change in morphology is used to distinguish premolars from molars. Usually the largest cusp of the last premolar is higher than cusps of the first molar. Butler and Krebs (1973) reviewed the available data on ontogeny of the dentitions of dryolestids and presented evidence of diphyodonty of some premolars in several species. Dryolestids have four premolars and seven to nine molars (note Prothero, 1981). In contrast, paurodontids have two to five premolars and three to six molars.

Criteria for distinguishing postcanines of *Amphitherium* are morphological. The first molar is the anteriormost cheek tooth with both paraconid and metaconid. Unlike most Jurassic mammals the highest cusp of the last premolar is approximately the same height as the highest cusp of the molars. One of the authors (W.A.C.) reviewed the available sample of *Amphitherium* (three mandibular fragments in the University Museum, Oxford, and one in collections of the British Museum (Natural History)) and concluded that the dental formula given by Simpson (1928) requires modification. His identification of the last premolar and first molar is not disputed. Additional study of the specimen he designated "Oxford III" (now J. 20.076) revealed six alveoli between the first of the two preserved premolars and the large alveolus thought to have contained the canine. The other specimens yield evidence in accord with the hypothesis that the dental formula is five premolars and six to seven molars.

The dental formula of *Peramus* is estimated and debated on morphological grounds; ontogenetic data are not available. Its first three postcanines are premolariform; the last three are fully molariform. Disagreement surrounds identification of the two intermediate teeth. The anterior of the two is a premolariform tooth; its main cusp is distinctly higher than those of the following tooth, which is submolariform. In several other groups of early mammals this difference in heights of principal cusps and degree of molariformity is employed to discriminate between the last premolar and first molar. On these bases Simpson (1928) and Clemens and Mills (1971) argued that the dental formula of *Peramus* was four premolars and four molars. McKenna (1975), in his attempt to clarify cladistic relationships of principal groups of therians, used comparisons with younger tribosphenic therians to assert that the dentition of *Peramus* probably included five premolars and three molars.

Similarity of lower molar structure (uppers are unknown) prompted Dashzeveg (1979) to suggest close phylogenetic relationship between *Peramus* and the Cretaceous Asian *Arguimus*, which has four premolars and at least three molars. However, as he noted, unlike *Peramus*, the "last premolar" of *Arguimus* is submolariform, with a relatively low principal cusp. Subsequently Dashzeveg and Kielan-Jaworowska (1984) explored the possibility that the tooth originally identified as the "first molar" might be the last premolar or P5. Morphological change from premolars to molars is neither great nor abrupt.

Data on differentiation of postcanine dentitions of early tribotheres are provided by the Early Cretaceous, Asian aegialodontid *Kielantherium* and an informally named taxon, "*Prokennalestes*." A mandibular fragment of *Kielantherium* contains four molariform teeth preceded by alveoli interpreted as having contained at least four, two rooted premolars and a root of a fifth premolar or the canine (Dashzeveg and Kielan-Jaworowska, 1984). The assertion that five premolars were present was bolstered with the observation that "*Prokennalestes*," which was found in the same deposits, has five premolars and three molars (also see Kielan-Jaworowska, Bown, and Lillegraven, 1979).

The only other tribotheres that provide direct information on numbers of premolars and molars are the Late Cretaceous, Asiatic deltatheridians. Changing interpretations were reviewed by Kielan-Jaworowska, Eaton, and Bown, 1979 (and see Kielan-Jaworowska, 1982). The current view is that the postcanine dentition included three premolars, four lower molars, and three or four upper molars.

Problems of determining the primitive number of premolars and molars of eutherians and morphology of the posterior premolars recently received the attention of a number of workers (*e. g.*, McKenna, 1975, Butler, 1977, Marshall, 1979, and Fox, 1984). Some data on morphology of the posterior premolars are presented in Table 1. The convention for upper premolars was to designate teeth with paracone, metacone, and protocone, as being molariform. Submolariform refers to teeth in which only the paracone and protocone are large, distinct cusps. The last lower premolar of *Gypsonictops*, which has all the talonid and, except for the paraconid, trigonid cusps well developed, was designated molariform. Submolariform was used to describe teeth with a basined, not necessarily closed, talonid. A small metaconid is present on some of these teeth.

Primitively, dentitions of eutherians probably contained at least five premolars. This interpretation is based primarily on evidence of presence of five premolars in the lower, but not the upper, dentition of *Gypsonictops* (see Fox, 1977), and in the upper dentition of a juvenile *Kennalestes*. Support also is drawn from compositions of dentitions of "*Prokennalestes*" and, following McKenna's (1975) interpretation, possibly *Peramus*. *Otlestes*, recently described by Nesov (1985a) and referred to the Palaeoryctidae, had five lower premolars. Butler (1980) as well as Luckett and Mair (1982) and Westergaard (1983) have voiced skepticism concerning the presence of five premolars in the dentitions of any early eutherians.

Recently, Domning (1982) has added support to the interpretation that five premolars were present (retained?) in the dentitions of some Eocene sirenians.

By the Late Cretaceous, the number of premolars in most eutherians was reduced to four or, rarely, three. Whether this reduction was simply the result of loss of an anterior premolar, or, in some lineages, also involved a change in ontogeny preserving the molariform DP5 in the "permanent" dentition and loss of the posterior molar (McKenna, 1975) remains a moot point. Among Late Cretaceous eutherians only specimens of *Kennalestes* unambiguously document diphyodonty of the last premolar. Isolated teeth thought to be deciduous premolars of *Gypsonictops* and *Cimolestes* have been described, suggesting that at least their posterior premolars were diphyodont.

A survey of premolar dentitions of Late Cretaceous eutherians reveals a trend for distinct morphological differentiation between the smaller, usually simpler, anterior premolars and the two larger posterior premolars (for ease of communication traditional designations of P3 and P4 are used). In many species, P3 is a large submolariform tooth, and its principal cusp is the highest of the postcanine dentition. Usually, P4 is either submolariform or, in some *(e. g., Gypsonictops)* almost fully molariform. In addition, the so far unvarying presence of three molars probably is a primitive trait.

Finally, known Late Cretaceous marsupials document a remarkably stable pattern of differentiation of the postcanine dentition. All described species have three premolars and four molars. The last premolar, although greatly enlarged in some genera *(e. g., Didelphodon)*, does not have a molariform or submolariform morphology. In many species its principal cusp is distinctly higher than those of the following molar.

Some evidence that the last "permanent" premolar replaced a molariform tooth has been found (see Clemens, 1979). Many workers interpreted this as an example of typical diphyodonty and designated the teeth in question as P3 and DP3. After study of ontogeny of dentitions of many (particularly Australian) marsupials, Archer (1978) concluded that incisors and canines were the only diphyodont teeth in dentitions of marsupials. He argued that the last premolar (P3) physically displaced M1, leaving four molars (M2-5). Archer's terminology has not been used in this paper.

In summary, the pattern of ontogeny and postcanine differentiation thought to be primitive for Cenozoic didelphoids and dasyuroids appears to have been firmly established in Late Cretaceous marsupials. If Archer (1978) is correct in his interpretation, then among tribosphenic therians this ontogenetic pattern is the most derived in reduction of polyphyodonty.

Probable primitive condition — Evolution of the mammalian dentition was characterized by increased precision and complexity in occlusion of upper and lower dentitions, probably consequent reduction in frequency of replacement of cheek teeth, and differentiation of the postcanine dentition into distinct, but sometimes intergrading, functional units. Attempts have been made to identify and describe these functional units in terms of the premolar/molar dichotomy found in marsupials and eutherians.

In most groups of Mesozoic mammals this division is placed at an abrupt morphological change in the postcanine teeth. Diphyodonty of the last premolar and monophyodonty of the first molar have been cited as supporting assertions of homology of these teeth. But, as the foregoing summary illustrates, ontogenetic data supporting identification of the division between premolars and molars of many Mesozoic groups are unknown, and in some (e. g., multituberculates and amphilestids), available ontogenetic data raises serious questions about current conventions for classification of elements of the dentitions (see Sloan, 1979, and Hahn, 1978). Also, in light of embryological studies summarized by Schwartz (1982 and see above) assertion of stability of position of the division between diphyodont and monophyodont teeth, particularly in a group undergoing selection for reduction in primitive polyphyodonty, is more than suspect.

With these caveats in mind, a few hypotheses concerning morphology of the last common ancestors of mammals and of therians with tribosphenic dentitions appear warranted.

Known non-therians and therians share a pattern in which some anterior postcanine teeth are diphyodont and the rest monophyodont. This pattern, unknown in cynodontians, might be a derived character of mammals. However, it must be stressed cynodontian-like patterns of loss without replacement of some anterior postcanines, diphyodonty of molariform teeth, as well as late eruption and individual variation in presence and morphology of the last molar occurred in some of the earliest mammals. Also, variation and high numbers of postcanine teeth in some therian and non-therian groups suggest the number of postcanines in early mammals was large, possibly 10 to 12, and addition or suppression of cheek teeth occurred frequently.

Known Late Cretaceous tribosphenic therians have no more than eight postcanine teeth, and this might be the primitive number for the group. If the division between diphyodont and monophyodont teeth is a stable guide to homologous teeth, then a primitive condition with at least five premolars, to account for the known maximum in eutherians, and four or five molars (see Archer, 1978), to account for the molars of marsupials, would be required. These assertions are based upon the probably unlikely hypothesis that molars or premolars were not added to the dentitions of these animals.

Among non-therians and early therians the last premolar in many groups (e. g., morganucodontids, triconodontids, some symmetrodonts, and most dryolestids) has a principal cusp higher than those of the following molar. This might be the primitive condition for mammals. In forms in which the principal cusp is lower (e. g., docodonts and *Amphitherium*), a distinct morphological break still separates the last premolar from the first molar. Examples of a graded transition with teeth that are not clearly identified as premolars or molars, e. g.,

note *Shuotherium*, *Peramus*, and *Arguimus*, are rare and are assumed to represent derived conditions.

At least two hypotheses can be advanced concerning the probable primitive morphology and ontogeny of posterior premolars and anterior molars of tribosphenic therians. First, it can be argued that deltatheridians and marsupials preserve the primitive condition for mammals in which the last premolar has a high crown, and a distinct morphological break separates premolars from molars. The more molariform posterior premolars and gradual transition from premolariformity to molariformity in eutherians would thereby be derived characters. If deltatheridians, marsupials, and eutherians shared a tribosphenic therian as their last common ancestor, then the eutherian-like patterns of morphological variation seen in the dentitions of *Peramus* and *Shuotherium*, which lack tribosphenic molars, must be interpreted as products of parallel evolution.

A second hypothesis suggests that the eutherian pattern of graded differentiation of postcanines evolved prior to origin of tribosphenic molars (i. e., in the common ancestral stock of *Shuotherium* and *Peramus* and tribosphenic therians), and was maintained in eutherians. Deltatheridians and marsupials thereby would be interpreted as groups that, probably independently, lost the eutherian-like pattern (either through loss of teeth or via modification of morphological complexity).

Given the complex functional pattern of tribosphenic dentitions involving many cusps and shearing blades plus the detailed similarities of tribosphenic molars of tribotheres, marsupials, and eutherians we strongly support hypotheses not involving multiple origins of the tribosphenic dentition. Of the two hypotheses noted above, which both meet this specification, we favor the former. It only requires assuming parallel evolution of a graded transition from premolariform to molariform postcanines in three distinct groups of therians: *Shuotherium*, *Peramus* (and the possibly closely related *Arguimus*), and eutherians. Depending upon the cladistic or phylogenetic interpretation of therian relationships adopted, to regard the graded morphological change between premolariforms and molariforms found in *Shuotherium* and all therians with which it shared a common ancestry requires invoking reversals to the assumed primitive condition in the following lineages: *Amphitherium*, dryolestids, paurodonts, deltatheridians, and marsupials. The consequent predictions concerning probable primitive conditions of the first hypothesis are listed in Table 2.

6-12, 14-16. Characters of individual tribosphenic molars

Lacking data on ontogeny of crowns of cheek teeth of Mesozoic mammals, similarity in size and position of cusps of fully developed teeth were the original bases for suggestions of cusp homology. Later, because of the apparently great stability of occlusal relationships, functional units of crests and cusps were recognized as providing more reliable indications of homology. Occlusal relationships of therian and non-therian dentitions differ significantly and, with few exceptions, only limited data are available on which to base hypotheses of homology. In contrast, therians are united by triangular disposition of the principal cusps plus particular occlusal relationships that provide the basis for hypotheses of cusp homology within the group. In our analysis the following conventions are used: (1) the immediate ancestors of therians are assumed to have had cheek teeth with a linear (anteroposterior) arrangement of cusps; (2) other than adopting the view (supported by embryological data) that the paracone and protoconid are homologs of the single cusp of early synapsid cheek teeth, no assertions concerning homologies of cusps of cynodontian, non-therian, and therian cheek teeth are made, although those suggested by Crompton and Sun (1985) certainly warrant consideration; (3) the question of whether triangular symmetry of therian cheek teeth arose by shift in position of preexisting cusps (cusp rotation) or evolution of new cusps is left unanswered; and (4) the hypothesis for evolution of tribosphenic molars developed by Patterson (1956) and reviewed in detail by Crompton (1971) is followed with only minor modifications.

Upper Cheek Teeth

6. Ratio of greatest transverse width to anterior-posterior length

Proportions of cheek teeth of advanced cynodontians, non-therians, and *Kuehneotherium* suggest that primitively for mammals anterior-posterior length of the crown significantly exceeded its width. Crowns of at least some upper cheek teeth of Middle and Late Jurassic therians *(e. g., Spalacotherium, Peramus, Palaeoxonodon;* and the "Forest Marble eupantotheres" described by Freeman, 1979) have length/width ratios approaching, or slightly exceeding, unity. Among dryolestids, width of some posterior molars significantly exceeds length, which is interpreted as a derived condition. A ratio of near unity (length approximates width) probably characterized the last common ancestor of tribosphenic therians.

Among tribosphenic therians, this ratio (based on M^2 where possible) is sensitive to changes in transverse widths of the stylar shelf and protocone. In Late Cretaceous tribotheres, usually with small protocones relative to those of marsupials and eutherians, the ratios are highly variable, ranging from 0.9:1 to 1.3:1. Marsupials, with larger protocones, have ratios in approximately the same range. Among Late Cretaceous eutherians the ratio usually is distinctly larger, 1.1:1 to 1.7:1. The ratio is only slightly above unity in those eutherian genera that would be classified as derived on the basis of large size of conules and presence of protoconular cingula *(e. g., Paranyctoides* and *Protungulatum).* Transverse widening of the protocone probably characterized the molars of this earliest eutherian.

7. Size and proportions of the protocone

Presence of the protocone and occluding talonid basin are used as the diagnostic characters of the tribosphenic dentition. The primitive condition for tribosphenic therians is, therefore, a small protocone whose crown is lower than those of the paracone and metacone (height measured from base of crown to apex).

8. Protoconular cingula

Basal pre- and postprotoconular cingula are lacking in deltatheridians, and are present only in those other tribotheres with relatively large protocones. In *Picopsis* and *Falepetrus* they are small structures; *Bistius,* in contrast, has narrow but distinct pre- and postprotoconular cingula. These genera are known from small samples of isolated teeth, and nothing can be said about patterns of individual variation.

In general, Late Cretaceous marsupials lack protoconular cingula. Molars of some species of *Alphadon* and *Pediomys* occasionally have a weak pre- or, more frequently, postprotoconular cingulum. Only in *Glasbius* is a preprotoconular cingulum consistently present and, in some individuals, a postprotoconular cingulum is developed on M^3. Some lineages of Late Cretaceous eutherians (e. g., leptictoids and condylarths) consistently have wide pre- and, particularly, postprotoconular cingula. In contrast, cingula are missing from molars of some species of the palaeoryctoid *Cimolestes* and present in others.

Absence of protoconular cingula is primitive for tribosphenic therians. The rare occurrence of these structures on molars of some tribotheres and marsupials and their limitation to some lineages of eutherians suggests parallel evolution of these cingula.

9. Conular development

Presence of paraconules and metaconules on at least some molars of late Early Cretaceous tribosphenic therians *(e. g., Pappotherium* and *Holoclemensia)* suggests they were added to the labially directed crests of the protocone soon after this cusp evolved (but note the contrary view expressed by Fox, 1975). Conules are present in deltatheridians and other tribotheres with transversely elongated protocones *(e. g., Bistius* and the unnamed upper molar described by Fox, 1982). Those tribotheres with small protocones lack conules *(e. g., Potamotelses* and *Picopsis),* and might preserve the primitive condition for tribosphenic therians. Presence of small, distinct para- and metaconules appears to be primitive for marsupials and eutherians.

10. Relative strengths of paracone and metacone

The paracone, thought to be homologous with the single cusp of postcanines of primitive synapsids (see Patterson, 1956), is the largest cusp of upper molars of *Kuehneotherium,* symmetrodonts, *Palaeoxonodon, Peramus,* and some dryolestids. If the metacone is defined as the cusp whose anterolabially directed crest (premetacrista) shears against the crest supported by hypoconid and hypoconulid, all these early therians except *Peramus* lack this cusp. Possibly one of the several cusps usually present on the crest linking the paracone with the posterior corner of the crown of these non-tribosphenic therians was the precursor of the metacone, but this is open to question (see Bown and Kraus, 1979, and the extensive discussion of a contrary view by Prothero, 1981).

The paracones of molars of *Peramus, Pappotherium,* and *Holoclemensia* are larger and higher than their metacones. These proportions, also found in all deltatheridians and other tribotheres, are regarded as primitive for tribosphenic therians. Although differences in height of paracone and metacone are muted in some species, in Late Cretaceous eutherians the paracone usually is the taller cusp. In a few Late Cretaceous marsupials *(e. g.,* species of *Alphadon* and *Albertatherium)* the paracone is slightly higher and larger than the metacone; in most, however, the proportions are reversed. Although not a diagnostic character of the Marsupialia, a trend for enlargement of metacone and reduction of paracone was established early.

11. Strength of stylar shelf

The term stylar shelf was coined to refer to the cusp-bearing part of the crown of tribosphenic molars labial to paracone and metacone, an area particularly well developed in many Cenozoic didelphoid and dasyuroid marsupials. Patterson (1956) recognized this area as homologous with the body of the crown of non-tribosphenic therian upper molars. A wide stylar shelf is maintained in all known Cretaceous deltatheridians, some other tribotheres, and most Cretaceous marsupials. Some Cretaceous eutherians that would be regarded as primitive on other dental criteria *(e. g.,* palaeoryctoids) have a wide stylar shelf. Its width is greatly reduced in other eutherian lineages. A wide stylar shelf probably is the primitive condition for tribosphenic therians.

12. Stylar cusp development

History of development of nomenclature and functional analyses of stylar cusps have been reviewed elsewhere (Clemens, 1979). Stylar cusps can be divided into three groups. The first includes only the stylocone (stylar cusp B), which is the anterolabial principal cusp of non-tribosphenic therian molars and the labial terminus of the preparacrista (note Prothero, 1981, for discussion of variation in non-tribosphenic therians). This crest, occluding with the crest between the protoconid and metaconid (protocristid), provided the primary shearing mechanism of molars of early therians and was maintained in primitive tribosphenic molars *(e. g.,* those of *Pappotherium* and *Holoclemensia).*

The parastyle (stylar cusp A) and metastyle (stylar cusp E) form the second group and functioned to interlock the molars as well as protect the gums by providing contacts between adjacent teeth. Also, to varying degrees, they contribute to shearing surfaces along the anterior and posterior edges of the crown. A parastyle is present on molars of *Kuehneotherium* and most other non-tribosphenic therians. Whether presence of a parastyle is a derived character of therian mammals or simply retention of a cusp providing a similar interlocking function in dentitions of cynodonts is open to question. Presence of a parastyle probably was a primitive character of tribosphenic molars.

A survey of therian dentitions illustrates great variability in absence or presence of one or more cusps at or near the posterolabial corner of the crown, the region of stylar cusp(s) E. On molars of *Kuehneotherium* a crest (postmetacrista) links the metacone with a small metastylar cusp. This crest is interrupted by a small cusp ("cusp c" of Crompton, 1971). In *Palaeoxonodon* and *Peramus* several small cusps are present in the metastylar region and along the posterolingual edge of the crown. Similarly, small cusps are present in this region on molars of *Pappotherium* and *Holoclemensia*. Presence of one or more cusps might be a primitive character of therian mammals and tribosphenic therians, but whether they are homologous cusps is an open question.

Finally, evolution of stylar cusps in the region between the stylocone and metastylar corner of the crown poses vexatious questions. In the nomenclature developed primarily for marsupials, these are stylar cusps C (mesostyle) and D. In all primitive tribosphenic molars with cusps in these positions, the crowns of the cusps show only apical wear, suggesting their major function was puncturing food and that they did not occlude with the lower dentition. Their homologies only can be assessed on the basis of size and position relative to other structures, both criteria of limited value.

Kuehneotherium, Palaeoxonodon, "Forest Marble eutherians" (see Freeman, 1979), *Peramus,* and symmetrodonts either lack cusps or have but minor rugosities in the C and D regions. Cusps are present posterior to the stylocone on the greatly shortened labial margins of molars of some dryolestids. Whether they should be considered metastylar cusps or placed in this third group is another question. The general pattern of absence or very small size of cusps on the labial margin of the crown posterior to the stylocone suggests that this was the primitive condition for therian mammals.

Tribosphenic therians, deltatheridians, some other tribotheres, and many Late Cretaceous eutherians either lack cusps in this region or have only minor marginal rugosities. The mammal represented by the unnamed molar described by Fox (1982), *Potamotelses, Picopsis, Falepetrus, Paranyctoides,* and *Zalambdalestes* might fall into this category, or be set apart as having small but distinct D and/or C cusps. Most Cretaceous marsupials have one or, in some instances, more stylar cusps D, and many have one or more stylar cusps C. Other combinations are illustrated by *Pappotherium* (with a small but distinct stylar cusp D but lacking a stylar cusp C), *Holoclemensia* (with a small stylar cusp D and very large, high stylar cusp C), and *Bistius* (with subequal, bulbous stylar cusps C and D).

Most workers agree that absence or only slight development of stylar cusps in the C region is the primitive condition for eutherians. Fox (1984) has argued that a cusp in the D region, smaller than the stylocone, was present on molars of the earliest eutherians.

Hypotheses concerning the primitive condition of development of stylar cusps in the C and D regions of molars of marsupials are hotly debated. For example, Slaughter (see 1981 and references cited) has argued strongly that presence of a large stylar cusp C is a diagnostic character of primitive marsupials. The higher frequency of occurrence of stylar cusp D than C in Cretaceous marsupials was the basis for speculation that presence of a stylar cusp D might be primitive for marsupials and stylar cusp C evolved later and independently in several lineages (Clemens, 1979). More recently, Fox (1983b), citing a report in preparation, argued that molars of early predidelphid marsupials had a stylar cusp D, but lacked a cusp C. He interpreted evolution of stylar cusp C as a derived feature of the Didelphidae.

Absence or slight development of cusps in the C and D regions of non-tribosphenic therians as well as many tribotheres suggests this was the primitive condition for tribosphenic therians. Was evolution of stylar cusps in the D and/or C region diagnostic of members of the Marsupialia? To argue that it was could result, for example, in inclusion within the Marsupialia of *Picopsis* (with a very primitive protoconal morphology), *Bistius* (which has a large protocone and pre- and postprotoconular cingula), and *Paranyctoides* (with even stronger protoconular cingula), as well as both *Holoclemensia* and *Pappotherium*. What little is known of the dentitions of these genera either offers no evidence that adds support to such an allocation or includes characters that argue against reference to the Marsupialia. If presence of a small cusp D is taken as a diagnostic character, then it could be argued that *Zalambdalestes* and *Kennalestes* are marsupials, references contradicted by a large number of other characters.

Because stylar cusps C and D of early tribosphenic therians did not occlude with the lower molars, functional constraints on their evolution might not have been as great as on modification of the principal cusps. Rare occurrences of stylar cusps in the C and D regions of molars of Cretaceous eutherians or tribosphenic therians that, on other criteria, appear unlikely candidates for reference to the Marsupialia, suggests stylar cusps evolved independently in these regions of molars of marsupials and less frequently in some tribotheres and eutherians.

Lower Cheek Teeth

14. Size and position of paraconid relative to metaconid

Kuehneotherium exhibits the probable primitive arrangement of principal cusps of the therian trigonid forming an obtuse angle, with the protoconid at its apex (Mills, 1984). The paraconid is situated anterior or slightly lingual to the protoconid; on any molar the metaconid commonly has a slightly more lingual position. Usually the paraconid is slightly higher than the metaconid; both are lower than the protoconid.

Other non-tribosphenic therians illustrate considerable variation in structure of the trigonid (note Prothero, 1981). Molars of the almost unicuspid amphidontid symmetrodonts and the paurodontids either lack a paraconid or the cusp is very small relative to other cusps of the trigonid. In trigonids of most of the other non-tribosphenic mammals, the paraconid ranges from somewhat smaller to approximately the same height as the metaconid. The metaconid is situated slightly farther lingually than the paraconid, and crests linking them to the protoconid (paracristid and protocristid) usually form an acute angle.

Among Early Cretaceous tribosphenic therians the range of variation encompasses the relatively low, open trigonid of *Kermackia*, an intermediate condition in *Trinititherium* (see Butler, 1978a), and the higher and more anteroposteriorly shortened trigonids of molars referred to *Holoclemensia* (see Slaughter, 1971) and *Kielantherium* (see Dashzeveg and Kielan-Jaworowska, 1984). Two adjacent molars preserved in the holotype of *Slaughteria* give a hint of the variation to be expected in the dentition of a tribothere. However, common occurrence of the following suggest that such characters constitute the primitive condition for tribosphenic therians: (1) the paraconid is distinctly smaller to about the same height as the metaconid; (2) the metaconid is slightly lingual to the paraconid; and (3) the cristids connecting the protoconid, paraconid, and metaconid form a large but still acute angle.

Among Late Cretaceous tribosphenic therians some appear to retain the primitive condition. *Kielantherium, Potamotelses,* deltatheridians, and stagodontid marsupials *(Eodelphis* and *Didelphodon)* stand apart in emphasis on postvallum/prevallid shearing that results in larger size and greater height of the paraconid relative to the metaconid. *Gallolestes* differs in that the paraconid is significantly smaller than the metaconid and set farther labially, characters cited by Lillegraven (1976) as indicative of eutherian affinities of this genus.

Late Cretaceous marsupials, excepting the stagodontids, maintain the probable primitive condition in relative heights of paraconid and metaconid. The more lingual placement of the paraconid so that it, metaconid, and entoconid are aligned, probably is a derived character of the group. In contrast, the paraconids of contemporaneous eutherians are uniformly lower than the metaconids. Their position can vary from being anterior (mesial) to the metaconid to positioned noticeably farther labially near the middle of the crown. Fox (1984) suggested that the primitive condition for eutherians was location of the paraconid slightly labial to the metaconid.

15. Position of hypoconulid relative to entoconid and hypoconid

The talonid region of molars of non-tribosphenic therians is a relatively small blade or spur extending distally to contact the trigonid of the following molar (note review by Prothero, 1981). As is the case with the parastyle (see character 12, above), whether presence of this spur is a derived character of therian mammals or simply retention of a projection providing a similar interlocking function in dentitions of advanced cynodontians is an open question. The cusp at the distal end of the talonid in molars of *Kuehneotherium* has been identified as the hypoconulid by some workers (Crompton, 1971). Slaughter (1971) reviewed the possible homologies of the terminal cusp of these simple trigonids and suggested it could be homologous with either the hypoconulid or hypoconid of tribosphenic molars. Freeman (1979) and Prothero

(1981) argued that it is homologous with the hypoconid. The closest approximations to the basined, cusp-ringed talonids of primitive tribosphenic therians are found in molars of *Palaeoxonodon, Peramus,* and *Arguimus.* Hypoconids and hypoconulids (incipient in *Palaeoxonodon*) are present in all three. *Peramus* and possibly *Palaeoxonodon* have rudimentary entoconids on some molars, and a groove or depression on the talonid might be the precursor of the basin of tribosphenic molars.

Tribosphenic (or near-tribosphenic) therians from the Early Cretaceous document several grades in evolution of a basined talonid that occluded with the protocone. *Kielantherium* and "Trinity molar type 6" (of Slaughter, 1971) both lack entoconids. Talonids of the former are closed basins, and that of the latter is open lingually. *Aegialodon* (see Kermack and others, 1965) and the majority of the molars of tribotheres from the Early Cretaceous of Texas have narrow, basined talonids that are approximately one-half to two-thirds the width of their trigonids with three or more cusps around their perimeters. The probable primitive condition for tribosphenic therians is presence of a talonid with hypoconid, hypoconulid, and possibly entoconid on some molars. The data suggest that the talonid might not have been closed.

Among Late Cretaceous tribotheres, the talonid is narrow in most species, and the three cusps are equally spaced around the perimeter of the talonid. The exception is *Gallolestes,* in which the entoconid is slightly closer to the hypoconulid than the hypoconid, a marsupial-like trait (Clemens, 1980).

In contrast, marsupials and eutherians of Late Cretaceous age usually have larger talonids; in some species they approximate the width of the trigonid. Talonid enlargement, correlated with enlargement of the protocone, followed at least two distinct patterns. In eutherians the primitive, approximately equidistant, disposition of the hypoconid, hypoconulid, and entoconid was maintained. In contrast, in marsupials the entoconid and hypoconulid remained or became more closely approximated while the hypoconid shifted labially. As noted by Slaughter (1971), the derived pattern found in marsupials could be functionally related to the early trend in the group for enlargement of the metacone and reduction of the paracone. This would require enlargmeent of the embrasure between the back of the talonid of one molar and the anterior surface of the trigonid of the following molar. Eutherians differ in tending to maintain the primitive proportions of paracone and metacone (see character 10, above) and enlargement of the protocone involves a distinct transverse widening of the trigon. In combination these would not require early, major enlargement of the embrasure between adjacent lower molars.

16. Relative strengths of antero- and posterolingual cingulids

Although not always complete, a cingulum or a cusp is present near the base of the anterolabial side of the trigonid of *Kuehneotherium* and some other non-tribosphenic therians. However, it is lost from molars of amphidontids and most dryolestids (Prothero, 1981). Likewise, presence of a cusp or cingulum in this region in most Early Cretaceous tribosphenic therians and Late Cretaceous tribotheres, excepting deltatheridians, suggests that the presence of one or the other of these structures might be the primitive condition for tribosphenic therians. However, available data are not sufficient to argue that the basal anterolabial cingulids or cusps of tribosphenic and non-tribosphenic therians are homologous structures.

A posterolabial cingulid is not present on talonids of early Cretaceous tribosphenic therians and is absent or short on talonids of molars of most Late Cretaceous tribotheres. Late Cretaceous eutherians usually lack or have only a weakly developed posterolabial cingulid. A major exception is *Paranyctoides*. In contrast, most Late Cretaceous marsupials have this cingulid on their molars, and frequently it is quite strong. This derived condition probably was correlated with enlargement of the metacone and consequent increase in size of the embrasure between the talonid and trigonid of adjacent molars.

CONCLUSIONS

Occurrences of standard morphological characters available to aid in the classification and evolutionary interpretation of advanced therian mammals from strata of Late Cretaceous age are presented in Table 1 (which, as noted above, does not include some Asian, Late Cretaceous mammals recently described by Nesov, 1985a, b). Inspection of the table shows profound gaps in dental anatomical information for most genera; full information is available for only a few. Even though analysis of commonly used characters highlights their inadequacy for purposes of classification, we suggest that four groups of tribosphenic therians can be recognized in the Late Cretaceous. We do not suggest formalized taxonomic categories for groups 3 and 4, but instead lump them informaly as tribotheres.

Groups "1" and "2"—Through comparison with Cenozoic species whose affinities can be assessed on a broader spectrum of dental, skeletal, and even reproductive features, some Cretaceous genera can be identified as members of the Marsupialia and Eutheria with a high degree of probability. Included among the marsupials are: *Alphadon, Peradectes, Glasbius, Albertatherium, Pediomys, Aquiladephis, Didelphodon,* and *Eodelphis*. Included among the eutherians are: *Gypsonictops, Kennalestes, Cimolestes, Batodon, Procerberus,* an unnamed palaeoryctid (see Fox, 1979c), *Asioryctes, Telacodon,* the so called "Champ-Garimond tooth" (see Ledoux and others, 1966), *Paranyctoides, Zalambdalestes,* ? *Daulestes, Barunlestes, Protungulatum, Oxyprimus, Ragnarok, Mimatuta,* and *Purgatorius*.

Group "3"—*Deltatheridium, Deltatheroides,* and possibly, *Hyotheridium* could represent a group (the "deltatheridians" of Van Valen, 1966) of equal rank to the marsupials or the eutherians and distinct from the remainder of the "tribotheres." They are distinguishable by the mixture of marsupial-like dental formula and differences in morphology of the premolars and molars combined with the eutherian-like characters of their molars.

Group "4"—This represents a taxonomic wastebasket of genera too poorly known to be referred with confidence to groups 1-3, and, beyond sharing a tribosphenic common ancestor, possibly is not closely related to any of them. Included within this group are: *Potamotelses, Beleutinus, Picopsis,* an unnamed upper molar (see Fox, 1982), *Falpetrus,* and *Bistius.* Also, although not cited in Table 1, we include the possibly tribosphenic therian known only from a premolar found in the Woodbine Formation (Cenomanian) of Texas (Krause and Baird, 1979) and the fragmentary tribosphenic lower molar discovered in the Eutaw Formation (late Santonian) of Mississippi (Emry and others, 1981).

Falepetrus shows reduction of the anterior part of the stylar shelf and presence of lingual cingula, features common to eutherians. Similar reduction in width of the stylar shelf, however, is known to have occurred independently in marsupials. The low ratio of measurements of transverse molar width to length is atypical of Cretaceous eutherians. We simply have no suite of features that allows confidence in assignment of *Falepetrus* to any particular major group of tribosphenic mammals.

In contrast, *Bistius* has a wide stylar shelf with marsupial-like stylar cusps, but the pre- and postprotoconular cingula are well developed in typically-eutherian fashion. Such a mosaic of characters precludes ready identification of *Bistius* either as a marsupial or eutherian.

Gallolestes is included here with some reservations. Lillegraven (1976) interpreted this mammal as a eutherian. Butler (1977) concurred, but stressed that it was very different from all other known Late Cretaceous mammals. Clemens (1980) stressed the combination of marsupial and eutherian characters of *Gallolestes* and suggested that reference to either the Marsupialia or Eutheria was not clearly warranted. Probably the best stance to take is that these are all appropriate working hypotheses to be tested when more material of this mammal is discovered.

The suite of characters presented in Table 1 (summarized in Table 2), and we know of no other generally utilitarian features, is wholly inadequate to allow a cladistically-based phylogenetic interpretation of the tribotheres. Dental formulae are virtually unknown, as are most elements of the antemolar dentition. Many common features of the molariform teeth of tribotheres (such as wide stylar shelves) are retentions of primitive characters. Other teeth are distinct in that they possess combinations of derived characters not seen in generally accepted marsupials or eutherians.

In spite of these difficulties, the tribotheres are of great interest for they provide hints to the occurrence of a major, diversification of tribosphenic therian mammals during the Cretaceous. The earlier interpretation that the Cretaceous history of evolution of tribosphenic therians simply involved differentiation of marsupial and eutherian stocks in the Northern Hemisphere followed by southward dispersal of some of their members must be abandoned. The deltatheridians and other tribotheres testify to a much more complex evolutionary radiation in the north. Additionally, recent discoveries indicate that groups of tribosphenic therians also differentiated on the various continents of the Southern Hemisphere adding to the diversity of mammalian lineages prior to the end of the Cretaceous.

ACKNOWLEDGMENTS

We thank the National Science Foundation (grants GB 39789 and DEB 81-19217 to W.A.C. and DEB 81-05452 and EAR 82-05211 to J.A.L.), the National Geographic Society (grants 2226-80 and 2353-81 to W.A.C.), and the Annie M. Alexander Endowment, Museum of Paleontology, University of California, Berkeley. We also gratefully acknowledge the help of Ms. Marisol Montellano, who permitted us to study the referred specimen of *Falepetrus* that is part of the sample of a local fauna that she is currently analyzing, and Dr. Malcolm C. McKenna, who discovered the holotype of this genus. Ms. Marisol Montellano, Ms. Nancy Simmons, Mr. Donald Lofgren, and Mr. Luo Zhexi read early drafts of the manuscript and provided valuable comments. We thank them, Mrs. Linda E. Lillegraven for her many activities in aid of development of the manuscript, and Ms. Augusta Lucas for illustrations of the fossils.

REFERENCES CITED

Archer, M., 1978, The nature of the molar-premolar boundary in marsupials and a reinterpretation of the homology of marsupial cheekteeth: Queensland Museum, Memoirs, v. 18, p. 157-164.

Archer, M., Flannery, T. F., Ritchie, A., and Molnar, R. E., 1985, First Mesozoic mammal from Australia — an early Cretaceous monotreme: Nature, v. 318, p. 363-366.

Archibald, J. D., 1982, A study of Mammalia and geology across the Cretaceous-Tertiary boundary in Garfield County, Montana: University of California, Publications in Geological Sciences, v. 122, *xvi* + 286 p.

Bashanov, V. S., 1972, First Mesozoic Mammalia (*Beleutinus orlovi* Bashanov) from the USSR: Teriologiya, Akademii Nauk SSSR, Sibirskoe Otdelenie, Novoskibirsk, v. 1, p. 74-80 (in Russian with English summary).

Berkovitz, B. K. B., 1968a, The early development of the incisor teeth of *Setonix brachyurus* (Macropodidae: Marsupialia) with special reference to the prelacteal teeth: Archives of Oral Biology, v. 13, p. 171-190.

―――― 1968b, Some stages in the early development of the postincisor dentition of *Trichosurus vulpecula* (Phalangeroidea, Marsupialia): Journal of Zoology, v. 154, p. 403-414.

Bown, T. M., and Kraus, M. J., 1979, Origin of the tribosphenic molar and metatherian and eutherian dental formulae, *in* Lillegraven, J. A., Kielan-Jaworowska, Z., and Clemens, W. A., eds., Mesozoic mammals: the first two-thirds of mammalian history: Berkeley, University of California Press, p. 172-181.

Butler, P. M., 1963, Tooth morphology and primate evolution, *in* Brothwell, D. R., ed., Dental anthropology: Oxford, Pergamon Press, p. 1-13.

———— 1977, Evolutionary radiation of the cheek teeth of Cretaceous placentals: Acta Palaeontologica Polonica, v. 22, p. 241-271.

———— 1978a, A new interpretation of the mammalian teeth of tribosphenic pattern from the Albian of Texas: Breviora, no. 44, 27 p.

———— 1978b, The ontogeny of mammalian heterodonty: Journal de Biologie Buccale, v. 6, p. 217-227.

———— 1980, The tupaiid dentition, *in* Luckett, W. P., ed., Comparative biology and evolutionary relationships of tree shrews: New York, Plenum Press, p. 171-204.

Butler, P. M., and Kielan-Jaworowska, Z., 1973, Is *Deltatheridium* a marsupial?: Nature, v. 245, p. 105-106.

Butler, P. M., and Krebs, B., 1973, A pantotherian milk dentition: Paläontologische Zeitschrift, v. 47, p. 256-258.

Cassiliano, M. L., and Clemens, W. A., 1979, Symmetrodonta, *in* Lillegraven, J. A., Kielan-Jaworowska, Z., and Clemens, W. A., eds., Mesozoic mammals: the first two-thirds of mammalian history: Berkeley, University of California Press, p. 150-161.

Chow, M., and Rich, T. H. W., 1982, *Shuotherium dongi*, n. gen. and sp., a therian with pseudo-tribosphenic molars from the Jurassic of Sichuan, China: Australian Mammalogy, v. 5, p. 127-142.

———— 1984, A new triconodontan (Mammalia) from the Jurassic of China: Journal of Vertebrate Paleontology, v. 3, p. 226-231.

Clemens, W. A., 1966, Fossil mammals of the type Lance Formation, Wyoming. Part II. Marsupialia: University of California, Publications in Geological Sciences, v. 62, vi + 122 p.

———— 1968, A mandible of *Didelphodon vorax* (Marsupialia, Mammalia): Natural History Museum of Los Angeles County, Contributions in Science, no. 133, 11 p.

———— 1973a, Fossil mammals of the type Lance Formation, Wyoming. Part III. Eutheria and summary: University of California, Publications in Geological Sciences, v. 94, vi + 102 p.

———— 1973b, The roles of fossil vertebrates in interpretation of Late Cretaceous stratigraphy of the San Juan Basin, New Mexico: Four Corners Geological Society Memoir Book, 1973, p. 154-167.

———— 1974, *Purgatorius*, an early paromomyid primate (Mammalia): Science, v. 184, p. 903-905.

———— 1979, Marsupialia, *in* Lillegraven, J. A., Kielan-Jaworowska, Z., and Clemens, W. A., eds., Mesozoic mammals: the first two-thirds of mammalian history: Berkeley, University of California Press, p. 192-220.

———— 1980, *Gallolestes pachymandibularis* (Theria, *incertae sedis*; Mammalia) from Late Cretaceous deposits in Baja California del Norte, Mexico: PaleoBios, no. 33, 10 p.

———— 1985, Jurassic mammals of North America: Ameghiniana, v. 22, p. 53-56.

Clemens, W. A., and Kielan-Jaworowska, Z., 1979, Multituberculata, *in* Lillegraven, J. A., Kielan-Jaworowska, Z., and Clemens, W. A., eds., Mesozoic mammals: the first two-thirds of mammalian history: Berkeley, University of California Press, p. 99-149.

Clemens, W. A., and Mills, J. R. E., 1971, Review of *Peramus tenuirostris* Owen (Eupantotheria, Mammalia): British Museum (Natural History) Bulletin (Geology), v. 20, p. 87-113.

Clemens, W. A., and Russell, L. S., 1965, Mammalian fossils from the Upper Edmonton Formation, *in* Vertebrate paleontology in Alberta: University of Alberta Bulletin (Geology 2), p. 32-40.

Crompton, A. W., 1963, Tooth replacement in the cynodont *Thrinaxodon liorhinus* Seeley: South African Museum, Annals, v. 46, p. 479-521.

———— 1964, A preliminary description of a new mammal from the Upper Triassic of South Africa: Zoological Society of London, Proceedings, v. 142, p. 441-452.

———— 1971, The origin of the tribosphenic molar, *in* Kermack, D. M., and Kermack, K. A., eds., Early mammals: Zoological Journal of the Linnean Society, v. 50, supplement 1, p. 65-87.

Crompton, A. W., and Jenkins, F. A., Jr., 1968, Molar occlusion in Late Triassic mammals: Biological Reviews, v. 43, p. 427-458.

———— 1979, Origin of mammals, *in* Lillegraven, J. A., Kielan-Jaworowska, Z., and Clemens, W. A., eds., Mesozoic mammals: the first two-thirds of mammalian history: Berkeley, University of California Press, p. 59-73.

Crompton, A. W., and Sun, A. -L., 1985, Cranial structure and relationships of the Liassic mammal *Sinoconodon*: Zoological Journal of the Linnean Society, v. 85, p. 99-119.

Dashzeveg, D., 1979, *Arguimus khosbajari* gen. n., sp. n. (Peramuridae, Eupantotheria) from the Lower Cretaceous of Mongolia: Acta Palaeontologica Polonica, v. 24, p. 199-204.

Dashzeveg, D., and Kielan-Jaworowska, Z., 1984, The lower jaw of an aegialodontid mammal from the Early Cretaceous of Mongolia: Zoological Journal of the Linnean Society, v. 82, p. 217-227.

Domning, D. P., Morgan, G. S., and Ray, C. E., 1982, North American Eocene sea cows (Mammalia, Sirenia): Smithsonian Contributions to Paleobiology, v. 52, p. 1-69.

Emry, R. J., Archibald, J. D., and Smith, C. C., 1981, A mammalian molar from the Late Cretaceous of northern Mississippi: Journal of Paleontology, v. 55, p. 953-956.

Fox, R. C., 1970, Eutherian mammal from the early Campanian (Late Cretaceous) of Alberta, Canada: Nature, v. 227, p. 630-631.

_____ 1971, Marsupial mammals from the early Campanian Milk River Formation, Alberta, Canada, *in* Kermack, D. M., and Kermack, K. A., eds., Early mammals: Zoological Journal of the Linnean Society, v. 50, supplement 1, p. 145-164.

_____ 1972, A primitive therian mammal from the Upper Cretaceous of Alberta: Canadian Journal of Earth Sciences, v. 9, p. 1479-1494.

_____ 1974, *Deltatheroides*-like mammals from the Upper Cretaceous of North America: Nature, v. 249, p. 392.

_____ 1975, Molar structure and function in the Early Cretaceous mammal *Pappotherium*; evolutionary implications for Mesozoic Theria: Canadian Journal of Earth Sciences, v. 12, p. 412-442.

_____ 1976a, Additions to the mammalian local fauna from the Upper Milk River Formation (Upper Cretaceous), Alberta: *ibid.*, v. 13, p. 1105-1118.

_____ 1976b, Cretaceous mammals (*Meniscoessus intermedius*, new species, and *Alphadon* sp.) from the lowermost Oldman Formation, Alberta: *ibid.*, v. 13, p. 1216-1222.

_____ 1977, Notes on the dentition and relationships of the Late Cretaceous insectivore *Gypsonictops* Simpson: *ibid.*, v. 14, p. 1823-1831.

_____ 1979a, Mammals from the Upper Cretaceous Oldman Formation, Alberta. I. *Alphadon* Simpson (Marsupialia): *ibid.*, v. 16, p. 91-102.

_____ 1979b, Mammals from the Upper Cretaceous Oldman Formation, Alberta. II. *Pediomys* Marsh (Marsupialia): *ibid.*, v. 16, p. 103-113.

_____ 1979c, Mammals from the Upper Cretaceous Oldman Formation, Alberta, III. Eutheria: *ibid.*, v. 16, p. 114-125.

_____ 1980, *Picopsis pattersoni*, n. gen. and sp., an unusual therian from the Upper Cretaceous of Alberta, and the classification of primitive tribosphenic mammals: *ibid.*, v. 17, p. 1489-1498.

_____ 1982, Evidence of a new lineage of tribosphenic therians (Mammalia) from the Upper Cretaceous of Alberta, Canada: Geobios, Mémoire spécial 6, p. 169-175.

_____ 1983a, Evolutionary implications of tooth replacement in the Paleocene mammal *Pararyctes*: Canadian Journal of Earth Sciences, v. 20, p. 19-22.

_____ 1983b, Notes on the North American Tertiary marsupials *Herpetotherium* and *Peradectes*: *ibid.*, v. 20, p. 1565-1578.

_____ 1984, *Paranyctoides maleficus* (new species), an early eutherian mammal from the Cretaceous of Alberta: Carnegie Museum of Natural History, Special Publication, no. 9, p. 9-20.

Freeman, E. F., 1976a, A mammalian fossil from the Forest Marble (Middle Jurassic) of Dorset: Geological Association, Proceedings, v. 87, p. 231-235.

_____ 1976b, Mammal teeth from the Forest Marble (Middle Jurassic) of Oxfordshire, England: Science, v. 194, p. 1053-1055.

_____ 1979, A Middle Jurassic mammal bed from Oxfordshire: Palaeontology, v. 22, p. 135-166.

Gaunt, W. A., and Miles, A. E. W., 1967, Fundamental aspects of tooth morphogenesis, *in* Miles, A. E. W., ed., Structural and chemical organization of teeth, Volume I: New York, Academic Press, p. 151-197.

Gill, P., 1974, Resorption of premolars in the early mammal *Kuehneotherium praecursoris*: Archives of Oral Biology, v. 19, p. 327-328.

Grambast, L., Martinez, M., Mattauer, M., and Thaler, L., 1967, *Perutherium altiplanense*, nov. gen., nov. sp., premier Mammifère mésozoïque d'Amérique du Sud: Comptes-rendus hebdomadaires des seances de l'Academie des Sciences, Paris, sér. D, v. 264, p. 707-710.

Gregory, W. K., 1947, The monotremes and the palimpsest theory: American Museum of Natural History, Bulletin, v. 88, p. 1-52.

Gregory, W. K., and Simpson, G. G., 1926, Cretaceous mammal skulls from Mongolia: American Museum Novitates, no. 225, 20 p.

Hahn, G., 1977, Neue Schädle-reste von Multituberculaten aus dem malm Portugals: Geologica et Palaeontologica, v. 11, p. 161-186.

_____ 1978, Milch Bezahnungen von Paulchoffatiidae (Multituberculata: Ober-Jura): Neues Jahrbuch Geologie und Paläontologie, Monatshefte, 1978, no. 1, p. 25-34.

Hopson, J. A., and Barghusen, H. R., *in press*, A cladistic analysis of the mammal-like reptiles, *in* Roth, J. J., Roth, F. C., Maclean, P. D., and Hotton, N., eds., The ecology and biology of mammal-like reptiles: Washington, D. C., Smithsonian Press.

Jenkins, F. A., Jr., 1984, A survey of mammalian origins, *in* Broadhead, T. W., ed., Mammals, notes for a short course: University of Tennessee, Department of Geological Sciences, Studies in Geology 8, p. 32-47.

Jenkins, F. A., Jr., Crompton, A. W., and Downs, W. R., 1983, Mesozoic mammals from Arizona: new evidence on mammalian evolution: Science, v. 222, p. 1233-1235.

Johnston, P. A., 1980, First record of Mesozoic mammals from Saskatchewan: Canadian Journal of Earth Sciences, v. 17, p. 512-519.

Johnston, P. A., and Fox, R. C., 1984, Paleocene and Late Cretaceous mammals from Saskatchewan, Canada: Palaeontographica, Abteil A, v. 186, p. 163-222.

Kemp, T. S., 1982, Mammal-like reptiles and the origin of mammals: London, Academic Press, *xiv* + 363 p.

_____ 1983, The relationships of mammals: Zoological Journal of the Linnean Society, v. 77, p. 353-384.

Kermack, D. M., and Kermack, K. A., 1984, The evolution of mammalian characters: Washington, D. C., Kapitan Szabo Publishers, x + 149 p.

Kermack, K. A., Lees, P. M., and Mussett, F., 1965, *Aegialodon dawsoni*, a new trituberculosectorial tooth from the lower Wealden: Royal Society (B), Proceedings, v. 162, p. 535-554.

Kermack, K. A., Mussett, F., and Rigney, H. W., 1981, The skull of *Morganucodon*: Zoological Journal of the Linnean Society, London, v. 71, p. 1-158.

Kielan-Jaworowska, Z., 1969, Preliminary data on the Upper Cretaceous eutherian mammals from Bayn Dzak, Gobi Desert, in Kielan-Jaworowska, Z., ed., Results of the Polish-Mongolian Palaeontological Expeditions, pt. 1: Palaeontologia Polonica, no. 19, p. 171-191.

―――― 1975a, Preliminary description of two new eutherian genera from the Late Cretaceous of Mongolia, in ibid., pt. VI: ibid., no. 33, p. 5-16.

―――― 1975b, Evolution of the therian mammals in the Late Cretaceous of Asia. Part I. Deltatheridiidae, in ibid., pt. VI: ibid., no. 33, p. 103-132.

―――― 1981, Evolution of the therian mammals in the Late Cretaceous of Asia. Part IV. Skull structure in *Kennalestes* and *Asioryctes*, in ibid., pt. IX: ibid., no. 42, p. 25-78.

―――― 1982, Marsupial-placental dichotomy and paleogeography of Cretaceous Theria, in Montanaro-Galitelli, E., ed., Proceedings of the international meeting, *Paleontology, essential of historical geology*: Modena, Italy, S. T. E. M. Mucchi, p. 367-383.

Kielan-Jaworowska, Z., Bown, T. M., and Lillegraven, J. A., 1979, Eutheria, in Lillegraven, J. A., Kielan-Jaworowska, Z., and Clemens, W. A., eds., Mesozoic mammals: the first two-thirds of mammalian history: Berkeley, University of California Press, p. 221-258.

Kielan-Jaworowska, Z., Eaton, J. G., and Bown, T. M., 1979, Theria of metatherian-eutherian grade, in ibid., p. 182-191.

Kielan-Jaworowska, Z., and Trofimov, B. A., 1980, Cranial morphology of the Cretaceous eutherian mammal *Barunlestes*: Acta Palaeontologica Polonica, v. 25, p. 167-185.

Kraus, M. J., 1979, Eupantotheria, in Lillegraven, J. A., Kielan-Jaworowska, Z., and Clemens, W. A., eds., Mesozoic mammals: the first two-thirds of mammalian history: Berkeley, University of California Press, p. 162-171.

Krause, D. W., and Baird, D., 1979, Late Cretaceous mammals east of the North American western interior seaway: Journal of Paleontology, v. 53, p. 562-565.

Kron, D. G., 1979, Docodonta, in Lillegraven, J. A., Kielan-Jaworowska, Z., and Clemens, W. A., eds., Mesozoic mammals: the first two-thirds of mammalian history: Berkeley, University of California Press, p. 91-98.

Krusat, G., 1969, Ein Pantotheria-Molar mit dreispitzigem Talonid aus dem Kimmeridge von Portugal: Palaontologische Zeitschrift, v. 43, p. 52-56.

―――― 1980, Contribuição para o Conhecimento da Fauna do Kimeridgiano da Mina de Lignito Guimarota (Leiria, Portugal), Parte IV, *Haldanodon exspectatus* Kühne and Krusat 1972 (Mammalia, Docodonta): Memórias dos Servicos Geologicos de Portugal, no. 27, 79 p.

Ledoux, J.-C., Hartenberger, J.-L., Michaux, J., Sudre, J., and Thaler, L., 1966, Découverte d'un Mammifère dans le Crétacé supérieur à Dinosaures de Champ-Garimond près de Fons (Gard): Comptes-rendus hebdomadaires des séances de l'Académie des Sciences, Paris, sér. D, v. 262, p. 1925-1928.

Lillegraven, J. A., 1969, Latest Cretaceous mammals of upper part of Edmonton Formation of Alberta, Canada, and review of marsupial-placental dichotomy in mammalian evolution: University of Kansas, Paleontological Contributions, Article 50 (Vertebrata 12), 122 p.

―――― 1972, Preliminary report on Late Cretaceous mammals from the El Gallo Formation, Baja California del Norte, Mexico: Natural History Museum of Los Angeles County, Contributions in Science, no. 232, 11 p.

―――― 1976, A new genus of therian mammal from the Late Cretaceous "El Gallo Formation," Baja California, Mexico: Journal of Paleontology, v. 50, p. 437-443.

―――― 1984, Why *was* there a "marsupial-placental dichotomy?", in Broadhead, T. W., ed., Mammals: notes for a short course: University of Tennessee, Department of Geological Sciences, Studies in Geology 8, p. 72-86.

―――― 1985, Use of the term "trophoblast" for tissues in therian mammals: Journal of Morphology, v. 183, p. 293-299.

Lillegraven, J. A., and Bieber, A. L., 1986, Reliability of measurements of small mammalian fossils with an industrial measuring microscope: Journal of Vertebrate Paleontology, v. 6, p. 96-100.

Lillegraven, J. A., and McKenna, M. C., 1986, Fossil mammals from the "Mesaverde" Formation (Late Cretaceous, Judithian) of the Bighorn and Wind River Basins, Wyoming, with definitions of Late Cretaceous, North American land-mammal "ages": American Museum Novitates, no. 2840, 68 p.

Luckett, W. P., and Maier, W., 1982, Development of deciduous and permanent dentition in *Tarsius* and its phylogenetic significance: Folia Primatologica, v. 37, p. 1-36.

McKenna, M. C., 1975, Toward a phylogenetic classification of the Mammalia, in Luckett, W. P., and Szalay, F. S., eds., Phylogeny of the Primates, a multidisciplinary approach: New York, Plenum Press, p. 21-46.

Marshall, L. G., 1979, Evolution of metatherian and eutherian (mammalian) characters: a review based on cladistic methodology: Zoological Journal of the Linnean Society, v. 66, p. 369-410.

Marshall, L. G., Muizon, C. de, and Sigé, B., 1983a, Late Cretaceous mammals (Marsupialia) from Bolivia: Geobios, no. 16, p. 739-745.

―――― 1983b, *Perutherium altiplanese,* un notongulé du Crétacé supérieur du Pérou: Palaeovertebrata, v. 13, p. 145-155.

Mills, J. R. E., 1971, The dentition of *Morganucodon, in* Kermack, D. M., and Kermack, K. A., eds., Early mammals: Zoological Journal of the Linnean Society, v. 50, supplement 1, p. 29-63.

―――― 1984, The molar dentition of a Welsh pantothere: Zoological Journal of the Linnean Society, v. 82, p. 189-205.

Muizon, C. de, Marshall, L. G., and Sigé, B., 1984, The mammal fauna from the El Molino Formation (Late Cretaceous, Maestrichtian) at Tiupampa, southcentral Bolivia: Bulletin du Muséum national d'Histoire naturelle, Paris, 4e sér., v. 6, section C, p. 327-351.

Nesov, L. A., 1982, The ancient mammals of the USSR: Ezhegodnik Vsesoivznogo Paleontologicheskogo Obshchestva (Yearbook of All-Union Palaeontological Society), Akademii nauk SSSR, v. 25, p. 228-243 (in Russian).

―――― 1984, Concerning some discoveries of the remains of mammals from the Cretaceous deposits of Middle Asia: Vestnik zoologii, no. 2, p. 60-65 (in Russian).

―――― 1985a, New Cretaceous mammals of the Kizylkum Desert: Vestnik Leningradskogo Universiteta, Geologiya, Geografiya, Leningrad, no. 17, p. 8-18 (in Russian).

―――― 1985b, Rare osteichthyans, terrestrial lizards, and mammals from the Cretaceous estuaries and coastal plane zone of the Kizylkum Desert: Ezhegodnik Vsesoiuznogo Paleontologicheskogo Obshchestva (Yearbook of All-Union Palaeontological Society), Akademii nauk SSSR, v. 28, p. 199-219 (in Russian).

Nesov, L. A., and Golovneva, L. B., 1983, Changes in Cenomanian-Santonian (Late Cretaceous) vertebrate communities of the Kizylkum Desert, *in* Palaeontology and evolution of the biosphere: Trudy XXV Sessii Vsesoiuznogo Paleontologicheskogo Obshchestva (Proceedings 25th Session, All-Union Palaeontological Society), Leningrad, Nauka Press, p. 126-134 (in Russian).

Nesov, L. A., and Gureyev, A. A., 1981, A discovery of a jaw of the most ancient shrew in the Upper Cretaceous of the Kizylkum: Doklady Akademii nauk SSSR, v. 257, p. 1002-1004 (in Russian, with English translation, v. 257 (1-6), p. 217-219).

Nesov, L. A., and Trofimov, B. A., 1979, The oldest insectivore of the Cretaceous of the Uzbek SSSR: Doklady Akademii nauk SSSR, v. 247, p. 952-954 (in Russian, with English translation, v. 247 (1-6), p. 237-239).

Novacek, M., 1982, *Diacodon alticuspis,* an erinaceomorph insectivore from the early Eocene of northern New Mexico: Contributions to Geology, University of Wyoming, v. 20, p. 135-149.

―――― 1984, Evolutionary stasis in the elephant-shrew, *Rhynchocyon, in* Eldredge, N., and Stanley, S. M., eds., Living fossils: New York, Springer-Verlag, p. 4-22.

Novacek, M., Bown, T. M., and Schankler, D., 1985, On the classification of the early Tertiary Erinaceomorpha (Insectivora, Mammalia): American Museum Novitates, no. 2813, 22 p.

Osborn, H. F., 1907, Evolution of mammalian molar teeth: New York, Macmillan Company, 250 p.

Osborn, J. W., 1978, Morphogenetic gradients: fields versus clones, *in* Butler, P. M., and Joysey, K. A., eds., Development, function and evolution of teeth: London, Academic Press, p. 171-201.

Osborn, J. W., and Crompton, A. W., 1973, The evolution of mammalian from reptilian dentitions: Breviora, no. 399, 18 p.

Parrington, F. R., 1971, On the Upper Triassic mammals: Royal Society of London, Philosophical Transactions, v. 261B, p. 231-272.

Patterson, B., 1956, Early Cretaceous mammals and the evolution of mammalian molar teeth: Fieldiana (Geology), v. 13, p. 1-105.

Presley, R., 1981, Alisphenoid equivalents in placentals, marsupials, monotremes and fossils: Nature, v. 294, p. 668-670.

Prothero, D. R., 1981, New Jurassic mammals from Como Bluff, Wyoming, and the interrelationships of non-tribosphenic Theria: American Museum of Natural History, Bulletin, v. 167, p. 277-326.

Roth, V. L., 1984, On homology: Biological Journal of the Linnean Society, v. 22, p. 13-29.

Sahni, A., 1972, The vertebrate fauna of the Judith River Formation, Montana: American Museum of Natural History, Bulletin, v. 147, p. 321-412.

Schwartz, J. H., 1982, Morphological approach to heterodonty and homology, *in* Kurten, B., ed., Teeth: form, function, and evolution: New York, Columbia University Press, p. 123-144.

Sigé, B., 1972, La faunule de mammifères du Crétacé supérieur de Laguna Umayo (Andes péruviennes): Bulletin du Museum national d'historie naturelle, Paris, 3e sér., no. 99, Sciences de la Terre 19, p. 375-409.

Simpson, G. G., 1925, Mesozoic Mammalia. I. American triconodonts: American Journal of Science, v. 10, p. 145-165 and 334-358.

―――― 1928, A Catalogue of the Mesozoic Mammalia in the Geological Department of the British Museum: London, British Museum of Natural History, *vii* + 215 p.

―――― 1929, American Mesozoic Mammalia: Peabody Museum (Yale University), Memoirs, v. 3, pt. 1, Yale University Press, *xv* + 171 p.

―― 1936, Studies of the earliest mammalian dentitions: Dental Cosmos, August-September 1936, p. 1-24.

―― 1961, Evolution of Mesozoic mammals, *in* Vanderbroek, G., ed., International colloquium on the evolution of lower and nonspecialized mammals: Koninklijke Vlaamse Academie voor Wetenschappen, Letteren en Schone Kunsten van Belgie, Part 1, p. 57-95.

―― 1971, Concluding remarks: Mesozoic mammals revisited, *in* Kermack, D. M., and Kermack, K. A., eds., Early mammals: Zoological Journal of the Linnean Society, v. 50, supplement 1, p. 181-198.

Slaughter, B. H., 1968a, Earliest known eutherian mammals and the evolution of premolar occlusion: Texas Journal of Science, v. 30, p. 3-12.

―― 1968b, Earliest known marsupials: Science, v. 162, p. 254-255.

―― 1971, Mid-Cretaceous (Albian) therians of the Butler Farm local fauna, Texas, *in* Kermack, D. M., and Kermack, K. A., eds., Early mammals: Zoological Journal of the Linnean Society, v. 50, supplement 1, p. 131-143.

―― 1981, The Trinity therians (Albian, mid-Cretaceous) as marsupials and placentals: Journal of Paleontology, v. 55, p. 682-683.

Sloan, R. E., 1979, Multituberculates, *in* Fairbridge, R. W., and Jablonski, D., eds., The encyclopedia of paleontology: Stroudsberg, Pennsylvania, Dowden, Hutchison, and Ross Inc., p. 492-498.

Sloan, R. E., and Van Valen, L., 1965, Cretaceous mammals from Montana: Science, v. 148, p. 220-227.

Szalay, F. S., 1965, First evidence of tooth replacement in the Subclass Allotheria (Mammalia): American Museum Novitates, no. 2226, 12 p.

―― 1977, Phylogenetic relationships and a classification of eutherian Mammalia, *in* Hecht, M. K., Goody, P. C., and Hecht, B. M., eds., Major patterns in vertebrate evolution: New York, Plenum Press, NATO Advanced Studies Institute Series no. 14, p. 315-374.

Turnbull, W. D., 1971, The Trinity therians: their bearing on evolution in marsupials and other therians, *in* Dahlberg, A. A., ed., Dental morphology and evolution: Chicago, University of Chicago Press, p. 151-179.

Van Valen, L., 1966, Deltatheridia, a new order of mammals: American Museum of Natural History, Bulletin, v. 132, p. 1-126.

―― 1978, The beginning of the age of mammals: Evolutionary Theory, v. 4, p. 45-80.

Van Valen, L., and Sloan, R. E., 1965, The earliest primates: Science, v. 148, p. 220-227.

Westergaard, B., 1980, Evolution of the mammalian dentition: Mémoires de la Société géologique de France, N.S., no. 139, p. 191-200.

―― 1983, A new detailed model for mammalian dentitional evolution: Zeitschrift für zoologische systematik Evolutionsforschung, v. 21, p. 68-78.

Wilson, J. T., and Hill, J. P., 1897, Observations upon the development and succession of the teeth in *Perameles;* together with a contribution to the discussion of the homologies of the teeth in marsupial animals: Quarterly Journal of Microscopal Science, v. 39, p. 427-588.

NOTE ADDED IN PROOF

Since completion of this study two pertinent papers have been published.

M. J. Novacek (1986) reviewed problems and data available concerning identification of the primitive eutherian postcanine dental formula; his results strengthen or expand some of our conclusions. He argued that eutherians primitively had at least five premolars. Reduction of the premolars to four, a condition probably primitive for a number of eutherian orders, is hypothesized to have occurred through loss of a premolar in the middle of the series. The hypothesis advanced by McKenna (1975) that reduction resulted from evolution of monophyodont ontogeny of the last (distal) premolar (P5) and loss of the last (distal) molar (M3) cannot be falsified, but appears less likely. Finally, Novacek's analysis emphasizes problems of establishing homologies between postcanine teeth of marsupials and eutherians, and the possibility of markedly different patterns of evolution of the postcanine dentitions of these groups from their last common, tribosphenic ancestor.

Domning and others (1986) added support to an earlier assertion (Domning and others, 1982) that early sirenians and, probably, the common ancestors of sirenians, proboscideans, and desmostylians had five premolars. Although the authors reserved full analysis of the polarity and distribution of this and other dental characters for a future paper, their initial findings add support to the view that presence of five premolars was the primitive condition of eutherians.

ADDITIONAL REFERENCES CITED

Domning, D. P., Ray, C. E., and McKenna, M. C., 1986, Two new Oligocene desmostylians and a discussion of tethytherian systematics: Smithsonian Contributions to Paleobiology, no. 59, 56 p.

Novacek, M. J., 1986, The primitive eutherian dental formula: Journal of Vertebrate Paleontology, v. 6, p. 191-196.

MANUSCRIPT RECEIVED NOVEMBER 20, 1985
REVISED MANUSCRIPT RECEIVED APRIL 1, 1986
MANUSCRIPT ACCEPTED JUNE 18, 1986

Paraphyly in *Catopsalis* (Mammalia: Multituberculata) and its biogeographic implications

NANCY B. SIMMONS — *Department of Paleontology, University of California, Berkeley, California 94720*

MIAO DESUI — *Department of Geology and Geophysics, The University of Wyoming, Laramie, Wyoming 82071-3006*

ABSTRACT

The genus *Catopsalis* Cope includes eight species *(C. matthewi, C. catopsaloides, C. joyneri, C. alexanderi, C. foliatus, C. utahensis, C. fissidens, C. calgariensis)* spanning Late Cretaceous through late Paleocene/early Eocene time on two continents, Asia (first two taxa) and North America (last six taxa). A cladistic analysis of dental and palatal features within the Taeniolabididae (which includes *Catopsalis, Kamptobaatar, Lambdopsalis, Prionessus, Sphenopsalis,* and *Taeniolabis)* indicates that *Catopsalis* is a paraphyletic taxon, composed of no fewer than five independent monophyletic groups. *Taeniolabis* is a monophyletic taxon, and *Lambdopsalis, Prionessus,* and *Sphenopsalis* (individually monophyletic by monotypy) together form another monophyletic group. These two clades appear to have evolved from ancestors within the paraphyletic taxon *Catopsalis;* accordingly, the smallest monophyletic group including all *Catopsalis* species also includes *Taeniolabis, Lambdopsalis, Prionessus,* and *Sphenopsalis. C. matthewi,* the most primitive member of this clade, is returned to *Djadochtatherium* Simpson, previously considered a junior subjective synonym of *Catopsalis.* The relationships demonstrated among various members of the Taeniolabididae support the hypothesis of a Late Cretaceous taeniolabidid dispersal from Asia to North America. The data additionally suggest a second dispersal event, probably in the middle to late Paleocene, in which the ancestors of the *Lambdopsalis/Prionessus/Sphenopsalis* lineage dispersed from North America back to Asia.

INTRODUCTION

The genus *Catopsalis* was erected by Cope (1882a), based on *C. foliatus* from the Puercan of New Mexico. Five North American species subsequently have been referred to the genus: *C. joyneri* Sloan and Van Valen, 1965 (Lancian, North America); *C. alexanderi* Middleton, 1982 (Puercan, North America); *C. utahensis* Gazin, 1939 (Puercan and Torrejonian, North America); *C. fissidens* Cope, 1884a,b (Torrejonian, North America); and *C. calgariensis* Russell, 1926 (Torrejonian/Tiffanian, North America). Kielan-Jaworowska and Sloan (1979) synonymized the Asian genus *Djadochtaterium* Simpson, 1925 with *Catopsalis,* thereby adding *Djadochtatherium matthewi* Simpson, 1925 ("early Upper Cretaceous," Mongolia) and *D. catopsaloides* Kielan-Jaworowska, 1974a (? Middle Campanian, Mongolia) to this otherwise exclusively North American genus. For reviews of the species thus referred to *Catopsalis* Cope see Kielan-Jaworowska and Sloan (1979), Middleton (1982), and Hahn and Hahn (1983).

According to Kielan-Jaworowska and Sloan (1979) and Middleton (1982), the eight species referred to *Catopsalis* can be placed in a chronological sequence characterized by a series of presumed evolutionary trends in morphology. These trends include an increase in relative size, reduction of the number of upper premolars, relative increase in the size of the first molars, decrease in size of the lower fourth premolar, and increases in numbers of cusps, cuspules, cingulae, and enamel infoldings on the molars (Kielan-Jaworowska and Sloan, 1979; Middleton, 1982). These apparent trends were used to link the various species together into a single genus (Kielan-Jaworowska and Sloan, 1979; Middleton 1982). Because the oldest and most primitive two members of the genus occur in Asia, Kielan-Jaworowska and Sloan (1979) hypothesized an Asian origin for *Catopsalis* as sometime before the late Santonian, followed by diversification and dispersal into North America in the Late Campanian or Early Maastrichtian.

Catopsalis has been included in the Taeniolabididae by all recent authorities (see Hahn and Hahn, 1983). The other members of the Taeniolabididae are: *Kamptobaatar kuczynskii* Kielan-Jaworowska, 1970 (Late Santonian/Early Campanian, Asia); *Taeniolabis taoensis* Cope, 1882b (Puercan, North America); *Taeniolabis* sp. (Puercan, North America [Simmons, manuscript]); *Lambdopsalis bulla* Zhou and Qi, 1978 (late Paleocene/early Eocene, Asia); *Prionessus lucifer* Matthew and Granger, 1925 (late Paleocene/early Eocene, Asia); and *Sphenopsalis nobilis* Matthew and others, 1928 (late Paleocene/early Eocene, Asia). The purpose of this study was to investigate relationships among the various taxa referred to the Taeniolabididae, with the specific goal of establishing relationships of the eight species referred to *Catopsalis* to one another and to other taeniolabidid species.

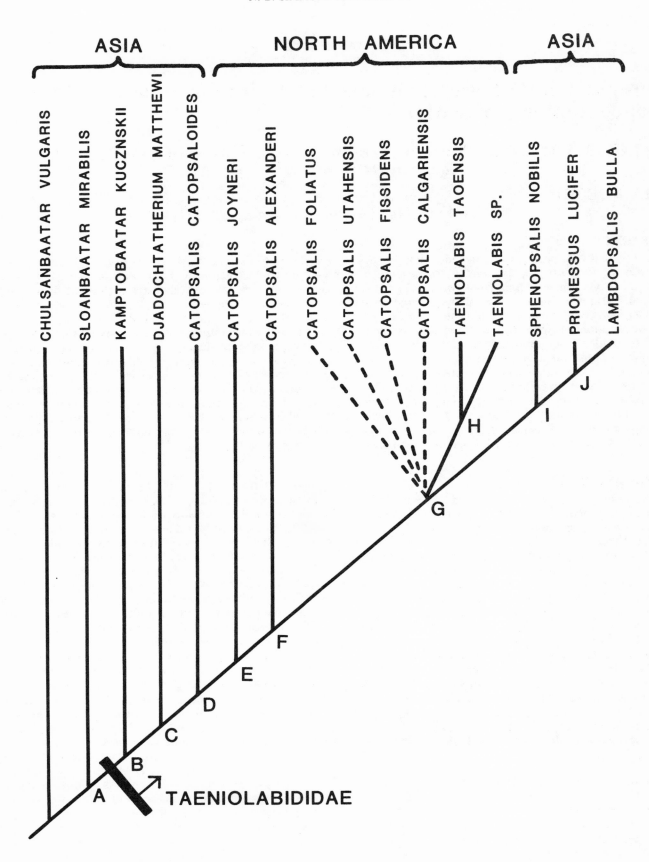

METHODS

Phylogenetic relationships among the various species referred to *Catopsalis* and other members of the Taeniolabididae were investigated through cladistic analysis; see Nelson and Platnick (1981) and Wiley (1981) for detailed discussions of this method. Two outgroups were used, *Chulsanbaatar vulgaris* Kielan-Jaworowska, 1974a (Eucosmodontidae; ?Middle Campanian, Asia) and *Sloanbaatar mirabilis* Kielan-Jaworowska, 1970 (Sloanbaataridae; Late Santonian/Early Campanian, Asia).

Choice of outgroups in multituberculate systematics is made difficult by lack knowledge of phylogenetic relationships at the familial level within the order (Clemens and Kielan-Jaworowska, 1979). *Chulsanbaatar vulgaris* and *Sloanbaatar mirabilis* were chosen as outgroups in this study for several reasons. First, each taxon represents one of the other two families currently included (along with the Taeniolabididae) in the potentially monophyletic suborder Taeniolabidoidea (see Hahn and Hahn, 1983; Kielan-Jaworowska, 1980). The Sloanbaataridae is a monotypic family, thus *Sloanbaatar mirabilis* was an obvious choice. *Chulsanbaatar vulgaris* was chosen as a representative of the Eucosmodontidae because of its lack of more extreme autapomorphies seen elsewhere in the family, the relatively large sample size available (36 specimens, mostly complete or nearly complete skulls with lower jaws [Kielan-Jaworowska, 1974a]), and the early Late Cretaceous age and Mongolian provenience of the species. A nearly complete data set (including information of most of the features utilized for this analysis) was available both for *Sloanbaatar mirabilis* and *Chulsanbaatar vulgaris,* enhancing usefulness of these taxa as outgroups.

The characters outlined below were analyzed using Wagner parsimony criteria, as adapted by J. Felsenstein in the MIX and PENNY algorithms of the PHYLIP (Version 2.6) Phylogenetic Inference Package. The inherent assumptions of the Wagner method as used in these algorithms are as follows (Felsenstein, 1973; 1978; 1979; 1981); (1) ancestral character states are unknown; (2) different characters evolve independently; (3) different lineages evolve independently; (4) character state changes from 'A' to 'B' are as equally probable as changes from 'B' to 'A'; (5) both kinds of character state changes are *(a priori)* improbable over the evolutionary time spans involved in the differentiation of the group in question; (6) other kinds of evolutionary events, such as retention of polymorphisms, are far less probable than character state changes; and (7) rates of evolution in different lineages are sufficiently low that two changes in one segment of the evolutionary tree are far less probable than one change in another segment. Use of this method results in production of an unrooted tree; rooting of the tree is left to discretion of the investigator. In the case of this study, the tree was rooted through outgroups based on the identified character polarities (see "Character Descriptions" below for discussions of polarities of individual characters).

The algorithms of the PHYLIP program package were run on Morrow MD-3 microcomputer. Multistate characters were translated into binary code using the FACTOR program included in the PHYLIP package. The MIX algorithm was run twenty times, with the order of the taxa considered randomized each time in order to force the algorithm to consider different approaches to constructing the most parsimonious cladogram. The PENNY algorithm was used to search 30,000 alternate trees in order to find the most parsimonious cladograms. The results of these analyses are presented in Figure 1 and are discussed in the "Results" section below.

CHARACTER DESCRIPTIONS AND DATA

Characters used in the analyses are drawn exclusively from dental and palatal morphology. Because most of the species referred to *Catopsalis* are known only from dental specimens, cranial and postcranial characters were not considered. Only characters exhibiting two or more states among the taxa considered are listed below, as single state characters contribute no information useful in assessing relationships of the taxa in question. The first seven characters concern the anterior (incisor and premolar) dentition and the palate; the second five characters describe cusp shape and cusp formulae for the molar dentition; the next two characters concern absolute size; and the final three characters deal with the relative proportions of consecutive teeth.

Character independence is of critical importance in any systematic study. This is particularly difficult to establish when dental features are considered, because of functional (and perhaps evolutionary) relationships among various elements of the dentition. However, as Kirsch (1982, p. 589) so aptly put it, "Yet the whole point of taxonomy is to discover correlations among features, so that it might be counter-productive to discount a dataset just for this reason." The characters utilized in this study were chosen and defined in such a way as to make identification of correlated characters easy in the final

---

Figure 1. Results of phylogenetic analysis of Taeniolabididae, based on dental and palatal characters; *Chulsanbaatar vulgaris* and *Sloanbaatar mirabilis* used as outgroups (note that *Catopsalis matthewi* has been returned to *Djadochtatherium* Simpson). Solid lines reflect relationships substantiated by this analysis; dotted lines indicate placement of taxa occurring in variable positions above node *F* and below nodes *H* and *I* in alternate most parsimonious cladograms generated (alternate positions of these taxa do not affect relative positions of taxa indicated by solid lines; see "Results" section for discussion). Key to character states pertaining at each node (see "Character Descriptions" for abbreviations): *A*, M_1 cusp formula 4:3 or higher. *B*, M^1 cusp formula 5:5:0 or higher. *C*, M_1 cusp formula 4:4 or higher. *D*, P^2 lost; M^1 cusp formula 5:5:4 or higher; $P^4/M^1 < 0.50$; $P_4/M_1 < 0.65$; $M_1/M_2 < 1.70$; M^1 length ≥ 5.5 mm; M_1 length ≥ 5.0. *E*, P^1 lost; P_4 triangular; M^1 cusp formula 6:6:5 or higher; M_1 cusp formula 5:4 or higher; M_2 cusp formula 3:2 or higher; $P^4/M^1 \leq 0.40$; $P_4/M_1 < 0.55$. *F*, I^3 marginal; P_3 lost. *G*, P^3 lost; P^4 single rooted; $P^4/M^1 < 0.35$; $P_4/M_1 < 0.45$. *H*, M^1 cusp formula 9:8:9 or higher; M_1 cusp formula 7:6 or higher; M_2 cusp formula 4:4 or higher; $P_4/M_1 < 0.40$. *I*, Crescentic cusps. *J*, P_4 peg-like; $P_4/M_1 < 0.30$.

analysis; no potentially useful characters were discarded because of possible functional and/or evolutionary correlations with other characters.

Definitions of the characters used in this analysis are presented below. In the case of characters exhibiting ranges of variation within species (such as cusp number and absolute tooth length), minimum values rather than mean values or full ranges of variation were used to describe the various states of each character. This choice was made for several reasons. First, the taxa considered in this study are known from numbers of specimens ranging from one *(Catopsalis matthewi)* to over 60 *(Taeniolabis taoensis)*. Mean values and full ranges of variation based on such widely different sample sizes are not strictly comparable, and the utility of an analysis based on characters so defined appears highly limited. Additionally, mean measurements and full ranges of variation are not characteristic of individual organisms, and hence are not useful for placing individual specimens either above or below a cladogram node characterized by such a value. A single new specimen either is, or is not, longer than 15 mm; but does it fall in the range 12-16 mm or the range 14-18 mm? Cladograms are particularly valuable for recognition of relationships of newly discovered specimens relative to previously described taxa, an advantage that is largely forfeited if mean values or ranges are used to define the nodes of the cladogram (mean values and ranges of variation are most useful when the goal is to describe specific taxa based on their autapomorphies). Because recognition of relationships (both well-known taxa and single specimens) is the goal of this study, minimum values were chosen to describe those features found to vary within species.

For each of the features for which minimum values were used to define character states (*e. g.*, cusp numbers, size measurements), arbitrary cut-off points were chosen in such a way as to allow placement of the complete known range of variation for every taxon either above or below that point. Round numbers (*e. g.*, 8.0 mm) were chosen whenever possible for the sake of convenience. It should be recognized that future changes in known ranges of variation for the taxa considered may necessitate alteration of these cut-off points, but this possibility does not detract from their current utility.

The references used to obtain data on the various characters for each taxon are as follows: *Chulsanbaatar vulgaris* (see Kielan-Jaworowska, 1974a); *Sloanbaatar mirabilis* (see Kielan-Jaworowska, 1970, 1971, 1974a); *Kamptobaatar kuczynskii* (see Kielan-Jaworowska, 1970, 1971, 1974a); *Catopsalis matthewi* (see Simpson, 1925, 1937; Kielan-Jaworowska, 1974a; Kielan-Jaworowska and Sloan, 1979); *C. catopsaloides* (see Kielan-Jaworowska, 1974a; Kielan-Jaworowska and Sloan, 1979); *C. joyneri* (see Sloan and Van Valen, 1965; Kielan-Jaworowska, 1974a; Kielan-Jaworowska and Sloan, 1979; Middleton, 1982); *C. alexanderi* (see Middleton, 1982); *C. foliatus* (see Cope, 1882a; Kielan-Jaworowska and Sloan, 1979; Middleton, 1982); *C. utahensis* (see Gazin, 1939; Kielan-Jaworowska and Sloan, 1979; Middleton, 1982); *C. fissidens* (see Cope, 1884a; Kielan-Jaworowska and Sloan, 1979; Middleton, 1982); *C. calgariensis* (see Russell, 1926; Middleton, 1982); *Taeniolabis taoensis* (see Granger and Simpson, 1929; Simmons, *personal observation*); *Taeniolabis* sp. (Simmons, *personal observation*); *Prionessus lucifer* (see Granger and Simpson, 1929; Miao, *personal observation*); *Sphenopsalis nobilis* (see Granger and Simpson, 1929; Miao, *personal observation*); and *Lambdopsalis bulla* (see Zhou and Qi, 1978; Miao, 1986).

Position of I^3.—Two positions of the I^3 relative to the palate and the remainder of the dentition have been identified among the taxa considered. The I^3 is located near the midline of the palate, just lateral to the palatine fissures and medial to the longitudinal axis of the cheek tooth row in *Chulsanbaatar, Sloanbaatar, Kamptobaatar, Catopsalis matthewi,* and *C. catopsaloides*. This is considered the primitive condition, based on appearance in the outgroups. The I^3 is located on the margin of the palate, posterior to the enlarged I^2 and in line with the longitudinal axis of the cheek tooth row, in *Catopsalis alexanderi, Taeniolabis taoensis,* and *Lambdopsalis;* this is considered the derived condition. The state of this character is unknown in other taxa examined.

Presence/absence of P_3.—Two premolars, identified as P_3 and P_4 (Clemens and Kielan-Jaworowska, 1979), are present in dentitions of *Chulsanbaatar, Sloanbaatar, Kamptobaatar, Catopsalis catopsaloides,* and, possibly, *C. matthewi*. This is considered the primitive condition based on occurrence in the outgroups. The derived condition consists of loss of P_3, leaving P_4 as the only lower premolar. The derived condition is seen in *Catopsalis alexanderi, C. foliatus, Taeniolabis taoensis, Taeniolabis* sp., *Prionessus,* and *Lambdopsalis*. The number of lower premolars is unknown in other taxa considered.

Shape of P_4.—The shape of P_4 is blade-like or parabolic in lateral outline in *Chulsanbaatar, Sloanbaatar, Kamptobaatar, Catopsalis matthewi,* and *C. catopsaloides*. This is considered the primitive condition based on appearance in the outgroups. Two derived conditions have been identified. The P_4 is triangular in lateral outline (apex directed dorsally) in *Catopsalis joyneri, C. alexanderi, C. foliatus, C. fissidens,* and *Taeniolabis*. The P_4 is peg-like (consisting of a single cusp only) in *Prionessus* and *Lambdopsalis*. The shape of P_4 is unknown in other taxa examined.

Presence/absence of P^2.—Four upper premolars (P^1, P^2, P^3, P^4) are present in *Chulsanbaatar, Sloanbaatar, Kamptobaatar,* and probably *Catopsalis matthewi* (only the anterior snout with alveoli for three upper premolars is preserved in this taxon, but morphologically they resemble those for P^{1-3}; P^4 was probably present, but not preserved). The presence of four upper premolars is considered the primitive condition, based on appearance in the outgroups. P^2 is absent in *Catopsalis catopsaloides, C. joyneri, Taeniolabis taoensis,* and *Lampdopsalis;* this is considered the derived condition. The number of upper premolars is unknown in other taxa considered.

Presence/absence of P^1.—The P^1 is present in *Chulsanbaatar, Sloanbaatar, Kamptobaatar, Catopsalis mat-*

thewi, and *C. catopsaloides;* this is considered the primitive condition. The derived condition consists of loss of P^1, which is seen in *C. joyneri, Taeniolabis taoensis,* and *Lambdopsalis.*

Presence/absence of P^3.—The P^3 is present in *Chulsanbaatar, Sloanbaatar, Kamptobaatar, Catopsalis matthewi, C. catopsaloides,* and *C. joyneri;* this is considered the primitive condition. All three of the anterior upper premolars are absent in *Taeniolabis taoensis* and *Lampdopsalis*, leaving only the P^4.

Number of roots on P^4.—The P^4 is double-rooted in *Chulsanbaatar, Sloanbaatar, Kamptobaatar, Catopsalis catopsaloides, C. joyneri,* and *C. alexanderi.* This is considered the primitive condition based on appearance in the outgroups. The P^4 is single-rooted in *C. calgariensis, Taeniolabis taoensis, Taeniolabis* sp., *Prionessus,* and *Lambdopsalis.* This is considered the derived condition. The number of roots on P^4 is unknown in other taxa examined.

Molar cusp shape.—The shape of the molar cusps is quadranglular in *Chulsanbaatar, Sloanbaatar, Kamptobaatar, Catopsalis matthewi, C. catopsaloides, C. joyneri, C. alexanderi, C. foliatus, C. utahensis, C. fissidens, C. calgariensis, Taeniolabis taoensis,* and *Taeniolabis* sp. This is considered the primitive condition, based on appearance in the outgroups. Molar cusp shape is crescentic in *Prionessus, Sphenopsalis,* and *Lambdopsalis;* this is considered the derived condition.

Cusp formula of M_1.—Following the pattern established by Simpson (1929), cusp numbers are expressed in the formulae (e. g., 4:5 or 1:2:3), in which the number of cusps in each longitudinal row is recorded, beginning with the labial and ending with the lingual row of cusps.

Based on cusp formulae seen in the outgroups, the primitive cusp formula for the M_1 is defined as '3:2 or fewer'; this is the condition seen in *Chulsanbaatar* (range = 2-3:2; N > 36 [see Kielan-Jaworowska, 1974a]). Five relatively derived states have been identified, each characterized by a higher minimum number of cusps in one or both cusp rows. These derived states are defined as '4:3 or higher,' '4:4 or higher,' '5:4 or higher,' '6:5 or higher,' and '7:6 or higher.' These states were defined on the basis of the following data: *Sloanbaatar* (4:3; N = 2); *Kamptobaatar* (4:3; N = 1); *Catopsalis matthewi* (4:4; N = 1); *C. catopsaloides* (4:4; N = 2); *C. joyneri* (5:4; N = 11); *C. alexanderi* (5:4; N = 9); *C. foliatus* (5:4; N = 1); *Prionessus* (5:4; N = 1); *Lambdopsalis* (5:4; N = 13); *C. calgariensis* (5-6:4-5; N = 10); *C. fissidens* (6:5; N = 1); *C. utahensis* (6-7:5; N = 2); *Taeniolabis* sp. (7:6; N = 1); and *Taeniolabis taoensis* (7-8:6-7; N = 35). The M_1 is unknown only in *Sphenopsalis*.

Cusp formula of M_2.—Based on cusp formulae seen in the outgroups, the primitive cusp formula for M_2 is defined as '2:2'; this is the state seen in *Sloanbaatar* (2:2; N = 2) and *Catopsalis catopsaloides* (2:2; N = 2). Three relatively derived states have been identified, each characterized by a higher minimum number of cusps in one or both cusp rows. These derived states are defined as '3:2 or higher,' '4:2 or higher,' and '4:4 or higher.' These states were defined on the basis of the following data: *Chulsanbaatar* (2-3:2; N > 36 [see Kielan-Jaworowska, 1974a]); *C. joyneri* (3:2; N = 1); *C. alexanderi* (3:2; N = 8); *C. fissidens* (3:2; N = 1); *Prionessus* (3:2; N = 1); *C. calgariensis* (3-4:2; N = 6); *C. foliatus* (4:2-3; N = 2); *Lambdopsalis* (4-5:2; N = 55); *Taeniolabis taoensis* (4-6:4-6; N = 40); and *Taeniolabis* sp. (6:4; N = 1). The M_2 is unknown in other taxa considered.

Cusp formula of M^1.—Based on cusp formulae seen in the outgroups, the primitive cusp formula for M^1 is defined as '4:4:0'; this is the state seen in *Sloanbaatar* (4:4:0; N = 2). Seven relatively derived states have been identified, each characterized by a higher minimum number of cusps in one or more cusp rows. These derived states are defined as '4:5:0 or higher,' '5:5:0 or higher,' '5:5:4 or higher,' '6:6:5 or higher,' '6:7:7 or higher,' '7:7:7 or higher,' and '9:8:9 or higher.' These states were defined on the basis of the following data *Chulsanbaatar* (4:5:1; N > 36 [see Kielan-Jaworowska, 1974a]); *Kamptobaatar* (5:5:0; N = 2); *Catopsalis catopsaloides* (5-6:5-6:4; N = 2); *Prionessus* (7:6:5; N = 1); *Lambdopsalis* (6-7:7-8:7; N = 35); *C. joyneri* (8:7:8; N = 1); *C. alexanderi* (7-8:7-8:9; N = 10); *C. calgariensis* (7-8:7:7-10; N = 4); *Taeniolabis taoensis* (9-10:8-11:9-11; N = 22); and *Taeniolabis* sp. (9:9:10; N = 1). The M^1 is unknown in other taxa considered.

Cusp formula of M^2.—Based on cusp formulae seen in the outgroups, the primitive cusp formula for M^2 is defined as '1:2:2'; this is the state seen in *Chulsanbaatar* (1:2:2; N > 36 [see Kielan-Jaworowska, 1974a]). Two relatively derived states have been identified, each characterized by a higher minimum number of cusps in one or more cusp rows. These derived states are defined as '1:2:3 or higher' and '1:3:3 or higher.' These states were defined on the basis of the following data: *Catopsalis catopsaloides* (2:2-3:2-3; N = 4); *Sloanbaatar* (1:2:3; N = 2); *Kamptobaatar* (1:2:3; N = 2); *Prionessus* (1:2:3; N = 2); *Lambdopsalis* (1:2:3-4; N = 24); *Sphenopsalis* (1:2:4); *C. joyneri* (1:3:3; N = 1); *C. calgariensis* (1:3:3; N = 6); *C. alexanderi* (1-2:3:3-4; N = 11); and *Taeniolabis taoensis* (1:4-5:4-6; N = 11). The M^2 is unknown in other taxa examined.

Length of M_1.—The primitive state for M_1 length is defined as length less than 2.0 mm; this state is seen in *Chulsanbaatar* (1.6-1.9 mm; N = 6), *Sloanbaatar* (1.7 mm; N = 2), and *Kamptobaatar* (1.8 mm; N = 2). Seven relatively derived states were defined, characterized by increases in the minimum tooth length. These derived character states are defined as '4.0 mm or more,' '5.0 mm or more,' '6.0 mm or more,' '7.0 mm or more,' '8.0 mm or more,' '10.0 mm or more,' '12.0 mm or more,' '14.0 mm or more,' and '18.0 mm or more.' These states were defined on the basis of the following data: *Catopsalis matthewi* (4.7 mm; N = 1); *Prionessus* (?5.6 mm; N = 1); *C. catopsaloides* (6.1 mm; N = 2); *C. joyneri* (6.4-7.4 mm; N = 11); *Lambdopsalis* (7.2-8.8 mm; N = 13; *C. alexanderi* (8.2-?9.3 mm; N = 9); *C. foliatus* (10.7 mm; N = 1); *C. utahensis* (12.0-13.0 mm; N = 2); *C. fissidens* (14.0 mm; N = 1); *C. calgariensis* (14.3-17.1 mm; N = 10); *Taeniolabis* sp. (16.0 mm; N = 1); and *Taeniolabis* *taoensis* (18.7-21.4 mm; N = 35). M_1 length is unknown only in *Sphenopsalis*.

Length of M^1.—The primitive state for M^1 length is defined as length less than 2.5 mm; this state is seen in *Chulsanbaatar* (1.8-2.5 mm; N = 6), *Sloanbaatar* (1.9 mm; N = 2), and *Kamptobaatar* (1.9 mm; N = 2). Six relatively derived states were defined, characterized by increases in the minimum tooth length. These derived character states are defined as '5.5 mm or more,' '6.5 mm or more,' '7.5 mm or more,' '9.0 mm or more,' '15.0 mm or more,' and '18.0 mm or more.' These states were defined on the basis of the following data: *Prionessus* (5.5-5.6 mm; N = 2); *Catopsalis catopsaloides* (6.9-7.9 mm; N = 4); *C. joyneri* (7.5-8.6 mm; N = 14); *Lambdopsalis* (9.0-11.0 mm; N = 35); *C. alexanderi* (9.9-10.7 mm; N = 7); *Sphenopsalis* (15.8 mm; N = 1); *C. calgariensis* (18.0-19.2 mm; N = 4); *Taeniolabis* sp. (18.9 mm; N = 1); and *Taeniolabis taoensis* (21.9-24.4 mm; N = 22). The M^1 is unknown in other taxa examined.

Ratio of length P_4 to length M_1 (P_4/M_1).—Ratios used in this study were calculated from consecutive teeth in individual specimens whenever possible; ratios calculated from collections of isolated teeth are designated '?', as these can be considered only as estimates.

Based on the P_4/M_1 ratio seen in outgroups, the primitive cusp P_4/M_1 ratio is defined as '0.9 or higher'; this is the condition seen in *Sloanbaatar* (1.58; N = 2), *Kamptobaatar* (1.28; N = 2), and *Chulsanbaatar* (0.9-1.0; N = 6). Six relatively derived states were identified, characterized by decreases in this ratio (indicating relative reduction of the P_4. These derived states are defined as 'less than 0.7,' 'less than 0.65,' 'less than 0.55,' 'less than 0.45,' 'less than 0.40,' and 'less than 0.30.' These states were defined on the basis of the following data: *Catopsalis matthewi* (0.68; N = 1); *C. catopsaloides* (0.62; N = 2); *C. alexanderi* (0.45-0.52; N = 3); *C. joyneri* (0.45; N = 1); *C. foliatus* (?0.39-?0.43; N = 2); *Taeniolabis taoensis* (0.29-0.38; N = 17); *Taeniolabis* sp. (0.35; N = 1); *Prionessus* (0.29; N = 1); and *Lambdopsalis* (0.18; N = 1). The P_4/M_1 ratio is unknown in other taxa examined.

Ratio of length M_1 to length M_2 (M_1/M_2).—Based on the M_1/M_2 ratio seen in outgroups, the primitive cusp M_1/M_2 ratio is defined as '1.75 or higher'; this is the condition seen in *Chulsanbaatar* (1.77-2.11; N = 6). Four relatively derived states were identified, characterized by decreases in this ratio (indicating relative increase in size of the M_2). These derived states are defined as 'less than 1.75,' 'less than 1.70,' 'less than 1.60,' and 'less than 1.30.' These states were defined on the basis of the following data: *Prionessus* (1.73; N = 1); *Catopsalis catopsaloides* (1.69; N = 2); *C. calgariensis* (?1.56-?1.64; N = 2); *C. foliatus* (?1.37-?1.62; N = 2); *Taeniolabis taoensis* (1.28-1.57; N = 24); *C. alexanderi* (1.19-1.49; N = 3); *C. joyneri* (1.27; N = 1); *Taeniolabis* sp. (1.23; N = 1); and *Lambdopsalis* (?1.05-?1.09; N = 2). The M_1/M_2 ratio is unavailable for other taxa examined.

Ratio of length P^4 to length M^1 (P^4/M^1).—Based on the P^4/M^1 ratio seen in outgroups, the primitive cusp P^4/M^1 ratio is defined as '0.85 or higher'; this is the condition seen in *Sloanbaatar* (0.89; N = 2) and *Kamptobaatar* (0.89; N = 2). Three relatively derived states were identified, characterized by decreases in this ratio (indicating relative reduction of the P^4). These derived states are defined as 'less than 0.8,' 'less than or equal to 0.40,' and 'less than 0.35.' These states were defined on the basis of the following data: *Chulsanbaatar* (0.46-0.78; N = 6); *Catopsalis catopsaloides* (0.38; N = 2); *C. joyneri* (0.40; N = 1); *C. alexanderi* (0.38; N = 1); *C. calgariensis* (?0.31-?0.33; N = 2); *Taeniolabis taoensis* (0.31-0.33; N = 2); and *Lambdopsalis* (0.15; N = 1). The P^4/M^1 ratio is unavailable for other taxa examined.

RESULTS

Hypothesized phylogenetic relationships among the taxa considered are illustrated in Figure 1. Character states pertaining at each node are listed in the caption to Figure 1; in each case, character states are assigned to the highest position on the cladogram that can be justified given available data.

Computer algorithms described in the "Methods" section above resulted in identification of a single most parsimonious cladogram expressing relationships among *Chulsanbaatar vulgaris*, *Sloanbaatar mirabilis*, *Kamptobaatar kuczynskii*, *Catopsalis matthewi*, *Catopsalis catopsaloides*, *Catopsalis joyneri*, *Catopsalis alexanderi*, *Taeniolabis taoensis*, *Taeniolabis* sp., *Sphenopsalis nobilis*, *Prionessus lucifer*, and *Lambdopsalis bulla*. Monophyly of the Taeniolabididae was not fully tested in this study, although our analysis indicates that the family does form a monophyletic group relative to the two outgroup species. The exact positions of *Catopsalis foliatus*, *Catopsalis utahensis*, *Catopsalis fissidens*, and *Catopsalis calgariensis* relative to other taxa could not be established; however, these species appeared above node 'F' and below nodes 'H' and 'I' in all fourteen alternate most parsimonious cladograms generated (differing positions of these four taxa never altered the relative positions of other taxa examined). Accordingly, *Catopsalis foliatus*, *C. utahensis*, *C. fissidens*, and *C. calgariensis* have been assigned to node 'G', as no other positioning could be justified given available data (see Fig. 1). Inability to resolve relationships of these taxa is due primarily to lack of data for these species; the upper dentition of *C. foliatus*, *C. utahensis*, and *C. fissidens* never has been described, and *C. calgariensis* is known only from isolated teeth.

Results of this analysis illustrate that the genus *Catopsalis* as currently recognized (Kielan-Jaworowska and Sloan, 1979; Hahn and Hahn, 1983) is a paraphyletic taxon, composed of no fewer than five monophyletic groups. *Catopsalis matthewi*, *C. catopsaloides*, *C. joyneri*, and *C. alexanderi* each constitute a separate monotypic monophyletic lineage within the Taeniolabididae. The status of *C. foliatus*, *C. utahensis*, *C. fissidens*, and *C. calgariensis* cannot be resolved given available data. At best, these four species may constitute a single monophyletic group and, at worst, each may represent another monotypic monophyletic branch within the Taeniolabididae.

Among other taxa considered, *Kamptobaatar, Sphenopsalis, Prionessus,* and *Lambdopsalis* are *(a priori)* monophyletic genera, because they are monotypic. Three additional significant monophyletic groups are recognized on the basis of this analysis: (1) *Taeniolabis,* which appears to be a monophyletic genus composed of two species; (2) *Prionessus* and *Lambdopsalis,* which together form another monophyletic group; and (3) *Sphenopsalis + Prionessus + Lambdopsalis,* together forming yet another clade. *Catopsalis foliatus, C. utahensis, C. fissidens,* and *C. calgariensis* may prove to be individually more or less closely related to one or the other of these two monophyletic groups; however, such relationships cannot yet be resolved.

CONCLUSIONS

The genus *Catopsalis* as currently recognized is a paraphyletic taxon; *Kamptobaatar, Taeniolabis, Sphenopsalis, Prionessus,* and *Lambdopsalis* all are monophyletic. In keeping with recognition that only monophyletic taxa are natural evolutionary units (Wiley, 1981), we recommend that species now referred to *Catopsalis* be split into genera that are at least potentially monophyletic. *C. foliatus, C. utahensis, C. fissidens,* and *C. calgariensis* form a potentially monophyletic group containing the type species of *Catopsalis* (*C. foliatus* Cope, 1882a); accordingly, these species reasonably may be retained in the genus *Catopsalis* until their relationships to other members of the family become clear. The generic epithet *Djadochtatherium* Simpson is available for *C. matthewi;* we recommend that the species be returned to that genus. No available names exist for *Catopsalis catopsaloides, C. joyneri,* or *C. alexanderi.* Because autopomorphies (uniquely derived character states) of each taxon have not been examined in this study, new generic epithets for these species will not be proposed, pending further study.

BIOGEOGRAPHIC IMPLICATIONS

Kielan-Jaworowska (1974a,b) was the first to hypothesize a taeniolabidid dispersal from Asia to North America during the Late Cretaceous. Based on morphological similarities, Kielan-Jaworowska (1974a,b) suggested that *Djadochtatherium catopsaloides* (as it was then known) was either ancestral to or close to the form which gave rise to *Catopsalis joyneri* in the Late Maastrichtian of North America. This relationship was formalized by Kielan-Jaworowska and Sloan (1979), who synonymized *Djadochtatherium* with *Catopsalis* and hypothesized a Late Campanian or Early Maastrichtian dispersal event from Asia to North America for the stock ancestral to *C. joyneri.*

Among taxa considered in this study, the five morphologically most primitive taxa (*Chulsanbaatar* and *Sloanbaatar* [both outgroups], and *Kamptobaatar, Djadochtatherium* and *Catopsalis catopsaloides* [taeniolabidids]) are exclusively Asian in distribution. This suggests an Asiatic origin for the Taeniolabididae.

All of these taxa are reputed to be Santonian or Campanian in age (all the European stages designated to Mongolian Late Cretaceous formations are disputable, see Lillegraven and McKenna, 1986), presumably somewhat older than the Lancian taxon *Catopsalis joyneri,* which morphologically the most primitive and oldest of the North American taeniolabidids. Accordingly, the current analysis supports Kielan-Jaworowska's (1974a,b) and Kielan-Jaworowska and Sloan's (1979) hypothesis of an Asian origin for the Taeniolabididae, with a dispersal from Asia to North America late in the Late Cretaceous.

One additional event, however, is necessary to account for the geographical distribution of taxa, given relationships illustrated in Figure 1. Of the Asian taxa examined, *Sphenopsalis, Prionessus,* and *Lambdopsalis* form a monophyletic group whose ancestry apparently is based within North American taeniolabidids. Two major explanations are possible for this distribution. First, that the stock ancestral to *Sphenopsalis + Prionessus + Lambdopsalis* evolved in North America and then dispersed to Asia before generic differentiation. A second possible explanation is that lineages ancestral to *Catopsalis joyneri, C. alexanderi,* and *C. foliatus + C. fissidens + C. utahensis + C. calgariensis + Taeniolabis* independently dispersed from Asia to North America (a minimum of three events, probably more), leaving the ancestral stock of *Sphenopsalis + Prionessus + Lambdopsalis* in Asia throughout its history. For the time being (given that this question cannot be resolved with available evidence, we favor the explanation requiring the fewest dispersal events, specifically the first hypothesis (*i. e.,* that ancestors of the *Sphenopsalis + Prionessus + Lambdopsalis* lineage originated in North America and then dispersed from North America back to Asia). This event probably took place in the middle to late Paleocene, after the North American taeniolabidid lineages were well established and before differentiation of *Sphenopsalis, Lambdopsalis,* and *Prionessus* from their ancestral stock.

ACKNOWLEDGMENTS

Thanks go to W. A. Clemens, J .A. Lillegraven, and M. C. McKenna for allowing access to specimens in their care. Thanks also to J. Felsenstein for developing and making available the PHYLIP program package, and to D. Lindberg for assistance in adapting these programs for the Morrow MD-3. W. A. Clemens, J. A. Lillegraven, and T. Rowe provided discussions and criticism crucial to development of this study, although their agreement with the ideas presented is not necessarily implied. Thanks also to D. W. Krause for his review, which stimulated important modifications in the manuscript. This work was partially supported by a University of California Graduate Opportunity Fellowship (1982-83) and a National Science Foundation Graduate Fellowship (1983-86) to Simmons, National Science Foundation Grant BSR-81-19217 to W. A. Clemens, and National Science Foundation Grant EAR 82-05211 to J. A. Lillegraven.

REFERENCES CITED

Clemens, W. A., and Kielan-Jaworowska, Z., 1979, Multituberculata, *in* J. A. Lillegraven, Z. Kielan-Jaworowska, and W. A. Clemens, eds, Mesozoic mammals: the first two-thirds of mammalian history: Berkeley, University of California Press, p. 99-149.

Cope, E. D., 1882a, A second genus of Eocene Plagiaulacidae: American Naturalist, v. 16, p. 416-417.

_____ 1882b, A new genus of Taeniodonta: *ibid.*, v. 16, p. 604-605.

_____ 1884a, Second addition to the knowledge of the Puerco epoch: American Philosophical Society, Proceedings, v. 21, p. 309-324.

_____ 1884b, The Tertiary Marsupialia: American Naturalist, v. 18, p. 686-697.

Felsenstein, J., 1973, Maximum likelihood and minimum-steps methods for estimating evolutionary trees from data on discrete characters: Systematic Zoology, v. 22, p. 240-249.

_____ 1978, Cases in which parsimony and compatibility methods will be positively misleading: *ibid.*, v. 27, p. 401-410.

_____ 1979, Alternate methods of phylogenetic inference and their interrelationship: *ibid.*, v. 28, p. 49-62.

_____ 1981, A likelihood approach to character weighting and what it tells us about parsimony and compatibility: Biological Journal of the Linnean Society, v. 16, p. 183-196.

Gazin, C. L., 1939, A further contribution to the Dragon Paleocene fauna of central Utah: Journal of the Washington Academy of Science, v. 29, p. 273-286.

Granger, W., and Simpson, G. G., 1929, The Tertiary Multituberculata: American Museum of Natural History, Bulletin, v. 56, p. 601-676.

Hahn, G., and Hahn, R., 1983, Multituberculata, *in* F. Westphal, ed., Fossilium Catalogus, I: Animalia: Amsterdam, Kugler Publications, v. 127, p. 1-409.

Kielan-Jaworowska, Z., 1970, New Upper Cretaceous multituberculate genera from Bayn Dzak, Gobi Desert, *in* Results of the Polish-Mongolian Palaeontological Expedition, Part II: Palaeontologica Polonica, v. 21, p. 35-49.

_____ 1971, Skull structure and affinities of the Multituberculata, *in ibid.*, Part II: *ibid.*, v. 25, p. 5-41.

_____ 1974a, Multituberculate succession in the Late Cretaceous of the Gobi Desert, Mongolia, *in ibid.*, Part V: *ibid.*, v. 30, p. 23-44.

_____ 1974b, Migrations of the Multituberculata and the Late Cretaceous connections between Asia and North America: South African Museum, Annals, v. 64, p. 231-243.

_____ 1980, Absence of ptilodontoidean multituberculates from Asia and its palaeogeographic implications: Lethaia, v. 13, p. 169-173.

Kielan-Jaworowska, Z., and Sloan, R. E., 1979, *Catopsalis* (Multituberculata) from Asia and North America and the problem of taeniolabidid dispersal in the Late Cretaceous: Acta Palaeontologica Polonica, v. 24, p. 187-197.

Kirsch, J. A. W., 1982, The builder and the bricks: toward a philosophy of characters, *in* Archer, M., ed., Carnivorous marsupials, Vol. 2: Mosman, Royal Zoological Society of New South Wales, p. 587-594.

Lillegraven, J. A., and McKenna, M. C., 1986, Fossil mammals from the "Mesaverde" Formation (Late Cretaceous, Judithian) of the Bighorn and Wind River Basins, Wyoming, with definitions of Late Cretaceous North American Land-Mammal "Ages": American Museum Novitates, no. 2840, 68 p.

Matthew, W. D., and Granger, W., 1925, Fauna and correlation of the Gashato Formation of Mongolia: American Museum Novitates, no. 189, 12 p.

Matthew, W. D., Granger, W., and Simpson, G. G., 1928, Paleocene multituberculates from Mongolia: *ibid.*, no. 331, 4 p.

Miao, D., 1986, Dental anatomy and ontogeny of *Lambdopsalis bulla* (Mammalia, Multituberculata): Contributions to Geology, University of Wyoming, v. 24, p. 65-76.

Middleton, M. D., 1982, A new species and additional material of *Catopsalis* (Mammalia, Multituberculata) from the western interior of North America: Journal of Paleontology, v. 56, p. 1197-1206.

Nelson, G., and Platnick, N., 1981, Systematics and biogeography: cladistics and vicariance: New York, Columbia University Press, 567 p.

Russell, L. S., 1926, A new species of the genus *Catopsalis* Cope from the Paskapoo Formation of Alberta: American Journal of Science, v. 12, p. 230-234.

Simpson, G. G., 1925, A Mesozoic mammal skull from Mongolia: American Museum Novitates, no. 201, 11 p.

_____ 1929, American Mesozoic mammals: Memoirs of the Peabody Museum, v. 3, pt. 1, p. 1-171.

_____ 1937, Skull structure of the Multituberculata: American Museum of Natural History, Bulletin, v. 73, p. 727-763.

Sloan, R. E., and Van Valen, L., 1965, Cretaceous mammals from Montana: Science, v. 148, p. 220-227.

Wiley, E. O., 1981, The theory and practice of phylogenetic systematics: New York, Wiley-Interscience, 439 p.

Zhou, M., and Qi, T., 1978, Paleocene mammalian fossils from the Nomogen Formation of Inner Mongolia: Vertebrata PalAsiatica, v. 16, p. 77-85.

MANUSCRIPT RECEIVED OCTOBER 11, 1985
REVISED MANUSCRIPT RECEIVED JUNE 3, 1986
MANUSCRIPT ACCEPTED JUNE 5, 1986

Competitive exclusion and taxonomic displacement in the fossil record: the case of rodents and multituberculates in North America

DAVID W. KRAUSE   *Department of Anatomical Sciences, Health Sciences Center, State University of New York, Stony Brook, New York 11794*

ABSTRACT

Competitive displacement of one taxon by another in the fossil record may be indicated when: (1) an inverse correlation in diversity and, particularly, relative abundance can be demonstrated between the two groups through time; (2) aspects of their paleobiology suggest utilization of common resources; and (3) it can be shown that the two taxa evolved in allopatry prior to their sympatric association. Data from recent collections of Paleocene and Eocene mammals in the Western Interior of North America show marked inverse correlations both of generic diversity and relative abundance between multituberculates and rodents. The largest diminution in multituberculate diversity occurred in the latest Paleocene, near the Tiffanian-Clarkforkian boundary, not in the early Eocene as suggested previously. Reconstruction of diets, diel activity patterns, locomotor habits, and body sizes of multituberculates and rodents suggests that both groups potentially utilized similar resources. The hypothesis that competitive exclusion may have played a role in the decline of multituberculates is strengthened by recent evidence that rodents evolved in Asia, immigrating to North America in latest Paleocene time. Evidence in support of alternative hypotheses employed to account for the decline and eventual extinction of multituberculates is wanting.

INTRODUCTION

Is there evidence for large-scale taxonomic displacement events in the fossil record? Or are the patterns interpreted as evidence of displacement events merely indicative of replacement? Kitchell (1985, p. 97) defined taxonomic replacement as "opportunistic diversification by surviving species following an extinction event," whereas taxonomic displacement most frequently is "linked causally with competitive interactions." Some of the most commonly cited examples of large-scale taxonomic displacement in the fossil record include the competitive exclusion of brachiopods by bivalves, mammal-like reptiles by dinosaurs, perissodactyls by artiodactyls, and South American mammals by North American mammals (Benton, 1983b, and references therein). However, recent analyses of each of these cases have led to an interpretation of replacement rather than displacement; the evidence for competitive interactions was found wanting, in most cases because of the role externally-driven mass extinction events seem to have played (e. g., Gould and Calloway, 1980; Benton, 1983a; Cifelli, 1981, 1985; see also Benton, 1985; Kitchell, 1985).

The sudden decline of multituberculates both in numbers and diversity in the early Cenozoic of North America is one of the more striking events in the history of the Mammalia, one that has long piqued the curiosity of paleobiologists. This paper presents evidence that North American multituberculates may have been actively displaced by rodents, supporting an interpretation of competitive exclusion in the fossil record. The evidence is derived from diversity and relative abundance data based on recent collections of Late Cretaceous and early Cenozoic mammals from the Western Interior, inferences of paleobiological attributes of the purported competitors, and paleobiogeographic data concerning centers of origin and intercontinental dispersal routes. Evidence in support of alternative hypotheses to account for the decline and eventual extinction of multituberculates, namely climatic fluctuations and predation pressure, is evaluated and found insufficient.

Multituberculates have one of the longest known histories (if not *the* longest) among mammalian orders. Their geological range extends from at least the Late Jurassic (and probably the Late Triassic) to the early Oligocene. Members of the order are abundantly represented in every well-sampled Cretaceous and Paleocene local fauna in western North America. The Multituberculata consist of three major suborders: Plagiaulacoidea, Ptilodontoidea, and Taeniolabidoidea. Plagiaulacoids ranged from Late Jurassic to Early Cretaceous time, are characterized by a greater number of incisors and premolars than later forms, and are generally thought to include, or be representative of, the stock that gave rise to ptilodontoids and taeniolabidoids (e. g., Clemens and Kielan-Jaworowska, 1979). Ptilodontoids first occur in the Late Cretaceous and extend into the early Oligocene of North America. They possess a pair of long, slender, procumbent lower incisors, the crowns of which are completely covered with enamel. The last lower premolar of ptilodontoids is much enlarged to form a slicing blade. Taeniolabidoids also occur first in the Late Cretaceous, and extend to the early Eocene. They are united as a group by the common possession of an enlarged pair of lower incisors bearing a ventrolabially restricted band of enamel. Unlike the condition in ptilodontoids, the last lower premolar of taeniolabidoids frequently is reduced in size. A fourth suborder, the poorly known Haramiyoidea from the Late Triassic and Early Jurassic (possibly Middle Jurassic) of Europe, is also frequently included in the Multituberculata.

PREVIOUS WORK

Historically, decline and eventual extinction of the Multituberculata has been attributed principally to competition with rodents. In addition, climatic fluctuations as well as predation pressure owing to the radiation of either mammalian or avian carnivores have been invoked as additional or alternative factors. Little attention has been paid, however, to distinction between the major decline of the Multituberculata as opposed to their eventual extinction; patterns in the fossil record rarely are sufficient to infer causality for termination of the last members of a clade (*e. g.*, Stanley, 1986).

Matthew (1897, p. 261) was the first to advance an hypothesis to account for final extinction of the multituberculates. Without further comment, he postulated that multituberculates died out in the middle Paleocene and were immediately replaced by rodents. As Landry (1965) has already noted, this suggestion was based, in part, on the mistaken assignment of *Mixodectes*, a proteutherian, to the Rodentia. Subsequent collections, primarily from Paleocene and Eocene deposits of the Bighorn Basin of northwestern Wyoming, persuaded Jepsen (1949) to speculate that multituberculates were the "economic ancestors" of rodents. He noted superficial resemblances between rodents and multituberculates in general morphology and inferred habitus and, more specifically, in gross structure of the lower incisors. These resemblances, Jepsen reasoned, placed multituberculates in direct competition with rodents. Rodents diversified while multituberculates declined in diversity for at least two possible reasons: (1) the lower incisors of ptilodontoid multituberculates were not ever-growing and self-sharpening as they are in rodents; and (2) multituberculates "had but a single set of teeth." [Szalay (1965), and others since, have found evidence for two sets of teeth in multituberculates.] Jepsen (1949), however, also compiled the Paleocene and Eocene diversities of multituberculates and rodents and documented a striking correlation; soon after rodents appeared in the fossil record in the late Paleocene of North America, multituberculates became extinct. In particular, Jepsen noted complete absence of multituberculates and presence of the first rodent in North America (*Acritoparamys atavus*—at that time assigned to the genus *Paramys*), at the Bear Creek locality of southern Montana, then thought to be Tiffanian (late Paleocene) in age. The immediacy of replacement of multituberculates by rodents implied a causal relationship to Jepsen; that is, that competition was the driving force (*i. e.*, that competitive displacement rather than opportunistic replacement was involved). Jepsen's views appear to have gained rapid general acceptance (*e. g.*, Simpson, 1949, 1953; Wilson, 1951; Wood, 1962).

Romer (1966, p. 200), for example, stated that multituberculates were "wiped out, presumably by the competition of advanced placentals—notably the rodents, which usurped the ecological niche long occupied by the Multituberculata."

However, there have been several critics of the competition hypothesis. McKenna (1961) noted that, in individual early Wasatchian (early Eocene) quarry samples, when multituberculates are common, rodents are not, and vice versa. McKenna preferred the interpretation that these differential abundances do not reflect differential success in competitive interactions but, instead, reflect two different environments, one inhabited by multituberculates, the other by rodents.

Landry (1965) also attempted to counter the competition hypothesis. He argued that multituberculates could not have been involved in direct competition with rodents because ptilodontoids, and possibly even taeniolabidoids, possessed incisors that were not adapted for gnawing. In support of this he noted that the rodent *Acritoparamys atavus* was much too small to have competed with the much larger *Ptilodus,* a common Paleocene multituberculate. Instead, Landry (1967) speculated that multituberculates (and plesiadapiform primates) were hunted to extinction by raptorial birds. To account for differential predation pressure, Landry (1967) proposed that multituberculates were primarily diurnal while early rodents were nocturnal.

Van Valen and Sloan (1966) elaborated considerably upon the competition hypothesis formulated by Jepsen (1949). They attributed the decline and extinction of multituberculates to competition with a series of herbivorous placental mammals: condylarths in the early Paleocene, plesiadapiform primates in the middle Paleocene, and rodents from the latest Paleocene onwards. This series of purported competitors was envisioned as causing a gradual and continuous decline in multituberculate abundance from a maximum in the Late Cretaceous until their demise (see Fig. 2 in Van Valen and Sloan, 1966), which is now known to have been as late as the early Oligocene. Van Valen and Sloan also showed that multituberculate diversity was highest in the Torrejonian (middle Paleocene), dropped slightly through the Tiffanian (late Paleocene) and Clarkforkian (latest Paleocene-earliest Eocene), and then declined sharply in the Wasatchian (early Eocene) (see Fig. 3 in Van Valen and Sloan, 1966). Using samples from various Wasatchian sites comprising the Four Mile local fauna, negative correlations in abundances between multituberculates and rodents and between multituberculates and plesiadapiform primates were also demonstrated by Van Valen and Sloan (but see McKenna, 1961). From this and other evidence, Van Valen and Sloan concluded that the decline in multituberculates probably involved both immigration and more or less sympatric evolution of "superior" competitors, sequentially limiting the resources available to multituberculates. Although fully cognizant of the fact that "competitive inferiority is not really demonstrated by apparent primitiveness in individual characters," Van Valen and Sloan (1966, p. 276-277; see also Simpson,

1953, p. 300) drew attention to several morphological features of multituberculates deemed potentially inferior to those of therians. Their list included primitive features of the inner ear, brain, pectoral and pelvic girdles, and bones of the forelimb. To this list Hopson (1967) added several more features, including inferred aspects of physiology, regarded by him as indicating inferiority in "general biological efficiency" (*i. e.,* thermoregulation, reproduction, behavior, locomotion). A recent analysis of the postcranial elements of North American multituberculates, however, led Krause and Jenkins (1983, p. 244) to conclude that there is "no evidence of features that might be considered significantly inferior to those of eutherians. Skeletal traits that are clearly divergent in multituberculates and eutherians cannot be assessed in terms of comparative locomotor ability, particularly because many such multituberculate features have no analogs among living mammals."

Finally, Ostrander (1984) reviewed some of the evidence concerning the decline and extinction of multituberculates. Like Landry (1965, 1967), he argued that rodents probably did not account for the eventual demise of the Multituberculata as a group, because the last known multituberculates, in the early Oligocene, postdate the first appearance of rodents in the North American fossil record by at least 15 million years. Ostrander (1984, p. 77) suggested that final extinction of multituberculates "was more likely caused by a combination of changes in climate, vegetation, and predatory pressure." In marshalling support for his argument, Ostrander (*ibid.,* p. 78) pointed to the change from "predominantly a forested ecosystem to that of predominantly a savanna ecosystem" from the late Eocene to the early Oligocene and its effect on other mammalian taxa. Ostrander also noted the diversification of miacid carnivorans during the early Cenozoic as another potential factor contributing to the demise of multituberculates.

THE COMPETITION HYPOTHESIS

Introduction

Hypotheses of competitive exclusion, even in modern ecological studies, are difficult to test. Recent discussions include Connell (1980, 1983), Connor and Simberloff (1979, 1984, 1986), Gilpin and Diamond (1984), Grant and Schluter (1984), Schoener (1983), Simberloff (1983), Strong (1984), Strong and others (1979), and Toft (1985). Central questions address: (1) whether patterns observed differ significantly from those expected under a random hypothesis; and (2) whether mechanisms other than competition strongly influence the structuring of communities. One of the most oft-cited alternative potential mechanisms is predation (*e. g.,* Connell, 1975; Toft, 1985).

Similar questions have been discussed in paleobiology as well (see review by Kitchell, 1985). Evidence for competition in the fossil record is restricted by inability to observe directly interactions between individuals or to experimentally manipulate populations and resources. Nevertheless, the fossil record, while restrictive, yields a distinct perspective not available to neo-ecologists, the dimension of time. The paleobiologist is permitted to observe, in temporal sequence, changes in relative abundance and diversity of two groups that may have been in coexistence over periods of millions of years. Of course inverse correlations of relative abundance or diversity between two groups through time do not "prove" competitive interaction. However, demonstration of such long-term patterns are essential to establishment of an hypothesis of competitive exclusion (Connell, 1980).

Incompleteness of the fossil record makes it difficult to differentiate reliably between patterns resulting from competitive displacement and those resulting from opportunistic replacement by another, adaptively similar taxon. Benton (1983a) recently outlined patterns that are claimed to differentiate between models of competitive displacement and opportunistic replacement in the fossil record. According to Benton (*ibid.,* p. 42), the following pattern pertains to instances of interspecific competition between groups A and B (and displacement of A by B):

"(1) A will tend to decrease in abundance, and B will increase in abundance over time;
(2) the rate of replacement ['displacement' in the terminology of this paper] should be gradual (in paleontological terms, this would imply a time span of more than one million years, say, and often more than 10 to 20 million years);
(3) A and B will be found together, and either could be dominant in any particular formation; and
(4) the replacement [i.e., 'displacement'] will not necessarily be associated with climatic or floral change."

Alternatively, opportunistic replacement of group A by group B is suggested by the following pattern:

"(1) B will appear or radiate only *after* the extinction of A;
(2) the rate of replacement should be rapid (in paleontological terms, this would imply a time span of less than one million years, and possibly a few thousands or tens of thousands of years, if such stratigraphic accuracy were possible);
(3) A and B will not be found together, or B may be unobtrusively present when A is dominant; and
(4) the replacement will be associated with climatic or floral changes."

In addition to consideration of diversities and relative abundances, evaluation of an hypothesis of competitive exclusion should entail consideration of the biology of purported competitors and purported resources (*e. g.,* James and Boecklen, 1984; Grant and Schluter, 1984) in an attempt to address the question: did taxa involved utilize the same resources (*e.g.,* space, food) at the same time in the same communities? As partly noted above, fossil evidence, because of its incomplete nature, cannot provide a conclusive answer to this question. For instance, similar morphological features often indicate similar food use or habits, but this is not always the case (Bock, 1977). Likewise, dissimilar features often indicate dissimilar ways of life, but again, exceptions are known to occur. Animals as different as insects, birds, and mammals have been shown to compete for similar food resources (Brown, 1978, and references therein; Davidson and others, 1984). To go beyond the inverse correlations of relative abundance or diversity observed in the fossil record one must *assume* that, among potential competitors, the more similar the morphology, the more similar the way of life and therefore, the greater the likelihood of competitive interaction (*e. g.*, Connor and Simberloff, 1986). Given this assumption, one can attempt to withdraw support from hypotheses of competition for similar resources by finding dissimilar adaptations. Such major differences in adaptation should reflect a substantial partitioning of resources. For instance, determination of specialized adaptations for browsing in one group of mammals (*e. g.*, mastodons) would eliminate it from consideration as a potential competitor for food resources of another group known to possess grazing adaptations (*e. g.*, mammoths). This, of course, does not preclude possibility that the taxa involved competed along some other axis; this must be determined independently.

Habitat, food, and diel activity patterns are the three major resource categories, or niche axes, traditionally examined by neo-ecologists (*e. g.*, Pianka, 1975), and some information concerning these categories can be obtained from the mammalian fossil record. Emphasis among ecologists is to use these categories to demonstrate resource partitioning, or niche shifts between potentially competing species. Body size often is employed as a measure of niche partitioning (*e. g.*, Bowers and Brown, 1982). Fortunately, body size is a parameter that can be estimated from dentitions of fossil mammals (*e. g.*, Gingerich and others, 1982).

Attempting to determine the actual cause for success and eventual domination of one group over another is an extremely speculative venture. Frequently, one or another character in the unsuccessful group is labelled as primitive, and therefore as potentially inferior; this potential inferiority is then used to explain the decline and/or extinction of the less successful group. As demonstrated in the historical review above, more attention has been paid to showing why multituberculates might have been competitively inferior than to why, adaptively, multituberculates and rodents might have competed for the same resources. Tests of such "hypotheses" of adaptive inferiority are difficult, if not impossible, without more knowledge about the biology of the organisms involved than is usually available in the fossil record. As emphasized by Benton (1983a, p. 43): "Scenarios that attempt to explain complex replacement [and displacement] processes by reference to single characters are likely to be gross oversimplifications." Simpson (1953, p. 300) was even more direct: "A frequent statement regarding broad competitive replacements ["displacements" in the terminology of this paper] is that the replacers ["displacers"] are more 'progressive' and the replaced ["displaced"] more 'primitive,' which even if it means something, certainly explains nothing."

Finally, geographical considerations can play a major role in evaluating hypotheses of competitive exclusion in the fossil record. To again quote Simpson (1953, p. 208): "occupation by a different group of a zone already occupied usually, perhaps always, involves changes in distribution; a geographically invading group, if the invasion is successful, ousts one already established in a region. . . . In most cases it was the invading group that survived, as would be expected because the ability to invade in the face of occupation implies probable competitive superiority" (but only under conditions of ecological saturation for a particular resource). It is appropriate, therefore, to determine likelihood that the purportedly successful competitor evolved elsewhere (geographically or ecologically) and subsequently migrated into the area (or niche) in which the ultimately unsuccessful group was already established.

The geographic aspect of competition is implicit in the work of field ecologists who introduce a species into an area occupied by a potential competitor and determine effects of the introduction by measuring various niche parameters (for summaries of field experiments on interspecific competition see: Connell, 1983; and Schoener, 1983). Given the nature of the fossil record, however, it is much more probable that large-scale migration patterns, rather than ecological shifts, would be detected.

In keeping with these theoretical considerations, three major sources of evidence will be considered in testing the competitive exclusion hypothesis for decline of multituberculates: (1) patterns of diversity and relative abundance in the fossil record; (2) inferred resource use; and (3) paleobiogeography. The last two aspects have received little or no detailed attention in previous literature on the decline and extinction of multituberculates.

Patterns of Diversity and Relative Abundance

Diversity

The pattern of inverse correlation in diversity between rodents and multituberculates established by Jepsen (1949) and more fully documented by Van Valen and Sloan (1966) can be refined further in light of new data. At the time of Jepsen's writing, the latest known multituberculates were early Wasatchian (early Eocene) in age. Multituberculates have since been discovered at late Wasatchian (Stucky and Krishtalka, 1982), Bridgerian (middle Eocene—Bown, 1982; Stucky, 1984a), Uintan and Duchesnean (late Eocene—Robinson and others, 1964; Sloan, 1966; Black, 1967; Krishtalka and Black, 1975), and Chadronian (Ostrander and others, 1979; Krishtalka and others, 1982; Ostrander, 1984) horizons. All post-early Wasatchian specimens have been referred to the neoplagiaulacid genus *Ectypodus,* with exception of a fragmentary Bridgerian tooth that, if from a multituberculate, may belong to *Neoliotomus* (see Bown, 1982). In addition, a specimen of a large multituberculate has been discovered recently in the late Wasatchian of the Wind River Basin, Wyoming (R. Stucky, *personal communication,* 1985).

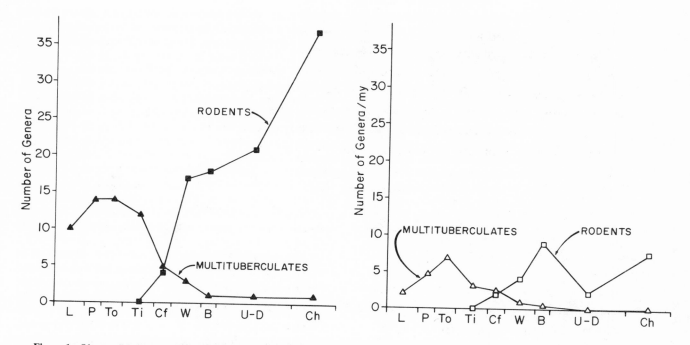

Figure 1. Observed (left) and standardized (right) generic diversities of multituberculates and rodents from Late Cretaceous to early Oligocene time. Multituberculates are indicated by triangles, rodents by squares. Land-mammal ages are (from oldest to youngest) as follows: L, Lancian; P, Puercan; To, Torrejonian; Ti, Tiffanian; Cf, Clarkforkian; W, Wasatchian; B, Bridgerian; U-D, Uintan-Duchesnean; and Ch, Chadronian.

Observed generic diversities of multituberculates and rodents from the Late Cretaceous to the early Oligocene, as well as diversities that have been standardized for time interval, are plotted in Figure 1. Because the land-mammal ages are not all of equal duration, the longer intervals are more likely to sample more genera than shorter ones. Standardized generic diversities have therefore been calculated by dividing the number of genera known for each land-mammal age by the duration of that age (in millions of years), thus yielding the number of genera per million years. Durations of the land-mammal ages correspond to estimates provided by Clemens and others (1979) for the Late Cretaceous, Gingerich (1983) for the Paleocene, and Savage and Russell (1983) for the Eocene and early Oligocene. It must be emphasized that durations of Late Cretaceous and Paleogene land-mammal ages are only rough estimates, since few radiometric dates are available for nonmarine sediments of these ages. Standardized diversity values may, therefore, be even less precise than observed diversity values. Observed generic diversity data are replotted in clade diversity diagrams in Figure 2, which shows, in addition, the proportion of the rodent generic diversity contributed by the Paramyidae. For both figures, generic ranges for multituberculates were derived primarily from Hahn and Hahn (1983) and Archibald and others (in press), with additions from Johnston and Fox (1984) and Krause (1982a, c; in press). Ranges for rodents were derived primarily from Black and Sutton (1984) and Wood (1980), with additions from Korth (1984, 1985) and Stucky (1984b). Although a tabulation of species, rather than generic, diversity would be preferable (e.g., Signor,

1985), this was not done; many species are known only from single localities, and several localities cannot be assigned to a specific land-mammal age (e. g., see Stucky, 1984b). This problem is mostly avoided by a consideration of generic ranges. As in most compendia of this sort, subjective decisions had to be made concerning taxonomic allocations; in almost all cases, the most recent allocation was accepted.

Both the observed and standardized plots indicate basically the same pattern from the Tiffanian to the Bridgerian, the interval of time most relevant to the present discussion. Van Valen and Sloan (1966), for lack of evidence to the contrary, suggested that the major decline in multituberculate diversity occurred in the Wasatchian. More recent collections indicate that dramatic reduction in multituberculate diversity occurred somewhat earlier, near the Tiffanian-Clarkforkian boundary (Krause, 1980, 1982a). Paramyid rodents appear in the North American fossil record in the Clarkforkian (Rose, 1980, 1981), and in some diversity (a minimum of four genera and five species; Korth, 1984). Multituberculates continue to decrease in generic diversity through the Clarkforkian and Wasatchian, and into the Bridgerian, during the time that rodents (including the Paramyidae) undergo an adaptive radiation. At every successive land-mammal age in which multituberculates and rodents co-occur, where there is an increase in observed rodent diversity there is a corresponding decrease, or equilibrium, in multituberculate diversity. Although the same pattern holds for the standardized diversities, after the Bridgerian there appears to be a significant drop in standardized generic diversity for rodents in the Uintan-Duchesnean. This may, in part,

TABLE 1. RELATIVE ABUNDANCES (PERCENTAGES OF MINIMUM NUMBER OF INDIVIDUALS (MNI) OF MULTITUBERCULATES, CONDYLARTHS, PLESIADAPIFORM PRIMATES, RODENTS, AND CREODONTS AND CARNIVORANS AT VARIOUS LOCALITIES IN THE WESTERN INTERIOR OF NORTH AMERICA SPANNING LATE CRETACEOUS (LANCIAN) THROUGH CHADRONIAN (EARLY OLIGOCENE) TIME. TO MINIMIZE SAMPLING BIASES, ONLY SAMPLES COLLECTED FROM LOCALITIES THAT WERE NOT EXCLUSIVELY SURFACE-COLLECTED AND WITH GREATER THAN 40 MNI WERE USED.

Locality	MULTITUBERCULATA Freq. (MNI)	PLESIADAPIFORM PRIMATES Freq. (MNI)	RODENTIA Freq. (MNI)	CONDYLARTHRA Freq. (MNI)	CREODONTA & CARNIVORA Freq. (MNI)
LANCIAN					
Lance[1]	97/182−.533	0/182−.000	0/182−.000	0/182−.000	0/182−.000
Flat Creek #5[2]	29/60−.483	0/60−.000	0/60−.000	0/60−.000	0/60−.000
Bug Creek Anthills[1]	196/234−.838	0/234−.000	0/234−.000	16/234−.068	0/234−.000
Bug Creek West[3]	90/108−.833	0/108−.000	0/108−.000	8/108−.074	0/108−.000
Harbicht Hill[3]	39/54−.722	0/54−.000	0/54−.000	12/54−.222	0/54−.000
PUERCAN					
Worm Coulee #1[2]	51/86−.593	0/86−.000	0/86−.000	10/86−.116	0/86−.000
Purgatory Hill[4]	20/43−.465	2/43−.047	0/43−.000	12/43−.279	0/43−.000
TORREJONIAN					
Gidley Quarry[5]	80/387−.206	89/387−.230	0/387−.000	106/387−.274	20/387−.052
Rock Bench Quarry[5]	91/497−.183	100/497−.201	0/497−.000	202/497−.406	15/497−.030
TIFFANIAN					
Douglass Quarry[6]	10/56−.179	6/56−.107	0/56−.000	25/56−.446	1/56−.018
Scarritt Quarry[5]	25/97−.257	10/97−.103	0/97−.000	3/97−.031	0/97−.000
Cedar Point Quarry[5]	101/503−.201	180/503−.358	0/503−.000	117/503−.233	18/503−.036
Circle[7]	17/46−.370	12/46−.261	0/46−.000	7/46−.152	0/46−.000
Brisbane[8]	16/54−.296	14/54−.259	0/54−.000	5/54−.093	2/54−.037
Olive[7]	28/75−.373	19/75−.253	0/75−.000	4/75−.053	6/75−.080
Princeton[5]	24/185−.130	50/185−.270	0/185−.000	41/185−.222	8/185−.043
CLARKFORKIAN					
Bear Creek[5]	0/75−.000	14/75−.187	4/75−.053	7/75−.093	2/75−.027
Holly's Microsite[9]	26/113−.230	16/113−.142	22/113−.195	8/113−.071	5/113−.044
Paint Creek[10]	7/46−.152	16/46−.348	0/46−.000	14/46−.304	1/46−.022

TABLE 1. (continued)

Locality	MULTITUBERCULATA Freq. (MNI)	PLESIADAPIFORM PRIMATES Freq. (MNI)	RODENTIA Freq. (MNI)	CONDYLARTHRA Freq. (MNI)	CREODONTA & CARNIVORA Freq. (MNI)
WASATCHIAN					
Bown's 150' level[5]	3/243=.012	24/243=.099	27/243=.111	61/243=.251	19/243=.078
Anthill Quarry[11]	0/63 =.000	7/63 =.111	20/63 =.317	16/63 =.254	4/63 =.063
Despair Quarry[11]	4/125=.032	12/125=.096	28/125=.224	35/125=.280	11/125=.088
West Alheit Quarry[11]	9/112=.080	9/112=.080	13/112=.116	32/112=.286	11/112=.098
East Alheit Quarry[11]	9/108=.083	8/108=.074	14/108=.130	31/108=.287	6/108=.056
Sand Quarry[11]	9/106=.085	8/106=.075	23/106=.217	30/106=.283	8/106=.075
Timberlake Quarry[11]	11/169=.065	11/169=.065	62/169=.367	40/169=.237	4/169=.024
BRIDGERIAN					
B-1 Fault[12]	0/111=.000	0/111=.000	38/111=.342	13/111=.117	3/111=.027
B-3 Hawk[12]	0/41 =.000	0/41 =.000	11/41 =.268	9/41 =.220	0/41 =.000
BS-1 Steele Butte[12]	0/41 =.000	0/41 =.000	1/41 =.024	6/41 =.146	6/41 =.146
BS-2 East Fork R.[12]	0/47 =.000	0/47 =.000	2/47 =.043	10/47 =.213	5/47 =.106
UINTAN-DUCHESNEAN					
Swift Current Creek[13]	3/154=.019	0/154=.000	42/154=.272	11/154=.071	5/154=.032

[1] Estes and Berberian (1970)
[2] Archibald (pers. comm., 1981)
[3] Sloan and Van Valen (1965)
[4] Van Valen and Sloan (1965)
[5] Rose (1981)
[6] Krause and Gingerich (1983)
[7] Wolberg (1978, 1979)
[8] Holtzman (1979)
[9] This paper - Appendix A
[10] This paper - Appendix B
[11] McKenna (1960)
[12] West (1973)
[13] Storer (1984)

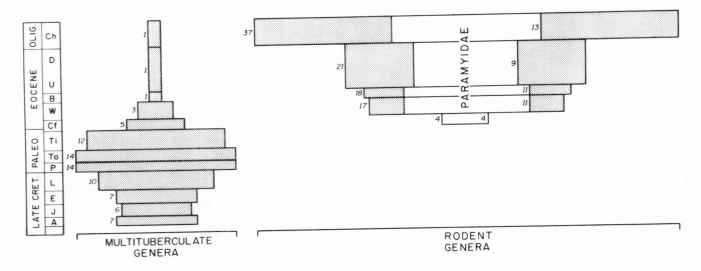

Figure 2. Observed generic diversity of multituberculates and paramyid plus other rodents from Late Cretaceous to early Oligocene time. Number of genera in each group in each land-mammal age is indicated opposite each bar. Paramyidae are indicated as subset (unstippled) of all rodents. Abbreviations for land-mammal ages as in Figure 2, with addition of A, Aquilan; J, Judithian; and E, Edmontonian for Late Cretaceous.

reflect biochronological problems attendant to definition of the Duchesnean (the most poorly defined land-mammal age for the entire Cenozoic period—see review in Savage and Russell, 1983), the different communities sampled in the Bridgerian (Black, 1967), or to poor sampling of the entire middle and late Eocene. In any case, rodents were much more diverse than multituberculates from Wasatchian time onwards.

Interestingly, taeniolabidoid multituberculates, the group of multituberculates most similar to rodents (at least in incisor structure), became extinct almost immediately after appearance of rodents in the North American fossil record; ptilodontoids persisted into the early Oligocene. The only taeniolabidoid to survive into the Wasatchain was *Neoliotomus ultimus*, the largest multituberculate present during that interval (Krause, 1982a). It must be noted, however, that taeniolabidoids, in general, were less diverse (and less abundant) in the early Cenozoic of North America than were ptilodontoids.

Of critical importance in past discussions of the decline and extinction of multituberculates has been the absence of multituberculates in the Bear Creek local fauna of southern Montana. The significance of the locality was thought to lie in the fact that it was the only North American Paleocene locality in which multituberculates were absent and that it contained the first record of rodents *(Acritoparamys atavus)* on this continent. While long thought to be Tiffanian (late Paleocene) in age (e. g., Simpson, 1929; Jepsen, 1937), typical Clarkforkian mammals have been identified in the fauna, and now it is generally regarded of Clarkforkian age (Van Valen and Sloan, 1966; Sloan, 1969; Rose, 1975, 1977, 1981; Gingerich, 1976; but see Korth, 1984). Absence of multituberculates at the Bear Creek locality is therefore not surprising, in that multituberculates were not common during the Clarkforkian and that the locality has not yielded a large sample of mammals (Krause, 1980).

Relative Abundance

Of potentially more relevance to studies of competition are relative abundances of individuals. Analyses of diversity alone do not emphasize that rare taxa should not be accorded the same level of competitive impact as common ones. Persistently rare taxa cannot dominate common ones in competition (Van Valen and Sloan, 1966). It is therefore of interest to compare the percent frequency of individual taxa as a test of Van Valen and Sloan's (1966) prediction that additional data should corroborate the inverse correlations of relative abundance between purported competitors.

The method most commonly employed to estimate relative abundances of taxa within Late Cretaceous and Paleogene mammalian assemblages is the minimum number of individuals (MNI); this usually is determined by the frequency of the most common identifiable element of each taxon in the assemblage. Although the MNI method is subject to a number of biases (e. g., Grayson, 1973, 1978; Holtzman, 1979), it is used here primarily because it is historically the most commonly employed method, and a much larger data base of relative abundance information is therefore available for the time period in question than would otherwise be the case. Because of the additional problem of numerous preservational and collecting biases, however, samples to be used in a comparative investigation must be chosen with care. To minimize these biases, I have restricted analysis to large samples (greater than 40 MNI, average of 136 MNI) that have been largely quarried and/or screen-washed (Table 1). Despite this restriction, the number of samples adequate for comparison is more than double the number available when Van Valen and Sloan (1966)

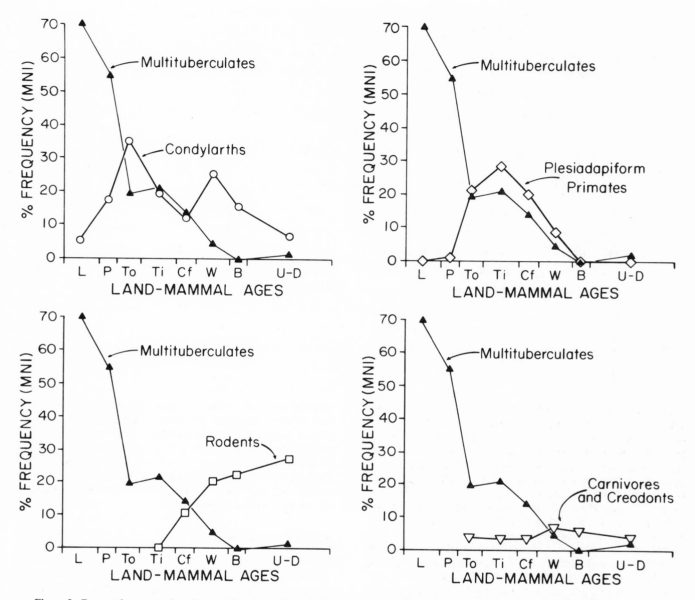

Figure 3. Percent frequency, based on minimum number of individuals (MNI), in total mammalian fauna of multituberculates, condylarths, plesiadapiform primates, and rodents from Late Cretaceous (Lancian - L) to late Eocene (Uintan-Duchesnean - U-D) time. Abundance of multituberculates is compared with that of condylarths in upper left, with that of plesiadapiform primates in upper right, with that of rodents in lower left, and with that of carnivorans and creodonts in lower right. Note particularly that the three major decreases in relative abundance of multituberculates (L-P, P-To, and Ti-W) coincide sequentially with increases in relative abundance of condylarths (L-To), plesiadapiform primates (P-To), and rodents (Ti-W). Abbreviations for land-mammal ages as in Figure 2. Data from Table 2.

performed their study. Most significantly, information on relative abundances of mammals during the Tiffanian and Clarkforkian Land-Mammal Ages was entirely lacking in the earlier analysis. The number of samples now available is, however, still extremely low considering the span of 40 or so million years involved from Lancian to Chadronian horizons. The samples within individual land-mammal ages were pooled (Table 2) to further reduce effects of paleoecological, preservational, and collecting biases. The least known intervals are the Puercan, Clarkforkian, Bridgerian, and Uintan-Duchesnean; no MNI information is available for the Chadronian. With exception of the samples from two Clarkforkian localities in the northern Bighorn Basin (Holly's Microsite and Paint Creek—see faunal lists in Appendices A and B), all samples are from localities that have been analyzed elsewhere (see footnote, Table 1).

Patterns of relative abundance, based on data presented in Table 2 and plotted as percent frequency of minimum number of individuals of the total mammalian fauna in Figure 3, are consistent with Van Valen and Sloan's hypothesis that the pattern of multituberculate

TABLE 2. SUMMARY STATISTICS FOR RELATIVE ABUNDANCES OF MULTITUBERCULATES, CONDYLARTHS (INCLUDING ARCTOCYONIDS BUT EXCLUDING MESONYCHIDS), PLESIADAPIFORM PRIMATES, RODENTS, AND CREODONTS AND CARNIVORES AT VARIOUS LOCALITIES IN THE WESTERN INTERIOR OF NORTH AMERICA SPANNING LANCIAN (LATE CRETACEOUS) THROUGH CHADRONIAN (EARLY OLIGOCENE) TIME. DATA DERIVED FROM TABLE 1.

Land-Mammal Age	MULTITUBERCULATA Freq. (MNI)	PLESIADAPIFORM PRIMATES Freq. (MNI)	RODENTIA Freq. (MNI)	CONDYLARTHRA Freq. (MNI)	CREODONTA & CARNIVORA Freq. (MNI)
LANCIAN	451/638 = .707	0/638 = .000	0/638 = .000	36/638 = .056	0/638 = .000
PUERCAN	71/129 = .550	2/129 = .016	0/129 = .000	22/129 = .171	0/129 = .000
TORREJONIAN	171/884 = .193	189/884 = .214	0/884 = .000	308/884 = .348	35/884 = .040
TIFFANIAN	221/1016 = .218	291/1016 = .286	0/1016 = .000	202/1016 = .199	35/1016 = .034
CLARKFORKIAN	33/234 = .141	46/234 = .197	26/234 = .111	29/234 = .124	8/234 = .034
WASATCHIAN	45/926 = .047	79/926 = .085	187/926 = .202	245/926 = .265	63/926 = .068
BRIDGERIAN	0/240 = .000	0/240 = .000	52/240 = .217	38/240 = .158	14/240 = .058
UINTAN-DUCHESNEAN	3/154 = .019	0/154 = .000	42/154 = .272	11/154 = .071	5/154 = .032

decline was brought about by a series of competitors. There are suggestive inverse correlations of relative abundance between multituberculates and condylarths from the Lancian to the Puercan (and extending into the Torrejonian), between multituberculates and plesiadapiform primates from the Puercan to the Torrejonian, and between multituberculates and rodents from the Tiffanian to the Clarkforkian. The two earlier stages of decreased multituberculate abundance are not, however, accompanied by correlative decreases in diversity. This suggested to Van Valen and Sloan that the adaptive zone of multituberculates became more finely partitioned during this time (see also Van Valen, 1978). This may be so, but it is pertinent to note also that condylarths, as a whole, did not continue to increase in abundance after the Torrejonian and that plesiadapiform primates did not continue to increase in abundance after the Tiffanian. A number of possible reasons exists for this, and it is therefore not clear that the earlier inverse correlations in relative abundance between multituberculates and condylarths and between multituberculates and plesiadapiform primates are due to interspecific competition.

Furthermore, in contrast to Van Valen and Sloan's Figure 2, the new data do not tend to support the perception of a gradual and continuous decline in relative abundance of multituberculates from Late Cretaceous to Oligocene time. Rather, multituberculates, after an initial major decline in numbers through the early Paleocene (possibly in response to the first major radiation of placentals), achieved some degree of constancy in abundance until introduction of rodents in the latest Paleocene. If multituberculates were earlier in competition with condylarths and plesiadapiform primates, as Van Valen and Sloan suggested, these pressures appear to have moderated through the middle and late Paleocene, perhaps to such a degree that the three groups were able to apportion resources. It is also implicit in these data that the range of adaptations of early rodents was sufficiently broad to overlap those *both* of multituberculates and plesiadapiform primates.

Inferred Resource Use

Diet

Unfortunately, precise estimates of dietary habits for either multituberculates or paramyid rodents are unavailable, and only the broadest of comparisons can be made. Nonetheless, superficial similarities in cranial and dental anatomy between multituberculates and early rodents cannot be denied (Fig. 4). As noted by Simpson (1953, p. 300): "The two groups are markedly different in ancestry and many features of anatomy, but strikingly similar in rodentlike adaptation." Both groups have skulls of similar shape, and their anterior dentitions are dominated by enlarged, procumbent upper and lower central incisors. In addition, the anterior teeth, particularly in the lower dentition, are separated from the cheek teeth by a sizeable diastema. The posterior dentitions also are superficially similar in multituberculates and rodents; both taxa have anteroposteriorly aligned grinding surfaces. In rodents the alignment is exploited by means of a proal power stroke of the chewing cycle, whereas in multituberculates the power stroke is palinally directed (Krause, 1982b).

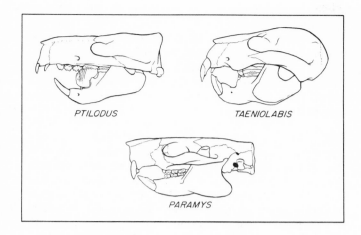

Figure 4. Comparison of cranial and dental morphology of multituberculates *Ptilodus* and *Taeniolabis* with paramyid rodent (*Paramys*). *Ptilodus* redrawn from Simpson (1937) and Krause (1982b); *Taeniolabis* redrawn from Granger and Simpson (1929); and *Paramys* redrawn from Wood (1962).

Probable dietary preferences of ptilodontoid multituberculates were evaluated recently by Krause (1982b); it appears unlikely that all members were strict folivores as was suggested by several earlier workers. By analogy with living arboreal folivorous mammals, the small size of most species of Ptilodontoidea suggests that they could not have subsisted on a totally folivorous diet. The length of striations on the sides of P_4 of *Ptilodus,* one of the largest of the Ptilodontoidea, indicates that large, hard food items, possibly seeds and nuts, were ingested. The presence of both smooth and highly-striated enamel on the same dental wear facets in different individuals of *Ptilodus* suggests a varied diet. The recent suggestion by Clemens and Kielan-Jaworowska (1979) that most multituberculates were probably omnivorous is supported by the above observations.

Taeniolabidoids apparently were even more like rodents in gross dental morphology. Structure of the lower incisor of rodents is convergent upon that seen in taeniolabidoids; in both taxa these teeth are enlarged, laterally compressed, and the enamel is restricted to the ventrolabial surface. This morphology produced, through wear, a self-sharpening tool for slicing or gnawing (however, unlike the condition in rodents, the roots of taeniolabidoid incisors were not evergrowing, but closed late in ontogeny). In addition, in most taeniolabidoids, as in rodents but unlike ptilodontoids, P_4 was not enlarged. Sloan (1979) suggested that while ptilodontoids utilized the enlarged P_4 for slicing, taeniolabidoids instead used their incisors for the same function.

Functional analyses of paramyid dentition and inferences concerning diet in the earliest North American rodents are virtually nonexistent. Rensberger (1982) regarded the paramyid dentition as specifically adapted to seed-eating. Butler (1985, p. 392), in an analysis of homologies of rodent molar cusps and crests, inferred that there was an evolutionary trend "toward increased efficiency in grinding hard or abrasive foods . . . ." While rodents have long been considered basically herbivorous mammals, all major lineages are now known to include omnivorous species (Landry, 1970; McLaughlin, 1984; Carleton and Musser, 1984; Klingener, 1984; Woods, 1984).

In sum, it appears that both multituberculates and rodents may have overlapped to a considerable degree in utilization of food resources, perhaps seeds and nuts. This is hardly a strong argument that the two groups were in competition with one another. Currently available data serve only to show that there is *no* evidence to suggest that rodents and multituberculates had markedly different diets.

Diel Activity Pattern

Multituberculates probably were nocturnal, not diurnal as suggested by Landry (1967). Kay and Cartmill (1977) examined a diversity of therian mammals and observed that, in mammals with skull lengths of less than 75 mm, three parallel but overlapping distributions obtain when skull length is plotted against orbital diameter.

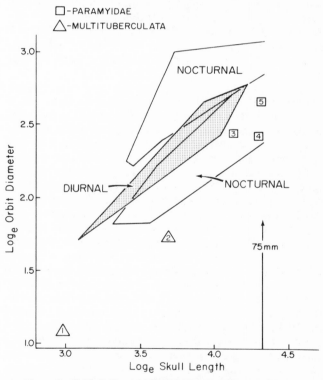

Figure 5. Orbital diameter and skull length (prosthion-inion) of various living (N = 83) and fossil (N = 5) mammals (measurement in mm). The three parallel, but overlapping, large polygons represent living mammals, and include large-eyed nocturnal forms (above), diurnal forms (stippled pattern) with eyes of intermediate size, and small-eyed, olfactory-dominated nocturnal forms (below). Addition of measurements of multituberculates (triangles - 1, *Ectypodus*; 2, *Ptilodus*) and paramyid rodents (squares - 3, *Reithroparamys*; 4, *Thisbemys*; and 5, *Paramys*) indicates that all probably were small-eyed nocturnal forms. Data for living mammals from Kay and Cartmill (1977). Measurements for *Ectypodus* taken from reconstruction in Sloan (1979), for *Ptilodus* from Simpson (1937), and for *Reithroparamys*, *Thisbemys*, and *Paramys* from Wood (1962).

Even with the most generous estimates, these multituberculates had eyes that were smaller than those of small-eyed, nocturnal living mammals (Fig. 5). Interestingly, while the eyes of *Ptilodus* are relatively small for living mammals, the olfactory bulbs are among the largest (Simpson, 1937; Jerison, 1973), suggesting that olfaction was indeed a dominant sense, moreso than vision (the same appears to be true of Asian Cretaceous multituberculates—see Kielan-Jaworowska, 1983). Sloan (1979, p. 493) suggested that taeniolabidoids also were nocturnal, based on observation that their orbits are "huge" ("about 30% of skull length"). However, the only North American taeniolabidoid for which such information is known is *Taeniolabis,* which has a skull length of approximately 150 mm. Sloan's conclusion about the nocturnal habits of taeniolabidoids must be tempered by the realization, as noted by Kay and Cartmill (1977), that interpretations of activity patterns for mammals with skulls greater than 75 mm in length on the basis of orbital diameter are less reliable than for those with skulls that are shorter.

Measurements obtained from reconstructed skulls of three genera of paramyid rodents (*Paramys, Reithroparamys,* and *Thisbemys*) fall entirely within the small-eyed, nocturnal distribution for living mammals (Fig. 5). Available data, therefore, do not support Landry's (1967) suggestion that activity patterns of multituberculates and rodents were appreciably different; both apparently were nocturnal.

Habitat and Locomotion

Functional analysis of the postcranial skeleton of North American multituberculates, particularly the hind foot and tail, indicates that the group probably was largely comprised of arboreal forms (Jenkins and Krause, 1983; Krause and Jenkins, 1983). This assessment is based on examination of skeletal remains of only 10 of the 27 recognized genera of Late Cretaceous and Early Tertiary North American multituberculates, but these genera are representative of both suborders and all six families (Suborder Ptilodontoidea, families Cimolodontidae, Neoplagiaulacidae, and Ptilodontidae; Suborder Taeniolabidoidea, families Eucosomodontidae and Taeniolabididae; suborder indeterminate, Family Cimolomyidae).

Detailed analyses of complete, or nearly complete, postcranial skeletons in early rodents have not yet been performed. Nonetheless, a good deal of information concerning locomotor habits of paramyids is known from selected regions of the skeleton, primarily the tarsus, as outlined below. Matthew (1910) reconstructed *Paramys* as an arboreal rodent. Wood (1962, p. 27), however, was not as confident about the presumed arboreal specializations of *Paramys,* stating that "it would seem probable that *Paramys* spent considerable periods of time on the ground as well as in the trees." Emry and Thorington (1982) concur with Wood's view. Szalay (1977; 1985, p. 111), on the other hand, has argued, based primarily on analyses of tarsal bones of *Paramys copei, P. delicatus,* and *Leptotomus parvus,* that the earliest paramyids were "capable arboreal forms" and that, in fact, they employed the same type of arboreal behavior practiced by multituberculates, namely headfirst descent down vertical supports.

Clearly, consensus regarding locomotor habits of paramyid rodents has yet to be achieved. Nonetheless, there is no available evidence to indicate that multituberculates and paramyid rodents were not cohabitants of trees, at least part of the time; and there is some evidence that they were.

Body Size

Body size is one of the most important parameters that can be utilized in evaluation of an animal's ecology (Peters, 1983; Fleagle, 1985). It is an attribute that has a profound influence on all of the above categories and, especially, diet (*e. g.,* Kay, 1975, 1978; Kay and Hylander, 1978; Fleagle, 1978), locomotion and substrate preference (*e. g.,* Fleagle and Mittermeier, 1980), and diel activity patterns (*e. g.,* Emmons and others, 1983).

Despite the fact that mammals of different size can and often do compete for the same resource (Grant, 1972), the most intense competition can be expected among mammals of similar body size (*e. g.,* Fleagle, 1978; Bowers and Brown, 1982). Based on this assumption, Landry (1965) asserted that rodents probably did not competitively exclude multituberculates because the earliest known rodent, *Acritoparamys* (= *Paramys*) *atavus,* was much smaller than *Ptilodus,* a common Paleocene multituberculate. Landry's comparison is inappropriate for at least two reasons: (1) *Ptilodus* was common in deposits considerably older than those in which the earliest rodents are found (there is but a single known specimen of *Ptilodus* that has been assigned a Clarkforkian age, and that only tentatively (Krause, 1980)); and (2) *Ptilodus* is the second largest of 13 recognized genera of ptilodontoid multituberculates.

There was, in fact, a broad overlap in size between Clarkforkian rodents and multituberculates. There are now at least five Clarkforkian species of rodents known (Korth, 1984), three of which are larger than *Acritoparamys atavus,* and five Clarkforkian species of multituberculates (Krause, 1980), three of which are smaller than *Ptilodus.* Reliable estimates of body size for all Clarkforkian multituberculates and rodents are not possible owing to the fragmentary nature of the fossils; comparisons are restricted further because of dissimilarity of their dentitions. Perhaps the best body size estimate attainable is from lengths of the lower dentitions in the smallest and largest representatives of each group. Length of the lower dentition (measured from the apex of the incisor to the distal end of the molar series) ranges from approximately 11 mm (*Acritoparamys atavus*—University of Michigan specimens 65244 and 69219) to 22 mm (*Paramys* cf. *excavatus* sensu Rose, 1981—UM specimens 65117 and 65120) in Clarkforkian rodents, and from about 10 mm (*Microcosmodon rosei*—UM 72662) to 38 mm (*Neoliotomus conventus*—estimated from specimens of *N. ultimus;* see Krause, 1982a) in Clarkforkian multituberculates. Based on this comparison, it appears that the entire size range of Clarkforkian rodents fits entirely within that of Clarkforkian multituberculates. At the very least, their body size distributions overlapped extensively. Arguments for improbability of competition between rodents and multituberculates on the basis of differences in body size are no longer tenable.

Paleobiogeography

Until recently, plesiadapiform primates were considered the most probable ancestors of rodents (Wood, 1962, 1977; McKenna, 1969; Van Valen and Sloan, 1966; Van Valen, 1966, 1971; Szalay, 1968), and Butler (1985) still appears to regard this hypothesis as unfalsified. Similarly, other workers (*e. g.,* McKenna, 1969; Szalay and Decker, 1974; Szalay, 1977, 1985) have suggested that leptictids or palaeoryctids were ancestral to rodents; but this suggestion has been disputed (*e. g.,* Novacek, 1982; Luckett and Hartenberger, 1985). Important paleontological discoveries in Paleocene deposits of Asia, particularly in the People's Republic of China, during the last decade document presence of a group of mammals, the Eurymyloidea, that most probably includes ancestors of rodents (Li, 1977; Gingerich and Gunnell, 1979; Hartenberger, 1977, 1980; Dawson and others, 1984; Li and Ting, 1985; Luckett and Hartenberger, 1985; Novacek, 1985; Li and others, *in press*). At the least, eurymyloids are close relatives of rodents. Rodents almost certainly originated in Asia. The sudden appearance of paramyid rodents in earliest Clarkforkian (latest Paleocene) deposits of western North America (Gingerich and Rose, 1977; Rose, 1980, 1981; Korth, 1984) therefore appears to represent an immigration event from Asia. The conclusion that rodents originated in Asia and dispersed to North America in the earliest Clarkforkian is in keeping with the requisite consideration that competitive exclusion of one group by another is preceded by a period of allopatry.

THE CLIMATE HYPOTHESIS

There were indeed significant climatic fluctuations in the Western Interior of North America during the early Cenozoic, as indicated by data from leaf floras (*e. g.,* Wolfe and Hopkins, 1967; Wolfe, 1978; Wolfe and Poore, 1982; Hickey, 1977, 1980, 1984), palynology (*e. g.,* Leffingwell, 1971; Tschudy, 1984), and oxygen isotopes (Buchardt, 1978). Nevertheless, the pattern of climatic fluctuations from the Late Cretaceous through the Eocene does not, in general, follow the changes in diversity and relative abundance of multituberculates. From the Late Cretaceous to the early Paleocene there appears to have been a decrease in mean annual temperature (MAT) in the continental interior, a decline that continued and reached a nadir in the Tiffanian (late Paleocene). This trend, however, was sharply reversed from the Tiffanian to the Clarkforkian and into the Wasatchian, when MATs again increased. In post-Wasatchian Eocene times, MATs fluctuated dramatically. As noted previously, the relative abundance of multituberculates decreased markedly from Lancian to Torrejonian horizons, the time during which MATs also decreased. Multituberculate diversity, however, increased during this time. From the Torrejonian to the Tiffanian, when MATs were continuing to decrease, multituberculate diversity decreased but relative abundance increased slightly. From the Tiffanian to the Wasatchian, when MATs increased, multituberculate diversity and relative abundance decreased dramatically. During the rest of the Eocene, MATs fluctuated considerably, but multituberculate diversity and relative abundance stayed uniformly low.

No clear pattern of relationship can therefore be established between broad patterns of MATs on the one hand and multituberculate diversity and relative abundance profiles on the other. This may simply reflect the inadequacy of data on early Cenozoic paleoclimates (*e. g.,* limitation to and analysis of MATs) or mammalian faunas, or both, but on the face of it there is nothing to suggest that changes in climate contributed significantly to the decline of multituberculates.

THE PREDATION HYPOTHESIS

Mammalian and avian predators both have been suggested as having had a significant impact on decline and extinction of multituberculates (Ostrander, 1984; Landry, 1967). As noted by Ostrander (1984), the most likely candidates among mammalian predators are miacid carnivorans, but data are insufficient to test whether the Miacidae had a significant effect on multituberculate diversity and abundance. Even in the broader context of all Carnivora plus all Creodonta, there appears to be little to support the predation hypothesis. The largest increase in relative abundance of mammalian carnivores appears to have been in the Wasatchian, well after multituberculates had begun to decline (Fig. 3). Relative abundances of carnivores at all levels is always low, however, and to place confidence in slight fluctuations in relative abundance estimates for such numerically insignificant components of the fauna is premature, especially in light of the numerous sampling and preservational biases involved.

Landry's (1967) suggestion that raptorial birds contributed to the extinction of multituberculates likewise is untestable at this time. Landry speculated that multituberculates were potentially vulnerable to predation by raptors owing to their diurnal habits but, as noted above, multituberculates probably were nocturnal mammals, not diurnal ones. Furthermore, raptorial birds are comprised both of nocturnal (Order Strigiformes—owls) and diurnal (Order Falconiformes—falcons, hawks, eagles, vultures, osprey, secretarybirds) forms. If either of these groups preyed substantially on multituberculates, it was most likely the Strigiformes, which appear in the North American fossil record in the late Paleocene (Rich and Bohaska, 1976, 1981), and not the Falconiformes, the earliest North American record of which is in the middle Oligocene (Feduccia, 1980). Nonetheless, on the basis of currently available information, Landry's hypothesis is untestable; the early Cenozoic fossil record of birds is wholly inadequate, making it impossible to determine diversity and relative abundance profiles, as well as aspects of life habits.

SUMMARY AND CONCLUSIONS

Although several other purported cases of competitive displacement in the fossil record have been reinterpreted recently as instances of mass extinction followed by opportunistic replacement (Benton, 1983b), three independent lines of evidence (diversity and relative abundance profiles, inferred resource use, and paleobiogeography) support the hypothesis that introduction of rodents into North America led to active displacement of multituberculates through competition. Patterns evinced from the fossil record are in keeping with the model of competition and differential survival developed by Benton (1983a), rather than his model of mass extinction and opportunistic replacement (see above).

North American multituberculates experienced significant reductions in diversity and relative abundance at the Tiffanian-Clarkforkian boundary. Recent collections from the Western Interior support earlier claims that there are marked inverse correlations in diversity and relative abundance between multituberculates and rodents from the Clarkforkian to the Chadronian, when multituberculates apparently finally became extinct. Rodents appeared, probably as immigrants from Asia, in the earliest Clarkforkian of North America. It seems probable that rodents were responsible for a major restructuring of early Cenozoic communities in North America. Insofar as aspects of their paleobiology can be determined, multituberculates and rodents probably overlapped extensively in utilization of resources. Evidence for these conclusions stems from a consideration of dietary and locomotor adaptations, diel activity patterns, and body size distributions.

Evidence for competition between condylarths and multituberculates from Lancian to Torrejonian and between multituberculates and plesiadapiform primates from Puercan to Torrejonian is not as strong as for that between multituberculates and rodents from the Tiffanian onwards. Similarly, current evidence to support the contention that decline of multituberculates was owing to climatic fluctuations or to predation pressure from mammalian or avian carnivores is either wanting or counter-indicative. I must make it explicit, however, that I do not claim that these alternative hypotheses are not possible; but the fact remains that there is no support for them in available data.

Despite major declines in multituberculate diversity and relative abundance at the Tiffanian-Clarkforkian boundary, multituberculates persisted at reduced levels for at least another 15 million years before becoming extinct. We know of only one multituberculate genus (possibly two) that survived into post-early Wasatchian times in western North America. Relative abundance data for post-early Wasatchian localities are few, but of those that are known, multituberculates did not comprise more than two percent of individuals in any particular local fauna. Such reduced levels make it all the more difficult to establish quantitative patterns (*i. e.,* correlations with diversity and relative abundance data of other taxa). Present data are insufficient to speculate realistically about a cause for the final termination of the Multituberculata.

ACKNOWLEDGMENTS

This project was initiated as one of the chapters of my doctoral dissertation (Krause, 1982c) and I thank members of my dissertation committee (P. D. Gingerich, J. A. Dorr, Jr., D. C. Fisher, P. Myers, and G. R. Smith) for feedback. For advice, discussions, reviews, and/or access to unpublished information pertinent to this paper since completion of my dissertation and a preliminary abstract (Krause, 1981), I am most grateful to J. D. Archibald, C. Badgley, S. J. Carlson, J. G. Fleagle, L. J. Flynn, F. E. Grine, W. L. Jungers, J. A. Kitchell, W. W. Korth, M. C. Maas, R. K. Stucky, L. Van Valen, and D. L. Wolberg. Collections from Holly's Microsite and the Paint Creek Locality were made with assistance from NSF grant DEB 80-10846 to P. D. Gingerich. This study was supported by National Science Foundation grant BSR-84-06707 and grants from EARTHWATCH and The Center for Field Research.

REFERENCES CITED

Archibald, J. D., Clemens, W. A., Gingerich, P. D., Krause, D. W., Lindsay, E. H., and Rose, K. D., *in press,* The Puercan, Torrejonian, Tiffanian, and Clarkforkian Land Mammal Ages, *in* Woodburne, M. O., ed., Cenozoic land-mammal ages: Berkeley, University of California Press.

Benton, M. J., 1983a, Dinosaur success in the Triassic: a noncompetitive ecological model: Quarterly Review of Biology, v. 58, p. 29-55

―――― 1983b, Large-scale replacements in the history of life: Nature, v. 302, p. 16-17.

―――― 1985, Mass extinction among non-marine tetrapods: *ibid,* v. 316, p. 811-814.

Black, C. C., 1967, Middle and late Eocene mammal communities: a major discrepancy: Science, v. 156, p. 62-64.

Black, C. C., and Sutton, J. F., 1984, Paleocene and Eocene rodents of North America, *in* Mengel, R. M., ed., Papers in vertebrate paleontology honoring Robert Warren Wilson: Pittsburgh, Carnegie Museum of Natural History, Special Publication no. 9, p. 67-84.

Bock, W. J., 1977, Adaptation and the comparative method, *in* Hecht, M. K., Goody, P. C., and Hecht, B. M., eds., Major patterns of vertebrate evolution: New York, Plenum Press, p. 57-82.

Bowers, M. A., and Brown, J. H., 1982, Body size and coexistence in desert rodents: chance or community structure?: Ecology, v. 63, p. 391-400.

Bown, T. M., 1982, Geology, paleontology, and correlations of Eocene volcaniclastic rocks, southeast Absaroka Range, Hot Springs County, Wyoming: U.S. Geological Survey, Professional Paper 1201A, p. 1-74.

Brown, J. H., 1978, Effects of mammalian competitors on the ecology and evolution of communities, in Snyder, D. P., ed., Populations of small mammals under natural conditions. The Pymatuning Symposia in Ecology, Pymatuning Laboratory of Ecology: University of Pittsburgh, Special Publication Series, v. 5, p. 52-57.

Buchardt, B., 1978, Oxygen isotope palaeotemperatures from the Tertiary period in the North Sea area: Nature, v. 275, p. 121-123.

Butler, P. M., 1985, Homologies of molar cusps and crests, and their bearing on assessments of rodent phylogeny, in Luckett, W. P., and Hartenberger, J.-L., eds., Evolutionary relationships among rodents: a multidisciplinary analysis: New York, Plenum Press, p. 381-401.

Carleton, M. D., and Musser, G. G., 1984, Muroid rodents, in Anderson, S., and Jones, J. K., Jr., eds., Orders and families of Recent mammals of the world: New York, John Wiley & Sons, p. 289-379.

Cifelli, R. L., 1981, Patterns of evolution among the Artiodactyla and Perissodactyla (Mammalia): Evolution, v. 35, p. 433-440.

———1985, South American ungulate evolution and extinction, in Stehli, F. G., and Webb, S. D., eds., The great American biotic interchange: New York, Plenum Press, p. 249-266.

Clemens, W. A., and Kielan-Jaworowska, Z., 1979, Multituberculata, in Lillegraven, J. A., Kielan-Jaworowska, Z., and Clemens, W. A., eds., Mesozoic mammals: the first two-thirds of mammalian history: Berkeley, University of California Press, p. 99-149.

Clemens, W. A., Lillegraven, J. A., Lindsay, E. H., and Simpson, G. G., 1979, Where, when, and what — a survey of known Mesozoic mammal distribution, in ibid., p. 7-58.

Connell, J. H., 1975, Some mechanisms producing structure in natural communities: a model and evidence from field experiments, in Cody, M. L., and Diamond, J. M., eds., Ecology and evolution of communities: Cambridge, The Belknap Press of Harvard University Press, p. 460-490.

——— 1980, Diversity and the co-evolution of competitors, or the ghost of competition past: Oikos, v. 35, p. 131-138.

——— 1983, On the prevalence and relative importance of interspecific competition: evidence from field experiments: The American Naturalist, v. 122, p. 661-696.

Connor, E. F., and Simberloff, D., 1979, The assembly of species communities: chance or competition?: Ecology, v. 60, p. 1132-1140.

——— 1984, Neutral models of species' co-occurrence patterns, in Strong, D. R., Jr., Simberloff, D., Abele, L. G., and Thistle, A. B., eds., Ecological communities: conceptual issues and the evidence: Princeton, Princeton University Press, p. 316-331.

——— 1986, Competition, scientific method, and null models in ecology: American Scientist, v. 74, p. 155-162.

Davidson, D. W., Inouye, R. S., and Brown, J. H., 1984, Granivory in a desert ecosystem: experimental evidence for indirect facilitation of ants by rodents: Ecology, v. 65, p. 1780-1786.

Dawson, M. R., Li, C.-K., and Qi, T., 1984, Eocene ctenodactyloid rodents (Mammalia) of eastern and central Asia, in Mengel, R. M., ed., Papers in vertebrate paleontology honoring Robert Warren Wilson: Pittsburgh, Carnegie Museum of Natural History, Special Publication, no. 9, p. 138-150.

Emmons, L. H., Gautier-Hion, A., and Dubost, G., 1983, Community structure of the frugivorous-folivorous forest mammals of Gabon: Journal of Zoology, London, v. 199, p. 209-222.

Emry, R. J., and Thorington, R. W., Jr., 1982, Descriptive and comparative osteology of the oldest fossil squirrel, *Protosciurus* (Rodentia: Sciuridae): Smithsonian Contributions to Paleobiology, no. 47, 35 p.

Estes, R., and Berberian, P., 1970, Paleoecology of a Late Cretaceous vertebrate community from Montana: Breviora, v. 343, p. 1-35.

Feduccia, A., 1980, The age of birds: Cambridge, Harvard University Press, 196 p.

Fleagle, J. G., 1978, Size distributions of living and fossil primate faunas: Paleobiology, v. 4, p. 67-76.

——— 1985, Size and adaptation in primates, in Jungers, W. L., ed., Size and scaling in primate biology: Plenum Press, New York, p. 1-19.

Fleagle, J. G., and Mittermeier, R. A., 1980, Locomotor behavior, body size, and comparative ecology of seven Surinam monkeys: American Journal of Physical Anthropology, v. 52, p. 301-314.

Gilpin, M. E., and Diamond, J. M., 1984, Are species co-occurrences on islands non-random, and are null hypotheses useful in community ecology?, in Strong, D. R., Jr., Simberloff, D., Abele, L. G., and Thistle, A. B., eds., Ecological communities: conceptual issues and the evidence: Princeton, Princeton University Press, p. 297-315.

Gingerich, P. D., 1976, Cranial anatomy and evolution of early Tertiary Plesiadapidae (Mammalia, Primates): University of Michigan Papers on Paleontology, v. 15, 140 p.

——— 1983, Paleocene-Eocene faunal zones and a preliminary analysis of Laramide structural deformation in the Clark's Fork Basin, Wyoming: Thirty-Fourth Annual Field Conference, Wyoming Geological Association Guidebook, p. 185-195.

Gingerich, P. D., and Gunnell, G. F., 1979, Systematics and evolution of the genus *Esthonyx* (Mammalia, Tillodontia) in the early Eocene of North America: The University of Michigan, Contributions from the Museum of Paleontology, v. 25 p. 125-153.

Gingerich, P. D., and Rose, K. D., 1977, Preliminary report on the American Clark Fork mammal fauna, and its correlation with similar faunas in Europe and Asia: Geobios, Memoire Special, v. 1, p. 39-45.

Gingerich, P. D., Smith, B. H., and Rosenberg, K., 1982, Allometric scaling in the dentition of primates and prediction of body weight from tooth size in fossils: American Journal of Physical Anthropology, v. 58, p. 81-100.

Gould, S. J., and Calloway, C. B., 1980, Clams and brachiopods — ships that pass in the night: Paleobiology, v. 6, p. 383-396.

Granger, W., and Simpson, G. G., 1929, A revision of the Tertiary Multituberculata: American Museum of Natural History, Bulletin, v. 56, p. 601-676.

Grant, P. R., 1972, Interspecific competition among rodents: Annual Review of Ecology and Systematics, v. 3, p. 79-106.

Grant, P. R., and Schluter, D., 1984, Interspecific competition inferred from patterns of guild structure, *in* Strong, D. R., Jr., Simberloff, D., Abele, L. G., and Thistle, A. B., eds., Ecological communities: conceptual issues and the evidence: Princeton, Princeton University Press, p. 202-231.

Grayson, D. K., 1973, On the methodology of faunal analysis: American Antiquity, v. 38, p. 432-439.

_____ 1978, Minimum numbers and sample size in vertebrate faunal analysis: *ibid.*, v. 43, p. 53-65.

Hahn, G., and Hahn, R., 1983, Fossilium Catalogus I: Animalia, Pars 127: Multituberculata: Amsterdam, Kugler Publications, 409 p.

Hartenberger, J.-L., 1977, A propos de l'origine des Rongeurs: Geobios, Memoire Special, v. 1, p. 183-193.

_____ 1980, Donnees et hypotheses sur la radiation initiale des Rongeurs: Palaeovertebrata, Memoire Jubilee R. Lavocat, p. 285-301.

Hickey, L., 1977, Stratigraphy and paleobotany of the Golden Valley Formation (early Tertiary) of western North Dakota: Geological Society of America, Memoir, no. 150, 181 p.

_____ 1980, Paleocene stratigraphy and flora of the Clark's Fork Basin, *in* Gingerich, P. D., ed., Early Cenozoic paleontology and stratigraphy of the Bighorn Basin, Wyoming: University of Michigan Papers in Paleontology, v. 24, p. 33-49.

_____ 1984, Changes in the angiosperm flora across the Cretaceous-Tertiary boundary, *in* Berggren, W. A., and Van Couvering, J. A., eds., Catastrophes and earth history: Princeton, Princeton University Press, p. 279-313.

Holtzman, R., 1979, maximum likelihood estimation of fossil assemblage composition: Paleobiology, v. 5, p. 77-89.

Hopson, J. A., 1967, Comments on the competitive inferiority of the multituberculates: Systematic Zoology, v. 16, p. 352-355.

James, F. C., and Boecklen, W. J., 1984, Interspecific morphological relationships and the densities of birds, *in* Strong, D. R., Jr., Simberloff, D., Abele, L. G., and Thistle, A. B., eds., Ecological communities: conceptual issues and the evidence: Princeton, Princeton University Press, p. 458-477.

Jenkins, F. A., Jr., and Krause, D. W., 1983, Adaptations for climbing in North American multituberculates (Mammalia): Science, v. 220, p. 712-715.

Jepsen, G. L., 1937, A Paleocene rodent, *Paramys atavus:* American Philosophical Society, Proceedings, v. 78, p. 291-301.

_____ 1949, Selection, "orthogenesis," and the fossil record: *ibid.*, v. 93, p. 479-500.

Jerison, H. J., 1973, Evolution of the brain and intelligence: New York, Academic Press, 482 p.

Johnston, P. A., and Fox, R. C., 1984, Paleocene and Late Cretaceous mammals from Saskatchewan, Canada: Palaeontographica, Abt. A, v. 186, p. 163-222.

Kay, R. F., 1975, The functional adaptations of primate molar teeth: American Journal of Physical Anthropology, v. 43, p. 195-216.

_____ 1978, Molar structure and diet in extant Cercopithecidae, *in* Butler, P. M., and K. A. Joysey, eds., Development, function and evolution of teeth: New York, Academic Press, p. 309-339.

Kay, R. F., and Cartmill, M., 1977, Cranial morphology and adaptations of *Palaechthon nacimienti* and other Paromomyidae (Plesiadapoidea, ?Primates) with a description of a new genus and species: Journal of Human Evolution, v. 6, p. 19-53.

Kay, R. F., and Hylander, W. L., 1978, The dental structure of mammalian folivores with special reference to Primates and Phalangeroidea (Mammalia), *in* Montgomery, G. G., ed., The biology of arboreal folivores: Washington, D. C., Smithsonian Institution Press, p. 173-191.

Kielan-Jaworowska, Z., 1983, Multituberculate endocranial casts: Palaeovertebrata, v. 13, p. 1-12.

Kitchell, J. A., 1985, Evolutionary paleoecology: recent contributions to evolutionary theory: Paleobiology, v. 11, p. 91-104.

Klingener, D., 1984, Gliroid and dipodoid rodents, *in* Anderson, S., and Jones, J. K., Jr., eds., Orders and families of Recent mammals of the world: New York, John Wiley & Sons, p. 381-388.

Korth, W. W., 1984, Earliest Tertiary evolution and radiation of rodents in North America: Carnegie Museum of Natural History, Bulletin, v. 24, p. 1-71.

_____ 1985, The rodents *Pseudotomus* and *Quadratotomus* and the content of the tribe Manitshini (Paramyinae, Ischyromyidae): Journal of Vertebrate Paleontology, v. 5, p. 139-152.

Krause, D. W., 1980, Multituberculates from the Clarkforkian Land-Mammal Age, late Paleocene-early Eocene, of western North America: Journal of Paleontology, v. 54, p. 1163-1183.

_____ 1981, Extinction of multituberculates and plesiadapiform primates: examples of competitive exclusion in the mammalian fossil record: Geological Society of America, Abstracts with Programs, v. 13, p. 491.

_____ 1982a, Multituberculates from the Wasatchian Land-Mammal Age, early Eocene, of western North America: Journal of Paleontology, v. 56, p. 271-294.

―――― 1982b, Jaw movement, dental function, and diet in the Paleocene multituberculate *Ptilodus*: Paleobiology, v. 8, p. 265-281.

―――― 1982c, Evolutionary history and paleobiology of early Cenozoic Multituberculata (Mammalia), with emphasis on the Family Ptilodontidae [Ph.D. thesis]: Ann Arbor, Michigan, The University of Michigan, 555 p.

―――― in press, *Baiotomeus*, a new ptilodontid multituberculate (Mammalia) from the middle Paleocene of western North America: Journal of Paleontology.

Krause, D. W., and Gingerich, P. D., 1983, Mammalian fauna from Douglass Quarry, earliest Tiffanian (late Paleocene) of the eastern Crazy Mountain Basin, Montana: The University of Michigan, Contributions from the Museum of Paleontology, v. 26, p. 157-196.

Krause, D. W. and Jenkins, F. A., Jr., 1983, The postcranial skeleton of North American multituberculates: Museum of Comparative Zoology, Harvard University, Bulletin, v. 150, p. 199-246.

Krishtalka, L., and Black, C. C., 1975, Paleontology and geology of the Badwater Creek area, central Wyoming. Part 12. Description and review of the late Eocene Multituberculata from Wyoming and Montana: Annals of Carnegie Museum, v. 45, p. 287-297.

Krishtalka, L., Emry, R. J., Storer, J. E., and Sutton, J. F., 1982, Oligocene multituberculates (Mammalia: Allotheria): youngest known record: Journal of Paleontology, v. 56, p. 791-794.

Landry, S. O., 1965, The status of the theory of the replacement of the Multituberculata by the Rodentia: Journal of Mammalogy, v. 46, p. 280-286.

―――― 1967, Disappearance of multituberculates: Systematic Zoology, v. 16, p. 172-173.

―――― 1970, The Rodentia as omnivores: Quarterly Review of Biology, v. 45, p. 351-372.

Leffingwell, H. A., 1971, Palynology of the Lance (Late Cretaceous) and Fort Union (Paleocene) formations of the type Lance area, Wyoming: Geological Society of America, Special Paper, no. 127, 64 p.

Li, C.-K., 1977, Paleocene eurymyloids (Anagalida, Mammalia) of Qianshan, Anhui: Vertebrata PalAsiatica, v. 15, p. 104-118.

Li, C.-K., and Ting, S.-Y., 1985, Possible phylogenetic relationship of Asiatic eurymylids and rodents, with comments on mimotonids, in Luckett, W. P., and Hartenberger, J.-L., eds., Evolutionary relationships among rodents: a multidisciplinary analysis: New York, Plenum Press, p. 35-58.

Li, C.-K., Wilson, R. W., and Dawson, M. R., *in press*, The origin of rodents and lagomorphs: Journal of Mammalogy.

Luckett, W. P., and Hartenberger, J.-L., 1985, Evolutionary relationships among rodents: comments and conclusions, in Luckett, W. P., and Hartenberger, J.-L., eds., Evolutionary relationships among rodents: a multidisciplinary analysis: New York, Plenum Press, p. 685-712.

McKenna, M. C., 1960, Fossil Mammalia from the early Wasatchian Four Mile fauna, Eocene of northwest Colorado: University of California, Publications in Geological Sciences, v. 37, p. 1-130.

―――― 1961, A note on the origin of rodents: American Museum Novitates, no. 2037, 5 p.

―――― 1969, The origin and early differentiation of therian mammals: New York Academy of Sciences, Annals, v. 167, p. 217-240.

McLaughlin, C. A., 1984, Protrogomorph, sciuromorph, castorimorph, myomorph (geomyoid, anomaluroid, pedetoid, and ctenodactyloid) rodents, in Anderson, S., and Jones, J. K., Jr., eds., Orders and families of Recent mammals of the world: New York, John Wiley & Sons, p. 267-288.

Matthew, W. D., 1897, A revision of the Puerco fauna: American Museum of Natural History, Bulletin, v. 9, p. 259-323.

―――― 1910, On the osteology and relationships of *Paramys*, and the affinities of the Ischyromyidae: *ibid.*, v. 28, p. 43-71.

Novacek, M. J., 1982, Information for molecular studies from anatomical and fossil evidence on higher eutherian phylogeny, in Goodman, M., ed., Macromolecular sequences in systematic and evolutionary biology: New York, Plenum Press, p. 3-41.

―――― 1985, Cranial evidence for rodent affinities, in Luckett, W. P., and Hartenberger, J.-L., eds., Evolutionary relationships among rodents: a multidisciplinary analysis: New York, Plenum Press, p. 59-81.

Ostrander, G., 1984, The early Oligocene (Chadronian) Raben Ranch local fauna, northwest Nebraska: Multituberculata; with comments on the extinction of the Allotheria: Nebraska Academy of Sciences, Transactions, v. 12, p. 71-80.

Ostrander, G., Jones, C. A., and Cape, R., 1979, The occurrence of a multituberculate in the lower Oligocene Chadron Formation of northwest Nebraska: Geological Society of America, Abstracts With Programs, v. 11, p. 299.

Peters, R. H., 1983, The ecological implications of body size: Cambridge, Cambridge University Press, 329 p.

Pianka, E. R., 1975, Niche relations of desert lizards, in Cody, M. L., and Diamond, J. M., eds., Ecology and evolution of communities: Cambridge, Harvard University Press, p. 292-314.

Rensberger, J. M., 1982, Patterns of dental change in two locally persistent successions of fossil aplodontid rodents, in Kurten, B., ed., Teeth: form, function, and evolution: New York, Columbia University Press, p. 323-349.

Rich, P. V., and Bohaska, D. J., 1976, The world's oldest owl: a new strigiform from the Paleocene of southwestern Colorado: Smithsonian Contributions to Paleobiology, v. 27, p. 87-93.

―――― 1981, The Ogygoptyngidae, a new family of owls from the Paleocene of North America: Alcheringa, v. 5, p. 95-102.

Robinson, P., Black, C., and Dawson, M. R., 1964, Late Eocene multituberculates and other mammals from Wyoming: Science, v. 145, p. 809-811.

Romer, A. S., 1966, Vertebrate paleontology: Chicago, University of Chicago Press, 468 p.

Rose, K. D., 1975, The Carpolestidae, early Tertiary primates from North America: Museum of Comparative Zoology, Harvard University, Bulletin, v. 147, p. 1-74.

―――― 1977, Evolution of carpolestid primates and chronology of the North American middle and late Paleocene: Journal of Paleontology, v. 51, p. 536-542.

―――― 1980, Clarkforkian Land-Mammal Age: revised definition, zonation, and tentative intercontinental correlations: Science, v. 208, p. 744-746.

―――― 1981, The Clarkforkian Land-Mammal Age and mammalian faunal composition across the Paleocene-Eocene boundary: University of Michigan Papers on Paleontology, v. 26, 196 p.

Savage, D. E., and Russell, D. E., 1983, Mammalian paleofaunas of the world: Reading, Massachusetts, Addison-Wesley Publishing Company, 432 p.

Schoener, T. W., 1983, Field experiments on interspecific competition: The American Naturalist, v. 122, p. 240-285.

Signor, P. W., III, 1985, Real and apparent trends in species richness through time, in Valentine, J. W., ed., Phanerozoic diversity patterns: profiles in macroevolution: Princeton, Princeton University Press, p. 129-150.

Simberloff, D., 1983, Competition theory, hypothesis testing, and other community ecological buzzwords: The American Naturalist, v. 122, p. 626-635.

Simpson, G. G., 1929, Third contribution to the Fort Union fauna at Bear Creek, Montana: American Museum Novitates, no. 345, 12 p.

―――― 1937, Skull structure of the Multituberculata: American Museum of Natural History, Bulletin, v. 73, p. 727-763.

―――― 1949, The meaning of evolution: New Haven, Yale University Press, 364 p.

―――― 1953, The major features of evolution: New York, Columbia University Press, 434 p.

Sloan, R. E., 1966, Paleontology and geology of the Badwater Creek area, central Wyoming. Part 2. The Badwater multituberculate: Annals of Carnegie Museum, v. 38, p. 309-315.

―――― 1969, Cretaceous and Paleocene terrestrial communities of western North America: North American Paleontological Convention, Proceedings, v. 1(E), p. 427-453.

―――― 1979, Multituberculata, in Fairbridge, R. W., and Jablonski, D., eds., The encyclopedia of paleontology: Stroudsberg, Dowden, Hutchinson & Ross, Inc., p. 492-498.

Sloan, R. E., and Van Valen, L., 1965, Cretaceous mammals from Montana: Science, v. 148, p. 220-227.

Stanley, S. M., 1986, Population size, extinction, and speciation: the fission effect in Neogene Bivalvia: Paleobiology, v. 12, p. 89-110.

Storer, J. E., 1984, Mammals of the Swift Current Creek local fauna (Eocene: Uintan), Saskatchewan: Saskatchewan Culture and Recreation, Museum of Natural History, Natural History Contributions, no. 7, 158 p.

Strong, D. R., Jr., 1984, Exorcising the ghost of competition past: phytophagous insects, in Strong, D. R., Jr., Simberloff, D., Abele, L. G., and Thistle, A. B., eds., Ecological communities: conceptual issues and the evidence: Princeton, Princeton University Press, p. 28-41.

Strong, D. R., Jr., Szyska, L. A., and Simberloff, D. W., 1979, Tests of community-wide character displacement against null hypotheses: Evolution, v. 33, p. 897-913.

Stucky, R. K., 1984a, Revision of the Wind River faunas, early Eocene of central Wyoming. Part 5. Geology and biostratigraphy of the upper part of the Wind River Formation, northeastern Wind River Basin: Annals of Carnegie Museum, v. 53, p. 231-294.

―――― 1984b, The Wasatchian-Bridgerian Land Mammal Age boundary (early to middle Eocene) in western North America: ibid., v. 53, p. 347-382.

Stucky, R. K., and Krishtalka, L., 1982, Revision of the Wind River faunas, early Eocene of central Wyoming. Part 1. Introduction and Multituberculata: ibid., v. 51, p. 39-56.

Szalay, F. S., 1965, First evidence of tooth replacement in the subclass Allotheria (Multituberculata): American Museum Novitates, no. 2226, 12 p.

―――― 1968, The beginnings of primates: Evolution, v. 22, p. 19-36.

―――― 1977, Phylogenetic relationships and a classification of the eutherian Mammalia, in Hecht, M. K., Goody, P. C., and Hecht, B. M., eds., Major patterns in vertebrate evolution: New York, Plenum Publishing Co., NATO Advanced Study Institute, Series A, v. 14, p. 315-374.

―――― 1985, Rodent and lagomorph morphotype adaptations, origins and relationships: some postcranial attributes analyzed, in Luckett, W. P., and Hartenberger, J.-L., eds., Evolutionary relationships among rodents: a multidisciplinary analysis: New York, Plenum Press, p. 83-132.

Szalay, F. S., and Decker, R. L., 1974, Origins, evolution and function of the tarsus in Late Cretaceous eutherians and Paleocene primates, in Jenkins, F. A., Jr., ed., Primate locomotion: New York, Academic Press, p. 223-259.

Toft, C. A., 1985, Resource partitioning in amphibians and reptiles: Copeia, v. 1985, p. 1-21.

Tschudy, R. H., 1984, Palynological evidence for change in continental floras at the Cretaceous Tertiary boundary, in Berggren, W. A., and Van Couvering, J. A., eds., Catastrophes and earth history: Princeton, Princeton University Press, p. 315-337.

Van Valen, L., 1966, Deltatheridia, a new order of mammals: American Museum of Natural History, Bulletin, v. 132, p. 1-126.

―――― 1971, Adaptive zones and the orders of mammals: Evolution, v. 25, p. 420-428.

―――― 1978, The beginning of the Age of Mammals: Evolutionary Theory, v. 4, p. 45-80.

Van Valen, L., and Sloan, R. E., 1965, The earliest primates: Science, v. 150, p. 743-745.

―――― 1966, The extinction of the multituberculates: Systematic Zoology, v. 15. p. 261-278.

West, R. M., 1973, Geology and mammalian paleontology of the New Fork-Big Sandy area, Sublette County, Wyoming: Field Museum of Natural History, Fieldiana: Geology, v. 29, 193 p.

Wilson, R. W., 1951, Evolution of the early Tertiary rodents: Evolution, v. 5, p. 207-215.

Wolberg, D. L., 1978, The mammalian paleontology of the late Paleocene (Tiffanian) Circle and Olive localities, McCone and Powder River counties, Montana [Ph.D. thesis]: Minneapolis, Minnesota, University of Minnesota, 385 p.

―――― 1979, Late Paleocene (Tiffanian) mammalian fauna of two localities in eastern Montana: Northwest Geology, v. 8, p. 83-93.

Wolfe, J. A., 1978, a paleobotanical interpretation of Tertiary climates in the Northern Hemisphere: American Scientist, v. 66, p. 694-703.

Wolfe, J. A., and Hopkins, D. M., 1967, Climatic changes recorded by Tertiary land floras in northwestern North America, *in* Hatai, K., ed., Tertiary correlation and climatic changes in the Pacific: 11th Pacific Scientific Congress, Symposium, v. 25, p. 67-76.

Wolfe, J. A., and Poore, R. Z., 1982, Tertiary marine and nonmarine climatic trends, *in* Geophysics Study Committee, Geophysics Research Board, Commission on Physical Sciences, Mathematics and Resources, National Research Council, compilers, Climate in earth history; Washington, D.C., National Academy Press, p. 154-158.

Wood, A. E., 1962, The early Tertiary rodents of the family Paramyidae: American Philosophical Society, Transactions, v. 52, p. 1-261.

―――― 1977, The Rodentia as clues to Cenozoic migrations between the Americas and Europe and Africa: Milwaukee Public Museum, Special Publications in Biology and Geology, v. 2, p. 95-109.

―――― 1980, The Oligocene rodents of North America: American Philosophical Society, Transactions, v. 70, p. 1-68.

Woods, C. A., 1984, Hystricognath rodents, *in* Anderson, S., and Jones, J. K., Jr., eds., Orders and families of Recent mammals of the world: New York, John Wiley & Sons, p. 389-446.

MANUSCRIPT RECEIVED DECEMBER 11, 1985
REVISED MANUSCRIPT RECEIVED JUNE 16, 1986
MANUSCRIPT ACCEPTED JUNE 18, 1986

APPENDIX A. RELATIVE ABUNDANCES OF MAMMALS AT HOLLY'S MICROSITE (UNIVERSITY OF MICHIGAN LOCALITY SC-188), MIDDLE CLARKFORKIAN Plesiadapis cookei ZONE, CLARK'S FORK BASIN, NORTHWESTERN WYOMING.

Taxon	No. of Specimens	MNI	Freq.(MNI)
Order MULTITUBERCULATA			
Family Eucosmodontidae			
Microcosmodon rosei	12	2	.018
Family Neoplagiaulacidae			
Ectypodus powelli	92	24	.212
Order POLYPROTODONTA			
Family Didelphidae			
Peradectes cf. P. chesteri	21	4	.035
Order PROTEUTHERIA			
Family Leptictidae			
cf. Prodiacodon tauricinerei	3	1	.009
Family Apatemyidae			
Labidolemur kayi	8	3	.027
Family Palaeoryctidae			
Palaeoryctes punctatus	2	1	.009
Order LIPOTYPHLA			
Family Nyctitheriidae			
cf. Leptacodon packi	22	7	.062
cf. Plagioctenacodon krausae	5	1	.009
Family Erinaceidae			
Leipsanolestes siegfriedti	1	1	.009
?Order LIPOTYPHLA			
cf. "Diacodon" minutus	10	4	.035
Order DERMOPTERA			
Family Plagiomenidae			
Worlandia inusitata	81	11	.097
Order PRIMATES			
Family Microsyopidae			
Niptomomys doreenae	2	1	.009
Tinimomys graybulliensis	2	1	.009

Appendix A. (continued)

Family Plesiadapidae			
Plesiadapis cookei	17	3	.027
Family Carpolestidae			
Carpolestes nigridens	31	6	.053
Family Paromomyidae			
Phenocolemur pagei	17	3	.027
Ignacius graybullianus	7	2	.018
Order CONDYLARTHRA			
Family Arctocyonidae			
Thryptacodon cf. antiquus	1	1	.009
Arctocyonidae incertae sedis	2	1	.009
Family Phenacodontidae			
Phenacodus primaevus	2	1	.009
Ectocion osbornianus	21	5	.044
Family Hyopsodontidae			
Haplomylus simpsoni	1	1	.009
Order PANTODONTA			
Family Coryphodontidae			
Coryphodon sp.	1	1	.009
Order NOTOUNGULATA			
Family Arctostylopidae			
Arctostylops steini	1	1	.009
Order CARNIVORA			
Family Miacidae			
Didymicitis proteus	5	2	.018
Viverravus cf. bowni	5	1	.009
Uintacyon rudis	1	1	.009
Order CREODONTA			
Family Oxyaenidae			
Oxyaena aequidens	2	1	.009
Order RODENTIA			
Family Paramyidae	163	22	.195
Totals =	538	113	1.003

APPENDIX B. RELATIVE ABUNDANCES OF MAMMALS AT THE PAINT CREEK LOCALITY (UNIVERSITY OF MICHIGAN LOCALITY SC-143), MIDDLE CLARKFORKIAN *Plesiadapis cookei* ZONE, CLARK'S FORK BASIN, NORTHWESTERN WYOMING.

Taxon	No. of Specimens	MNI	Freq.(MNI)
Order MULTITUBERCULATA			
Family Eucosmodontidae			
Microcosmodon rosei	16	4	.087
Family Neoplagiaulacidae			
Ectypodus powelli	3	1	.022
Parectypodus laytoni	6	2	.043
Order PROTEUTHERIA			
Family Apatemyidae			
Labidolemur kayi	1	1	.022
Order LIPOTYPHLA			
Family Nyctitheriidae			
cf. *Pontifactor bestiola*	1	1	.022
cf. *Plagioctenodon krausae*	4	1	.022
?Order LIPOTYPHLA			
cf. "*Diacodon*" *minutus*	1	1	.022
Order PRIMATES			
Family Microsyopidae			
Microsyops simplicidens	2	1	.022
Family Plesiadapidae			
Plesiadapis dubius	36	5	.109
Plesiadapis cookei	3	2	.043
Chiromyoides major	3	1	.022
Family Carpolestidae			
Carpolestes nigridens	14	2	.043
Family Paromomyidae			
Phenacolemur pagei	18	4	.087
Phenacolemur praecox	2	1	.022

Appendix B. (continued)

Order CONDYLARTHRA
  Family Arctocyonidae
     Thryptacodon cf. pseudarctos   2   1   .022
     cf. Tricentes sp.   4   1   .022
  Family Phenacodontidae
     Phenacodus sp.   4   1   .022
     Ectocion osbornianus   21   4   .087
  Family Hyopsodontidae
     Aletodon gunnelli   5   2   .043
     Apheliscus nitidus   38   5   .109
Order MESONYCHIA
  Family Mesonychidae
     Dissacus sp.   1   1   .022
Order DINOCERATA
  Family Uintatheriidae
     Probathyopsis sp.   2   1   .022
Order CARNIVORA
  Family Miacidae
     Didymictis?   2   1   .022
Order Uncertain
  Family Metacheiromyidae
     Palaeanodon ?parvulus   2   2   .043

                                191   46   1.002

Gradual evolution and species discrimination in the fossil record

KENNETH D. ROSE — *Department of Cell Biology and Anatomy, Johns Hopkins University School of Medicine, Baltimore, Maryland 21205*

THOMAS M. BOWN — *Paleontology and Stratigraphy Branch, U.S. Geological Survey, Denver, Colorado 80225*

ABSTRACT

Although species are the basic units of many paleontological and evolutionary studies, the term "species" applied to the fossil record does not convey the same concept to all workers. Simpson's "evolutionary species" incorporated time into the species concept, but considered each non-branching lineage as a separate species; longer lineages with more continuous fossil records may require subdivision into successional species. One's perception of paleontological species affects, and is affected by, evolutionary philosophy and models of how new species form and evolve. For example, if species actually arise abruptly and persist for much longer periods essentially unchanged (punctuated equilibria), discrimination of paleontological species should be a relatively simple matter. Alternatively, if there is continuous change within and between successive species (gradualism), species boundaries would be nebulous, and would have to be imposed arbitrarily. We summarize our study of omomyid primates and cite other supportive evidence which suggests that, where the record is sufficiently dense, gradual evolution (requiring arbitrary boundaries) is common between species and even genera.

INTRODUCTION

The nature and recognition of species in the fossil record—long a topic of special interest to Professor Simpson—has spawned voluminous discussion and debate (*e. g.,* Simpson, 1943, 1961; Sylvester-Bradley, 1956; Imbrie, 1957; Mayr, 1969; Wiley, 1978; Levinton and Simon, 1980; Thaler, 1983; Gingerich, 1985). Species have posed special problems for paleontologists owing to certain factors peculiar to the fossil record, namely the limited evidence with which we must work, and the element of time. As a result, one focus of controversy has been on how closely paleontological species approximate biological species. Eldredge and Gould (1972) suggested that this issue cannot be resolved; nonetheless, much current debate over the tempo and mode of evolution concerns how species form and evolve, so it is vital to understand what we mean when we speak of species in the fossil record (see also Thaler, 1983). This debate and the advent of cladistic methodology have stimulated renewed interest in the nature of paleontological species.

As the fossil record has improved, species boundaries in many lineages have become increasingly nebulous—particularly in the presence of a stratigraphically dense and relatively continuous record. This has important implications for the definition of paleontological species and the interpretation of evolutionary modes. In this paper we consider the meaning of paleontological species and the intimate relationship between species recognition and evolutionary models and philosophy. We close with an example from Eocene primates, illustrating how an improved fossil record can make species discrimination very difficult, while at the same time markedly enhancing our understanding of their evolution.

PALEONTOLOGICAL SPECIES AND EVOLUTIONARY MODELS

Numerous definitions of species have been proposed (see for example Cain, 1960; Simpson, 1961; Mayr, 1970), underscoring at least one sure conclusion: the term species means different things to different people. However it is applied to fossils, a paleontological species is recognized in practice by its characteristic morphology, encompassing a range of variability comparable to that seen in extant species (*e. g.,* Gingerich, 1974). Thus we tend to consider as distinct species those samples that display a certain definable variability, and which at the same time are sufficiently distinct from other samples. That the characters involved and the degree of difference are to some extent subjective is obvious, but most paleontologists use generally similar guidelines for recognizing fossil species. Considered only in this context, paleontological species are morphospecies (Cain, 1960).

The incomplete fossil record of many groups makes it easier to delineate samples that can be reasonably considered as species—particularly those known from only one horizon or a restricted interval (Fig. 1). But as the fossil record improves, discrimination between species becomes increasingly difficult (see also Cain, 1960; Tobias, 1978). Stated succinctly, "species are in many cases objective evolutionary units on a time plane and at the same time arbitrary units crossing time planes" (Gingerich, 1985, p. 29).

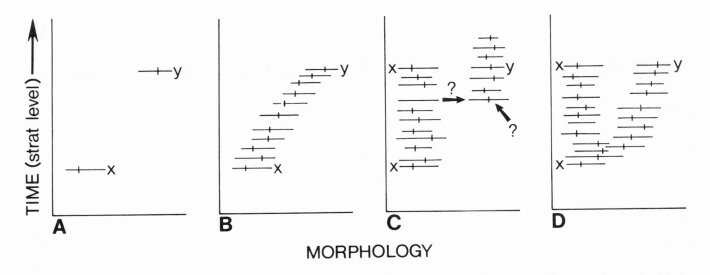

Figure 1. Different interpretations of two species in fossil record, dependent on relative completeness of record and evolutionary philosophy. *A*, two species, x and y, separated in time, are distinguished on morphological criteria (horizontal bars indicate range, vertical slash denotes mean). This is the state of knowledge for most paleontological species. *B*, Intermediates are found that document continuity from x to y (gradualism), with no apparent branching. Cladistically, x and y may now be considered one species, although they are no less distinct simply because the fossil record has improved. *C*, Species x shows stasis through time (only minor oscillations of mean) and species y appears abruptly (punctuated equilibria), either as an immigrant or evolving rapidly from x, without preserved intermediates. In the latter case, if x survives after appearance of y, a branching event has occurred. In any case, two species are still recognized. *D*, Species x shows stasis, but a cladogenetic event early in its history results in species y, which diverges gradually. Two species are recognized.

Time is the unique component of species in the fossil record. The element of time, coupled with the inability to demonstrate reproductive isolation definitively, makes it impossible (and probably inappropriate) to impose a strict biological species concept to the fossil record. Many authors have emphasized the difficulty of recognizing the equivalent of biospecies in the fossil record. Some suggest, for example, that our concept of paleontological species underestimates the number of biological species that were present, grouping similar forms that are indistinguishable on the basis of fossil remains (*e. g.*, Thaler, 1983; Hecht, 1983; Jaeger, 1983). However, as Gould (1983) points out, the nature of many fossil samples suggests that they are indeed close approximations to biological species, and the evolutionary history of these fossil samples is of great interest.

To incorporate time into the species concept, Simpson (1961, p. 153; 1951) proposed the "evolutionary species," defining it as "a lineage (an ancestral-descendant sequence of populations) evolving separately from others and with its own unitary evolutionary role and tendencies." By this definition, each branch or lineage of a branching phylogeny is an evolutionary species. The weakness of this concept becomes obvious as the fossil record of a lineage improves:

> If you start at any point in the sequence and follow the line backward through time, there is no place where the definition ceases to apply. You never leave an uninterrupted, separate, unitary lineage and therefore never leave the species with which you started unless some other criterion of definition can be brought in. If the fossil record were complete, you could start with man and run back to a protist still in the species *Homo sapiens*. Such classification is manifestly both useless and somehow wrong in principle. Certainly the lineage must be chopped into segments for purposes of classification, and this must be done arbitrarily because there is no nonarbitrary way to subdivide a continuous line.
> Simpson, 1961, p. 165.

Simpson suggested that such an evolutionary species be divided into segments (separated by morphological differences comparable to those that separate biospecies) called successive or successional species (see also Imbrie, 1957); others have referred to them as paleospecies (*e. g.*, Cain, 1960) or chronospecies (Thomas, 1956; Rhodes, 1956; Stanley, 1979). Although Simpson did not consider successional species to be the equivalent of biospecies, this concept of the species is widely used in paleontology.

Some paleontologists still adhere to Simpson's evolutionary species concept as a better approximation to biological species in the fossil record, maintaining that if a continuum can be established between successive units then they represent a single species (Wiley, 1978; Eldredge and Cracraft, 1980). This view excludes the possibility of origination of new species within a lineage ("phyletic speciation"), restricting the appearance of new species to cladogenetic (lineage-splitting or speciation) events. We consider this potentially misleading from both a systematic and an evolutionary standpoint (see below).

Thus it is obvious that evolutionary philosophy, in addition to morphological criteria, has a significant impact on the recognition of paleontological species. The concept of what constitutes a species in the fossil record is influenced by, and may affect, interpretations of rates and processes of species transformation. The latter are, in turn, influenced by the relative completeness of the stratigraphic and fossil records.

Considerable attention has focused on two models of species transformation: gradualism and punctuated equilibria. Different interpretations of the terms "gradualism," "punctuation," and "equilibrium" have led to confusion. The time-frame in which evolution is observed is important here, for the scale we employ may determine whether we infer a gradual or punctuated pattern (see, e. g., Ginzburg, 1981; Jones, 1981; Schindel, 1982; Gingerich, 1983; Penny, 1985). But these terms generally have been applied to the fossil record and are most appropriate for evolutionary patterns or events viewed on a geologic time-scale. Gradualism, in our view, implies continuous change within and between species, as evidenced by intermediate stages, but it does not require constant or slow rates—only rates slow enough to be detected in the fossil record. It may involve one or more character complexes (probably functionally related), but in any gradual transformation most characters remain stable; in this sense, any example of gradualism is a composite of gradualism and equilibrium.

Equilibrium, as envisioned in the punctuated equilibrium model (e. g., Gould and Eldredge, 1977), is stasis or only minor non-directional fluctuation in all characters. This is impossible to test definitively, due to obvious limitations of fossil evidence, but it is not unreasonable to argue for equilibrium if no significant changes can be detected in any preserved parts. Stasis in one character (e. g., molar size: see West, 1979; Gingerich, 1985), however, is not a demonstration of equilibrium (see also Martin, 1984), since other traits may have changed gradually during the same interval (e. g., the Eocene mammal *Diacodexis*: compare Krishtalka and Stucky, 1985, with Gingerich, 1985).

Since both gradualism and stasis have relatively long durations, any reasonably continuous sequence of adequate samples should allow a test of these two modes.

Punctuation has been characterized as a "geologically instantaneous" event associated with branching speciation, although within-lineage punctuations also have been reported (e. g., Bookstein and others, 1978; Malmgren and others, 1984). Since "geologically instantaneous" may be ambiguous, Gould (1982b, p. 84) suggested that this might be defined as "1 percent or less of later existence in stasis." For most species, this is so brief that the event rarely would be sampled in the fossil record. Punctuation must be inferred from a sharp phenotypic break (without preserved intermediates) at which ancestors are abruptly replaced by descendants, interrupting much longer periods during which little or no phenotypic change occurred (Gould and Eldredge, 1977).

It is thus based on negative evidence, and is untestable (Schopf, 1982; Boucot, 1983; Hecht, 1983; Gingerich, 1984) because it cannot resolve the mode of species transformation (*i. e.*, saltational or gradual).

It is important to appreciate that each model influences our perception of paleontological species. According to the gradual model of evolution, successive species would be expected to intergrade imperceptibly, and boundaries between them would be nebulous. By the punctuated equilibria model, discrete species should persist unchanged for periods on the order of a million years or more and be separated by abrupt breaks or punctuations of much shorter duration ($\sim 10^4$ years).

Morphological continuity between species—especially, but not restricted to, successional species within a lineage—is the hallmark of gradual evolution. Although the occurrence of phyletic gradualism has been and still is widely accepted (e. g., Vrba, 1980; Hoffman, 1982; Gould, 1982b; Chaline [ed.], 1983), some now consider it a relatively rare or insignificant mode of evolution (Gould and Eldredge, 1977; Stanley, 1979; Gould, 1985). Simpson (1961, p. 163-164) clearly regarded it as common, noting "numerous cases of temporally successive, appreciably changing samples from a single lineage or closely related lineages." We believe that today there is even more persuasive documentation of continuity between successional species, and cumulative change within many such lineages far exceeds that observed in any extant species (e. g., Ozawa, 1975; Bookstein and others, 1978; Gingerich, 1980, 1985; Raup and Crick, 1981; Chaline, 1983, 1985; Enay, 1983; Jaeger, 1983; Thaler, 1983). To regard the successional forms in these cases as a single species simply because intermediates linking them are now known would greatly amplify the morphologic limits of paleontological species compared to biospecies and would, in effect, reject widely-held criteria and procedures for establishing species in the fossil record. Moreover, it would obscure important information on evolution. After all, gaps in the fossil record are biologically no more meaningful for separating such species than are any other arbitrary guidelines (Maglio, 1973). We emphasize that species considered distinct by established criteria are no less distinct if evolutionary intermediates can be demonstrated (Fig. 1).

It follows that the classification of lineages with a relatively continuous fossil record into successional species must, if evolution has been gradual, involve subjective decisions:

> One must somewhere draw a completely arbitrary line, representing a point in time, across some steadily evolving lineage and say, "Here one taxon ends and another begins." Simpson, 1961, p. 117.

The absence of discontinuities between species and the consequent need for arbitrary boundaries does not diminish the reality of these taxa (e. g., Newell, 1956). Application of arbitrary time-lines as species boundaries has been advocated by Campell (1974) and Gingerich (e. g., 1976; 1980; Gingerich and Simons, 1977), among others. Even this method of separation between species may pose a dilemma, however, as illustrated in our example below.

Well documented lineages that show continuity and gradual change pose an additional taxonomic problem. It may not be sufficient simply to divide the lineage into successional species, for assignment of each sample to one or another of the successional species could mask the transitional nature of some samples, giving the false impression of morphologically static species separated by real discontinuities (see also Schopf, 1981). For example, this may explain Stanley's (1982) conclusion—despite substantial documentation to the contrary—that all early Eocene mammalian species from the Bighorn Basin remained essentially in stasis over periods of two to three million years. Both formal and informal means of recognizing intermediate forms have been proposed. Some authors have advocated use of temporally-successive subspecies (*e. g.*, Simpson, 1943, 1961; Tintant, 1980; Beden, 1983). Others have defended more cumbersome compound names (*e. g.*, Ewer, 1967; Tobias, 1969; Crusafont-Pairó and Reguant, 1970; Krishtalka and Stucky, 1985), usually involving some combination of the names of successive taxa bridged by the intermediate. Maglio (1971) proposed the use of informal numbered or named stages. This procedure has the advantage of not further burdening the formal taxonomy, hence we have adopted it for omomyid primates (Rose and Bown, 1984a, and below).

The preceding paragraphs have focused on problems of species recognition in relatively continuous, gradually evolving lineages. However, if most species "generally do not change substantially in phenotype over a lifetime that may encompass many million years (stasis), and most evolutionary change is concentrated in geologically instantaneous events of branching speciation" (Gould, 1982a, p. 383), as predicted by the punctuated equilibria model (Eldredge and Gould, 1972; Gould and Eldredge, 1977; Stanley, 1979; Gould, 1983), then species should be easy to distinguish in the fossil record and no arbitrary boundaries need be drawn. In the face of a good fossil record, however, often this is not the case (see also Boucot, 1983). Simpson (1944, 1953, 1961) considered true discontinuities between taxa to be unusual, suggesting that apparent discontinuities (particularly those associated with the origin of new higher level taxa) might result from periods of very rapid evolution. Even in these cases, however, the exact point of separation during a period of rapid transition would still be arbitrary. The evolution of molluscs in the Turkana Basin (Williamson, 1981) and of certain planktonic foraminifera (Malmgren and others, 1984) appear to represent cases of this sort—*i. e.*, relatively rapid, but still gradual, transformation of species.

Although gradualism tends to be associated with phyletic (within-lineage) evolution and punctuated equilibria with cladogenetic (branching) speciation, these associations are not universal and can be misleading. Because the term speciation has been defined as the formation of new species by branching (cladogenesis), strict application of the definition, as noted earlier, excludes the origin of new species by phyletic evolution—a point with which we disagree. As Tintant (1983) has pointed out, both phyletic evolution and cladogenesis may, in theory, be gradual or punctuated. Many apparent cases of gradual cladogenesis are now known (*e. g.*, Prothero and Lazarus, 1980; Lazarus and others, 1985; Godinot, 1985), and proposed examples of within-lineage punctuation and punctuated equilibria already have been cited. Chaline (1983, 1985) cites evidence from the fossil record of rodents for both gradual and punctuated patterns of species origination associated with both phyletic evolution and cladogenesis.

Examples adduced to support any model of evolution are compelling only if we can be confident that the fossil record is relatively complete. Several authors have questioned whether the stratigraphic record is complete enough to allow meaningful tests of evolutionary models (Schindel, 1980; van Andel, 1981; Sadler, 1981; Dingus and Sadler, 1982), and many supposed cases of both gradualism and punctuation have been challenged, often justifiably, on the basis that the stratigraphic record on which they are founded contains too many gaps to yield a reliable evolutionary record. Clearly, the most reliable evolutionary patterns will come from sequences in which the temporal record is most nearly complete.

Many recent studies of evolutionary patterns, including the one we summarize below, are based on fossils from alluvial deposits. In the alluvial regime, episodes of deposition and erosion are sporadic, and may occupy only a relatively small amount of the geologic time represented compared to periods of relative stability (Kraus and Bown, *in press*). This has led some to conclude that the fossil record is equally imperfect; however, periods of stability in the alluvial regime nearly always are occupied by soil formation, and fossils are commonly incorporated into the soils throughout this period (Bown and Kraus, 1981b). Thus, although paleosols represent gaps in the record of sediment accumulation, they do not necessarily reflect gaps in time. On the contrary, aside from periods of erosion (which often can be restored by reference to sections lateral to erosional scours), the time record may be virtually complete (Behrensmeyer, 1982; Kraus and Bown, *in press*); hence ancient alluvial sequences may provide an excellent testing ground for patterns of evolution. Admittedly, fossil samples from paleosols are usually time-averaged, and their temporal resolution can be no finer than the duration of soil formation. But this is relatively brief in geologic terms (2×10^3-1.4×10^4 years for similar modern soils: Bown and Kraus, 1981a; Kraus and Bown, *in press*). Using Gould's guidelines cited above, we should expect to see evidence of, or actually to record punctuations (if they occur) between adjacent paleosols, or possibly within a single paleosol.

A CASE FROM FOSSIL MAMMALS: PALEONTOLOGICAL SPECIES AND EVOLUTION OF EARLY EOCENE OMOMYID PRIMATES

The early Eocene Willwood Formation of the Bighorn Basin, Wyoming, is an alluvial deposit consisting largely of well exposed, richly fossiliferous superposed paleosols. It is ~770 m thick and spans about 3.5-4.0 million years. Few significant erosional gaps are known, hence the Willwood is temporally very complete (Kraus and Bown, in press). With few exceptions, fossils occur at intervals of 5 m or less throughout the sequence, and most can be determined to come from specific paleosols (~2,000-14,000 years duration, see above). Only rarely is it not possible to ascertain the exact paleosol source, and nearly all fossils can be bracketed within intervals of <10m, a thickness not known to contain more than four paleosols in the Willwood Formation. For these reasons, vertebrate fossils from the Willwood Formation constitute a major source of empirical evidence on tempo and mode in evolution.

Tarsier-like primates of the family Omomyidae have been found throughout this stratigraphic section. Over 700 specimens of omomyids (jaws and teeth) are now known from the Bighorn Basin and the adjoining Clark's Fork Basin—more than from any other basin in the world. They are among the most intensively studied Willwood fossils, and they provide an excellent illustration of how an improved fossil record can complicate the process of species discrimination.

Earlier researchers, dealing with only a small number of omomyid specimens from these beds, recognized several species based on differences of size and morphology. Most are still considered distinct (though some under different names), and several new taxa have been added in the last decade. Although precise stratigraphic occurrences of these taxa were unknown until recently, it has become clear that many species differ from others in age as well as morphology. But as a result of the large number of specimens collected over the last 20 years, the morphological and stratigraphic gaps that once existed between certain species have been filled in (Rose and Bown, 1984a). This has substantially improved our understanding of the phylogeny and interrelationships of early Eocene omomyids, but it has made discrimination between species—and, in some cases, genera—much more difficult.

Several lineages of omomyids can now be identified in the Bighorn Basin. The two best documented ones are: (1) small omomyids belonging to the genus *Teilhardina;* and (2) larger omomyids assigned to *Tetonius* and *Pseudotetonius*. Documented evolutionary changes in both lineages involve antemolar mandibular teeth, especially the last premolar (P_4) in the *Teilhardina* lineage, but teeth anterior to P_4 in the *Tetonius-Pseudotetonius* lineage (Rose and Bown, 1984a).

Teilhardina americana (Fig. 2), the oldest known North American omomyid, is characterized by a relatively tall and narrow P_4 dominated by the protoconid, and a simple P_3 lacking paraconid and metaconid. It appears to have descended from the slightly older European *T. belgica*. The two are quite similar, and if both were known from successive strata in the same local section they certainly would be difficult to separate. However, minor morphological differences (e. g., somewhat taller P_4 with slightly lower metaconid) justify their specific distinction pending more data, and this is strengthened by their geographic separation. Specimens clearly belonging to *T. americana* occur only in the lowest 50 m of section in the southern and central Bighorn Basin, but the lineage persists throughout almost all of the lowest 200 m. In the upper part of this sequence (180-190 m), specimens usually have been considered to represent a distinct species (placed in a different genus, *Tetonoides*, by Rose and Bown [1984a] and most previous authors, but here considered a new species of *Teilhardina*), characterized by broader molars, a broader, lower-crowned, and more molarized P_4 (Fig. 3), and development of the paraconid and metaconid on P_3. If only the samples from these two disjunct intervals were known, there would be little dispute or difficulty in recognizing two species.

In recent years an increasing number of specimens has been discovered that bridge the morphological gap between these two species. All intermediates come from stratigraphic levels in between the two species, and they display a progressive increase in molarization of P_4, trigonid development of P_3, and breadth of the cheek teeth with higher stratigraphic level (Fig. 3), thus documenting the gradual transition between the two. Although the differences between stratigraphically successive samples are extremely subtle, cumulative change through the entire sequence (between the end members) is easily perceived, and is at least of species magnitude. Unless

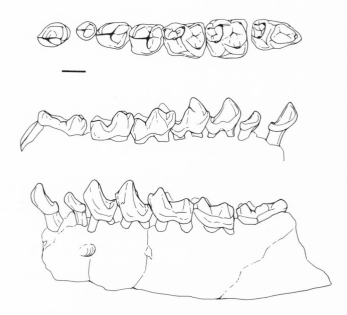

Figure 2. Holotype specimen of *Teilhardina americana* (University of Wyoming 6896), in crown, lingual, and buccal views (top to bottom), to illustrate primitive morphology of omomyids in this study. Scale bar is one millimeter.

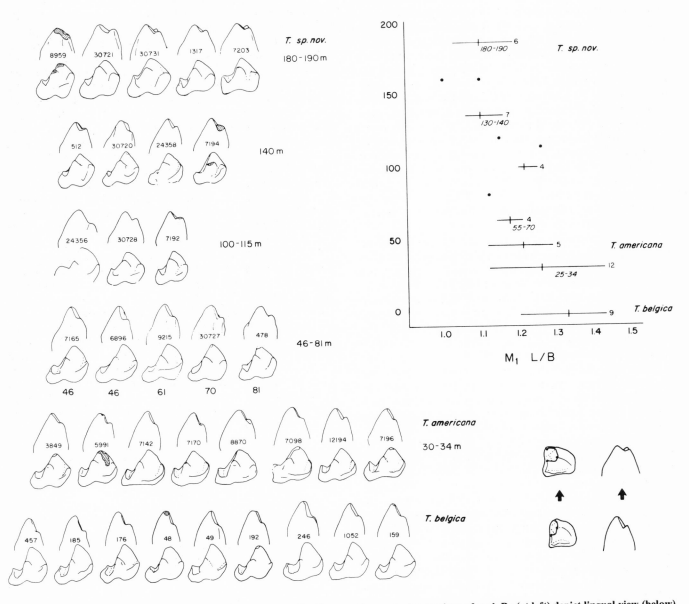

Figure 3. Morphological changes in P_4 and M_1 in *Teilhardina* lineage. Camera lucida tracings of each P_4 (at left) depict lingual view (below) and posterior view of trigonid (above). All are drawn to same scale, and specimen numbers (various institutions) and meter levels are shown; complete documentation and discussion will be presented elsewhere. Figure at lower right compares crown view (left) and posterior aspect of trigonid of *T. americana* (below) and *T.* new species (above). Stratigraphic distribution of M_1 length:breadth ratios (upper right) shows apparent decrease in ratio (*i. e.*, increase in relative breadth) up-section. Horizontal bar is range, vertical slash is mean; sample size shown at right (dots are single specimens). Numbers in italics are meter levels.

we are to ignore the obvious evolution that occurred, the boundary between the two species must be drawn somewhere in the interval containing intermediates, but its precise position is wholly arbitrary. Notably, molar crown area reveals no significant difference between the two species (highest and lowest samples of *Teilhardina* in Fig. 4), despite some minor oscillations and an apparent shift toward slightly smaller molars in the middle of the sequence. If only this attribute were examined, one might conclude that these samples represent a single species in stasis—quite a different interpretation than that based on all available evidence.

Considerably more data are available for the second lineage, consisting of *Tetonius* and *Pseudotetonius*.

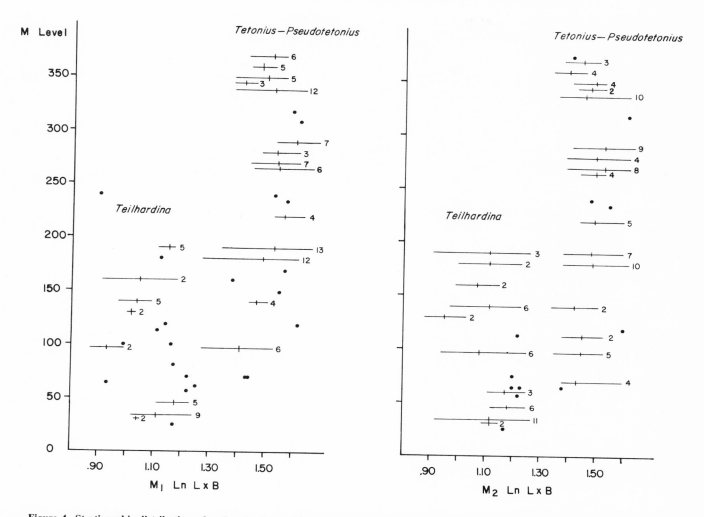

Figure 4. Stratigraphic distribution of molar size (natural logarithm of length x breadth) in both lineages of omomyids, showing apparent stasis (minor oscillations of mean). Samples indicated as in Figure 3.

Tetonius coexisted in the lower part of its range with *Teilhardina,* and probably descended directly from the latter, but perhaps not in the Bighorn Basin. This lineage endured much longer than *Teilhardina,* occurring in the stratigraphic interval from ~70 m to 370 m in the study area. Since 1915, two species have been distinguished by their antemolar dentitions, *Tetonius homunculus* and *"Tetonius" ambiguus* (the latter placed in the new genus *Pseudotetonius* by Bown, 1974).

The earliest well known representatives of *Tetonius,* traditionally referred to *T. homunculus*[1], closely resemble *Teilhardina americana* but are somewhat larger, with relatively broader cheek teeth and relatively more reduced canine and P_2. *Pseudotetonius ambiguus* is readily separated from *T. homunculus* by its conspicuously reduced and compacted anterior teeth, coupled with greater hypertrophy of the medial incisor (Matthew, 1915; Bown, 1974)—although its upper and lower P4 and molars are virtually indistinguishable from *Tetonius*. As in the *Teilhardina* lineage, recently it has become clear that these two disparate morphologies are widely separated stratigraphically, *P. ambiguus* being restricted to the upper 20-30 m of the stratigraphic range of the lineage and *T. homunculus* to the lower 100 m or so. (As an aside, we know of no evidence to support Stanley's (1985, p. 21) claim that teeth of primitive *Tetonius* have been found "in sediments about 1.5 ma younger than the alleged interval of transition." Even if this could be substantiated, it would not invalidate our interpretation of phyletic evolution in this lineage (contrary to Stanley's assertion). In fact, it would suggest instead that a branching event had occurred at about the time of stage 1—thereby strengthening our contention that the two forms must be assigned to separate species.)

From the intervening strata there are now numerous specimens, all of which display morphology intermediate between *Tetonius homunculus* and *Pseudotetonius ambiguus*—e. g., absence of P_2 associated with variably

---

[1]. As detailed elsewhere (Bown and Rose, *in preparation*), *Tetonius homunculus* technically is not the valid name for these early members of *Tetonius;* they require a new species name. However, pending our justification of this assessment, we use the name *T. homunculus* here.

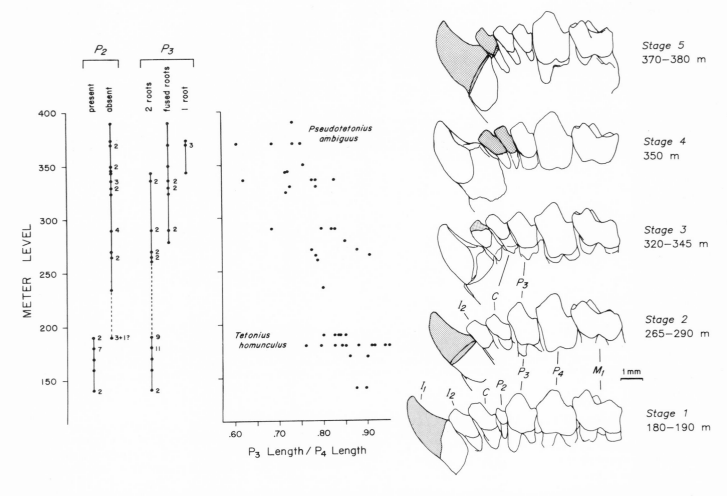

Figure 5. Stratigraphic distribution of some lower antemolar traits in *Tetonius-Pseudotetonius* lineage. Dental configurations at right depict individuals or composites representing five stages, with intervals of their most common occurrence; in fact, each stage overlaps stratigraphically with preceding and succeeding ones. Modified from Rose and Bown (1984).

smaller anterior teeth, compression or fusion of the roots of P_3, and enlargement of I_1 (Fig. 5). These traits are progressively better expressed as stratigraphic level increases, documenting a smooth transition between the two. To recognize the gradual and mosaic nature of the species transformation, we divided the best documented part of the lineage into five successive morphological stages (Rose and Bown, 1984a). The choice of five stages was arbitrary, each representing a landmark in temporally stratified variability based on our present state of knowledge. There are, in fact, no discrete breaks between successive stages, and many specimens straddle stages rather than fitting conveniently into one of them. Variation, including overlap of successive stages, occurs at each level (Fig. 6). For example, in the 180-190 m interval most specimens represent stage 1, but some individuals lack P_2, loss of which is the principal distinction of stage 2. Thus *T. homunculus* includes individuals at stage 1 and some at stage 2 (those from below 200 m), and *P. ambiguus* includes specimens at stage 5 and most at stage 4 (350 m and above); intermediates include specimens at stages 2-4.

Although the end points of this lineage show morphologic differences consistent with specific (we believe generic) separation, the boundary between the two species must be arbitrary. In this case, however, even a boundary based on an arbitrary time line is unsatisfactory because of overlapping successive morphologies. To place the boundary in the "convenient" gap in the record just above 190 m (where we still lack data on anterior teeth) would result in stage 2 individuals, which occur on both sides of the gap, being classified as both species. Conversely, any morphologic species definition—(e. g., presence of P_2 defines *Tetonius homunculus*) would require assigning only slightly different, essentially contemporaneous individuals from a single horizon to separate species.

Here again, molars remained conservative, showing essentially no change in size or crown morphology, apart from minor oscillations about mean size, through 300 m

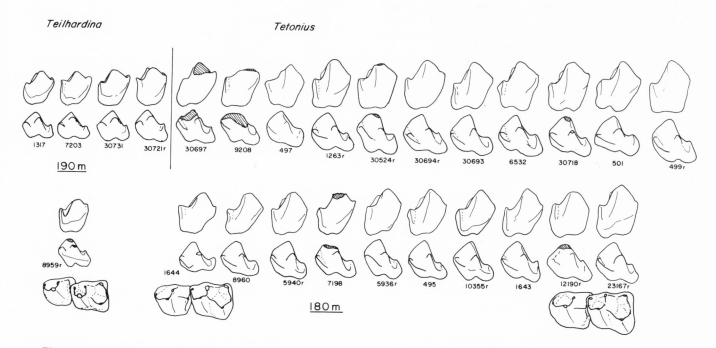

Figure 6. Variability in P_4 morphology in omomyids at 180-190 m interval. At each level, most of sample derives from a single paleosol. Two genera are represented. *Tetonius* sample includes both stage 1 and stage 2 individuals, and size variation in this interval exceeds that at any other, indicating either one highly variable population or two species inseparable (except arbitrarily) by available evidence (stages 1 and 2 do not segregate by size).

of section or ~1.5 my (Fig. 4). Thus, while the anterior dentition underwent a gradual but marked modification, the molars appear to have been in stasis.

Smaller samples of both omomyid lineages from a thicker section in the Clark's Fork Basin show the same changes, in the same sequence, as just described for these lineages in the central Bighorn Basin. Continuous and gradual transition between species also has been documented in adapid primates (Gingerich and Simons, 1977; Bookstein and others, 1978; Rose and Bown, 1984b) and other mammals from the Willwood Formation (Gingerich, 1976, 1980; Bown, 1979; Bown and Rose, *unpublished data*), although these examples have been based mainly (or only) on molar size and merit more thorough substantiation. In all of these cases, where the record is relatively dense, species boundaries have become ambiguous. These data strongly contradict Stanley's (1982, p. 472) recent statement that among virtually all lineages of Willwood mammals there is "no example of significant phyletic evolution over spans of 2-3 Myr." On the contrary, in each of the few well-studied lineages, gradual and continuous change is the rule rather than the exception.

CONCLUSIONS

Disagreements over what constitutes a species in the fossil record and how to discriminate between species are not trivial concerns for taxonomists only. Paleontological species (whether comparable to biological species or not) are the basic units of most paleontological and many evolutionary studies. In this essay we have attempted to show that species discrimination is closely related to more fundamental questions on how evolution has occurred. Substantial improvements in the fossil record have been made in recent years, particularly accumulation of many new collections with precise stratigraphic control. These are providing significant new data for testing evolutionary models and reevaluating concepts of species viewed through time.

Examples discussed here indicate that a dense fossil record often reveals gradual transformations between species, both within lineages and at branching events, thus complicating species recognition. In the best documented lineages, this compels us either to impose arbitrary species boundaries or none at all. We suspect that where circumstances permit more intensive sampling, examples like these will increase, forcing paleontologists to look more carefully at how they define paleontological species. From a taxonomic outlook, this will make our work more difficult, but from an evolutionary viewpoint it will undoubtedly be more rewarding.

ACKNOWLEDGMENTS

For access to and information about omomyid primates used in the study summarized here we are grateful to Drs. D. Baird, R. T. Bakker, M. R. Dawson, R. J. Emry, P. D. Gingerich, M. Godinot, F. A. Jenkins, Jr., J. A. Lillegraven, L. D. Martin, M. C. McKenna,

G. E. Meyer, J. H. Ostrom, D. E. Savage, C. R. Schaff, E. L. Simons, and R. W. Wilson. We thank Elaine Kasmer for preparing Figures 1 and 2, and T. Urquhart for photography. Drs. P. D. Gingerich, D. W. Krause, B. Van Valkenburgh, A. C. Walker, and especially D. B. Weishampel provided constructive comments that improved the manuscript. This research has been supported by grants from the American Philosophical Society (Penrose #9269) and the National Science Foundation (BSR-8215099 and BSR-8500732) to KDR. Finally, we thank Ms. Kathryn Flanagan and Mrs. Anne Roe Simpson for the invitation to contribute to this volume in memory of George Gaylord Simpson.

REFERENCES CITED

Beden, M., 1983, Family Elephantidae, in Harris, J. M., ed., Koobi Fora Research Project, v. 2, The fossil ungulates: Proboscidea, Perissodactyla, and Suidae: Oxford, Clarendon Press, p. 40-129.

Behrensmeyer, A. K., 1982, Time resolution in fluvial vertebrate assemblages: Paleobiology, v. 8, p. 211-227.

Bookstein, F.L., Gingerich, P. D., and Kluge, A. G., 1978, Hierarchical linear modeling of the tempo and mode of evolution: ibid., v. 4, p. 120-134.

Boucot, A., 1983, Rates of behavioral evolution, community evolution, biogeographic history, and the constraints on the evolution of species imposed by these classes of data obtained from the fossil record, in Chaline, J., ed., 1983, p. 73-77.

Bown, T. M., 1974, Notes on some early Eocene anaptomorphine primates: Contributions to Geology, University of Wyoming, v. 13, p. 19-26.

―――― 1979, Geology and mammalian paleontology of the Sand Creek Facies, lower Willwood Formation (lower Eocene), Washakie County, Wyoming: Geological Survey of Wyoming, Memoir 2, 151 p.

Bown, T. M., and Kraus, M. J., 1981a, Lower Eocene alluvial paleosols (Willwood Formation, northwest Wyoming, U.S.A.) and their significance for paleoecology, paleoclimatology, and basin analysis: Palaeogeography, Palaeoclimatology, Palaeoecology, v. 34, p. 1-30.

―――― 1981b, Vertebrate fossil-bearing paleosol units (Willwood Formation, lower Eocene, northwest Wyoming, U.S.A.): implications for taphonomy, biostratigraphy, and assemblage analysis: ibid., v. 34, p. 31-56.

Cain, A. J., 1960, Animal species and their evolution: New York, Harper & Row, 190 p.

Campbell, B. G., 1974, A new taxonomy of fossil man: Yearbook of Physical Anthropology, 1973, p. 194-201.

Chaline, J., 1983, Les rôles respectifs de la spéciation quantique et diachronique dans la radiation des Arvicolidés (Arvicolidae, Rodentia), conséquences au niveau des concepts, in Chaline, J., ed., 1983, p. 83-89.

Chaline, J. (ed.), 1983, Modalités, rythmes, mécanismes de l'évolution biologique: Paris, Colloques internationaux du Centre national de la recherche scientifique, no. 330, 335 p.

―――― 1985, Evolutionary data on steppe lemmings (Arvicolidae, Rodentia), in Luckett, W. P., and Hartenberger, J.-L., eds., Evolutionary relationships among rodents: a multidisciplinary analysis: New York, Plenum, p. 631-641.

Crusafont-Pairó, M., and Reguant, S., 1970, The nomenclature of intermediate forms: Systematic Zoology, v. 19, p. 254-257.

Dingus, L., and Sadler, P. M., 1982, The effects of stratigraphic completeness on estimates of evolutionary rates: ibid., v. 31, p. 400-412.

Eldredge, N., and Cracraft, J., 1980, Phylogenetic patterns and the evolutionary process: New York, Columbia University Press, 349 p.

Eldredge, N., and Gould, S. J., 1972, Punctuated equilibria: an alternative to phyletic gradualism, in Schopf, T. J. M., ed., Models in paleobiology: San Francisco, Freeman, Cooper & Co., p. 82-115.

Enay, R., 1983, Spéciation phylétique dans le genre d'ammonite téthysien *Semiformiceras* Spath, du Tithonique inférieur des chaînes bétiques (Andalousie, Espagne), in Chaline, J., ed., 1983, p. 115-123.

Ewer, R. F., 1967, Professor Tobias's new nomenclature: South African Journal of Science, v. 63, p. 281.

Gingerich, P. D., 1974, Size variability of the teeth in living mammals and the diagnosis of closely related sympatric fossil species: Journal of Paleontology, v. 48, p. 895-903.

―――― 1976, Paleontology and phylogeny: patterns of evolution at the species level in early Tertiary mammals: American Journal of Science, v. 276, p. 1-28.

―――― 1980, Evolutionary patterns in early Cenozoic mammals: Annual Review of Earth and Planetary Sciences, v. 8, p. 407-424.

―――― 1983, Rates of evolution: effects of time and temporal scaling: Science, v. 222, p. 159-161.

―――― 1984, Punctuated equilibria—where is the evidence?: Systematic Zoology, v. 33, p. 335-338.

―――― 1985, Species in the fossil record: concepts, trends, and transitions: Paleobiology, v. 11, p. 27-41.

Gingerich, P. D., and Simons, E. L., 1977, Systematics, phylogeny, and evolution of early Eocene Adapidae (Mammalia, Primates) in North America: Museum of Paleontology, University of Michigan, Contributions, v. 24, p. 245-279.

Ginzburg, L. R., 1981, Bimodality of evolutionary rates: Paleobiology, v. 7, p. 426-429.

Godinot, M., 1985, Evolutionary implications of morphological changes in Palaeogene primates: Special Papers in Palaeontology, no. 33, p. 39-47.

Gould, S. J., 1982a, Darwinism and the expansion of evolutionary theory: Science, v. 216, p. 380-387.

———— 1982b, The meaning of punctuated equilibrium and its role in validating a hierarchical approach to macroevolution, in Milkman, R., ed., Perspectives on evolution: Sunderland, Massachusetts, Sinauer, p. 83-104.

———— 1983, Dix-huit points au sujet des équilibres ponctués, in Chaline, J., ed., 1983, p. 39-41.

———— 1985, The paradox of the first tier: an agenda for paleobiology: Paleobiology, v. 11, p. 2-12.

Gould, S. J., and Eldredge, N., 1977, Punctuated equilibria: the tempo and mode of evolution reconsidered, ibid., v. 3, p. 115-151.

Hecht, M., 1983, Microevolution, developmental processes, paleontology and the origin of vertebrate higher categories, in Chaline, J., ed., 1983, p. 289-294.

Hoffman, A., 1982, Punctuated versus gradual mode of evolution: Evolutionary Biology, v. 15, p. 411-436.

Imbrie, J., 1957, The species problem with fossil animals, in Mayr, E., ed., The species problem: Washington, D. C., American Association for the Advancement of Science, Publ. 50, p. 125-153.

Jaeger, J.-J., 1983, Equilibres ponctués et gradualisme phylétique: un faux débat?, in Chaline, J., ed., 1983, p. 145-153.

Jones, J. S., 1981, An uncensored page of fossil history: Nature, v. 293, p. 427-428.

Kraus, M. J., and Bown, T. M., in press, Paleosols and time resolution in alluvial stratigraphy, in Wright, P. V., ed., Paleosols, their origin, classification, and interpretation: London, Blackwell and Princeton University Geological Series.

Krishtalka, L., and Stucky, R. K., 1985, Revision of the Wind River faunas, early Eocene of central Wyoming. Part 7. Revision of *Diacodexis* (Mammalia, Artiodactyla): Annals of Carnegie Museum, v. 54, p. 413-486.

Lazarus, D., Scherer, R. P., and Prothero, D. B., 1985, Evolution of the radiolarian species-complex *Pterocanium:* a preliminary survey: Journal of Paleontology, v. 59, p. 183-220.

Levinton, J. S., and Simon, C. M., 1980, A critique of the punctuated equilibria model and implications for the detection of speciation in the fossil record: Systematic Zoology, v. 29, p. 130-142.

Maglio, V. J., 1971, The nomenclature of intermediate forms: an opinion, ibid., v. 20, p. 370-373.

———— 1973, Origin and evolution of the Elephantidae: American Philosophical Society, Transactions, new series, v. 63, part 3, p. 1-149.

Malmgren, B. A., Berggren, W. A., and Lohmann, G. P., 1984, Species formation through punctuated gradualism in planktonic foraminifera: Science, v. 225, p. 317-319.

Martin, L. D., 1984, Phyletic trends and evolutionary rates: Carnegie Museum of Natural History, Special Publication, no. 8, p. 526-538.

Matthew, W. D., 1915, A revision of the lower Eocene Wasatch and Wind River faunas. Part IV.—Entelonychia, Primates, Insectivora (part): American Museum of Natural History, Bulletin, v. 34, p. 429-483.

Mayr, E., 1969, Principles of systematic zoology: New York, McGraw-Hill, 428 p.

———— 1970, Populations, species, and evolution: Cambridge, Belknap Press, 453 p.

Newell, N. D., 1956, Fossil populations: Systematics Association (London), Publ. 2, p. 63-82.

Ozawa, T., 1975, Evolution of *Lepidolina multiseptata* (Permian foraminifer) in East Asia: Kyushu University, Memoirs of the Faculty of Science, Series D, Geology, v. 23, p. 117-164.

Penny, D., 1985, Two hypotheses on Darwin's gradualism: Systematic Zoology, v. 34, p. 201-205.

Prothero, D. R., and Lazarus, D. B., 1980, Planktonic microfossils and the recognition of ancestors: ibid., v. 29, p. 119-129.

Raup, D. M., and Crick, R. E., 1981, Evolution of single characters in the Jurassic ammonite *Kosmoceras:* Paleobiology, v. 7, p. 200-215.

Rhodes, F. H. T., 1956, The time factor in taxonomy: Systematics Association (London), Publ. 2, p. 33-52.

Rose, K. D., and Bown, T. M., 1984a, Gradual phyletic evolution at the generic level in early Eocene omomyid primates: Nature, v. 309, p. 250-252.

———— 1984b, Early Eocene *Pelycodus jarrovii* (Primates: Adapidae) from Wyoming: phylogenetic and biostratigraphic implications: Journal of Paleontology, v. 58, p. 1532-1535.

Sadler, P. M., 1981, Sediment accumulation rates and the completeness of stratigraphic sections: Journal of Geology, v. 89, p. 569-584.

Schindel, D. E., 1980, Microstatigraphic sampling and the limits of paleontologic resolution: Paleobiology, v. 6, p. 408-426.

———— 1982, The gaps in the fossil record: Nature, v. 297, p. 282-284.

Schopf, T. J. M., 1981, Punctuated equilibrium and evolutionary stasis: Paleobiology, v. 7, p. 156-166.

———— 1982, A critical assessment of punctuated equilibria. I. Duration of taxa: Evolution, v. 36, p. 1144-1157.

Simpson, G. G., 1943, Criteria for genera, species, and subspecies in zoology and paleozoology: New York Academy of Sciences, Annals, v. 44, p. 145-178.

———— 1944, Tempo and mode in evolution: New York, Columbia University Press, 237 p.

———— 1951, The species concept: Evolution, v. 5, p. 285-298.

———— 1953, The major features of evolution: New York, Columbia University Press, 434 p.

———— 1961, Principles of animal taxonomy: New York, Columbia University Press, 247 p.

Stanley, S. M., 1979, Macroevolution: San Francisco, W. H. Freeman and Co., 332 p.

———— 1982, Macroevolution and the fossil record: Evolution, v. 36, p. 460-473.

———— 1985, Rates of evolution: Paleobiology, v. 11, p. 13-26.

Sylvester-Bradley, P. C. (ed.), 1956, The species concept in paleontology: Systematics Association (London), Publ. 2, 156 p.

Thaler, L., 1983, Image paléontologique et contenu biologique des lignées évolutives, in Chaline, J., ed., 1983, p. 327-335.

Thomas, G., 1956, The species conflict—abstractions and their applicability: Systematics Association (London), Publ. 2, p. 17-31.

Tintant, H., 1980, Problématique de l'espèce en Paléozoologie: Mémoires de Societe Zoologique de France, v. 3, no. 40, p. 321-372.

———— 1983, Cent ans après Darwin, continuité ou discontinuité dans l'évolution, in Chaline, J., ed., 1983, p. 25-37.

Tobias, P. V., 1969, Bigeneric nomina: a proposal for modification of the rules of nomenclature: American Journal of Physical Anthropology, v. 31, p. 103-106.

———— 1978, The earliest Transvaal members of the genus *Homo* with another look at some problems of hominid taxonomy and systematics: Zeitschrift für Morphologie und Anthropologie, v. 69, p. 225-265.

van Andel, T. H., 1981, Consider the incompleteness of the fossil record: Nature, v. 294, p. 397-398.

Vrba, E. S., 1980, Evolution, species, and fossils: how does life evolve?: South African Journal of Science, v. 76, p. 61-84.

West, R. M., 1979, Apparent prolonged evolutionary stasis in the middle Eocene hoofed mammal *Hyopsodus:* Paleobiology, v. 5, p. 252-260.

Wiley, E. O., 1978, The evolutionary species concept reconsidered: Systematic Zoology, v. 27, p. 17-26.

Williamson, P. G., 1981, Palaeontological documentation of speciation in Cenozoic molluscs from Turkana Basin: Nature, v. 293, p. 437-443.

MANUSCRIPT RECEIVED DECEMBER 17, 1985
REVISED MANUSCIRPT RECEIVED MARCH 12, 1986
MANUSCRIPT ACCEPTED JUNE 18, 1986

Nycticeboides simpsoni and the morphology, adaptations, and relationships of Miocene Siwalik Lorisidae

R. D. E. MacPHEE — *Department of Anatomy, Duke University Medical Center, Durham, North Carolina 27710*

LOUIS L. JACOBS — *Department of Geological Sciences, Southern Methodist University, Dallas, Texas 75275*

ABSTRACT

The fossil record of lorisiforms in Asia is currently restricted to specimens recovered from a half-dozen localities of Miocene age (13 Ma to 7 Ma) in the Siwalik Group of northern Pakistan and in related deposits of India. More than one lorisid taxon is represented in the Pakistan material, but *Nycticeboides simpsoni* Jacobs, 1981 is currently the only named species. A partial skeleton of *Nycticeboides,* although poorly preserved, possesses diagnostic lorisid synapomorphies of the auditory region and the vertebral column. The fact that *Nycticeboides* was a small animal is important for understanding its ecology. A primate frugivore with the M_1 dimensions of *Nycticeboides* should have a body weight of only about 500 g according to commonly-used regression statistics. However, if *Nycticeboides* was mostly insectivorous, and its molar teeth scaled to body size in the manner characteristic of highly insectivorous primates and non-zalambdodont insectivores, then it may have weighed much less than this estimate.

INTRODUCTION

In a very real sense, George Gaylord Simpson was responsible for initiating the research that led to this paper. One of Dr. Simpson's special and continuing interests was the evolution of primates, particularly "prosimian" primates (*e. g.,* Simpson, 1967). He was therefore the logical person to ask for help when one of us (LLJ) was presented with the task of identifying a small, primatelike molar collected by the joint Yale/Geological Survey of Pakistan (YGSP) expedition to the Siwaliks in 1976. After examining the specimen, Dr. Simpson sent back a brief note outlining his provisional interpretation. His note and accompanying sketch, reproduced here as Figure 1, were informal and not meant for publication or citation. They were, however, meant to be helpful and encouraging to a graduate student, and it is in this spirit that we include them here, to add a personal touch to this paper.

Dr. Simpson reached no definite conclusion about the taxonomic allocation of the tooth (YGSP 26008). Indeed, he was not even willing to grant that it was positively primate, although he was clearly impressed by the tooth's general resemblance to molars of some Paleogene plesiadapoids. As things turned out, his hunch that YGSP 26008 had a "primatish feel" was correct: additional remains, including a partial skeleton of an incontestible lorisid with a similar LM2, were later found at other YGSP localities. The skeleton became the holotype of *Nycticeboides simpsoni,* a name chosen in recognition of Simpson's contributions to primate paleontology (Jacobs, 1981).

The present paper has several purposes. The first is to describe all heretofore unpublished lorisid (or lorisidlike) material in collections made by the Siwalik expeditions organized by the Geological Survey of Pakistan and by Yale and Harvard Universities. This includes not only isolated finds like YGSP 26008, but also the associated elements of the *Nycticeboides* holotype, not described in the original report on this taxon. Our second purpose is to evaluate what new light the Siwalik forms shed on the origins of Lorisidae and their complex biogeography. Our third and last aim is to illustrate some of the difficulties involved in interpreting diet and body size in small extinct primates, currently an important topic in paleoprimatology.

All osteological measurements in tables are in millimeters.

MIOCENE LORISIFORMS FROM THE SIWALIKS OF PAKISTAN

Nycticeboides simpsoni is still the only named taxon of Tertiary Asian primates that unquestionably belongs in Lorisidae. (Other described Siwalik primates of "prosimian" grade— *Indraloris himalayensis* [= *I. lulli*], *Sivaladapis palaeindicus,* and *S. nagrii*— are adapids according to Gingerich and Sahni [1979, 1984].) Some of the new Siwalik specimens (Table 1), described in ascending chronostratigraphic order in the following paragraphs, may eventually deserve their own nomina. However, much better material will have to be found before the erection of new taxa is warranted, and for the present we simply assign these remains to Lorisidae *gen. et spec. indet.*

Fossil localities in the Siwalik Group of Pakistan are of two principal kinds: red bed localities, and those

31 Mar 76

Small primatelike tooth from the Siwaliks. Right lower M, probably M_2.

This has a primatish feel to it, but it does not belong to any genus of primates that I know or can locate. It is quite distinct from any Recent or described fossil lorisid. The only fossil lorisid hitherto described from the Siwaliks, or from Asia as far as I know, is *Indraloris** Lewis, 1933, which is definitely unlike this specimen. Curiously enough the most similar single teeth I know are from the American & European Paleocene & Eocene, for example some Paromomyidae, but the distinction is still at least generic. I am not suggesting that it is a paromomyid, but just that it is a funny thing to find in the Siwalik late Tertiary.

I have compared it only with primates. I once mistook bat teeth for lorisids, so I'm gun-shy. I'd want to compare insectivores (& bats!) before saying positively that this is a new primate.

G.G.S.

*Pseudoloris is a tarsiid, not a lorisid, & this tooth is not tarsiid.

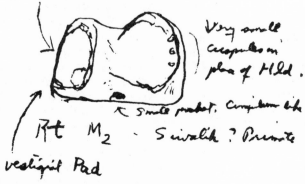

Figure 1. Dr. Simpson's comments and sketch of first lorisid molar discovered in Siwalik Group rocks of Indian subcontinent.

TABLE 1. CATALOG OF SIWALIK LORISIDAE.

Accession No.	Specimen	Maximum length	Maximum width	Yale/Geological Survey of Pakistan	Approximate Age (1)
YGSP 8091	holotype, N. simpsoni (2)			Locality 363	7.0-7.5 Ma (T)
YGSP 26009	left LP4	2.3	1.3	Locality 182A	ca. 8.0 Ma (D)
YGSP 26008	left LM2	2.5	2.0	Locality 182A	ca. 8.0 Ma (D)
YGSP 26007	left UM3	1.7	2.7	Locality 450	9.0 Ma (Fn)
YGSP 26006	left UM2	2.1	3.1	Locality 450	9.0 Ma (Fn)
YGSP 26005	LM3 talonid	—	1.4	Locality 259	9.4 Ma (D)
YGSP 26004	right UP2	2.4	1.2	Locality 491	13.2 Ma (Fc)

(1) As determined by inference from the magnetic polarity stratigraphy sequence for the Siwaliks, according to the following key:

  (D) Locality directly in magnetic section (Barry et al., 1982; Tauxe, 1980).

  (Fc) Faunal correlation with Chinji Formation localities dated by Johnson et al. (1985).

  (Fn) Faunal correlation with Nagri Formation localities dated by Johnson et al. (1985)

  (T) Lithologic unit traced into measured section dated by Tauxe and Opdyke (1982).

(2) For dental measurements of holotype, see Jacobs (1981).

formed in channel systems (Badgley, *in press*). Channel system localities are characterized by having low representations (by element) of a large number of taxa, in contrast to red bed localities, which tend to have high representations of one or a few taxa. Red bed contexts have been interpreted as representing accumulations made on land surfaces. All Siwalik localities that have produced articulated or reasonably complete specimens have been red beds, although not all red bed localities have produced articulated specimens.

Except for the holotype of *Nycticeboides simpsoni*, which was collected at a red bed locality, all other Siwalik lorisids were found while washing matrix taken from channel system localities (Fig. 2). The partial skeleton of *Nycticeboides simpsoni* is separately described in the section which follows these brief notes on isolated finds.

In this paper, we place the living lorisiforms in two families: Lorisidae, containing *Loris* and *Nycticebus* of Asia and *Arctocebus* and *Perodicticus* of Africa; and Galagidae, containing *Galago* and *Euoticus,* both of which are restricted to Africa (see Schwartz and Tattersall, 1985).

ISOLATED TEETH

Right UP2 (YGSP 26004, loc. y491): This specimen exhibits a single high cusp, with a sharp anterior ridge extending to the mesiobuccal stylar area and a blunter posterior ridge (Fig. *3A,B*). The tooth is morphologically rather *Loris*-like, but its cusps are more acute and trenchant than those of the UP2 of the living slender loris. YGSP 26004 has two roots and in that character resembles homologous teeth of *Loris, Arctocebus,* and *Galago,* but differs from the single-rooted UP2 of *Nycticeboides, Nycticebus,* and *Perodicticus*. All other extant lorisiforms have triple-rooted UP3s and UP4s, although Walker (1978) reports that these teeth are double-rooted in *Mioeuoticus* (East Africa, Miocene).

LM3 (YGSP 26005, loc. y259): This very small specimen (Fig. *4D*), consisting only of the talonid, is considered to be lorisiform because of the asymmetrical arrangement of the entoconid, hypoconid, and hypoconulid. This trait is also seen in extant lorisids, but not in any of the lipotyphlan insectivores so far recovered from the Siwaliks. The LM3 talonid width (1.44 mm) is similar to that of *Galago demidovii* (1.48 mm, N = 1), but is substantially smaller than widths recorded for *Nycticeboides simpsoni* (1.96 mm, holotype) and *Loris tardigradus* (2.2 mm, N = 1). Wear on YGSP 26005 is similar to that seen on the homologous tooth of the holotype of *N. simpsoni*.

Left UM2 (YGSP 26006) and left UM3 (YGSP 26007, both loc. y450): These specimens (Fig. 4*A,C*) seem to be within the size range of the taxon represented by YGSP 26005. They were found together at locality y450,

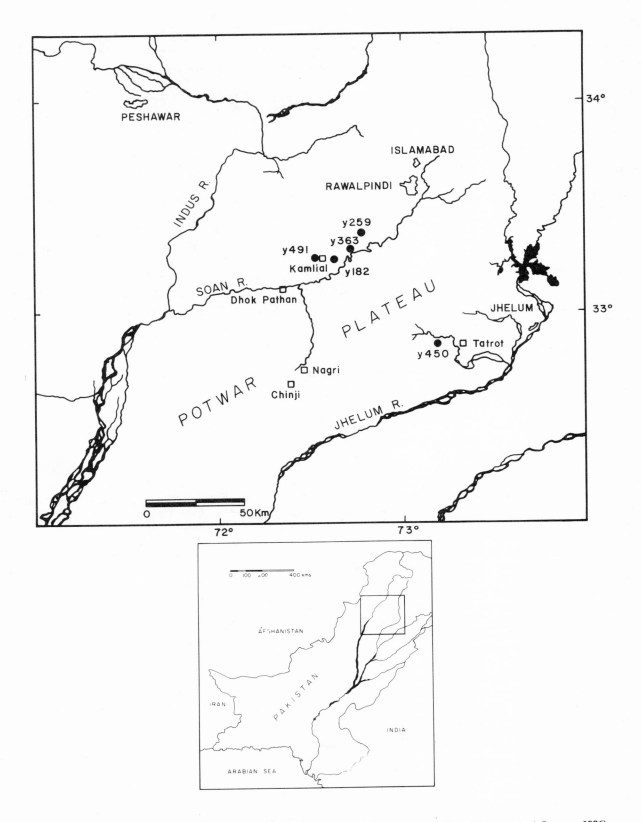

Figure 2. Potwar Plateau area of northern Pakistan, with fossil lorisid localities indicated (current to 1 January 1986).

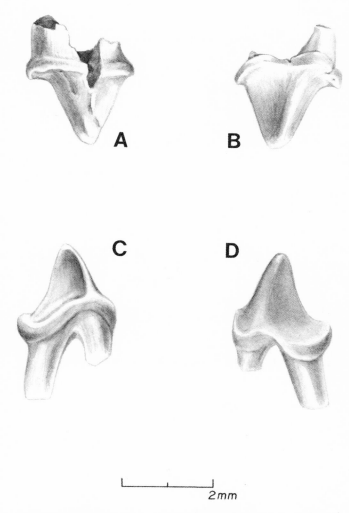

Figure 3. Isolated fossil lorisid teeth from Miocene sites in Siwaliks. A, Lorisid right UP2 (YGSP 26004), buccal view; B, same, lingual view; C, LP4 (YGSP 26009), lingual view; and D, same, buccal view.

tinct posteriorly. YGSP 26009 is distinct from the homologous tooth of *Loris tardigradus* in being premolariform rather than submolariform or molariform, and differs from that of *Nycticebus coucang* and *Perodicticus potto* in being smaller but more trenchant. Cingula are somewhat smaller than in *Nycticeboides*, and much smaller than in *Nycticebus*.

SKELETON OF *Nycticeboides simpsoni* (YGSP 8091)

Introduction

Bruce J. MacFadden found YGSP 8091 weathering out of red siltstone exposed along the Ratha Kas, a creek on the Potwar Plateau of northern Pakistan. The place of discovery, now numbered as locality y363, is stratigraphically the youngest of the Siwalik lorisiform-bearing localities (Table 1). *Nycticeboides simpsoni* is the only species represented at this locality, and all of the bones collected there apparently belong to a single individual.

All of the identifiable cranial and postcranial elements comprising YGSP 8091 are listed in Table 2. Although the matrix immediately surrounding the main concentration of bones was carefully scooped up for later screening, much of the skeleton had apparently washed away by the time of discovery. That the entire skeleton was originally preserved is suggested by several facts, including the large number of bone fragments with complementary, weathered margins (suggesting *in situ* breakage before discovery), and the co-occurrence of the right scapula and humerus (suggesting *in situ* decomposition; cf. Hill, 1979). There are no taphonomic indications, such as tooth marks, that might imply carnivore or scavenger activity. All long bones with preserved articular surfaces show epiphyseal-diaphyseal fusion, indicating that the animal was fully adult at death.

Since the familial affinities of *Nycticeboides* are not in issue, we emphasize other members of Lorisidae in the anatomical comparisons that follow.

Cranium

Portions of the frontal's contributions to both orbital rims and postorbital bars can be pieced together from the available fragments (Fig. 5). Judging from these remains, *Nycticeboides* had large eyes for its body size, as do living lorisids. The upper part of the postorbital bar is comparatively wide, rather robust, triangular in cross section, and exhibits no descending orbital plates. The width of this part of the bar might be taken as a specific resemblance to *Loris*, but there is no evidence that the rest of the orbit's lateral wall was apomorphously expanded (as it is in the living slender loris). *Nycticeboides* seems to have differed from *Loris* in the form of the interorbital area. In *Loris*, the orbits are so closely appressed that the interorbital area is reduced to a thin septum; other lorisids lack this extreme specialization. Unfortunately, the single fragment of the *Nycticeboides* skeleton that bears a recognizable part of the interorbital area is so incomplete that interorbital width cannot be reliably determined.

have a similar degree of wear, and may therefore represent a single individual. Cusps are markedly acute, as in *Loris, Nycticeboides,* and small species of *Galago*. The hypocone, paraconule, and metaconule of UM2 are smaller than the homologous features of *Nycticeboides* and *Loris*. A hypocone is present on UM3.

Left LM2 (YGSP 26008, loc. y182A): This tooth (Figs. 1, 4B), the one worked on by Dr. Simpson, came from a site situated 12 m stratigraphically below a *Sivapithecus* locality (locality y182), previously dated at 8 Ma by magnetic polarity (Tauxe, 1980). The specimen is closer in size to the holotype of *Nycticeboides simpsoni* than are the other specimens described here, although some differences exist in small details of crown morphology (e. g., in shearing blade lengths; see Discussion).

Left LP4 (YGSP 26009, loc. y182A): Locality y182A also yielded a double-rooted LP4 (Fig. 3C,D) similar to that of *Nycticeboides,* but more trenchant. This tooth has a single cusp, with a lingual cingulum that extends from the mesiobuccal area and becomes broader and less dis-

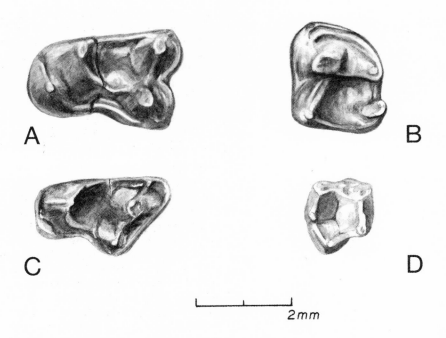

Figure 4. Isolated fossil lorisid teeth from Miocene sites in Siwaliks. *A*, Lorisid left UM2 (YGSP 26006); *B*, left LM2 (YGSP 26008); *C*, left UM3 (YGSP 26007); and *D*, LM3 talonid (YGSP 26005).

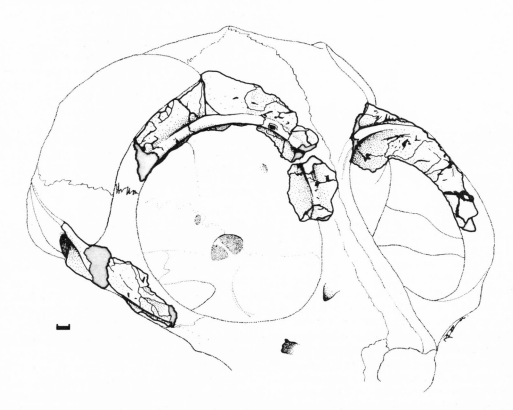

Figure 5. Known parts of orbital region of holotype of *Nycticeboides simpsoni,* projected against imaginary reconstruction of skull. Fragments of other craniofacial bones (nasal, temporal, parietal) are known, but they cannot be precisely placed on reconstruction, and have been omitted. Scale = 1 mm.

TABLE 2. SKELETON OF <u>Nycticeboides simpsoni</u> YGSP 8091.

Region/Element	Parts preserved
1. Cranium	maxillary and mandibular fragments with teeth; right petrosal, with parts of tympanic floor and mastoid; parts of orbits and cranial vault
2. Vertebral Column	
cervicals	axis (centrum and dens only); some neural arches and spinous processes
thoracics	one centrum; some neural arches and spinous processes
lumbars	one neural arch (?)
sacrals, coccygeals	---
3. Thorax and Pectoral Girdle	
ribs	none complete, some large fragments
sternal elements	----
clavicles	----
scapulae	glenoid, acromion, and part of blade of right scapula
4. Forelimb	
arms and forearms	parts of both humeri; proximal articular end of one radius and left ulna; diaphyseal fragments of radii and ulnae
hands	----

Small fragments of both zygomatic arches are preserved, but except for their marked robusticity and fluted external aspect (both general lorisid traits) they are not systematically informative. The posterior edge of the maxillary root of the zygomatic arch lies vertically above the middle of UM3 (Fig. 6), or more posteriorly than in *Loris*, *Nycticebus*, and *Perodicticus* (but cf. *Galago demidovii*). Elevated ridges on small, isolated pieces of neurocranium establish that *Nycticeboides* must have had well developed temporal lines, but whether they were situated high or low on the neurocranium cannot be determined.

Both maxillary alveolar processes and most of the upper dentition were recovered (Fig. 6), but the palatal portion is incomplete and the premaxillaries are missing. As a result the number of incisors and the position of incisive foramina cannot be determined. Part of one of the nasal bones is preserved, but it sheds no light on the size or orientation of the rostrum.

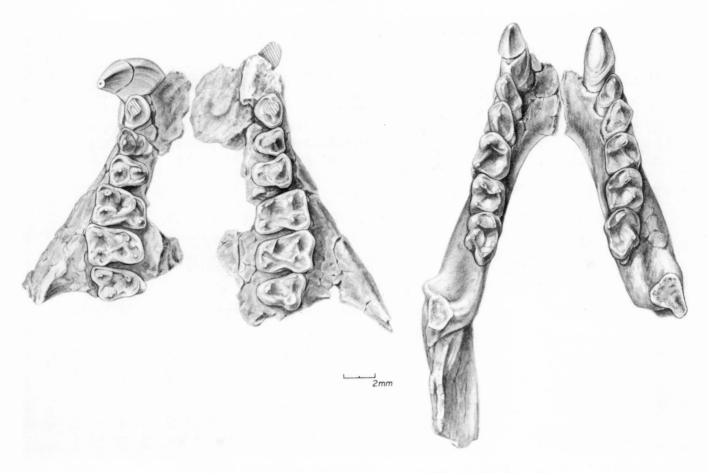

Figure 6. Upper and lower dentition of the holotype of *Nycticeboides simpsoni*.

A substantial portion of the occipital is preserved, including most of the right exoccipital, condyle, supraoccipital, and a small portion of the basioccipital (all fused together). Small denticles along the lambdoidal edge of the supraoccipital section indicate that the sutural margin is intact, allowing accurate estimation of the size of the squama. The foramen magnum is far from complete, although portions of its right lateral and posterosuperior margins are present.

The orientation of the foramen varies considerably among living prosimians (Seth, 1966). In the absence of a complete skull of *Nycticeboides*, orientation has to be inferred from other evidence. In living prosimians the proximity of the nuchal line to the foramen magnum is a good guide to orientation. Table 3 expresses the distance between the nuchal line and the posterosuperior margin of the foramen magnum as a percentage of the distance between the latter and the lambdoidal suture. As can be seen, species in which the foramen magnum is noticeably displaced anteriorly on the basicranium are characterized by a high value (more than 0.7), while those in which the foramen magnum is in the primitive, posteriorly-facing position have lower indices (less than 0.5). *Nycticeboides* places well within the latter group. Locomotor behavior cannot be determined from the position of the foramen magnum: for example, *Tarsius* and *Galago* are both energetic leapers with anteriorly displaced foramina, but *Loris*—in which displacement has also occurred—is a slow climber.

Because of abrasion, the preserved occipital condyle cannot be accurately measured, although its transverse diameter is unlikely to have exceeded 1.8 mm (cf. 1.8 mm in *Cheirogaleus medius,* 2.0 mm in *Galago senegalensis,* 2.7 mm in *Loris tardigradus,* and 3.3 mm in *Perodicticus potto*).

The right temporal is represented only by its petrosal portion (Fig. 7A). The bulla is shorn off at its origin from the petrosal, although a collection of angular fragments pasted onto the mastoid aspect of the petrosal appears to be the collapsed walls of the mastoid cavity. The size of the mastoid cavity cannot be ascertained from the material at hand, although presumably it was large as in all other lorisiforms.

The promontorium bears a fine network of grooves. In modern lorises and galagos, these grooves carry nerves and the functionless remnants of the internal carotid/promontorial artery (MacPhee, 1981). During lorisiform ontogeny, the internal carotid and its divisions involute, and the job of feeding the brain and its meninges is taken over by other cephalic arteries (principally the ascending pha-

TABLE 3. OCCIPITAL DIMENSIONS OF PROSIMIANS.

	I(1)	II(2)	Ratio
Tarsius bancanus	9.4	7.5	0.80
Loris tardigradus	12.2	9.7	0.80
Galago senegalensis	9.9	7.3	0.74
Perodicticus potto	16.0	7.9	0.49
Nycticebus coucang	16.1	8.5	0.53
Cheirogaleus medius	10.5	4.9	0.47
Nycticeboides simpsoni	7.5	3.2	0.43

(1) Distance (in mm) between posterosuperior margin of foramen magnum to lambdoidal suture, in midsagittal plane. N = 1 for each species.

(2) Distance (in mm) between posterosuperior margin of foramen magnum to nuchal line (or greatest eminence on planum nuchale) in midsagittal plane. N = 1 for each species.

ryngeal and external carotid arteries). *Nycticeboides* almost certainly possessed the same developmental sequence.

Enough is left of the root of the bulla to be certain that *Nycticeboides* possessed a "double" medial wall, with an interposed medial accessory cavity produced during bullar inflation. The ascending pharyngeal artery is also found in chierogaleid lemurs (MacPhee and Cartmill, 1986), but the medial accessory cavity is a diagnostic feature of lorisiforms. No part of the ectotympanic is identifiable, although presumably it was phaneric as in living lorisiforms. Living lorisids differ from living galagids in having a less inflated mastoid region (MacPhee, 1981), but too little is left of the auditory region of *Nycticeboides* to ascertain whether it possessed this trait as well.

On the whole, the basicranial evidence confirms the identification of *Nycticeboides* as a lorisiform, but the traits that would identify it as specifically lorisid as opposed to galagid are not preserved.

In *Nycticeboides*, as in all other lorisiforms, the mandibular rami are not fused at the symphysis. In the Siwalik species, jaw depth (Fig. 8) increases posteriorly in the same manner as in *Nycticebus* (i. e., abruptly, between LM1 and LM2). By contrast, in *Mioeuoticus bishopi* the increase in jaw depth at this position is slight (Walker, 1978), while in *Loris tardigradus, Arctocebus calabarensis, Galago crassicaudatus,* and *G. demidovii* the depth actually decreases.

Some aspects of the dentition of the type specimen of *Nycticeboides simpsoni* have been described already (Jacobs, 1981). The anterior dentition of *Nycticeboides* formed a functional toothcomb used for grooming (Rose and others, 1981). Cusps on the premolars and molars are more acute than in *Nycticebus coucang*, and in this regard resemble the homologous teeth of *Loris tardigradus*. This difference between *Nycticeboides* and *Nycticebus* may be related to diet, a topic we explore in detail later (see Discussion). The last two taxa agree with each other and differ from *Loris* in possessing a single-rooted UP2 and a less molariform U/LP4, and in lacking a well-developed UM3 hypocone (Fig. 6)

In order to get a sense of the probable length of the skull of *Nycticeboides*, we regressed UC-UM3 length on maximum skull length for two samples of lorisids. The first sample, consisting exclusively of *Nycticebus* (N = 20; 5 each of *N. pygmaeus, N. coucang bengalensis, N. c. javanicus* and *N. c. coucang*), yielded an estimate of 46.1 mm using a value of 15.6 mm for UC-UM3 length in *Nycticeboides* ($r = 0.92$, $a = 10.35$, $b = 2.29$). The second sample, consisting of insectivorous lorisids (N = 9; 6 specimens of *Loris tardigradus lydekkerianus*, 1 of *L.t. tardigradus*, and 2 of *Arctocebus calabarensis*), yielded a slightly higher estimate of 47.8 mm ($r = 0.85$, $a = 8.99$, $b = 2.49$). These estimates suggest that skull length in *Nycticeboides* may have been within the range of that of

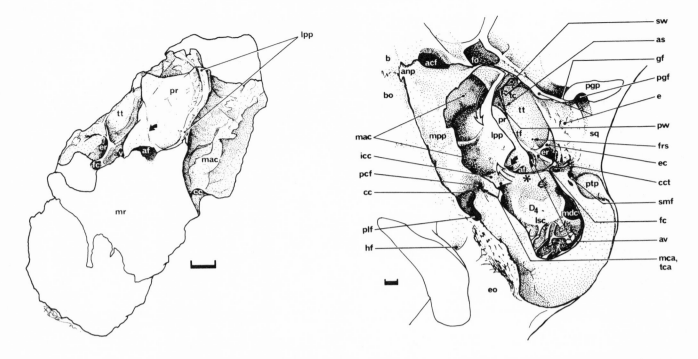

Figure 7. Petrosal anatomy of holotype of *Nycticeboides simpsoni* (left). Drawing on left is ventral view of right petrosal of YGSP 8091; drawing on right is similarly-oriented view of left petrosal of a specimen of *Galago crassicaudatus* (from MacPhee [1981], with permission of Karger AG). The fossil is severely battered, and only the least damaged portions are depicted. Nevertheless, structural correspondences in petrosal anatomy between *Nycticeboides* and a typical living lorisiform are obvious and striking. In diagram on left, arrow points to shallow sulcus which probably carried internal carotid nerve and involuted remnant of promontorial artery. In diagram on right, arrow points to small sinus in lateral lamella of petrosal plate (lpp); a homologous feature (not separately identified) exists in the fossil. Key to features that can be identified in both specimens: *af*, aperture of fossula of cochlear fenestra; *av*, aperture of vestibular fenestra; *cc*, cochlear canaliculus; *er*, epitympanic recess; *fc*, facial canal; *lpp*, lateral lamella of petrosal plate; *mac*, medial accessory cavity; *pr*, promontory of cochlea; and *tt*, tegmen tympani. Crushed mastoid region (*mr*) of fossil clearly possessed a large mastoid cavity; floor of homologous region of galago has been cut away. Scales = 1 mm.

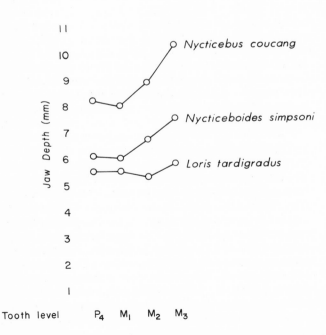

Figure 8. Depths of mandibles at various tooth levels in some Asian lorisids (compare to Walker, 1978, Fig. 6.1).

the smallest living lorisid, *Loris tardigradus tardigradus* (mean length 48.6 mm [Hill, 1933]).

Vertebral Column

Other than one fragmentary and uninformative neural arch, the only identifiable cervical element is a piece consisting of the centrum and dens of the axis (Table 4). The dens has a noticeable dorsal tilt, as in lorisiforms generally, but the atlantal facets are more vertical than in living taxa and lack wide "shoulders."

Only one other vertebral centrum (?thoracic) was found. Deformation *post mortem* has affected its appearance, but its narrowness and marked median keel may be original features. Its sagittal depth is 3.7 mm, which compares favorably to upper thoracic centra of *Arctocebus* and *Loris*.

Although no thoracolumbar vertebrae could be completely restored, the YGSP 8091 collection includes several partial neural arches which can be assigned confidently to this part of the spine (Fig. 9). In *Nycticeboides*, the laminae of the best-preserved partial arches are wide craniocaudally, as are the spinous processes. Unfortunately, none of the recovered spinous processes is complete, but judging from the least damaged specimens they cannot have been very long. Abbreviation of spinous processes is presumably correlated with a relatively small

TABLE 4. SOME POSTCRANIAL DIMENSIONS OF Nycticeboides simpsoni
COMPARED TO OTHER LORISIDAE (1).

	AxisH	AxisW	HumTroD	UlnTroW
Nycticeboides simpsoni	7.1	5.1	2.0	2.0
Nycticebus coucang	9.3	7.8	3.5	3.3
Perodicticus potto	11.5	8.0	3.6	4.0
Arctocebus calabarensis	7.6	5.6	2.4	2.2
Loris tardigradus	5.7	5.9	2.0	2.0

(1) AxisH - Maximum height of axis, from dens to caudal margin, ventral aspect.

AxisW - Maximum width of axis, lateral dimension through bases of atlantal facets.

HumTroD - Minimum diameter of trochlea of humerus (superior border--inferior border, on posterior side).

UlnTroW - Maximum width of trochlear notch of ulna.

N = 1 for each species.

epaxial muscle mass. Preserved thoracolumbar processes have a uniform inclination relative to their pedicles, and were clearly directed caudad in the living animal. This is circumstantial evidence for the absence of an anticlinal vertebra in *Nycticeboides*. In all these features *Nycticeboides* resembles slow-climbing and hanging mammals (cf. Straus and Wislocki, 1932).

Given the material at hand it is impossible to predict how many separate elements comprised the thoracolumbar part of the axial skeleton, but if *Nycticeboides* resembles living lorisids, the number probably was large.

Nycticeboides possesses transpedicular foramina (Hill, 1947) in at least the thoracic part of its spine (Fig. 9). Transpedicular foramina are found in lorisids but not in galagids (or any other known primates), and therefore comprise an important autapomorphy of family Lorisidae (Ankel-Simons, 1983). Hill (1947) states that transpedicular foramina are venous ports, although in other mammals possessing apertures in similar situations (*e. g., Sus*) the transpedicular foramina transmit dorsal rami of spinal nerves. In the course of making a preliminary dissection of an adult *Nycticebus coucang*, one of us (RDEM) noted several errors in Hill's description of nervous and vascular organization in the vertebral region of the slow loris. Most of these are not significant, but there is one osteological discrepancy which should be mentioned here.

Hill (1947) describes two transpedicular foramina in Lorisidae, the "main" and the "accessory." In general, the main foramen is separated from the intervertebral notch by a bridge of bone that connects the vertebral centrum with the base of the ipsilateral transverse process (Fig. 9). In *Arctocebus, Nycticebus,* and *Loris,* a second, smaller bridge runs out from the pedicle to join with the middle portion of the first bridge. This creates a small aperture directed upward and backward, where the pedicle merges with the transverse process. This foramen can be seen in posterior or caudal view; it cannot be seen from the cranial side, *contra* Hill's Figure 1*B*. In lower thoracics and in lumbars the smaller bridge often is incomplete, indicating that its formation is ontogenetically independent of that of the larger. *Contra* Hill (1947), the ribs do not hide the "accessory" foramen in *Perodicticus,* because none exists in the specimens we have

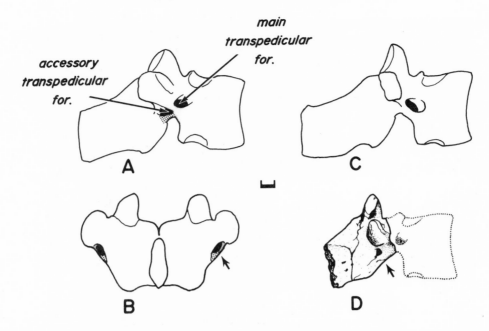

Figure 9. Vertebrae of holotype of *Nycticeboides simpsoni* and extant lorisids, illustrating "main" and "accessory" transpedicular foramina. *A, Nycticebus coucang* (T13, lateral aspect); *B, N. coucang* (T3, dorsal aspect); *C, Perodicticus potto* (T12, lateral aspect); and *D, Nycticeboides simpsoni* (a lower thoracic vertebra, lateral aspect). In *A* and *B*, faint stipple indicates bridge of bone which connects vertebral centrum and base of transverse process, and thereby delimits main foramen. Dark stipple indicates second bridge running from pedicle to first bridge, which floors the accessory foramen (arrow in *B*). This foramen is present in all extant lorisids except *Perodicticus* (*C*), and is also present in *Nycticeboides* (*D*, arrow). Outline of centrum and position of main foramen in *D* based on crushed specimen in YGSP 8091 sample. Scale = 1 mm.

examined: the smaller bridge is absent both in the thoracic and lumbar regions, hence no foramen. *Nycticeboides* agrees with the majority of Lorisidae in having both bridges and both foramina.

In living lorisids, the osseous bridges which subdivide the spaces related to the intervertebral foramen possibly originate through the ossification of tissues that remain ligamentous in other primates (cf. position of accessory superior costotransverse ligaments in humans; Warwick and Williams, 1973, p. 419). Their adaptive value is unknown, although Hill (1947) believes that the complexity of vertebral vascular arrangements in Lorisidae is related to their locomotor habits.

Thorax and Pectoral Girdle

One interesting skeletal adaptation of *Nycticeboides* for which there is reasonably good evidence is rib expansion (Fig. 10). Lorisids, like several other groups of fossorial and arboreal mammals, have developed ribs larger than would be expected for animals in their size range (see Jenkins, 1970). Two conditions are seen in living lorisids. In *Arctocebus* and, to a lesser extent, in *Perodicticus,* the ribs have retained a primitive flattened cross-section, but have acquired platelike flanges on their blades. When especially pronounced, as in some but not all specimens of *Arctocebus* (see Jenkins, 1970), adjacent flanges may even imbricate. By contrast, the ribs of *Nycticebus* lack noticeable flanges but have greatly thickened shafts with semicircular cross-sections. Those of *Loris* are flat and flangeless. So far as we can tell, *Nycticeboides* lacked pronounced costal flanges, but it clearly possessed cross-sectionally enlarged ribs. Indeed, in the dorsal one-third of the largest ribs, cross-sectional area nearly matches that of the humerus at mid-shaft.

As far as can be determined, the articular facets of the tubercle and head were not joined as they are in *Arctocebus*. Chest shape varies in lorisids (Hill 1936, 1953), but the ribs of *Nycticeboides* are too fragmentary for a valid estimation of its probable thoracic breadth.

The glenoid cavity, the base of the coracoid process, and a small part of the blade of the right scapula are preserved (Fig. 11). Although too little of the scapula of *Nycticeboides* is left to be certain that it had the characteristically long vertebral border of lorisids, the scapular neck is unquestionably wider than in *Galago*. The margin of the fossil's glenoid labrum is abraded, but in all essential features it agrees with that of living lorises. The root of the coracoid process bears a pronounced eminence which is apparently the homolog of the tubercle that provides, in *Homo,* attachment for the conoid part of the coracoclavicular ligament. A less likely possibility is that it served as the insertion area for the pectoralis minor. A similar eminence is seen in all lorisids; the galagid version is much less pronounced. Since the purpose of the coracoclavicular ligament is to prevent dislocation of the acromioclavicular joint, its enlargement or strengthening in animals that spend much of their time in hanging postures is expected.

Roberts (1974) contrasts primate glenoids having a "hemispherical outline" with those having a "pear-shaped outline" (as in living lorisids). In the case of the latter shape, he holds that the lip on the upper border

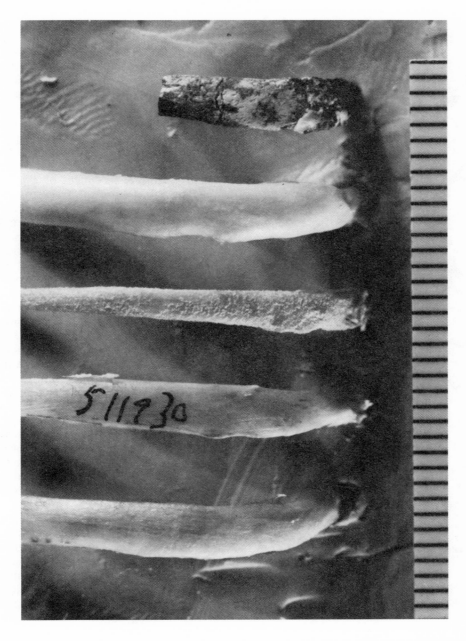

Figure 10. Ribs of holotype of *Nycticeboides simpsoni* and extant lorisids (lateral aspect, left side). In descending order: *Nycticeboides simpsoni*, lower thoracic rib; *Nycticebus coucang*, 13th rib; *Loris tardigradus*, 13th rib; *Arctocebus calabarensis*, 10th rib; and *Perodicticus potto*, 10th rib. Scale gradations in millimeters. Although *Nycticeboides* was a small animal, its ribs were as wide and thick as those of the largest extant lorisids.

of the glenoid functions to prevent dislocation of the head of the humerus when the humerus is retracted. Generally speaking, ligaments and muscles are far more important for that purpose, although it is possible that the lip provides extra surface for connective tissue attachment. Considering their style of slow, deliberate locomotion, it is not obvious why lorisids have need of an osseous adaptation against humeral dislocation, if that is in fact the advantage of having a pear-shaped glenoid. The wide occurrence of this shape (Roberts, 1974) indicates that it is effectively integrated with a number of different locomotor styles.

Forelimb

Humerus length in *Nycticeboides* could not be directly ascertained because both humeri of the type specimen lack their midsections (Fig. 12). We found that maximum humerus length was moderately highly correlated with minimum midshaft width in living Lorisidae ($r = 0.877$), and using a multispecies sample we developed a regres-

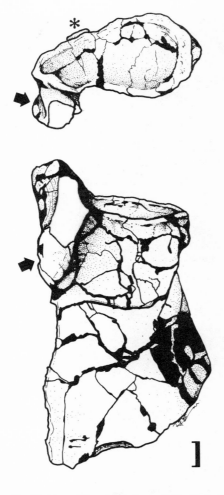

Figure 11. Right scapula of holotype of *Nycticeboides*, ventral (above) and lateral (below) aspects. Scale = 1 mm. Stippled area identifies broken base of coracoid process. Arrows indicate position of tuberosity for attachment of coracoclavicular ligament (or possibly for pectoralis minor m.). Asterisk identifies base of scapular spine.

sion equation appropriate for predicting length on the basis of width (Table 5). According to this equation, maximum humerus length of *Nycticeboides* would have been about 53 ± 1 mm, which is within the recorded ranges of *Arctocebus calabarensis* and *Nycticebus pygmaeus,* and just below that of *Loris tardigradus.*

Although crushing and breakage *post mortem* have noticeably affected the morphology of the YGSP 8091 humeral fragments, a few details deserve comment. As noted by Jacobs (1981), *Nycticeboides* has the humeral morphology of a slow climber, and appreciably differs from *Galago* in many diagnostic features.

The right proximal humeral fragment is almost unrecognizable, and will not be considered further. The lateral surface of the left fragment (including the external part of the humeral head) is crushed flat, but other surfaces seem to be only moderately damaged. The lesser tuberosity is present and well defined, *contra* Jacobs (1981), as in lorisids generally; the greater tuberosity is damaged, but it clearly did not extend craniad beyond the humeral head. The intertubercular sulcus is deep, also as in lorisids generally, and the margins of the sulcus are prominent (although less so than in *Galago,* as noted by Jacobs [1981] in relation to the insertion area for m. teres major). Crushing probably increased the definition of the deltoid ridge, although equally prominent ridges can be found in normal specimens of *Loris.*

The distal humeral fragments appear to be undistorted, although they differ in their minimum widths (Tables 4, 5). The left fragment establishes that *Nycticeboides* had a remarkably deep, sharply bordered olecranon fossa, as do *Nycticebus* and *Loris.* The medial epicondyle evidently was robust, although its original extent cannot be determined because of abrasion. A small, uninformative part of the middle section of the trochlea is preserved, but the distolateral portion of the distal articular end, including all of the capitulum, is missing. The medial supracondylar area is perforated above by a large entepicondylar foramen for the brachial artery and median nerve, as in lorisiforms other than *Arctocebus.* The lateral supracondylar line is unemphasized (as in living lorisids other than *Perodicticus*), which suggests small origins for mm. brachioradialis and extensor carpi radialis longus.

There are numerous small diaphyseal fragments that must relate either to the ulnae or the radii, but they are too fragmentary for reassembly. The only definite ulnar fragment is a piece from the left side bearing the radial notch and the lower half of the trochlear notch (Table 4, Fig. 13).

Because no other useful limb bone measurements can be taken on YGSP 8091, it is not possible to comment on the body proportions of *Nycticeboides.* In view of the small size of this lorisid, it is worth noting that the small extant species *Arctocebus calabarensis* and *Loris tardigradus* have proportionately longer limbs and limb segments than do the larger *Perodicticus potto* and *Nycticebus coucang* (see McArdle, 1978). Interestingly, *Nycticebus pygmaeus* has intermembral and crural indices which resemble those of *Loris tardigradus* more than they do those of *Nycticebus coucang* (although the brachial index is like that of the latter).

DISCUSSION

Relationships of *Nycticeboides simpsoni* and Other Siwalik Lorisiform Primates

Tables 6 and 7 are listings of character distributions found in fossil and recent Lorisiformes. All lorisiforms are united by a shared pattern of bullar inflation,

Figure 12. Humeri (opposite page) of holotype of *Nycticeboides simpsoni* and extant lorisids. *Top series*— Anterior aspect of humerus, right side (except *D* and *E*): A, *Nycticebus coucang*; B, *Perodicticus potto*; C and D, *Nycticeboides simpsoni*; E, *Loris tardigradus*; and F, *Arctocebus calabarensis*. Arrow in *D* points to superior aperture of entepicondylar foramen. *Bottom series*— Posterior aspect of humerus: side lettering as in top photograph. Arrow in *D* points to lesser tuberosity, pushed laterally by deformation. Both humeri of *Nycticeboides* are missing portions of their diaphyses, but the complete bones probably were comparatively short (see text). Scale units in millimeters.

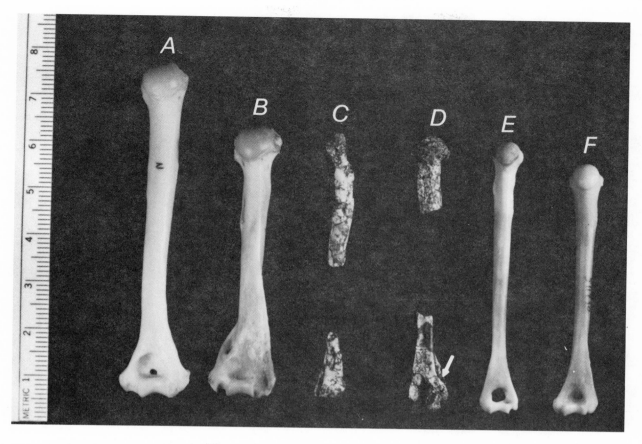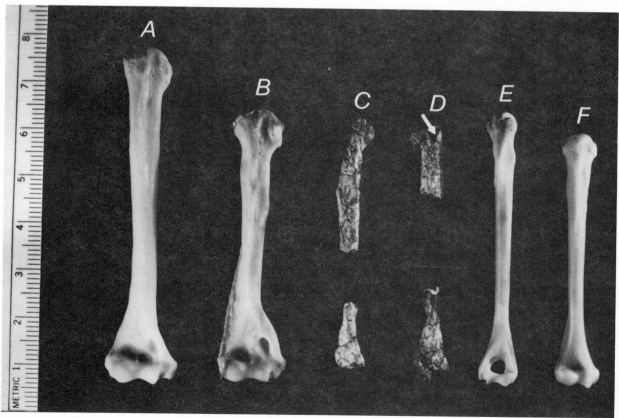

TABLE 5. HUMERUS DIMENSIONS OF LIVING LORISIDAE AND Nycticeboides.

	Maximum length (MaxHL), in mm				Minimum width (MinHW), in mm			
	N	X	1SD	Range	N	X	1SD	Range
Perodicticus potto	16	71.2	4.7	63.5-79.4	16	4.0	0.5	3.5-5.3
Nycticebus coucang	14	67.4	5.4	60.3-77.8	14	3.7	0.6	3.0-4.8
Nycticebus pygmaeus	3	55.3	1.4	53.8-56.5	3	3.0	0.2	2.8-3.2
Loris tardigradus	6	58.0	2.2	56.5-61.2	6	2.3	0.2	1.9-2.5
Arctocebus calabarensis	8	51.2	3.5	46.0-55.1	8	2.4	0.3	1.9-2.5
Nycticeboides simpsoni(1)	1	~52.6 ±0.96	--	--	1	2.2	---	---

(1) Mean estimate for *N. simpsoni* based on following regression equation (r = 0.877):

$$MaxHL = 10.06\{MinHW\} + 30.48$$

Sample used for computation of equation was composed of 7 *Perodicticus potto*, 7 *Nycticebus* (4 *N. coucang* + 3 *N. pygmaeus*), 6 *Loris tardigradus*, and 8 *Arctocebus* (N=28).

MinHW for *Nycticeboides* is the average for right (2.3 mm) and left (2.1 mm) sides.

---

although Lorisidae and Galagidae differ in details of the pneumatization process. All lorisids are distinct from all galagids in possessing skeletal adaptations for slow climbing, traits which we believe are derived. The molariform shapes of both the UP4 and LP4 distinguish living galagids from all known lorisids, but some extinct galagids lack these character states. *Nycticeboides* consistently groups with other lorisids for these and other features considered in this report.

As we noted in our introduction, more than one lorisiform species is present in the Siwalik sample, but lack of adequate material at present precludes systematic disposition of this material. The oldest specimen, the double-rooted UP2 (YGSP 26004) from locality y491 (Table 1), clearly is not allocatable to *Nycticeboides simpsoni* (or any other named taxon). We strongly suspect that the tooth belongs to a lorisiform, but the salutary lesson provided by Dr. Simpson's (1967) own misinterpretation of the chiropteran *Propotto leakyi* as a primate (Walker, 1969) teaches us to proceed cautiously. We are more confident about the two upper molars from locality y450 (YGSP 26006, 26007): they are lorisid, although they are smaller than and different from those of other extant or extinct Asian lorisids— except, perhaps, the taxon represented by the LM3 talonid from locality y259 (YGSP 26005), which is approximately half a million years older.

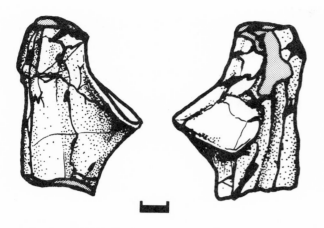

Figure 13. Left ulna of holotype of *Nycticeboides simpsoni*, proximal part, medial (left) and lateral (right) aspects. Olecranon and shaft broken where stippled. Scale = 1 mm.

The two specimens from locality y182A (YGSP 26008, 26009) are similar enough to the holotype of *Nycticeboides* to warrant placing them in *N. simpsoni* without a question mark.

Determination of the affinities of *Nycticeboides simpsoni* within Lorisidae is difficult. The four living lorisid genera each possess a number of distinctive autapomorphies, but these traits either cannot be compared because of the incompleteness of the skeleton of *Nycticeboides*, or they are present in differently derived versions. *Nycticeboides* agrees with *Nycticebus* in having cross-sectionally expanded ribs, single-rooted UP2s, and a mandible which deepens posteriorly. We regard each of these traits as derived within Lorisidae, although we do not know how the similar jaw shape of *Progalago* fits into this picture. The premolar and mandibular traits just listed are also found in *Perodicticus*; if these apomorphies are shared by virtue of common ancestry, then a special relationship seems to link this genus with *Nycticeboides* and *Nycticebus*. *Perodicticus* lacks a distinctive trait present in other Lorisidae (including *Nycticeboides*), the "accessory" transpedicular foramen (Table 7, character 10). We have noted that the bony bridges defining the main and accessory foramina may originate through ossification of ligamentous tissue, and that they tend to be less complete in the lower thoracic and lumbar regions. In the absence of decisive evidence, we cannot determine if the one- or the two-foramen arrangement is primitive for Lorisidae.

Paleobiogeography of Lorisidae

Where did the Lorisidae originate, and how did they achieve their present distribution? Little worthwhile evidence pertains to either of these questions, since the lorisid fossil record is exceptionally poor. An African origin has been presumed (*e. g.*, by Walker, 1972) because Africa is the continent on which the oldest indisputable fossil lorisids have been found, and because the sister-group of Lorisiformes *sensu stricto* is either Lemuriformes generally or Cheirogaleidae specifically, all of which are resident in Madagascar. However, these reasons are not very compelling. The oldest East African evidence is only Early Miocene, by which time the lorisid and galagid lineages were already quite separate. An earlier origin, not necessarily African, is therefore indicated for Lorisiformes. Since no true lemurs are known either from mainland Africa or South Asia, their exact relationship to lorisiforms is irrelevant to solving the puzzle of the latter's geographical origin. "Solving" the puzzle by claiming that lorisiforms arose from an immigrant Malagasy primate which managed to cross the Mozambique Channel only begs a further currently unanswerable question, that is, where did lemuriforms originate?

The earliest documented indication of a Neogene faunal interchange between South Asia and Africa comes from Miocene (>18.3 Ma) Bugti Beds of Pakistan (Barry and others, 1985), although earlier interchanges are implied by the presence of similar tethytheres and rodents in the Eocene of Africa and South Asia (Wells and Gingerich, 1983; Jaeger and others, 1985; Flynn and others, 1986). No lorisids or other primates have been recovered from Bugti rocks, but a slightly younger (16.1 Ma) Siwalik locality has yielded a single tooth very similar to African *Micropithecus* and Chinese *Dionysopithecus*. This fossil is the earliest dated Eurasian catarrhine. The Asian record for Lorisiformes starts at least 9 Ma and possibly as much as 13 Ma ago, or only a few million years later than the earliest African record for the infraorder (Walker, 1978). Since adapids were present in South Asia as well as the Fayum of Africa, and since adapids are widely regarded as the sister-group of the modern strepsirhine infraorders, it is interesting to ask whether lorisids could have originated in Asia rather than Africa. Nothing precludes this possibility. If African and Asian lorises constitute a holophyletic taxon (but see Dene and others, 1976; Schwartz and Tattersall, 1985), at least one migration event is required to place a lorisid initiator on the continent where the family *did not* originate.

It is also perfectly possible that multiple dispersal events occurred. Barry and others (1985) have documented successive faunal events (*i. e.*, compositional changes in a fauna due to speciations, novel introductions, and extinctions) in the Siwalik section. These faunal events are strongly correlated with changes in sea level and 1_8O curves, and hiatuses in deep sea deposition. According to their assessment of this evidence, a major faunal event occurred at 9.5 Ma, and a minor one at about 13.2 Ma. If the Siwalik record reflects the biogeographical history of early Asian lorises, and if Siwalik lorisid lineages originally entered Asia from Africa, they probably did so during one or both of these faunal episodes.

If Afroasian faunal interchange was repeatedly possible during the period between 20 Ma and 7.4 Ma, as Barry and others (1985) have suggested, it is curious that no member of Galagidae has been recognized in Asian Neogene sites. Assuming that this is not due to any form of collecting bias, the failure of galagos to enter or prosper in Asia may have had something to do with niche availability. In modern African forests, galagos and

TABLE 6. DISTRIBUTION OF SOME SELECTED CHARACTERS IN EXTINCT AND EXTANT LORISIFORMS (+ INDICATES CHARACTER PRESENT, - INDICATES CHARACTER ABSENT).

Taxa	(1)	(2)	(3)	(4)	(5)	(6)
YGSP 26004	?	?	?	?	?	-
Progalago	?	-	?	-	+	-
Komba	+	-	?	-	-	-
Galago	+	-	-	+	-	C
Mioeuoticus	+	?	?	?	B	-
Loris	+	+	+	A	B	-
Arctocebus	+	+	+	A	-	-
Nycticeboides	+	+	+	-	+	+
Nycticebus	+	+	+	-	+	+
Perodicticus	+	+	+	-	+	+

(1) Inflated bullae

(2) Skeletal modifications for slow climbing

(3) Transpedicular foramina present

(4) UP4/LP4 molariform

(5) Depth of mandible increases posteriorly

(6) UP2 single-rooted

(A) LP4 in Loris and Arctocebus are not fully molariform and almost certainly developed apomorphic submolariform shape independently from galagos and perhaps from each other.

(B) Depth increases only slightly, and the pattern of increase, at least in Loris, is distinct from that seen in Nycticeboides, Nycticebus, and Perodicticus.

(C) Galago (Euoticus) elegantulus has a single-rooted caniniform UP2, perhaps useful in gummivory. Other species of Galago have a double-rooted U P2.

TABLE 7. DISTRIBUTION OF SELECTED CHARACTERS IN EXTANT LORISIDS AND Nycticeboides. (+ INDICATES CHARACTER PRESENT, - INDICATES CHARACTER ABSENT).

Taxa	(1)	(2)	(3)	(4)	(5)	(6)	(7)	(8)	(9)	(10)	(11)
Loris	-	-	-	-	+	+	-	+	-	-	+
Arctocebus	-	-	-	+	+	-	+	-	+	-	+
Nycticeboides	+	+	+	-	-	-	-	+	-	-	+
Nycticebus	+	+	+	-	-	-	-	+	-	-	+
Perodicticus	+	+	-	±	-	-	-	+	-	+	-

(1) UP2 single-rooted
(2) Depth of mandible increases strongly posteriorly
(3) Ribs cross-sectionally enlarged
(4) Ribs with flange
(5) LP4 submolariform
(6) Interorbital area reduced
(7) Tubercle and head of rib joined
(8) Entepicondylar foramen present
(9) Cusps, cingula and crests on molars acute and distinct
(10) Strong supracondylar line on humerus
(11) Two bridges present in thoracic intervertebral foramina (accessory transpedicular foramen present)

lorises with broadly similar diets do not compete for the same resources. Charles-Dominique (1977) points out that insectivorous galagids seem to have outcompeted their lorisid counterparts, which tend to concentrate on foul-smelling or otherwise undesirable prey items avoided by other predators (but see Rasmussen and Izard, in press). Competition may also help to explain why *Perodicticus potto* looks for fruit in trees while *Galago alleni* forages on the ground, although in this case it is hard to argue that it is the lorisid that has been outcompeted or driven into the less favorable environment.

Adaptations and Body Size of *Nycticeboides* and Other Asian Lorisids

Introduction

Since body size is an important physiological attribute of organisms (Bourlière, 1975), knowing where *Nycticeboides* ranks in body size among primates could provide some insight into its behavioral ecology, especially its diet. While it is clear from evidence discussed already that *Nycticeboides* was a very small animal, establishing that it was as small as *Loris* may have very different implications from showing that it was as large as *Nycticebus*. Obviously, the chief difficulty involved in assigning a fossil taxon to a specific size/diet category is that body mass cannot be established empirically but must be predicted according to some inferential standard, such as a regression line relating body size to some other parameter (e.g., long bone measurements, total skeletal weight). There are numerous problems, both practical and conceptual, in making such predictions (see papers in volume edited by Jungers, 1984).

The dentition of *Nycticeboides* is sufficiently well preserved to permit use of the log-transformed regression equations developed by Gingerich and coworkers (Gingerich and others, 1982; Gingerich and Smith, 1984). The predictive strength of these statistics rests on the argument that, among primates generally, a high interspecific correlation exists between cheektooth crown area and body weight (but see Smith, 1984). Despite their evident promise, the adequacy of these equations as estimators of body size within specific phylogenetic or dietary groups is still on test. For example, Gingerich and others (1982) show that crown area and body size regress quite dif-

TABLE 8. SHEARING QUOTIENTS (SQ) OF PROSIMIANS (DATA OF COVERT, 1985).

	N (species)	Mean SQ(1)	SD	Range of means of SQ values
Frugivores	11	−9.92	6.75	−22.10 to −0.80
Folivores	8	+8.68	9.73	−0.70 to +28.79
Insectivores	6	+14.60	9.92	+2.40 to +27.95

(1) Using data collected by R. F. Kay, Covert (1985) calculated shearing quotients as follows. Two variables were used to derive a least-squares regression for calculating 'expected shear' (ES): (1), Ln of sum of lengths of specified M_2 shearing crests ('observed shear', OS); and (2) Ln of mesiodistal length of M_2. Regressing (1) on (2) yielded the equation (H. Covert, pers. comm.)

$$\text{Ln ES} = 0.989\{\text{Ln L}(M_2)\} + 0.684.$$

The expected shear for each species is derived by inserting into the equation the known M_2 lengths from the original data set. Relative concordance between OS and ES is expressed as the shearing quotient, SQ = (OS − ES)/ES × 100. The folivore group excludes <u>Lepilemur mustelinus</u>.

ferently in insectivorous and folivorous/frugivorous primates. Compared to folivores/frugivores (or dietary "generalists"), insectivores have larger crown areas than would be "expected" for their body size, and regression equations appropriate for computing body weight in plant feeders consistently overestimate body weights of insectivorous primates (Gingerich and Smith, 1984).

Gingerich and coworkers have not published equations appropriate for body-weight estimation in primate insectivores. However, they have shown graphically that a separate equation, developed from data for nonzalambdodont lipotyphlan insectivores (Table 12), roughly "predicts" the body weights of small insectivorous primates (Gingerich and Smith, 1984 and *personal communication*). Whether crown area vs. body size regress similarly in insectivorous primates and lipotyphlans over the entire range of comparable body sizes has not been investigated.

Estimating Body Size in Nycticeboides

Of the equations listed by Gingerich and others (1982), the one having the highest correlation coefficient ($r = 0.967$; Table 12) predicts that a primate taxon having the M_1 dimensions of *Nycticeboides* and a "generalist" diet ought to weigh about 530 g (95% C.I. = 464-610 g). This is remarkably close to the weight (493 g) predicted by the equation developed by Kay and Simons (1980) for estimating body weight in primates on the basis of M_2 length regardless of diet. By contrast, the same tooth crown dimensions applied to the lipotyphlan equation (Table 12) yield a weight estimate only one-quarter as much— about 110 g (95% C.I. = 94-125 g). Which is the better estimate? In order to decide this, we need some valid extrinsic criterion applicable to fossil taxa that can be used to reject the less acceptable estimate.

Since the "generalist" and lipotyphlan equations yield markedly different predictions, it is tempting to proceed intuitively, by identifying and rejecting one of the estimates as improbable. This procedure is permissible only when one is dealing with absolutely large body sizes (over 1000 g), as may be illustrated by two examples. Using body weight predictions alone, insectivory could be rejected as the dietary specialization of the mantled howler *Alouatta villosa*, even if the only facts known about this species were the length and width of its M_1.

TABLE 9. SHEARING QUOTIENTS (SQ) OF Nycticeboides simpsoni (1).

	YGSP 26008	YGSP 8091 (right and left sides)		Mean
1. Ln summed length of shearing crests 1-6 (in mm) (OS)	1.5433	1.5261	1.4996	1.5281
2. Ln mesiodistal length	0.9083	1.0716	1.0438	0.9830
3. Expected shear (ES)	1.5823	1.7438	1.7163	1.6561
4. SQ	-2.46	-12.48	-12.63	-7.65

(1) For method, see Table 8. 'Mean' is the average of right and left LM2 of holotype added to the value for YGSP 26008 and divided by 2; if the three teeth are treated as separate specimens (N=3), the mean is higher (-9.19). By either computation, N. simpsoni obviously falls within the frugivore range.

The lipotyphlan regression predicts an adult weight in excess of 2200 g for *A. villosa* (using data of Swindler [1976], sexes averaged), and there are no unequivocally full-time insectivores among living primates which are this large (Kay, 1975). However, the same pair of measurements would not permit a decision about the guild affiliation of the red-backed squirrel monkey, *Saimiri oerstedii* (data of Swindler [1976], as above). The "generalist" equation yields a value of about 730 g for this species, which is in fact close to its recorded weight (800-900 g) and strongly implies a large amount of plant food in its diet. The lipotyphlan equation, on the other hand, predicts a weight of 150 g, which is well within the range of primate insectivores. If actual diet and weight were unknown for this species, in order to reject the insectivore value it would be necessary to have evidence other than crown dimensions; to assume that the "generalist" equation is correct without such evidence is tantamount to assuming what needs to be proven. The same consideration applies to the interpretation of *Nycticeboides;* since this Miocene lorisid was not so large that it "must" have been a plant eater, we are forced to consider the possibility that it might have been primarily insectivorous.

Obviously, for interpreting habits of extinct taxa, an analytical procedure is needed which allows one to predict diet without having to estimate body weight, at least initially. Building on work by Kay and Simons (1980), Covert (1985 and *personal communication*) has developed a candidate method involving the calculation of "shearing quotients" (SQ). The utility of this method (outlined in Table 8) depends on the observation that frugivores have considerably less cheektooth shearing capacity than do either insectivores or folivores. Frugivores tend to display SQ values well below the zero value (less shear than "expected"), while folivores display the reverse tendency (more shear than "expected"). Insectivores also display more shear than expected, but this does not inhibit interpretation since all living folivores are much larger than all living primate insectivores.

Although Covert's method shows promise, interpretative and analytical difficulties attend its application. First, although SQ grand means for the three primate dietary guilds recognized by Covert are notably different, within-guild standard deviations are large. This reflects the fact that some within-species SQ ranges are extraordinarily wide: the range for *Lemur catta,* for example, is 26 percentage points (-5.71 to $+20.17$; mean, $+6.01$), and that for *Nycticebus coucang* is more than 22 points (-13.46 to $+9.10$; mean, $+3.34$). Secondly, resolution around the critical zero point is poor. Only 3.2 percentage points separate the mean SQ of the frugivore *Galago alleni* (-0.80; range, -5.43 to $+5.64$) from that of the insectivore *Loris tardigradus* ($+2.40$; range, -2.43 to $+7.76$), and their respective ranges markedly overlap. Thirdly, only six shearing blades are measured in Covert's procedure, despite the fact that in some groups additional blades can be involved in "total" shearing capacity. In

TABLE 10. RECORDED BODY WEIGHTS OF LIVING LORISIDAE.

Species/Subspecies (1)	Source of Data (2)	C (3)	N	'Total' Range and Mean(g)(4)	N	Male Range and Mean(g)	N	Female Range and Mean (g)	Notes
1. *Loris tardigradus*									
subsp. indet. or mixed sample	Prater ('65)	?	?	225-340	?	280-340	?	'225'	
	Bauchot and Stephan ('66)	c	5	195-347(322)	3	195-347-(264)	2	300+320(310)	
	Napier and Napier ('67)	?	23	85-348	16	85-348	7	85-270	
	Kavanagh ('83)	?	?	'300'					
	Jungers ('84)	?	16	(275)					
tardigradus	Hill ('33)	c	4	85-128(103)	2	85+128(107)	2	113+85(99)	
	Phillips ('35)	?	5	(ca.130)	4	(128)	1	135	
malabaricus	Ryley ('13b)	?	1	170					
	Rasmussen (unpubl.)	c	7	161-202(186)	3	169-202(187)	4	161-201(185)	
grandis	Phillips ('35)	?	2	204+238(221)	1	204	1	238	
nycticeboides	Hill ('42)	c	2	140-190(165)	1	140	1	190	
nordicus	Hill ('33)	c	4	198-340(236)	1	212	3	198-40(245)	(5)
lydekkerianus	Shortridge ('14)	?	?	269-354					
	Hill ('33)	c	4	226-347(279)	1	347	3	247-297(256)	
	This paper	f	5	259-347(295)					
2. *Nycticebus coucang*									
subsp. indet. or mixed sample	Stephan and Bauchot ('65)	c	2	570+645(600)	1	570	1	64	
	Napier and Napier ('67)	?	12	1012-1675	9	1012-1675	3	1105-1370	
	Medway ('69)	?	14	375-900					
	Roonwal and Mohnot ('77)	?	?	850-1675	?	850-1675	?	900-1320	
	Gingerich et al. ('82)	?	30	'1250'	21	(1300)	9	(1200)	(6)
	Kavanagh ('83)	?	?	'1400'					
	Jungers ('84)	?	27	(920)					
menagensis	Banks ('31)	f	1	418	1	418			(7)
	Davis ('62)	f	1	230	1	230			
coucang	Manouvrier (1888)	c	1	210					(8)
	Weber (1896)	c	2	416+500(458)	2	416+500(458)			
	Dubois (1897)	c	1	500	1	500			
	Spitzka ('03)	c	1	612	1	612			
	Warncke ('08)	c	2	365+572(469)					
	Fooden ('76)	f	2	575+720(648)	1	720	1	575	
	Muul (unpubl.)	f	?	700-965	?	740-965	?	700-830	
	This paper	f	1	610			1	610	
javanicus	no data available								
bengalensis	Manley ('66)	c	?	up to 2000					(9)
	Fooden ('71)	f	3	850-920(890)	2	850-920(885)	1	900	
	Acharjyo and Misra ('73)	c	2	1400+1594(1497)	1	1594	1	1400	(10)
3. *Nycticebus pygmaeus*	no data available								
4. *Arctocebus calabarensis*									
subsp. indet. or mixed sample	Napier and Napier ('67)	c	?	266-465					
	Kavanagh ('83)	?	?	'210'					
	Jungers ('84)	?	22	(265)					
	Bourliere ('85)	?	?	'210'					
calabarensis	Hill ('53)	?	?	'larger than aureus'					
aureus	Charles-Dominique ('77)	f	30	150-270(210)					

TABLE 10. (CONTINUED).

Species/Subspecies	Source of Data	C	N	'Total' Range and Mean(g)	N	Male Range and Mean(g)	N	Female Range and Mean(g)	Notes
5. *Perodicticus potto*									
subsp. indet. or mixed sample	Weber (1896)	?	3	710-756(733)					
	Warncke ('08)	?	1	538					
	von Bonin ('37)	?	4	684					
	Kennard and Willner ('41)	c	5	550-1250(930)	2	700+1200(950)	3	550-1250(1917)	
	Kingdon (1971)	?	--	1000-1500					
	Charles-Dominique ('77)	f	--	'1000'					
	Gingerich et al. ('82)	?	22	'1150'	12	(1200)	10	(1100)	(6)
	Kavanagh ('83)	?	?	'1300'					
	Jungers ('84)	?	20	(1150)					
	Bourliere ('85)	?	?	'1100'					
potto	Baudenon ('49)	f	15	720-1050(840)	11	720-1050(835)	4	820-910(855)	
	Rahm ('60)	f	3	700-900					
	Rahm ('60)	c	1	1000	1	1000			
ibeanus	Stephan and Bauchot ('65)	f	7	750-1250(1093)	3	750-1250(1067)	3	1000-1250(1150)	
edwardsi	Malbrant and Maclatchy ('49)	f	4	1000-1400(1156)	2	1025+1400(1213)	2	1000+1200(1100)	(11)
	Charles-Dominique ('77)	f	33	850-1600(1100)					

(1) Subspecies of *Loris tardigradus* and *Arctocebus calabarensis*, after Hill (1953); of *Nycticebus coucang* (with *pygmaeus* excluded), after Groves (1971). For *Perodicticus potto*, nominal subspecies *P. p. faustus*, *batesi*, and *juju* are considered part of *edwardsi* (but see Hayman in Sanderson [1940]). Schwartz (1974), who did not study *potto*, implausibly regards *ibeanus* as belonging to a separate species (*P. ibeanus*) because of its dental eruption sequence.

(2) Authors cited here are ones providing 'original' data sets. However, in many cases precisely where the authors gathered their information cannot be determined, and hence some duplication is likely. Data identified as 'this paper' refer to body weight information collected from specimen labels at the British Museum (Natural History) by W. Downs.

(3) Animal(s) measured: f = field-caught specimen(s); c = laboratory/zoo/other captive specimen(s); ? = not known.

(4) As given by author or reconstructed from author's data set, for adult animals only when subadults also listed. Computed means are enclosed in parentheses; single figures that are not described by author as computed means are enclosed in quotation marks. Small computational errors in original sources are not corrected. English measures (lb, oz) in original works have been converted into gram equivalents. All measures rounded to nearest gram. True (computed) means of total samples (original measurements) as reported by Bauchot and Stephan (1966) are usually different from their stated means: true means are, *Loris*, 283 g; *Nycticebus*, 608 g; *Perodicticus*, 1044 g. In table, total range and mean for *Perodicticus* includes 1 specimen of unknown sex weighting 1000 g. Field-caught specimens of *Perodicticus* are from eastern Zaire, and therefore represent *ibeanus*.

(5) Does not include estimates by Mayor, which were described by Hill (1933) as 'guesswork'.

(6) Gingerich et al. (1982) do not provide means for total samples; total sample 'means' reported here are means of their separate male and female averages.

(7) It is not clear from Davis' (1962) text whether this extremely low weight is an average of the 2 specimens captured, or the weight of the 'young' male only.

(8) Specimens named *N.* 'tardigradus' in older sources most likely belong to this subspecies (see Elliot, 1913; Hill, 1953). Fooden (1971) suspects that *N. c. tenasserimensis* may be part of *N. c. coucang* rather than *bengalensis*, as Groves (1971) believes.

(9) Manley (1966:15) states that '...healthy adults of the large race *bengalensis* may weigh up to 2 kg'. It is not clear whether this information refers to wild-caught or captive animals.

(10) Subspecies not named by Acharjyo and Misra (1973), but female was caught in Assam and is therefore *bengalensis*. Weight for female taken 3 wk after capture (and after birth of male recorded in next line). Male weight is average of weeks 44-52 (animal reached maximum weight of 1605 g at 44 wk).

(11) Larger female was skinned before weighing; smaller female listed as 'subadult'.

Lorisidae, for example, the paracristid (crest mesial to the metaconid) is well defined, and must contribute to the shearing activity of the molars.

These problems seriously affect interpretation of the SQ derived from *Nycticeboides simpsoni* (Table 9). The mean SQ estimate, -7.65 (range, -2.46 to -12.63), is close to the grand mean for living primate frugivores, and suggests that *Nycticeboides* relied on fruits (or foods of similar texture) rather than insects. If correct, this is a striking finding; among living lorisids, high fruit intake is seemingly characteristic only of those species whose "average" body weight (as usually reported) exceeds 1000 g (Table 10). According to the weight estimators used here, *Nycticeboides* weighed half that much at the most, and possibly weighed considerably less.

Just because no living lorisid is both very small and a frugivore need not mean very much, since there are frugivores of very small size in Galagidae and Lemuriformes. However, it seems to us that Covert's method will yield reliable results only when diet/body size correlations form reasonably discrete clusters within the evolutionary group in question. If overlaps among clusters are considerable, then there are no good grounds for assuming that improbable diets can be rejected on the basis of the sign of an SQ. The issue, then, is to establish whether lorisid frugivores and insectivores occupy distinct, non-overlapping ranges in body size. So-called "average" species weights cited in the recent literature (*e. g.*, by Gingerich and others, 1982; Jungers, 1984) imply that overlap does not occur. However, the meaningfulness of a "species average" depends on the nature of intertaxon variability for the measure in question (cf. Ford and Corruccini, 1984) as well as comparative invariance . . . diet within species. If intrataxon body size/diet correlations vary in any significant manner, then confidence in inductive pyramids based on supposed invariance in such measures must be correspondingly reduced. Thus the questions that have to be answered before any reliable extrapolation could be made about the behavioral ecology of *Nycticeboides* are: (1) how variable is body size within living lorisid species; and (2) does intrataxon variability in these species correlate with variation in diet?

Body Size Variation in Living Lorisidae

There are no comparative studies of body size variation within living lorisid species. As an approximation, not meant to replace the proper study which still needs to be undertaken, we assembled body weight data for each well-defined *subspecies* currently recognized (Table 10). These data are quite unsatisfactory in many respects. Nevertheless, we are struck by the fact that subspecies of *Nycticebus coucang* and (to a lesser extent) *Loris tardigradus* seem to differ in reported weight by much greater margins than do subspecies of either *Arctocebus* or *Perodicticus*. This contrast cannot be explained by reference to the absolute size of the species concerned or to sexual dimorphism (which is low in lorisiforms in any event; cf. McArdle, 1978; Seth, 1966).

A sense of the scale of variability involved can be gained by comparing the weight ranges of Asian lorisids with a sampling of ranges reported for other primates and *Tupaia* (Table 11). Logarithmic ratios for most species in this sample are in the range 0.5-1.0, which we subjectively regard as "normal." Taxa that exceed this limit include *Galago senegalensis, Presbytis entellus, Macaca fascicularis, M. arctoides,* and both species of Asian lorises. No well-investigated anthropoid species exceeds genus *Macaca* in body-size variability (Roonwal and Mohnot, 1977), but *Nycticebus coucang* is nearly as variable if the published extremes are reliable indicators. Even if extreme values are discarded (*e. g.*, all weights under 400 g), *Nycticebus* is still apparently more variable in size than are most other Asian primates. This is not the impression that one gets from the thin literature on the systematics of the slow loris (*e. g.*, Lyon, 1907; Elliot, 1913; Hill, 1953; Groves, 1971).

Another unexpected finding is that body-size predictions for component taxa of *Nycticebus*, computed by means of the equations in Table 12, do not accord well with some reported values. Thus the average estimate for *N. coucang bengalensis* (1506 g) seems rather high in view of the few wild-caught weights available for this subspecies. Captive animals reach and surpass this value (Manley, 1966; Acharjyo and Misra, 1973); these outliers may represent the clinical equivalent of obesity in humans, because slow lorises have a tendency to store quantities of fat (Banks, 1931). Correspondence between predicted and published weights for the subspecies *coucang* also is poor, but many of the relevant data are frankly suspect. Unpublished data (I. Muul, *personal communication*) on field-caught *coucang* indicate that mean body weight for males and nonpregnant, nonlactating females lies between 750-800 g.

Perhaps the most difficult member of *Nycticebus* to evaluate is the pygmy slow loris, *N. pygmaeus*. There do not appear to be any published empirical body weights for this taxon, once considered to represent a subspecies of *N. coucang* (see Ellerman and Morrison-Scott, 1951; Groves, 1971). The estimated "generalist" weight, 691 g, (95% C.I. = 611-783 g) is below many recently-reported empirical values for slow loris taxa. Hill's (1953) statement that *N. pygmaeus* is "only half the size" of *N. c. coucang* is probably based on a reading of Bonhote (1907), rather than original observation (see also Kolar, 1984).

Dietary Variation in Living Lorisidae

Although a certain amount of information has been collected on the diets of free-ranging African lorisids (see especially Charles-Dominique, 1977), the dietary preferences of *Nycticebus* and *Loris* have been investigated only in the most general terms. Thus it is quite impossible to test, for the present, the hypothesis that lorisid subspecies of different body size have different food preferences. The following notes are designed to draw attention to some observations that may imply such differences.

TABLE 11. VARIABILITY IN ADULT BODY WEIGHT IN LORISIFORMS, SOME SOUTHEAST ASIAN ANTHROPOIDS, AND Tupaia.

Species	Body Weight(kg)(1)	Proportionate Ratio(2)
Macaca min/max	1.5-18.0	2.48
Nycticebus coucang	0.21-2.0/0.37-2.0	2.25/1.69
Macaca fascicularis	1.5-8.3	1.71
Presbytis min/max	3.8-20.9	1.70
Loris tardigradus	0.09-0.35	1.36
Macaca arctoides	6.0-18.0	1.10
Presbytis entellus	7.5-20.9	1.02
Galago senegalensis	0.11-0.30	1.02
Perodicticus potto	0.54-1.40	0.95
Presbytis senex	3.8-9.3	0.90
Arctocebus calabarensis	0.20-0.47	0.85
Pongo pygmaeus	81.0-189.0/37.0-69.0	0.84/0.62
Tupaia glis	0.09-0.19	0.75
Galago demidovii	0.05-0.10	0.69
Galago crassicaudatus	1.0-2.0	0.69
Tarsius sp.	0.09-0.16	0.64
Galago alleni	0.19-0.34	0.58
Hylobates lar	4.4-7.6	0.55
Euoticus elegantulus	0.27-0.36	0.29

(1) Sources for body weight data: Table 10 of this paper, Kingdon (1971), Stephen and Bauchot (1965), Roonwal and Mohnot (1977), Napier and Napier (1967), and Charles-Dominique (1977). Macaca and Presbytis min/max, least and greatest weights on record for genus in question, regardless of species. Weight range for Pongo pygmaeus pygmaeus (first range) may be too large; second range represents P. p. abelii (see Napier and Napier, 1967). For Nycticebus, first range utilizes weight recorded by Manouvrier (1888), while second uses Warncke's (1908) and Medway's (1969) minima.

(2) P.R. = Ln X_i(largest variate) - Ln X_j(smallest variate).

TABLE 12. BODY WEIGHTS OF LORISIDAE PREDICTED FROM DIMENSIONS OF FIRST LOWER MOLAR (1).

	Species/Subspecies	N	L	W	LnY	LnX{1}	Wt(95% CI)	LnX{2}	Wt(95% CI)
G E N E R A L I S T S	Perodicticus potto								
	P. p. edwardsi	9	4.03	3.23	2.5662	7.3737	1593(1460-1739)	5.8716	354(300-419)
	P. p. ibeanus	6	3.48	2.60	2.2025	6.8317	927(830-1035)	5.2860	198(170-230)
	P. p. potto	8	3.78	2.59	2.2814	6.9493	1042(939-1158)	5.4131	224(192-262)
	Nycticebus coucang								
	N. c. bengalensis	6	4.07	3.08	2.5286	7.3175	1506(1377-1648)	5.8110	334(283-394)
	N. c. javanicus	8	3.74	2.86	2.3699	7.0812	1189(1077-1313)	5.5555	259(221-303)
	N. c. coucang	11	3.70	2.86	2.3592	7.0651	1170(1059-1293)	5.5383	254(217-298)
	N. c. menagensis	8	3.46	2.63	2.2083	6.8403	935(838-1043)	5.2953	199(171-232)
	Nycticebus pygmaeus	10	3.11	2.39	2.0059	6.5388	691(611-783)	4.9695	144(125-166)
I N S E C T I V O R E S	Loris tardigradus								
	L. t. lydekkerianus	12	3.06	2.46	2.0186	6.5577	704(623-797)	4.9899	147(127-170)
	L. t. nordicus	1	3.10	2.40	2.0069	6.5402	692(612-784)	4.9711	144(125-167)
	L. t. tardigradus	3	2.90	2.23	1.8667	6.3314	562(491-643)	4.7454	115(100-133)
	Arctocebus calabarensis	7	3.50	2.60	2.2083	6.8403	935(838-1043)	5.2954	199(177-232)
	Nycticeboides simpsoni	1	2.70	2.30	1.83	6.277	532(464-610)	4.686	108(94-125)

(1) Body weight predictions based on log-transformed regression equations developed by Gingerich et al. (1982) and Gingerich and Smith (pers. comm.). In this table, LnY is the natural log of length (L) x width (W) of M_1 (in mm). The columns LnX{1} and LnX{2} list values for dependent variable LnX (natural log of body weight, in g) according to the following least-squares regression equations:

Eq. {1}: LnX = 1.49(LnY) + 3.55   (r = 0.967)

Eq. {2}: LnX = 1.61(LnY) + 1.74   (r = 0.966)

The slope and intercept of Eq. 1 were taken from Table 3 of Gingerich et al. (1982). This equation is based on data from a broad sampling of primate frugivores and folivores (dietary "generalists"). Eq. 2 is based on M_1 and body weight data for nonzalambdodont lipotyphlan insectivores, not primates (see Gingerich and Smith, 1984). The slope and intercept for Eq. 2, heretofore unpublished, were taken from Gingerich and Smith's work-in-progress on tooth size and body scaling in insectivores.

The 95% CI of each predicted mean weight (Wt) in g was computed by means of the statistic recommended by Gingerich et al. (1982), i.e.,

$$95\% \text{ CI for Ln X} = \text{Ln X} \pm 1.96 \sqrt{C_1 + (\text{Ln Y} - C_2)^2 (C_3)}.$$

The C values used for Eq. 1 were taken from Table 3 of Gingerich et al. (1982). Those for Eq. 2 were computed from the data supplied to us by these authors (C_1 = .00473, C_2 = 1.3498, and C_3 = .00173).

Arctocebus calabarensis. Charles-Dominique (1977) studied the subspecies *aureus* in Gabon, which was found to rely on insects for about 85 percent of its food intake. The diet of the slightly larger *A. c. calabarensis* has not been reported.

Most weight estimates for the angwantibo range between 200-260 g, although there is at least one record of a captive animal weighing twice this amount (Napier and Napier, 1967). The misevaluation of insectivore body weight by the "generalist" equation is exhibited plainly by the entry for *Arctocebus* in Table 12 (more than 900 g); conversely, the lipotyphlan equation yields a mean value of 199 g (95% C.I. = 171-232 g), which is notably close to empirically-based estimates.

Perodicticus potto. About 10 percent of the food intake of Gabonèse *P. p. edwardsi* (Charles-Dominique,

1977) in Gabon is insect matter; this rises to 30% in East African *P. p. ibeanus* according to Kingdon (1971). The notable taste for social insects evinced by *edwardsi* (Charles-Dominique, 1977) has not been reported for the other two subspecies recognized here (see also Jewell and Oates, 1969; Baudenon, 1949; Rahm, 1960).

"Large-toothed" (*edwardsi*) and "small-toothed" (*potto*) subspecies of *Perodicticus potto* were distinguished by Schwarz (1931). In Table 12 the "generalist" value for *potto* (1042 g) is much less than that for *edwardsi* (1593 g). In fact, the second estimate is larger than most empirical weights for *Nycticebus coucang bengalensis* (Table 10), usually regarded as the largest living lorisid. This may be evidence that *P. p. edwardsi* has "larger" teeth than would be expected for its body size (empirical mean, about 1100 g), although if this is the case its tooth size is not linked to pronounced insectivory.

Loris tardigradus. Although in most dietary classifications *Loris tardigradus* is stated to be "chiefly insectivorous" (e.g., Hill, 1953; Roonwal and Mohnot, 1977), there are many records of Indian slender lorises eating other foods, including birds' eggs, small birds, lizards, shoots, leaves, and "almost every fruit with a hard rind" (Roonwal and Mohnot, 1977; see also Phillips, 1931; Ryley, 1913a; Subramonian, 1957). The bulk of the information on the dietary habits of *Loris tardigradus* seem to concern the largest subspecies, *L. t. lydekkerianus* of southern India, in which dietary catholicism would not be unexpected in view of its comparatively large size (Table 10). Feces of the slightly smaller *L. t. nordicus* from Sri Lanka have been reported as containing nothing but insect parts (Petter and Hladik, 1970), but in another paper dealing with the same forest study site (Polonnaruwa), Hladik and Hladik (1972) indicate that a considerable amount of fruit may contribute to the "diététogram" of this slender loris. Diet may, of course, vary according to the time of year; Petter and Hladik (1970) point out that the Polonnaruwa appeared to be nearly empty of visible insect life in December. Interestingly, the lipotyphlan equation accurately estimates the body weight of the smallest subspecies, *L. t. tardigradus,* but underestimates by an unacceptably wide margin the weights of the two largest, *L. t. nordicus* and *lydekkerianus*.

Nycticebus coucang and *N. pygmaeus.* As in the case of the slender loris, most of the rather anecdotal information about the diet of *Nycticebus coucang* concerns the largest subspecies. *Nycticebus c. bengalensis* evidently is quite opportunistic, taking a variety of insects and other animal foods in addition to fruit (MacKenzie, 1929; Roonwal and Mohnot, 1977; see also Fooden [1976] on stomach contents of *N. c.* cf. *coucang*).

The Bengalese and pygmy slow lorises are sympatric in Vietnam and Laos (Osgood, 1932; Dao, 1960), and are the only members of their genus that have overlapping distributions (Groves, 1971). No behavioral or dietary information exists for *N. pygmaeus,* but it is reasonable to expect that this species has some mechanism for avoiding competition with its larger relative. One possibility is that *pygmaeus* is much more insectivorous than *bengalensis,* or, to put it another way, *pygmaeus* is small enough so that it does not have to greatly supplement its diet with plant food. All African lorisiforms, for example, have gut and behavioral adaptations that permit insectivory; taxa with comparatively large body size, however, have greater nutritional requirements which they meet by eating other food items (Charles-Dominique, 1977, 1978). If both *bengalensis* and *pygmaeus* are predominantly frugivorous, then they presumably utilize different plants at different forest levels, in order to avoid direct competition for the same food resources. If they occupy similar levels of the same forests, it must be inferred that they utilize *quite* different resources. In either case they may subdivide their ecological space in a manner analogous to sympatric *Galago demidovii* and *G. alleni* in equatorial West African forests; highly insectivorous *demidovii* hunts its prey in the canopy, while moderately frugivorous *G. alleni* gathers fruit in forest undergrowth (Charles-Dominique, 1977). If *N. pygmaeus* is in fact predominantly insectivorous and shows the expected relationship of tooth size to body size, then 691 g should overestimate its weight by a considerable margin. A weight of 144 g, as predicted by the lipotyphlan equation (Table 12), seems much too small; however, the actual weight of the pygmy slow loris may yet prove to be comparatively low if 600 g is a reasonable estimate of the average weight of the smallest subspecies of *Nycticebus coucang* (Table 10).

CONCLUSION

We started this section by expressing our subjective opinion that *Nycticeboides* was a small animal, and we must conclude it nearly as subjectively. Although it would be epistemologically convenient if predominantly animalivorous primates could always be reliably segregated from those that are predominantly plant feeders, nature is less tractable than that.

Nycticeboides probably was not a folivore. Whether it relied on animal protein or non-leafy plant parts or exudates for most of its dietary intake cannot be settled at present, because existing methods for discriminating the guild affiliation of small nonfolivores are too imprecise to warrant our confidence. Although improvements in methods of inference may be possible, Lorisidae is likely to remain a difficult group. In this family, the average body-weight difference between populations or races of the largest nominal insectivores and the smallest nominal frugivores may be less than 200 g, with marked overlap in shearing capabilities. So narrow a range exceeds the limits of resolution of any of the methods considered here, which in turn means that extinct lorisid taxa with indeterminately small body sizes will have to continue to be assigned to indeterminately broad dietary guilds ("nonfolivorous generalists" or "frugi-animalivores").

A last point deserves special emphasis. If the relationship of *Nycticeboides* and *Nycticebus* is as close as we believe, the exceedingly small size of this fossil and its even smaller Siwalik relatives may be our first good hint that Asian lorisids have experimented with body size in a manner analogous to African Galagidae. Reported size contrasts in subspecies of living *Loris tardigradus* and *Nycticebus coucang* could imply that the experiment continues. The analogy is not exact: with the exception of *Nycticebus coucang* and *N. pygmaeus* in Southeast Asia, surviving lorisid taxa do not display the overlapping geographical distributions and concomitant dietary specializations seen in species of *Galago* (see Charles-Dominique, 1977). Instead, the adaptive strategy of Asian lorisids seems to have involved range and habitat extension. Populations of *Loris* manage to live in habitats ranging from sea level to the highest peaks in Sri Lanka and southern India, in diverse branch settings in both humid evergreen forests and dry deciduous forests (Hill, 1933; Phillips, 1935, Subramoniam, 1957; Hladik and Hladik, 1972; Eisenberg and McKay, 1970). *Nycticebus* is even more widespread, although less is known about its utilization of different forest types and branch settings within its range (but see captive study by Dykyj, 1980). The point now most in need of testing is whether populations in different parts of each species' range have diets that differ as sharply as their body sizes do. This would be unlikely judging by the standard of other primate species (but see Kay and others, 1978). However, as we have shown, some such inference has to be made in the case of the Bengalese and pygmy slow lorises, which do have overlapping distributions. The relevant hypotheses could be easily tested by appropriate fieldwork.

ACKNOWLEDGMENTS

We thank P. Andrews, R. Leaky, G. Musser, M. Rutzmoser, and R. Thorington for access to collections in their care; P. Wright, T. Rasmussen, I. Muul, J. C. Barry, W. Downs, and the late L. Radinsky, for certain data; P. D. Gingerich and B. Holly Smith, for allowing us access to their unpublished analysis of tooth and body size variation in insectivores; Veronica M. MacPhee for editorial assistance; and J. C. Barry, H. H. Covert, L. J. Flynn, P. D. Gingerich, R. F. Kay, D. Pilbeam, and T. Rasmussen for discussions and criticism. Special thanks are due the Government of Pakistan and members of the cooperative expeditions that collected the material described here, and the Government of Kenya and the National Museums of Kenya for permitting access to fossil specimens. This work was supported in part by NSF grants BSR 82-08797 (to RDEM) and BSR 85-00145 (to LLJ). Illustrations of teeth are by Lewis Sadler.

REFERENCES CITED

Acharjyo, L. N., and Misra, R., 1973, Notes on the birth and growth of a slow loris (*Nycticebus coucang*) in captivity: Bombay Natural History Society, Journal, v. 70, p. 193-194.

Ankel-Simons, F., 1983, A survey of living primates and their anatomy: New York, MacMillan, 313 p.

Badgley, C., *in press,* Taphonomy of mammalian fossil remains from Siwalik rocks of Pakistan: Paleobiology.

Banks, E., 1931, A popular account of the mammals of Borneo: Malayan Branch of the Royal Asiatic Society, Journal, Singapore, v. 9, p. 1-139.

Barry, J. C., Johnson, N. M., Raza, S. M., and Jacobs, L. L., 1985, Neogene mammalian faunal change in Southern Asia: correlations with climatic, tectonic, and eustatic events: Geology, v. 13, p. 637-640.

Barry, J. C., Lindsay, E. H., and Jacobs, L. L., 1982, A biostratigraphic zonation of the middle and upper Siwaliks of the Potwar Plateau of northern Pakistan: Palaeogeography, Paleoclimatology, and Paloeoecology, v. 37, p. 95-130.

Bauchot, R. and Stephan, H., 1966, Données nouvelles sur l'encéphalisation des insectivores et des prosimiens: Mammalia, v. 30, p. 160-196.

Baudenon, P., 1949, Contribution á la connaissance du potto de Bosman dans le Togo-Sud: *ibid.,* v. 13, p. 76-99.

Bonhote, J. L., 1907, On a collection of mammals made by Dr. Vassal in Annam: Zoological Society of London, Proceedings, p. 3-11.

Bonin, G. von, 1937, Brain-weight and body-weight of mammals: Journal of General Psychology, v. 16, p. 379-389.

Bourlière, F., 1975, Mammals, small and large: the ecological implications of size, in Golley, F. B., Petrusewicz, K., and Ryszkowski, L., eds., Small mammals: their productivity and population dynamics: Cambridge, Cambridge University Press, p. 1-8.

―――― 1985, Primate communities: their structure and role in tropical ecosystems: International Journal of Primatology, v. 6, p. 1-26.

Charles-Dominique, P., 1977, Ecology and behavior of nocturnal primates: New York, Columbia University Press, 277 p.

―――― 1978, Ecological position of the family Lorisidae compared to other mammalian families: Carnegie Museum of Natural History, Bulletin, v. 6, p. 26-30.

Covert, H. H., 1985, Adaptations and evolutionary relationships of the Eocene primate subfamily Notharctinae [Ph.D. thesis]: Durham, North Carolina, Duke University.

Dao Van Thien, 1960, Sur une nouvelle espéce de *Nycticebus* au Vietnam: Zoologischer Anzeiger, v. 164, p. 240-243.

Davis, D. D., 1962, Mammals of the low-land rain forest of North Borneo: National Museum of the Straits of Malaya, Bulletin, Singapore, v. 31, p. 1-129.

Dene, H. T., Goodman, M., Prychodko, W., and Moore, G. W., 1976, Immunodiffusion systematics of the Primates. III. The Strepsirhini: Folia primatologica, v. 25, p. 35-61.

Dubois, E., 1897, Sur le rapport du poids de l'encéphale avec le grandeur du corps chez les mammifères: Bulletin de la Société d'Anthropologie de Paris, série 4, v. 8, p. 337-376.

Dykyj, D., 1980, Locomotion of the slow loris in a designated substrate context: American Journal of Physical Anthropology, v. 52, p. 577-586.

Eisenberg, J. F., and McKay, G. M., 1970, An annotated check list of the Recent mammals of Ceylon, with keys to species: Ceylon Journal of Science (Biological Sciences), v. 8, p. 69-99.

Ellerman, J. R., and Morrison-Scott, T. C. S., 1951, Checklist of Palaearctic and Indian mammals, 1758 to 1946: London, British Museum of Natural History, 810 p.

Elliot, D. G., 1913, A review of the primates: New York, American Museum of Natural History, 3 vols.

Flynn, L. J., Jacobs, L. L., and Cheema, I. U., 1986, Baluchimyinae, a new ctenodactyloid rodent subfamily from the Miocene of Baluchistan: American Museum Novitates, no. 2841, 58 p.

Fooden, J., 1971, Report on primates collected in western Thailand, January-April, 1967: Field Museum of Natural History, Fieldiana: Zoology, v. 59, p. 1-62.

―――― 1976, Primates obtained in peninsular Thailand June-July, 1973, with notes on the distribution of continental Southeast Asian leaf-monkeys (*Presbytis*): Primates, v. 17, p. 95-118.

Ford, S. M., and Corruccini, R. S., 1984, Intraspecific, interspecific, metabolic, and phylogenetic scaling in platyrrhine primates, in Jungers, W. L., ed., Size and scaling in primate biology: New York, Plenum Press, p. 383-436.

Gingerich, P. D., and Sahni, A., 1979, *Indraloris* and *Sivaladapis*: Miocene adapid primates from the Siwaliks of India and Pakistan: Nature, v. 279, p. 415-416.

―――― 1984, Dentition of *Sivaladapis nagrii* (Adapidae) from the late Miocene of India: International Journal of Primatology, v. 5, p. 63-79.

Gingerich, P. D., and Smith, B. H., 1984, Allometric scaling in the primates and insectivores, in Jungers, W. L., ed., Size and scaling in primate biology: New York, Plenum Press, p. 257-272.

Gingerich, P. D., Smith, B. H., and Rosenberg, K., 1982, Allometric scaling in the dentition of primates and prediction of body weight from tooth size in fossils: American Journal of Physical Anthropology, v. 58, p. 81-100.

Groves, C. P., 1971, Systematics of the genus *Nycticebus*, in Proceedings of the Third International Congress of Primatology: Basel, Karger, vol. 1, p. 44-53.

Hill, A., 1979, Disarticulation and scattering of mammal skeletons: Paleobiology, v. 5, p. 261-274.

Hill, W. C. O., 1933, A monograph on the genus *Loris*, with an account of the external, cranial and dental characters of the genus: a revision of the known forms; and the description of a new form from northern Ceylon: Ceylon Journal of Science, v. 18B, p. 89-132.

―――― 1936, The affinities of lorisoids: *ibid.*, v. 19B, p. 287-314.

―――― 1942, The slender loris of the Horton Plains, Ceylon. *Loris tardigradus nycticeboides,* subsp. nov.: Bombay Natural History Society, Journal, v. 43, p. 73-78.

―――― 1947, Thoracic transpedicular foramina in the Lorisidae: Zoological Society of London, Proceedings, p. 525-530.

―――― 1953, Primates: comparative anatomy and taxonomy, vol. 1, Strepsirhini: Edinburgh, Edinburgh University Press, 798 p.

Hladik, C. M., and Hladik, A., 1972, Disponibilités alimentaires et domaines vitaux des primates à Ceylan: Terre et Vie, v. 2, p. 149-215.

Jacobs, L. L., 1981, Miocene lorisid primates from the Pakistan Siwaliks: Nature, v. 289, p. 585-587.

Jaeger, J.-J., Denys, C., and Coiffait, B., 1985, New Phiomorpha and Anomaluridae from the late Eocene of north-west Africa: phylogenetic implications, in Luckett, W. P., and Hartenberger, J.-L., eds., Evolutionary relationships among rodents, a multidisciplinary analysis: New York, Plenum Press, p. 567-588.

Jenkins, F. A., 1970, Anatomy and function of expanded ribs in certain edentates and primates: Journal of Mammalogy, v. 51, p. 288-301.

Jewell, P. A., and Oates, J. F., 1969, Ecological observations on the lorisoid primates of African lowland forest: Zoologica Africana, v. 4, p. 231-248.

Johnson, N. M., Stix, J., Tauxe, L., Cerveny, P. F., and Tahirkheli, R. A. K., 1985, Paleomagnetic chronology, fluvial processes, and tectonic implications of the Siwalik deposits near Chinji Village, Pakistan: Journal of Geology, v. 93, p. 27-40.

Jungers, W. L., 1984, Body size and scaling of limb proportions in primates, in Jungers, W. L., ed., Size and scaling in primate biology: New York, Plenum Press, p. 345-382.

Kavanagh, M., 1983, A complete guide to monkeys, apes, and other primates: London, Cape, 224 p.

Kay, R. F., 1975, The functional adaptations of primate molar teeth: American Journal of Physical Anthropology, v. 43, p. 195-216.

Kay, R. F., and Simons, E. L., 1980, The ecology of Oligocene African Anthropoidea: International Journal of Primatology, v. 1, p. 21-38.

Kay, R. F., Sussman, R. W., and Tattersall, I., 1978, Dietary and dental variations in the genus *Lemur,* with comments concerning dietary-dental correlations among Malagasy primates: American Journal of Physical Anthropology, v. 49, p. 119-128.

Kennard, M. A., and Willner, M. D., 1941, Weights of brains and organs of 132 New and Old World monkeys: Endocrinology, v. 28, p. 977-984.

Kingdon, J., 1971, East African mammals, an atlas of evolution in Africa: London, Academic Press, 2 vols.

Kolar, K., 1984, Tree shrews and prosimians, in Grzimek, B., ed., Grzimek's animal life encyclopedia: New York, Van Nostrand Reinhold, v. 10, p. 270-311.

Lyon, M. W., 1907, Notes on the slow lemurs: U.S. National Museum, Proceedings, v. 31, p. 527-538.

McArdle, J. E., 1978, The functional morphology of the hip and thigh of the Lorisiformes [Ph.D. thesis]: Chicago, Illinois, University of Chicago, 271 p.

MacKenzie, J. M. D., 1929, Food of the slow loris (*Nycticebus coucang*): Bombay Natural History Society, Journal, v. 33, p. 971.

MacPhee, R. D. E., 1981, Auditory regions of primates and eutherian insectivores: morphology, ontogeny and character analysis: Contributions to Primatology, v. 18, p. 1-282.

MacPhee, R. D. E., and Cartmill, M., 1986, Basicranial structures and primate systematics, *in* Swindler, D. R., ed., Comparative primate biology, vol. 1, Systematics, evolution, and anatomy: New York, Alan R. Liss, p. 219-275.

Malbrant, R., and Maclatchy, A., 1949, Faune de l'équateur africain français, II, Mammiféres: Encyclopédie de biologie, v. 36, p. 1-323.

Manley, G. H., 1966, Prosimians as laboratory animals: Zoological Society of London, Symposium, no. 17, p. 11-39.

Manouvrier, L., 1888, Sur l'interpretation de la quantité dans l'encéphale et dans le cerveau en particulier: Mémoires de la Société d'Anthropologie de Paris, série 2, v. 3, p. 157-326.

Lord Medway, 1969, The wild mammals of Malaya and offshore islands including Singapore: London, Oxford University Press, 128 p.

Napier, J. R., and Napier, P. H., 1967, A handbook of living primates: New York, Academic Press, 456 p.

Osgood, W. H., 1932, Mammals of the Kelley-Roosevelts and Delacour Asiatic Expeditions: Field Museum of Natural History, Fieldiana: Zoology, v. 18, p. 191-339.

Petter, J. J., and Hladik, C. M., 1970, Observations sur le domaine vital et la densité de population de *Loris tardigradus* dans les forêts de Ceylan: Mammalia, v. 34, p. 395-409.

Phillips, W. W. A., 1931, The food of the Ceylon slender loris (*Loris tardigradus*) in capitivity: Ceylon Journal of Science, v. 16, p. 205-208.

―――― 1935, Manual of the mammals of Ceylon: Colombo, Colombo Museum, issued by Dulau and Co., London.

Prater, S. H., 1965, The book of Indian mammals, 2nd rev. ed.: Bombay, Bombay Natural History Society and Prince of Wales Museum of Western India, 323 p.

Rahm, U., 1960, Quelques notes sur le potto de Bosman: Bulletin de l'Institut français d'Afrique noire, v. 22, p. 331-342.

Rasmussen, D. T., and Izard, M. K., *in press,* Scaling of growth and life history traits relative to body size, brain size, and metabolic rate in lorises and galagos (Lorisidae, Primates): Evolution.

Roberts, D., 1974, Structure and function of the primate scapula, *in* Jenkins, F. A., ed., Primate locomotion: New York, Academic Press, p. 171-199.

Roonwal, M. L., and Mohnot, S. M., 1977, Primates of South Asia: ecology, sociobiology, and behavior. Cambridge, Harvard University Press, 421 p.

Rose, K. D., Walker, A., and Jacobs, L. L., 1981, Function of the mandibular tooth comb in living and extinct mammals: Nature, v. 289, p. 583-585.

Ryley, K. V., 1913a, Bombay Natural History Society's mammal survey of India: Bombay Natural History Society, Journal, v. 22, p. 283-295.

―――― 1913b, Bombay Natural History Society's mammal survey of India, Report No. 10: *ibid.,* v. 22, p. 464-513.

Sanderson, I. T., 1940, The mammals of the North Cameroons forest area, being the results of the Percy Sladen expedition to the Mamfe division of the British Cameroons: Zoological Society of London, Transactions, v. 24, p. 623-725.

Schwartz, J. H., 1974, Dental development and eruption in the prosimians and its bearing on their evolution [Ph.D. thesis]: New York, New York, Columbia University, 423 p.

Schwartz, J. H., and Tattersall, I., 1985, Evolutionary relationships of living lemurs and lorises (Mammalia, Primates) and their potential affinities with European Eocene Adapidae: American Museum of Natural History, Anthropological Papers, v. 60, p. 1-100.

Schwarz, E., 1931, On the African short-tailed lemurs or pottos: Annals and Magazine of Natural History, series 10, v. 8, p. 249-256.

Seth, P. K., 1966, Cranial variability in the Indian lorises *Loris tardigradus lydekkerianus* and *Nycticebus coucang:* Zeitschrift für Morphologie und Anthropologie, v. 7, p. 179-191.

Shortridge, G. C., 1914, Notes on the weights of animals: Bombay Natural History Society, Journal, v. 22, p. 793-794.

Simpson, G. G., 1967, The Tertiary lorisiform primates of Africa: Museum of Comparative Zoology of Harvard University, Bulletin, v. 136, p. 39-62.

Smith, R. J., 1984, The present as a key to the past: body weight of Miocene hominoids as a test of allometric methods for paleontological inference, *in* Jungers, W. L., ed., Size and scaling in primate biology: New York, Plenum Press, p. 437-448.

Spitzka, E. A., 1903, Brain-weights of animals with special reference to the weight of the brain in the macaque monkey: Journal of Comparative Neurology, v. 13, p. 9-17.

Stephan, H., and Bauchot, R., 1965, Hirn-Körpergewichtsbeziehungen bei den Halbaffen (Prosimii): Acta Zoologica (Stockholm), v. 46, p. 209-231.

Straus, W. L., and Wislocki, G. B., 1932, On certain similarities between sloths and slow lemurs: Museum of Comparative Zoology of Harvard College, Bulletin, v. 74, p. 45-56.

Subramoniam, S., 1957, Some observations on the habits of the slender loris, *Loris tardigradus* (Linnaeus): Bombay Natural History Society, Journal, v. 54, p. 387-398.

Swindler, D. R., 1976, Dentition of living primates: New York, Academic Press, 308 p.

Tauxe, L., 1980, A new date for *Ramapithecus*: Nature, v. 282, p. 399-401.

Tauxe, L., and Opdyke, N. D., 1982, A time framework based on magnetostratigraphy for the Siwalik sediments of the Khaur area, northern Pakistan: Paleogeography, Palaeoclimatology, and Palaeoecology, v. 37, p. 43-61.

Walker, A., 1969, True affinities of *Propotto leakyi* Simpson, 1967: Nature, v. 223, p. 647-648.

―――― 1972, The dissemination and segregation of early primates in relation to continental configuration, *in* Bishop, W. W., and Miller, J. A., eds., Calibration of hominid evolution: Edinburgh, Scottish Academic Press, p. 192-218.

―――― 1978, Prosimian primates, *in* Maglio, V. J., and Cooke, H. B. S., eds., Evolution of African mammals: Cambridge, Harvard University Press, p. 90-99.

Warncke, P., 1908, Mitteilungen neuer Gehirn- und Körpergewichtsbestimmungen bei Säugern, nebst Zusammenstellung der gesamten bisher beobachteten absoluten und relativen Gehirngewichte bei den verschiedenen Spezies: Zeitschrift für Psychologie und Neurologie, v. 13, p. 355-403.

Warwick, R., and Williams, P. L., eds., 1973, Gray's anatomy (35th British edition): Philadelphia, Saunders.

Weber, M., 1896, Vorstudien über das Hirngewicht der Säugetiere, *in* Festschrift Carl Gegenbaur, v. 3, p. 103-123. (Cited by Bauchot and Stephan, 1966.)

Wells, N. A., and Gingerich, P. D., 1983, Review of Eocene Anthracobunidae (Mammalia, Proboscidea) with a new genus and species, *Jozaria palustris,* from the Kuldana Formation of Kohat (Pakistan): Museum of Paleontology, University of Michigan, Contributions, v. 26, p. 117-139.

MANUSCRIPT RECEIVED JANUARY 12, 1985
REVISED MANUSCRIPT RECEIVED JUNE 2, 1986
MANUSCRIPT ACCEPTED JUNE 24, 1986

The Paleogene record of the rodents: fact and interpretation

ROBERT W. WILSON *Museum of Natural History and Department of Systematics and Ecology, The University of Kansas, Lawrence, Kansas 66045*

ABSTRACT

Work of the past fifty years, or so, has shed much additional light on the phylogeny and history of the Order Rodentia. Classical views have been challenged, and new concepts invoked, or older views rediscovered. Of these new concepts, hystricomorphy, rather than protrogomorphy, as a primitive state for rodents seems difficult to accept, and is not really new. Hystricognathy versus sciurognathy as the fundamental division of the Rodentia seems perilous if parallelism is as important a phenomenon as is frequently suggested. The argument for multiserial incisor enamel, rather than pauciserial, as the primitive incisor kind is very persuasive, but perhaps more work is needed on Eocene rodent enamel. Punctuated equilibrium, if really a new idea, seems promising in explaining the obscure origin of most rodent groups, but gradualism is evident in many specific lines of descent in rodents. Virtually excluding temporal consideration from phylogenetic studies seems extreme, as does cladistic analysis when it excludes parallelisms and paraphyletic groups. In spite of recent work, the gap between Eocene groups such as the Paramyidae and Ctenodactyloidea, and the Oligocene and later families remains considerable, and largely unexplained. Extraterrestrial collision events in this case can hardly be regarded as pertinent for rodents.

INTRODUCTION

In the last fifty years or so, the number of recovered rodent fossils has increased enormously, and the number of workers with them also has multiplied correspondingly. Then too, in the past few years, areas such as China and Outer Mongolia have been the sites of discovery of Eocene rodents, a group previously essentially unknown from there. The present paper is a review which includes these new discoveries, and which is followed by a review of various interpretations of the Paleogene record, both classical and recent.

The oldest known rodent may be *Acritoparamys atavus* (= *Paramys atavus* Jepsen, 1937). This paramyid from the Eagle Coal Mine, Bear Creek, Montana, is either from the latest Paleocene or earliest Eocene. European paramyids are known from a nearly comparable, but possibly slightly younger age. In China, the only known early Eocene rodent is *Cocomys*, a ctenodactyloid. At slightly higher levels appear the first sciuravids in North America, the earliest possible theridomyids in Europe, followed by cylindrodonts in the middle Eocene of North America, and glirids in Europe. In the middle Eocene of China and India, ctenodactyloids (Cocomyidae/Yuomyidae/Chapattamyidae) are dominant, but paramyids do appear before the close of this age, if not earlier in Pakistan (de Bruijn and others, 1982).

The earliest rodents had a chewing and gnawing musculature generally termed protrogomorphous. The infraorbital foramen is relatively large, say compared to the Sciuridae, but small compared with such as the Zapodidae. There is no indication of passage through it by a significant slip of the medial masseter. The origin of the masseter seems to have been confined to the zygomatic arch, and to lie below the infraorbital foramen; its origin had not advanced onto the rostrum. Several rodent groups (among others ctenodactyloids, theridomyids) show enlargement of the infraorbital canal during the course of the Eocene.

In the late Eocene of North America appear the problematical genera *Simimys, Griphomys, Prolapsus,* and *Protoptychus.* These show some modification of the zygomasseteric structures seen in the paramyids and early Eocene cocomyids. In addition, the first eomyids appear, although we do not know much (if anything) about their zygomasseteric structure. With the early Oligocene, the older groups of rodents were rapidly replaced by such modern families as the Sciuridae, Castoridae, and Cricetidae. This change was striking enough for European workers to refer to the *grande coupure,* a marker which probably correlates with some part of the later Chadronian of North America. In North America, the changeover seemingly started in late Uintan and was largely completed before the beginning of the Orellan (medial Oligocene). Unfortunately, in Africa and South America, no determinable rodents are known from the Eocene. Phiomyids appear in the early Oligocene (or late Eocene?) of Africa, and four families of caviomorphs in the late Oligocene of South American (Patterson and Wood, 1982; MacFadden and others, 1985). In only a few cases, at best, has it been possible to trace lines of descent across the Eocene-Oligocene "boundary."

Basic data for the following distribution tables (Tables 1-4) were obtained from: Wilson (1980), Wood (1980), Black and Sutton (1984), Golz and Lillegraven (1977), J. A. Wilson (1978) for North America; Savage and Russell (1983), Dawson (1977) for Europe; Li and Ting (1983), Dawson (1977), Shevyreva (1983) for eastern Asia; Wood (1968), Lavocat (1973) for Africa; and Patterson and Wood (1982) for South America.

TABLE 1. THE NORTH AMERICAN RECORD OF RODENT FAMILIES.

Families	Range						
	Early Eocene	Middle Eocene	Late Eocene	Duchesnean	Early Oligocene	Middle Oligocene	Late Oligocene
Paramyidae	X X	X	X	X X	X	X	?(Manitsha tanka)
Sciuravidae		X	X	X X	?		
Cylindrodontidae	X	X	X X	X	X		
Ischyromyidae				?	X	X	?
Aplodontidae			? ?		X	X	X
Protoptychidae			X				
Eomyidae			X		X	X	X
Zapodidae			?(Simimys)	?			
Heteromyidae				?(Griphomys)	X	X	X
Cricetidae					?	X	X
Sciuridae					X	X	X
Castoridae					X	X	X
Eutypomyidae	?(Mattimus kalicola)			?(Janimus)	X		

The question arises as to how this Paleogene rodent record supports or does not support various ideas introduced in the last twenty years or so which relate to fossils and their classification, phylogeny, geographic relationships, and extinction. I propose to discuss briefly the following: (1) classical interpretation of the record; (2) hystricomorphy as a primitive feature; (3) hystricognathy; (4) enamel structure of incisors; (5) punctuated equilibrium and the depositional record; (6) collision events at the Eocene-Oligocene boundary; (7) validity of the fossil record in phylogeny; and (8) cladistic analysis.

INTERPRETATIONS

The Paleogene Rodent Record

Most mammalian paleontologists think that the fossil record has much to contribute to the understanding of the evolutionary history of mammals. So far as rodents are concerned, the following contributions are valuable, either as representing older views, now open to modification, or as paleontological views of the past ten years or so:

   Black and Sutton, 1984
   Chaline and Mein, 1979
   Dawson, 1961, 1977
   Fahlbusch, 1970, 1979, 1983
   Korth, 1984
   Matthew, 1910
   Wahlert, 1968, 1973, 1978
   Wilson, 1949c, 1972, 1980
   Wood, 1955, 1959, 1962, 1980.

As early as 1910, W. D. Matthew proposed that the Ischyromyidae (essentially the Paramyidae of many authors) were a group ancestral to all later rodents. Miller and Gidley (1918), however, in a preliminary statement which was never followed up by them, stated that there was no known rodent which could be considered ancestral to "any considerable number of subsequent forms." Since then, there has been mostly support for the Matthew thesis, but with few individual lines which could be traced from this presumed ancestral base into modern families of rodents.

In 1949c, I presented a review of the early Tertiary rodents of North America. Much information on these rodents has accumulated since that date. This increased knowledge is the result of several factors: (1) a considerable increase in the number of workers investigating small fossil mammals; (2) improvements in recovery methods (screen washing); and (3) important discoveries in Asia, especially China, a *terra incognito* in 1949. Black and Sutton (1984) have updated my review so far as North America is concerned. Li and Ting (1983) have summarized the Chinese stratigraphic record for the Paleogene. Modern work also has vastly improved the Oligocene record of Africa (Wood, 1968) and South America (Patterson and Wood, 1982). Finally, various French, Ger-

THE PALEOGENE RECORD OF THE RODENTS

TABLE 2. THE EUROPEAN RECORD OF RODENT FAMILIES.

Families — Range

Grande Coupure

Families	Early Eocene	Middle Eocene	Late Eocene	Early Oligocene	Middle Oligocene	Late Oligocene
Paramyidae	X	X	X			
Theridomyidae	X	X	X	X	X	X
Gliridae		X	X	X	X	X
Eomyidae				X	X	X
Aplodontidae				X	X	X
Sciuridae				X	X	X
Castoridae				X	X	X
Cricetidae				X	X	X
Zapodidae					X	X

Approximate Time Equivalents

Europe — North America

Early Eocene = Wasatchian and early Bridgerian
Middle Eocene = late Bridgerian and early Uintan
Late Eocene = Uintan, Duchesnean, earliest Chadronian
Early Oligocene = typical middle Chadronian
Middle Oligocene = late Chadronian, Orellan, early Whitneyan
Late Oligocene = late Whitneyan, early Arikareean

TABLE 3. THE EASTERN ASIAN AND INDIAN RECORD OF RODENT FAMILIES.

Families — Range

Families	Early Eocene	Middle Eocene	Late Eocene	early Oligocene*	Middle Oligocene	Late Oligocene
Paramyidae	?	X	X	X?		
Cocomyidae	X	X	X			
Yuomyidae			X			
Ctenodactylidae					X	X
Cylindrodontidae				X	X	X
Sciuridae						X
Cricetidae				X	X	
Zapodiae				X	X	X
Rhizomyidae					X	X
Castoridae					X	X
Aplodontidae					X	
Thryonomyidae					X	

*and late Eocene

TABLE 4. SOUTH AMERICAN FIRST RECORD OF RODENT FAMILIES.

Families

	Late Oligocene*	Mid Miocene
Erethizontidae		X
Cephalomyidae		X
Chinchillidae		X
Octodontidae	X	
Echimyidae	X	
Dasyproctidae	X	
Eocardidae		X
Dinomyidae	X	

*see MacFadden, et al., 1985 for this revised dating of the Deseadan

African Record of Rodent Families
(Only indets from Eocene)

Family	Range		
	Early Oligocene**	Middle Oligocene	Late Oligocene
Phiomyidae	X	?	?

**if not late Eocene

man, Austrian, Dutch, and Spanish workers have revolutionized the entire Tertiary rodent record of Europe.

My present attempt is to bring an updated rodent record of the various continents into one article. This factual assemblage stresses suprageneric groups (families) at the expense of genera and species. One problem with my earlier (1949c) interpretation was that it was based on the record as known from North America and Europe. The critical Asiatic record was missing, a point that troubled me at the time. For many years this could be minimized because the records were more or less comparable in the two known areas, and the North Atlantic was thought to have been a barrier for direct communication. Hence, the Asiatic fauna was assumed to be similar to account for the high faunal resemblance between North America and Europe in the early Eocene. About the time we were discovering that there was a North Atlantic land connection between North America and Europe in the early Eocene, the long sought Eocene record of rodents from eastern Asia was emerging as the result of discoveries, especially in China.

It now appears that in Asia the dominance of paramyids characteristic of the North American record was substituted by a dominance of ctenodactyloids. The earliest ctenodactyloid, *Cocomys,* is near enough to the earliest paramyid, *Acritoparamys,* to have been identified as *Microparamys,* when the genus was known only by fragments. If *Heomys* is near to the ancestry of the Order Rodentia, *Cocomys* may lie closer than typical paramyids in that *Cocomys* and *Heomys* both have similar P^4, with undivided paracone and metacone. P$_4$ in *Cocomys* lacks a hypoconid. In *Cocomys, Heomys,* and paramyids generally, the infraorbital foramen is not enlarged to any extent, and there is no visible origin of the *M. medialis* anterior to the infraorbital foramen. In the Asiatic ctenodactyloids, and the glirids and theridomyids, as mentioned earlier, the infraorbital foramen enlarges in the course of the Eocene; but in paramyids, and hence typically in the North American Eocene, there is no such enlargement, nor do we see evidence of enlargement in the Cylindrodontidae. The sciuravids (restricted to North America) are usually similar, but Dawson *(oral communication)* reports what seems to be enlargement in a specimen from the latest early Eocene (Lost Cabin). *Prolapsus,* another possible sciuravid, has an infraorbital foramen hardly any larger than usual for sciuravids, but is hystricognathus. In Africa, the phiomyids appear in the early Oligocene, or possibly late Eocene, depending on correlation, and are hystricomorphous. In South America, the late Oligocene cavioids are hystricomorphous, although there is doubt about the size of the infraorbital foramen in *Platypittimys* (see Wood, 1949 versus Landry, 1957).

As previously mentioned, there exists a considerable gap between Eocene rodent groups and those of the typical Oligocene, and few lines of Eocene rodents can be traced across this gap with any certainty. The paramyids, as now defined, seem to have given rise to the glirids and theridomyoids of Europe, and probably to the sciurids and aplodontids. The Sciuravidae, probably to be derived from microparamyids, may be ancestral to the zapodids and cricetids on the one hand, and geomyoids on the other. At least there are no other known Eocene rodents which seem as promising. In Asia, the early ctenodactyloids are close to common ancestry with the paramyids, and the cylindrodontids may have a relationship to caviomorphs and/or bathyergoids.

Among North American late Eocene rodents are several of uncertain position such as *Simimys, Griphomys,* and *Eohaplomys. Simimys* could occupy one of several taxonomic positions. It could be a peculiar sciuravid with loss of P4/4, a rodent close to the common ancestor both of zapodids and cricetids without being either, or a zapodid. Wood (1980) solved this problem by establishing the family Simimyidae. *Griphomys* is still something of a riddle also, but has been assigned to the Geomyoidea with a query (Wilson, 1940; Lillegraven, 1977), and to the Eomyidae by Wood (1980). *Eohaplomys* was assigned by Stock (1935) to the Ischyromyidae, although he recognized its relationship to primitive members of the Aplodontidae. Transfer to the latter family was made by Wilson (1949a), if not actually by McGrew (1941). More recently, Rensberger (1975) has cast much doubt on this assignment.[1] If *Eohaplomys* is not an aplodontid, however, it is even less similar to most paramyids. It might be interesting to compare *Eohaplomys* to certain tropical flying squirrels such as *Belomys*.

Once again, these late and latest Eocene rodents serve to emphasize the break between Eocene and typical Oligocene rodents. The situation is not very different from eutherians as a whole (see, for example, Black and Dawson, 1966).

In the past few years papers have appeared which have offered non-traditional or non-classical approaches to interpretations of the fossil record, some applying to rodents (hystricomorphy, hystricognathy, incisor enamel), and some to the Tertiary record more generally (punctuated equilibrium, impact extinction, cladistic analysis). These may now be considered *in seriatum*.

---

1. I might add that Rensberger (1975, Fig. 2*b*) identifies as *Prosciurus relictus* a specimen which surely is not this species, as numerous specimens from South Dakota (Slim Buttes) and northeastern Colorado testify.

Hystricomorphy

Charles B. Eastman (1982, p. 163) has argued that hystricomorphy was possessed by all primitive rodents. "Protrogomorphy is simply the primitive mammalian condition, and is not part of the derived morphology that defines the earliest rodents." The argument is based on Eastman's study of *Aplodontia* (three specimens), which revealed to him that part of the medial masseter muscle did pass through the infraorbital foramen and originated on the side of the rostrum. He assumed that the aplodont condition prevailed in the paramyids and other early rodents as well. He thought that the qualitative aspects of the muscle origin was more important than the quantitative (percentage of muscle passing through) aspects. It would seem that this would imply that invasion took place before development of the gnawing incisors, or at least in tandem with it. Moreover, insertion of the masseter on the mandible is usually under the second molar, rather than under the first molar, or last premolar, or even anterior to that in those rodents having some branch of the masseter originating on the muzzle. The coronoid process is large, and the temporal muscle correspondingly strong. These characters suggest a morphology with gnawing incisors, but no effective invasion of the infraorbital foramen and attachment on the rostrum by the masseter. It also should be noted that so far as I know, the origin of the *medialis* cannot be detected on the muzzle, which suggests a slip not of much importance in the chewing-gnawing aspects of these rodents. Eastman's figure, however, shows a considerable bit of *medialis* passing through the foramen.

We have evidence in the fossil record that enlargement of the infraorbital foramen does occur as time goes on in the Eocene in some rodent groups (theridomyids, ctenodactyloids, and possible some sciuravids). I, for one, would consider hystricomorphy to involve some real functional significance, and not merely a tiny amount of muscle tissue. The *medialis* originates on the zygomatic arch at its anterior and medial end in protrogomorphous rodents, and thus close to the posterior entrance to the foramen—hence also the possibility of a strand being anchored still farther forward. Moreover, what is the case in the Recent *Aplodontia rufa* may not have been so in early Eocene rodents, or even early aplodontids. Eastman (1982, p. 163) says, of *Aplodontia,* "The masticatory mechanism of the mountain beaver was previously described by Tullberg (1899) *among others* [italics mine]. In his description, however, he did not note that a portion of the medial masseter runs through the infraorbital foramen and originates on the rostrum." The "and others" deserves comment, especially because the references given are, with the exception of Tullberg, later than 1930.

In 1877, Coues (Coues and Allen, 1877) reported that in *Aplodontia,* the *masseter medialis* sends a small branch through the infraorbital foramen. Ten years later, in 1887, Winge proposed that the primitive condition in all rodents was hystricomorphous save the Ischyromyidae, placing reliance on Coues' statement. By 1924, Winge had modified this arrangement somewhat. In his words (English translation, 1941, p. 115), "After Tullburg discovered that in *Haplodon* no part of Masseter passes through the infraorbital foramen, and I myself had occasion to make the same observation, all grounds for placing these rodents in different families (*i. e.,* Haplodontidae and Ischyromyidae) disappeared." He continued to regard all other rodents as primitively hystricomorphous. Thus, Eastman's conclusions follow Coues, and a somewhat modified Winge. If Tullberg and Winge actually dissected the masseter, it almost suggests individual variation in *Aplodontia.* If they did not, it still suggests the mass of muscle tissue penetrating the infraorbital foramen must be slight. In passing, it should be noted that Miller and Gidley (1918) state that in the Paramyidae and Aplodontidae, the infraorbital foramen does not transmit muscle. Miller, at least, must have been aware of Coues' statement. Finally, it would seem that in cases where there is significant invasion of the infraorbital foramen, some means is provided to protect the trigeminal nerve from muscle pressure. That seems not to be the case in aplodontids and paramyids.

Problems in Hystricognathy

Hystricognathy in rodents is an important feature of the jaw mechanism which has been used by most specialists to separate those rodents having it from all others. Its opposite number is sciurognathy. The latter is certainly the more primitive condition, although the earliest rodents may have been slightly different (see Landry, 1957, p. 82-83; Korth, 1984, p. 7) from a fully sciurognath type. Be that as may be, it can be assumed that a condition exists in which jaws first became slightly modified away fom the ancestral type toward hystricognathy. A. E. Wood has thought he could see this condition in certain North American Eocene rodents such as *Franimys,* although Korth does not.

The hystricognath condition, as usually defined, is one in which the inner edge of the angular process is lateral to the external edge of the incisor, or the incisor sheath. Tullberg was a strong advocate of the fundamental division of rodents on the basis of the two conditions. In his classic work (1899, p. 69), however, he writes, "Samtliche *Hystricognathen* zeichnen sich dadurch aus, dass der Angularforsetz des Unterkiefers vorn seithwarts verschoben ist, so dass sein Unterteil von das ausseren Seite des Corpus ausgeht, wie auch dadurch dass seine Margo inferior nahezu mit dem Jochbogen parallel und in seitliche Ansicht, wenigstens im grossten Teil seiner Lange, ganz horizontal verlauft." Nothing in this statement speaks directly of the course of the incisor, whatever Tullberg's intent was.

It is not necessary, however, to assume that every "incipient" condition marches steadily toward a full-blown hystricognathy in which the whole angular process is lifted upward and lateral to the body of the ramus. Further, neither is it necessary to assume that the possession of incipient, or even fully developed, hystricognathy relates rodents to each other, and excludes relationship with others. Hystricognathy is perhaps more valuable as a taxonomic character than most such in rodents, but to use it alone and not in combination with other derived features is perhaps to make a key, not a classification.

Years ago now (Wilson, 1949c, p. 149 and 152), I commented on the fact that the molar patterns of Bridgerian rodents were of three kinds: paramyid, cylindrodontid, and sciuravid, and that these patterns in early, brachydont kinds of rodents were of taxonomic value, and modified from Wasatchian patterns. According to various statements of A. E. Wood, rodents having these patterns all have some members which have incipient to fully developed hystricognathy: that is, *Franimys* and *Reithroparamys* among paramyids, *Cylindrodon* among cylindrodonts, and *Prolapsus* among sciuravids. He also thinks that *Mysops boskeyi* could be ancestral to South American caviomorphs (*Mysops* was probably ancestral to *Cylindrodon*). It seems to me on the other hand, that the distribution suggests that incipiency, if it exists, is not necessarily of taxonomic worth at a supergeneric level, and even when fully developed as is said to be the case in *Prolapsus,* does not as a single character standing alone suggest a natural group of the importance indicated by Wood (1980, p. 6) in his review of the Oligocene rodents of North America (but see Wood, 1984, for a more ambivalent view). I would even speculate that until late in the Eocene real hystricognathy had not yet started in the Hystricomorpha, and the ctenodactyloids may be closer to the Hystricomorpha than most have thought. The latter was suggested by Hussain and others (1978), de Bruijn and others (1982), and Wahlert (1984), and by Landry (1957) many years ago. Admittedly, facts to back this possibility are almost lacking. If rodent history, however, is as shot through with parallelism as several of my colleagues claim—seemingly in endless amounts to prove a point—why not extend it to the kind of jaw modification we call hystricognathy? Also, there seem to be cases of "incipient" hystricognathy other than Wood has cited, the exact significance of which eludes me.

In a description of *Ischyrotomus littoralis* (see Wilson, 1949a, p. 5), I commented on the close resemblance of the jaw to that of *Erethizon,* but ascribed the resemblance largley to deformation of the jaw by crushing, and stated, "there is little likelihood that any significant relationship to the Hystricomorpha exists." Curiously, I also find two Chinese Eocene ctenodactyloids which, strictly speaking, seem also to be incipiently hystricognathous. These are : (1) *Cocomys lingchaensis,* right ramus, No. 78006, early Eocene (Dawson and others, 1984); and (2) *Yuomys cavioides,* right ramus, No. V4796.2, late Eocene (Li, 1975)—both in the collections of the Institute of Vertebrate Paleontology and Paleoanthropology, Academia Sinica, Beijing. These jaws have the inside of the angular process lateral to the outside of the incisor. They do not show, however, grooves for the reflexed part of the *M. m. superficialis.* In both cases there are associated left jaws which, for whatever reason, do not exhibit this lateral position of the angular process. Neither do other jaw fragments of these species which are available, but the jaws which do are in reasonably good shape. One might conclude that distortion, or incompleteness, or both, are responsible for this apparent variation, whatever the true jaw shape.

Ellerman (1940, p. 42) wrote that in the Caviidae the mandible is not by any means typically hystricid. He further stated that some approach to the caviid mandible was made by the Chinchillidae and Ctenodactylidae. Moreover, he said the jaw instead was modified in shape by the *medialis* muscle (hence corresponding to the *medialis* series grouping of the Miller and Gidley classification of 1918). These statements of Ellerman have been strongly criticized by Landry and by Woods (1972), but at least partly because of, I think, misinterpretation of Ellerman's nomenclature. For example, Woods (p. 124) says, "Ellerman (1940) incorrectly reported that the hystricognath jaw is correlated with the insertion of the lateral masseter muscle." In fact, Ellerman says (p. 42), "Tullberg made this his major division of the Order, and divided it into "Hystricognathi"—forms in which the mandible has the angular process lifted outwards by the specialized limb of *masseter lateralis superficialis,* and the "Sciurognathi" in which this does not take place." Elsewhere, when Ellerman writes of the *m. lateralis,* he, I think, is referring to the *M. m. superficialis*. Moreover, he has a point when he distinguishes the Caviidae (including Hydrochoeridae) as not showing this. If one refers to page 69 of Tullberg, the reference is to the body of the ramus, and not to the incisor position. In caviids, the angular process does not originate external to the body of the ramus, and jaws are different from most other hystricognaths. This is not to say, however, that caviids are not Hystricomorpha in the taxonomic sense, only that almost any meaningful definition will be open to exceptions. If jaws of the caviids were to be broken at a point behind the posterior termination of the incisor, it would be difficult to demonstrate hystricognathy.

If we use the incisors as a point of reference rather than the body of the lower jaw, we may still have trouble as previously mentioned for individuals of some ctenodactyloids such as *Cocomys* and *Yuomys*. I think in this regard that Landry's observation (1957, p. 82-83) is pertinent that *Reithroparamys delicatismus* [sic] is "not entirely sciurognath for the angle did not come off directly from the bottom of the incisor alveolus," and again, "In spite of this, the paramyine jaw does have one evident hystricognath feature. The masseteric ridge runs almost entirely parallel to the zygomatic arch. There is neither the in-pushing at the anterior end of the angle, nor the up-turning at the posterior end of the angle seen in sciurognaths." It seems to be this condition that Wood describes in various publications as incipiently hystricognath, and that I have seen in the jaws of Eocene rodents from Asia. I have not been able to see this "incipiency," however, in any eurymyloids save possibly *Rhombomylus*. The earliest eurymyloids, because of the fragmentary nature of the lower jaws, do not offer much information, but can hardly be described as having even incipient hystricognathy. Korth (1984, p. 7) suggests that the condition which Wood described as incipient may be frequent in early rodents, and in fact be the primitive condition.

If the hystricognathus jaw is as correlated with anteroposterior motion as has been thought (see Woods, 1972, p. 175-176), one would on the one hand not expect pronounced hystricognathy to develop with cuspate, short-crowned teeth, nor on the other hand expect hystricognathy to be limited totally to one taxonomic group. It is obvious, however, that anteroposterior chewing does not necessarily produce hystricognathy.

Among North American rodents, *Prolapsus sibilatoris* is described by Wood (1973, p. 22) as fully hystricognathous, although the cheekteeth are still relatively short-crowned. Wood's figures (5B and 6B) confirm the description. I have examined a plaster cast, and perhaps owing to the cast, the degree of hystricognathy seems somewhat less than the figure shows, but the jaw is hystricognathous. Korth (1984, p. 7), however, denies that the type specimen is fully hystricognathous, and states that the "angle of the mandible of *Prolapsus* is no more laterally displaced than that of the early Eocene sciuravid *Knightomys*." *Guanajuatomys hibbardi* is also described by Black and Stephens (1973) as hystricognathous. A cast of the lower jaw of this genus is available to me also. The original was set in plasticine prior to casting, and presumably owing to this fact, the inferior border of the ramus is obscured, and I can not see the described hystricognathy. Korth again states that the mandible should not be considered hystricognathous. *Protoptychus hatcheri* also has been described as probably hystricognathous. It is reported by Korth that additional preparation of lower jaw material from the Washakie Basin indicates a sciurognath rather than hystricognath condition. A fragmentary lower jaw of *Protoptychus* from the Washakie Basin which I have seen would confirm Korth's report.

Thus, the best evidence for a fully hystricognathous jaw from the Eocene anywhere in the world seems to be *Prolapsus,* which otherwise indicates sciuravid affinities. The skull of *Prolapsus* does not appear to be very hystricomorphous. Seemingly there are no specimens of Eocene age which are really hystricomorphous and even incipiently hystricognathous except possibly *Yuomys* of the late Eocene of Honan and Inner Mongolia, which genus is supposedly ctenodactyloid.

As a speculation, I suggest that the combination of hystricomorphy—hystricognathy found in the South American and Old World Hystricomorpha developed neither earlier nor later than for other kinds of zygomasseteric structures such as are found in advanced sciuromorphs or muroids. Development of an hystricomorphous condition in Eocene rodents may be related more to jaw stabilization than to more complex chewing and gnawing as is the case in advanced rodents. If this be so, it leaves open the possibility that the early ctenodactyloids are not already on a side branch but, as Hussain and others (1978), de Bruijn and others (1982), and Korth (1984) have suggested, may occupy a position in the Old World similar to the paramyids in the New World. Possibly, early ctenodactyloids gave rise to various later groups in addition to typical ctenodactyloids, and perhaps especially to hystricoids and caviomorphs, depending on what turns out to be so in the argument of pauciserial versus multiserial as the primitive type of incisor enamel (see following section). If multiserial enamel is primitive, it removes a "derived" character from a list holding these groups together.

Structures of Incisor Enamel

The inner enamel layer of rodent incisors shows one of three kinds of enamel structure: pauciserial, multiserial, or uniserial. Pauciserial has been thought to be primitive, with the other two types derived from it (Korvenkontio, 1934; Wahlert, 1968). It is the least organized, occurs almost entirely in Eocene rodent populations, is associated with protrogomorphous masseter structures, and transitions occur to the uniserial structure (Theridomyidae, Gliridae?), and to the multiserial (Ctenodactyloidea?; *Cyclomylus* and *Tsaganomys* of the Tsaganomyidae, Patterson and Wood, 1982).

In recent years, however, W. V. Koenigswald has argued that multiserial is primitive, and that pauciserial enamel is transitional between multiserial and uniserial (1980). He has gained a convert to this point of view with Wahlert (1984). I am not expert in this field, and bow to those who are. It still seems curious to me, however, that a transitional stage lasts so long (duration of Eocene), and that no fully multiserial types have been recorded from the Eocene. Possibly it is owing to the difficulty of distinguishing pauciserial from primitive multiserial. If multiserial is primitive, as outgroup comparison suggests, it would perhaps better account for its distribution in South America and Africa with most all areas inbetween represented by rodents of uniserial type. With refinement in examination techniques (scanning electron photomicroscopy versus light photography) it might also pay to redo the earlier work on Eocene rodent enamel. Examination of such rodents as *Cocomys,* and perhaps the eurymyloid *Heomys,* for incisor type seems necessary. The polarity of transition types such as is reported from the early Eocene of Pakistan (de Bruijn and others, 1982) needs positive establishment.

PUNCTUATED EQUILIBRIUM AND THE STRATIGRAPHIC RECORD

Eldredge and Gould (1972) have proposed that lineages remain stable for long periods of time, and if changes take place they do so in small populations and very quickly, perhaps too quickly for the resolving power of the stratigraphic record. Gould, especially, has regarded the stratigraphic record as complete enough to be taken at face value. They contrast their theory with gradualism, in which lineages are supposed to evolve slowly, in large populations, and over a large area. Gingerich has defended this second theory in a number of papers (see, for example, 1979). Simpson (1983), however, has pointed out that one must view with suspicion any "either or" hypothesis. In any case, it seems to me that Simpson (1953) and others (Wilson, 1951, 1962, for example) have largely anticipated either side some time back, except that such terms as bradytely and tachytely were employed.

As for the completeness of the stratigraphic record, Gould seems to imply nearly continuous, if slow, deposition. Many continental Tertiary records hardly fit this pattern. Take, for example, the Chadron Formation. In the Big Badlands of South Dakota, this formation reaches a maximum thickness of only about 62 m (200 ft). The usual section is nearer 31 m (100 ft). The total duration for the Chadron is some five million years. Much of the observed sediment was deposited rapidly. For example, titanothere skulls are frequently buried without weathering of the bone surface. Thus, diastems are more important by far than the time represented by actual deposition. The thickest section, with fossil control, is in Wyoming (Flagstaff Rim) where the Chadron measures 248 m (800 ft). Even this works out to be about a foot of sediment per six thousand years, hardly enough to cover a titanothere skull. It may be noted that the evolution of early Oligocene titanotheres as visualized by H. F. Osborn (1929) is a fiction based on essentially contemporaneous specimens of the Middle Chadron, or at best on specimens lacking stratigraphic control.

So far as rodents are concerned, it seems that the late Eocene-early Oligocene break in these mammals is the result of: (1) improvement of incisor strength which permitted the development of more powerful jaw mechanisms; (2) loss of Eocene habitat; (3) in North America, at least, a shift in environment of deposition which permitted a different sampling of the record; and (4) dispersal of certain progressive families (Wilson, 1972). No "either or" explanation fits this complex of events.

Nevertheless, in regard to this last point, in the past twenty years or so we have been obtaining a record of Eocene rodents from China, Outer Mongolia, and the Indian Peninsula. Although this record has produced one surprise, the "dominance" of ctenodactyloids, ancestors of the modern families of rodents are still as obscure as ever. Further, Late Eocene-early Oligocene records are continuous enough to indicate no gaps of any great length. So the record as it now stands seems to favor some kind of rapid evolution at the family level. There are, however, rather considerable areas of land in Siberia and northern North America without much of a fossil record in the critical time periods. Especially interesting in this regard is the idea put forward by Hickey and others (1983, 1984) that equivalent biologic faunas in the far north of North America appear to be distinctly younger than they really are in respect to those of Rocky Mountain basins. This view has been challenged by others for poor paleomagnetic data (see, for example, Flynn and others, 1984). The major conclusion, however, of a latitudinal shift of faunas southward is in any case not really new. W. D. Matthew, as early as 1915, thought that secular changes of climate produced such a shift in mammalian geography. At present, Dr. Larry D. Martin of The University of Kansas *(oral communication)* favors the idea of climatic changes producing sudden arrival of more advanced species in mid-temperate latitudes, and I, myself, have been troubled about this possibility causing errors in geologic correlations (Wilson, 1967, p. 596). A southward shift, then might help to account for the "sudden" appearance of more advanced groups of rodents. On the other side of the coin, at a lower taxonomic level we do have slowly evolving lines such as *Ondatra* (see L. Martin, 1979), castorids (Wilson, 1949c), and various cricetids and eomyids in Europe (Fahlbusch, 1964, 1970, 1979, 1983) which show gradualism in the evolution of rodents.

COLLISION EVENTS AND THE EOCENE-OLIGOCENE BOUNDARY

It has been suggested that an extraterrestrial impact event, possibly two or several, caused catastrophic extinctions at the Eocene-Oligocene boundary (Alvarez and others, 1982). The transformation of the rodent world at this time does not support an impact theory. A certain amount of gradual change occurs, as for example in the decline of sciuravids in the late Eocene (extinction by evolution?), and a similar but later decline in paramyids. Also a number of rodent genera such as *Simimys, Protadjidaumo,* and the like, show modernization in the late Eocene. More important, however, is the fact that impact on a large scale might wipe out rodents, but would hardly permit survival of developing families of similar body size at the same time. Also, the largest known paramyid, *Manitsha tanka,* comes from beds which post-date the break. It is hardly to be thought that the modern kinds had a greatly different habitat which would protect them.

VALIDITY OF THE FOSSIL RECORD IN PHYLOGENY

(1) Socrates is supposed to have said that "stones and trees can teach me nothing." Circa 400 B.C.

(2) ". . . it is pointless either to look for ancestors or to consider fossils to be ancestors. Ancestral-descendant relationships are unknowable and unprovable; no criteria exist for recognizing a particular fossil as an ancestor. Naturally the loss of the idea that fossils are ancestors (in the sense that they can be recognized as ancestors) will be difficult for some. But in doing so we will increase the precision of the working methods in paleontology, and eliminate a considerable number of irresponsible statements, . . ." Cracraft (1972, p. 390).

(3) "The primary record of the history of life is written in the successive strata of rocks as in the pages of a book." ". . . fossils may be called the writing on these pages. They represent once living things and should be seen as such. They must be put in their sequence in time. They are materials for the study of many factors of organic evolution." ". . . combining historical geology and historical biology into one great synthesis." Simpson (1983, p. 1).

Socrates has his followers today in those who think that the fossil record has no role to play in systematics, and refuse to allow them even the equal status granted to them by the international rules of zoological nomenclature. On this it seems fitting to quote W. D. Matthew (1937, p. 309): "Such methods are analogous to those of the sociologist who, utterly ignorant of history, and classing Napoleon and Julius Caesar together as "historical personages" in his discussion, would attempt to trace the origin and evolution of European civilization solely upon the basis of modern sociology." Cracraft does not go nearly this far, judging from his own research. His statement that no criteria exist for recognizing a particular fossil as an ancestor is true in a very strict sense, although it might be argued that temporal position is a criterion. Simpson (1950, p. 55) wrote that: "Perhaps the most truly basic of all historical principles is that no one can be ancestral to his grandfather. Applied to the interpretation of contemporaneous data, the law means flatly that these are not historical in nature." The statement by Cracraft that it is "pointless to consider fossils to be ancestors" is, of course, an opinion. What he regards as pointless is obviously far from the views of Simpson.

So far as Eocene rodents are concerned, we might very well not have any actual ancestors at the species level, but the families Cocomyidae (and relatives) and Paramyidae together almost certainly do contain the ancestry of the rodents. Although the oldest members of a group as now known are not necessarily ancestral, the probability is that in many cases they are. There are no known undoubted Paleocene rodents, and the above two groups are dominant in Asia, Europe, and North America for some twenty million years. Morphological characters seem to indicate the same ancestral position. Korth (1984) has argued for the Cocomyidae as being ancestral to the paramyids. If so, *Cocomys* (early Eocene) is the closest rodent we have for an ancestor, and can not be far removed from a real ancestor.

Although not altogether relevant, I can not help remarking on Schaeffer and others (1972, p. 37). They state that even with horses, the temporal element is secondary to outgroup comparison. "The resulting morphocline might initially be interpreted as either *Hyracotherium-Equus* or *Equus-Hyracotherium*. By comparing this spectrum of equid morphology with similar spectra exhibited by related perissodactyl families, it is evident that the one digit character state is derived. *Hyracotherium,* with its four front and three rear functional toes, is not the most primitive equid merely because it is the oldest taxon referred to the family, but because many of its character states are shared with other perissodactyl groups." If, however, one assumes a graded series and polarity is the only consideration, the temporal factor is decisive because we know the direction of a closely graded sequence in time. That five is the primitive number of digits in mammals does not depend on modern cladistic analysis, but on what might be termed inherited knowledge in comparative anatomy, if I may be this unscientific. In this connection, I remember a lecture by Thomas Hunt Morgan in which he stated that such a sequence could run either way (see above) except that temporal consideration gave the answer. If by *Hyracotherium-Equus* is meant the whole equid sequence from early Eocene to Recent, I would argue that polarity is more strongly determined by the temporal sequence than by outgroup comparison. In this sequence, the oldest temporal position can not be disputed; its equid position is determined by the sequence leading up to *Orohippus* of the middle Eocene. If on the other hand, reference is only to typical *Hyracotherium* and *Equus,* the former probably would not have been placed in the Equidae at all—it was not by Richard Owen—as the name itself suggests. It is essentially at the base of the whole Perissodactyla, even with such detailed analysis as given by MacFadden (1976), and only in its evolution toward *Orohippus,* is it the ancestral horse. *Hyracotherium* can be distinguished from the contemporary tapiroid *Homogalax* only with difficulty. In fact, many fragmentary and worn dentitions can not be placed with certainty, even by experts. It would be united, if we lived in the Wasatchian, with *Homogalax,* and possibly believed to be a peculiar phenacodontid.

CLADISTICS

Sometimes it would appear that vertebrate paleontology is about to be revolutionized by the new systematics of Willi Hennig. I am not a cladist, but even in my home institution, I feel as if I were one of a very few King Canutes resisting the engulfing tide of the future. I think it is appropriate at this point to quote George Gaylord Simpson (1978, p. 271—I could also quote Leigh Van Valen, "Why not to be a cladist," 1978) who says in his autobiography, "The main principles of the Hennigian system are: first, that the basic process of organic evolution (phylogeny) is the splitting (dichotomy) of an ancestral species into two descendant species; second, that each dichotomy should be taken as marking the origin of two new units (taxa) of classification; and third, that the hierarchic level of such units (whether species, genera, families, *etc.*) should be determined by the geologic time when the dichotomy occurred, the earlier the time, the higher the level—I will just say that the first principle, as given above, an apparent statement of fact, is not true and that the second and third principles, statements of opinion, are inane."

Beyond this there is cladistic analysis with total reliance on derived characters and their distribution. Cladists are quite reluctant to grant a role to parallelism and convergence, citing parsimony to justify this reluctance, and frequently reluctant to grant much of a role to fossils. The use of derived character states to determine phylogenetic relationships is not new with Hennigians, nor have vertebrate paleontologists refused to see the light. W. D. Matthew (1937, but written almost entirely in the period 1916-1930) made use of derived characters in determining relationships in his great monograph on Paleocene mammals (see p. 103-105, for example), as did George Gaylord Simpson (1928, p. 167-169), or even Wilson (1949b, p. 430-448) in pre-Hennigian times. Wahlert has published formal cladograms (as, for example, 1978, p. 11-15) for some rodent groups. Cladists, however, seem to have difficulties with fossils, as seen in the attempt by some to ignore fossils or at best to relegate them to special places in analyses.

Convergences and parallelisms obviously cause difficulties with cladograms. On the other hand, A. E. Wood thinks one of the outstanding characteristics of the Order Rodentia lies in the repeated parallelisms. One example that may be cited is that of the zygomasseteric structure. Does the sciuromorph structure present in eomyids and heteromyids relate these groups not only to each other but also to sciurids and castorids, or is zygomasseteric structure subject to parallelism from a protrogomorph type as Wilson (1949b, 1949c) has suggested, and Wood (1980), Wahlert (1978), and Black and Sutton (1984) have accepted or confirmed?

At times it seems to me that cladists are ardent followers of Descartes. He thought that the human mind had sufficient reasoning powers to solve most problems providing *proper* methods were followed. Proper methods seem to be Hennigian methodology, with pre-Hennigian work assigned to the trash heap of history and poor science.

In any case, it is not so much individual lines that make for some kinds of paleontological evidence, but faunas as a whole. Thus, for example, the Paleocene mammalian faunas of the northern hemisphere contain no rodents at all (except perhaps very late in the epoch). The conclusion from this makes it fairly certain that the Order Rodentia did not exist in a morphological sense in the Cretaceous. This has not prevented, however, Landry (1957, p. 90), in his excellent study of hystricomorphs, from suggesting that the order originated in Late Cretaceous time, or at latest in the earliest Paleocene. Croizat (1979) even thought that the ancestor of the Hystricomorpha negotiated the South Atlantic in the Cretaceous or even earlier. Appeals to these deficiencies in the fossil record are simply not supported by the increasing data from Asia, Europe, and North America.

There was a time when Aristotle was the great authority in matters scientific. Only when Aristotle conflicted with faith and the Bible did faith win. Nowadays it would seem that Aristotle is replaced by Hennig, with a liberal dose of Cartesian philosophy as to the importance of methodology. I too have faith, but in the historical record and in historical geology, and the use of fossils and temporal control. I think in many cases we can point to ancestral groups, paraphyletic or otherwise, if not ancestors. I feel, moreover, that cladistic analysis at times comes close to resulting in old-fashioned keys, although this may be more a result of practitional failure than methodology. At least for rodents, parallelism seems an important factor in their evolution, and a difficulty with some cladistic analyses. Finally, I say Prosit to those old-fashioned, out-of-date researches of A. S. Romer, W. D. Matthew, and George Gaylord Simpson.

REFERENCES CITED

Alvarez, W., Asaro, F., Michel, M. V., and Alvarez, L. W., 1982, Iridium anomaly approximately synchronous with terminal Eocene extinctions: Science, v. 216, p. 886-888.

Black, C. C., and Dawson, M. R., 1966, A review of late Eocene mammalian faunas from North America: American Journal Science, v. 264, p. 321-349.

Black, C. C., and Stephens, J. J., III, 1973, Rodents from the Paleogene of Guanajuato, Mexico: Occasional Papers of the Museum, Texas Tech University, v. 14, 10 p.

Black, C. C., and Sutton, J. F., 1984, Paleocene and Eocene rodents of North America: Carnegie Museum of Natural History, Special Publication no. 9, p. 67-84.

de Bruijn, H., Hussain, S. T., and Leinders, J. J. M., 1982, On some early Eocene rodent remains from Barbara Banda, Kohat, Pakistan, and the early history of the order Rodentia: Proceedings Koninklijke Nederlandse Akademie van Wetenschappen, Series B, v. 85, p. 249-258.

Chaline, J., and Mein, P., 1979, Les rongeurs et l'evolution: Paris, Doin, *xii* + 236 p.

Coues, E., 1877, *in* Coues, E., and Allen, J. A., Monographs of North American rodents, Family Haplodontidae: Report of United States Geological Survey of Territories (F. V. Hayden), v. *xi*, no. *ix*, Washington, D.C., Government Printing Office, p. 543-600.

Cracraft, J., 1972, The relationships of the higher taxa of birds: problems in phylogenetic reasoning: The Condor, v. 74, p. 379-392.

Croizat, L., 1979, Review of: Biogeographie: Fauna und Flora der Erde und ihre geschichtliche Entwicklung, by Banarescu, P., and Boscain, N.: Systematic Zoology, v. 28, p. 250-252.

Dawson, M. R., 1961, The skull of *Sciuravus nitidus,* a middle Eocene rodent: Postilla, Yale Peabody Museum, no. 53, 13 p.

―――― 1977, Late Eocene rodent radiations: North America, Europe and Asia: Geobios Memoire, Special 1, p. 195-209.

Dawson, M. R., Li, C.-k., and Qi, T., 1984, Eocene ctenodactyloid rodents (Mammalia) of eastern and central Asia: Carnegie Museum of Natural History, Special Publication no. 9, p. 138-150.

Eastman, C. B., 1982, Hystricomorphy as the primitive condition of the rodent masticatory apparatus: Evolutionary Theory, v. 6, p. 163-165.

Eldredge, N., and Gould, S. J., 1972, Punctuated equilibria: an alternative to phyletic gradualism, *in* Schopf, T. J. M., ed., Models in paleobiology: San Francisco, Freeman, Cooper and Co., p. 82-115.

Ellerman, J. R., 1940, The families and genera of living rodents: British Museum (Natural History), v. 1, *xxvi* + 689 p.

Fahlbusch, V., 1964, Die Cricetiden (Mamm.) der oberen Susswasser-Molasse Bayerns: Bayerische Akademie der Wissenschaften, Mathematisch-Naturwissenschaftliche Klasse, Abhandlungen Neue Folge, Heft 118, 136 p.

―――― 1970, Populationsverschiebungen bei tertiaren Nagetieren, eine Studie an oligozanen und miozanen Eomyidae Europas: *ibid.,* Heft 145, 136 p.

―――― 1979, Eomyidae-Geschichte einer Saugertierfamilie: Palaontologische Zeitschrift, v. 53, p. 88-97.

―――― 1983, Mikroevolution—Makroevolution—Punktualismus ein Diskussionsbeitrag am Beispiele miozaner Eomyidae (Mammalia, Rodentia): *ibid.,* v. 57, 314, p. 213-230.

Flynn, J. J., MacFadden, B. J., and McKenna, M. C., 1984, Land-mammal ages, faunal heterochrony, and temporal resolution in Cenozoic terrestrial sequences: Journal of Geology, v. 92, p. 687-705.

Gingerich, P. D., 1979, The stratophenetic approach to phylogeny reconstruction in vertebrate paleontology, *in* Cracraft, J., and Eldredge, N., eds., Phylogenetic analyses and paleontology: New York, Columbia University Press, p. 41-77.

Golz, D. J., and Lillegraven, J. A., 1977, Summary of known occurrences of terrestrial vertebrates from Eocene strata of southern California: Contributions to Geology, University of Wyoming, v. 15, p. 43-65.

Hickey, L. J., West, R. M., and Dawson, M. R., 1984, Response to arctic biostratigraphic heterochronicity: Science, v. 224, p. 175-176.

Hickey, L. J., West, R. M., Dawson, M. R., and Chai, D. K., 1983, Arctic terrestrial biota: paleomagnetic evidence of age disparity with mid-northern latitudes during the Late Cretaceous and Early Tertiary: *ibid.,* v. 221, p. 1153-1156.

Hussain, T., de Bruijn, H., and Leinders, J. M., 1978, Middle Eocene rodents from the Kala Chitta Range (Punjab, Pakistan): Proceedings Koninklijke Nederlandse Akademie van Wetenschappen, series B, v. 81, p. 74-112.

Jepsen, G. L., 1937, A Paleocene rodent, *Paramys atavus:* American Philosophical Society, Proceedings, v. 78, p. 291-301.

Koenigswald, W., von, 1980, Schmelzstruktur und Morphologie in den Molaren der Arvicolidae (Rodentia): Abhandlungen der Senckenburgischen Naturforschenden Gesellschaft, v. 539, 129 p.

Korth, W. W., 1984, Earliest Tertiary evolution and radiation of rodents in North America: Bulletin of Carnegie Museum of Natural History, no. 24, 71 p.

Korvenkontio, V. A., 1934, Mikroskopische Untersuchungen am Nagerincisiven unter Hinweis auf die Schmelzstruktur der Backenzahne: Annals Zoological Society, Zoology—Botany Fennicae Vanamos, v. 2, *xiv* + 274 p.

Landry, S. O., Jr., 1957, The interrelationships of the New and Old World hystricomorph rodents: University of California, Publications in Zoology, v. 56, p. 1-118.

Lavocat, R., 1973, Les rongeurs du Miocene d'Afrique Orientale: Memoires et Travaux, l'Ecole Pratique des Hautes Etudes, Institut de Montpellier, no. 1, 284 p.

Li, C.-k., 1975, *Youmys,* a new ischyromyoid rodent genus from the upper Eocene of North China: Vertebrata PalAsiatica, v. 13, p. 58-70.

Li, C.-k., and Ting, S.-y., 1983, The Paleogene mammals of China: Bulletin Carnegie Museum of Natural History, no. 21, 98 p.

Lillegraven, J. A., 1977, Small rodents (Mammalia) from Eocene deposits of San Diego County, California: American Museum of Natural History, Bulletin, v. 158, p. 221-262.

MacFadden, B. J., 1976, Cladistic analysis of primitive equids, with notes on other perissodactyls: Systematic Zoology, v. 25, p. 1-14.

MacFadden, B. J., Campbell, K. E., Jr., Cifelli, R. Z., Siles, O., Johnson, N. M., Naeser, C. W., and Zeitler, P. K., 1985, Magnetic polarity stratigraphy and mammalian fauna of the Deseadan (Late Oligocene-early Miocene) Salla beds of Northern Bolivia: Journal of Geology, v. 93, p. 223-250.

McGrew, P. O., 1941, The Aplodontoidea: Field Museum of Natural History, Geological Series, v. 9, p. 3-30.

Martin, L. D., 1979, The biostratigraphy of arvicoline rodents in North America: Nebraska Academy of Sciences, Transactions, v. 7, p. 91-100.

Matthew, W. D., 1910, On the osteology and relationships of *Paramys,* and the affinities of the Ischyromyidae: American Museum of Natural History, Bulletin, v. 28, p. 43-71.

_____ 1915, Climate and evolution: New York Academy of Sciences, Annals, v. 24, p. 171-318.

_____ 1937, Paleocene faunas of the San Juan Basin, New Mexico: American Philosophical Society, Transactions, New Series, v. 30, *viii* + 372 p.

Miller, G. S., Jr., and Gidley, J. W., 1918, Synopsis of the supergeneric groups of rodents: Journal of the Washington Academy of Sciences, v. 8, p. 431-448.

Osborn, H. F., 1929, The titanotheres of ancient Wyoming, Dakota, and Nebraska: U.S. Geological Survey, Monograph 55, Washington, D.C., Government Printing Office, 2 vols., *xxiv* + *xi,* 953 p.

Patterson, B., and Wood, A. E., 1982, Rodents from the Deseadan Oligocene of Bolivia and the relationships of the Caviomorpha: Harvard University, Museum of Comparative Zoology, Bulletin, v. 149, p. 371-543.

Rensberger, J. M., 1975, *Haplomys* and its bearing on the origin of the aplodontoid rodents: Journal of Mammalogy, v. 56, p. 1-14.

Savage, D. E., and Russell, D. E., 1983, Mammalian paleofaunas of the world: Reading, Massachusetts, Addison—Wesley Publishing Company, *xvii* + 432 p.

Schaeffer, B., Hecht, M. K., and Eldredge, N., 1972, Phylogeny and paleontology, *in* Dobzhansky, T., Hecht, M. K., and Steere, W., eds., Evolutionary biology, v. 6: New York, Appleton, Century, Crofts, p. 31-46.

Shevyreva, N. S., 1983, Rodents (Rodentia, Mammalia) of the Neogene of Eurasia and North Africa: the evolutionary foundation of the Pleistocene and Recent faunas of Palearctic rodents, *in* Gromov, I. M., ed., History and evolution of the Recent rodent fauna (in Russian): Moscow, Nauka, p. 9-145.

Simpson, G. G., 1928, A catalogue of the Mesozoic Mammalia in the Geological Department of the British Museum: British Museum (Natural History), 225 p.

_____ 1950, Some principles of historical biology bearing on human origins: Cold Spring Harbor Symposia on Quantitative Biology, v. 15, p. 55-66.

_____ 1953, The major features of evolution: New York, Columbia University Press, *xii* + 434 p.

_____ 1978, Concession to the improbable: an unconventional autobiography: New Haven, Yale University Press, *xi* + 291 p.

_____ 1983, Fossils and the history of life: New York, Scientific American Library, 239 p.

Stock, C., 1935, New genus of rodent from the Sespe Eocene: Geological Society of America, Bulletin, v. 46, p. 61-68.

Tullberg, T., 1899, Ueber das System der Nagethiere: eine phylogenetische Studie: Upsala, Akademischen Buchdruckerei, *v* + 514 p.

Van Valen, L., 1978, Why not to be a cladist: Evolutionary Theory, v. 3, p. 285-299.

Wahlert, J. H., 1968, Variability of rodent incisor enamel as viewed in thin section, and the microstructure of the enamel in fossil and recent rodent groups: Museum of Comparative Zoology, Breviora, no. 309, 18 p.

_____ 1973, *Protoptychus,* a hystricomorphous rodent from the late Eocene of North America: *ibid.,* no. 419, 14 p.

_____ 1978, Cranial foramina and relationships of the Eomyoidea (Rodentia, Geomorpha), skull and upper teeth of *Kansasimys:* American Museum Novitates, no. 2645, 16 p.

_____ 1984, Hystricomorphs, the oldest branch of the Rodentia: Colloquium in Biological Sciences, Annals of the New York Academy of Sciences, p. 356-357.

Wilson, J. A., 1978, Stratigraphic occurrence and correlation of Early Tertiary vertebrate faunas, Trans-Pecos Texas: Texas Memorial Museum, Bulletin 25, *v* + 42 p.

Wilson, R. W., 1940, Two new Eocene rodents from California: Carnegie Institution of Washington, Contributions to Paleontology, no. 514, p. 85-95.

_____ 1949a, Additional Eocene rodent material from southern California: *ibid.,* no. 584, art. 1, p. 1-25.

_____ 1949b, On some White River fossil rodents: *ibid.,* no. 584, art. 2, p. 27-50.

_____ 1949c, Early Tertiary rodents of North America: *ibid.,* no. 584, art. 4, p. 67-164.

_____ 1951, Evolution of the early Tertiary rodents: Evolution, v. 5, p. 207-215.

_____ 1962, The significance of the geological succession of organic beings: 1859-1959: University of Kansas, Science Bulletin, v. 42 (supplement), p. 157-178.

_____ 1972, Evolution and extinction in early Tertiary rodents: 24th International Geological Congress, Montreal, section 7, p. 217-224.

——— 1980, The stratigraphic sequence of North American rodent faunas: Palaeovertebrata, Memoire Jubilaire Rene Lavocat, p. 273-283.

——— 1967, Fossil mammals in Tertiary correlations, in Teichert, C., and Yochelson, E. L., eds., Essays in paleontology and stratigraphy: Lawrence, University of Kansas Press, p. 590-606.

Winge, H., 1887, Jordfundne og nulevende Gnavere (Rodentia) fra Lagoa Santa, Minas Geraes, Brasilien, Med Udsigt over Gnavernes indbyrdes Slaegtskab: Museo Lundii, v. 1, 178 p.

——— 1924, Pattedyr-Slaegter, v. 2, Rodentia, Carnivora, Primates: Copenhagen, H. Hagerups Forlag, 321 p.

——— 1941, The interrelationships of the mammalian genera, II: translated from Danish by E. Deichmann and G. M. Allen, Copenhagen, 367 p.

Wood, A. E., 1949, A new Oligocene rodent genus from Patagonia: American Museum Novitates, no. 1435, 54 p.

——— 1955, A revised classification of the rodents: Journal of Mammalogy, v. 36, p. 165-187.

——— 1959, Eocene radiation and phylogeny of the rodents: Evolution, v. 13, p. 354-361.

——— 1962, The Early Tertiary rodents of the family Paramyidae: American Philosophical Society, Transactions, New Series, v. 52, part 1, 261 p.

——— 1968, Early Cenozoic mammalian faunas, Fayum Province, Egypt, Part II, The African Oligocene Rodentia: Yale University, Peabody Museum Natural History, Bulletin 28, p. 23-105.

——— 1973, Eocene rodents, Pruett Formation, Southwest Texas, their pertinence to the origin of the South American Caviomorpha: Pearce-Sellards Series, no. 20, Texas Memorial Museum, 41 p.

——— 1980, The Oligocene rodents of North America: American Philosophical Society, Transactions, v. 70, part 5, 68 p.

——— 1984, Hystricognathy in the North American Oligocene rodent *Cylindrodon* and the origin of the Caviomorpha: Carnegie Museum Natural History, Special Publication no. 9, p. 151-160.

Woods, C. A., 1972, Comparative myology of jaw, hyoid, and pectoral appendicular regions of New and Old World hystricomorph rodents: American Museum of Natural History, Bulletin, v. 147, p. 115-198.

ADDENDUM

The present paper has a cutoff date of August, 1985. Not considered were papers in the important volume, W. P. Luckett and J.-L. Hartenberger, eds., 1985, Evolutionary relationships among rodents: a multidisciplinary analysis: New York and London, Plenum Press, NATO ASI Series. Series A: Life Sciences, v. 92, *xiii* + 721 p.

I also wish to thank an anonymous reviewer for calling my attention to MacFadden and others, 1985, with its important conclusion on the geologic age of the Deseadan.

MANUSCRIPT RECEIVED AUGUST 29, 1985
REVISED MANUSCRIPT RECEIVED MAY 9, 1986
MANUSCRIPT ACCEPTED JUNE 17, 1986

Machaeroides simpsoni, new species, oldest known sabertooth creodont (Mammalia), of the Lost Cabin Eocene

MARY R. DAWSON
RICHARD K. STUCKY
LEONARD KRISHTALKA

Section of Vertebrate Fossils, Carnegie Museum of Natural History, Pittsburgh, Pennsylvania 15213

CRAIG C. BLACK

Natural History Museum of Los Angeles County, Los Angeles, California 90007

ABSTRACT

A new species of machaeroidine creodont, *Machaeroides simpsoni,* is described from the Lostcabinian and Gardnerbuttean of the Wind River Formation, Natrona County, Wyoming. The sabertooth adaptation is fully developed in this species, which has a combination both of more primitive characters and a more derived sabertooth condition than in the Bridgerian species, *M. eothen.* Some morphological features of *Machaeroides* indicate that the machaeroidines are more closely allied with the oxyaenids than with the limnocyonids.

INTRODUCTION

"I also made an extensive study of some sabertooth cats and how they used their sabers. A Swedish paleontologist disagreed and we had a friendly controversy for some time. I still think I was right!" In this modest way, George Gaylord Simpson (1978, p. 117) described his important paper (1941) on the function of the saber in some carnivorous mammals having this adaptation. Simpson (1944) returned to the sabertooths in his discussion of using an adaptive grid to illustrate the evolutionary history of the Felidae, especially how the felid structure was made possible by creodont extinction and herbivore evolution. Thus, it seems appropriate in a volume dedicated to the memory of our mentor and friend to describe the oldest known carnivorous mammal to have the sabertooth adaptation. This is a creodont, related to the middle Eocene *Machaeroides eothen* Matthew. It occurs in the Lostcabinian to Gardnerbuttean, late early Eocene to early middle Eocene, Wind River Formation of the Wind River Basin, Wyoming.

Abbreviations for collections from which specimens referred to in the text have come are: CM, Carnegie Museum of Natural History, and USNM, National Museum of Natural History,

SYSTEMATIC PALEONTOLOGY

Class MAMMALIA Linnaeus, 1758
Order CREODONTA Cope, 1875
?Family OXYAENIDAE Cope, 1877
Machaeroides Matthew, 1909
Machaeroides simpsoni, new species
Figures 1 and 2; Table 1

Holotype: CM 45115, fragments of a skull including parts of the palate, with incisors, canines, right P^1-M^2 and left P^4-M^2.

Hypodigm: CM 36397, left lower jaw with P_1, P_4-M_2; CM 37342, left maxilla with P^3-M^2.

Horizon and locality: Natrona County, Wind River Basin, Wyoming. CM 45115, locality 1042, and CM 36397, locality 27/78, Lostcabinian; CM 37342, locality 34, Gardnerbuttean (detailed locality information is on file at Carnegie Museum). All material from the Wind River Formation.

Diagnosis: Differs from *M. eothen* in having P^2 more obliquely oriented, P^3 two-rooted with weaker protocone, M^2 more robust. Enamel on anterior and posterior ridges of upper canine and crests of cheek teeth serrated. Jaw more vertical at symphysis. P_4 protoconid inclined slightly more posteriorly.

Etymology: in honor of George Gaylord Simpson, whose many accomplishments included analysis of the function of the saber in carnivorous mammals.

Description: Machaeroides eothen is known from excellent material from the Bridger Basin (Gazin, 1946). Although the Wind River Basin material is not as complete, it clearly shows similarities that indicate reference to the same genus. In the following description and comparisons, differences from *M. eothen* are emphasized. Where characters are not mentioned, they are similar to those in *M. eothen* or are not known from the less complete Wind River Basin material.

Skull fragments of CM 45115 include a piece of the skull roof that has a distinct sagittal crest, and bases of both glenoid fossae, which are deeply cupped, with

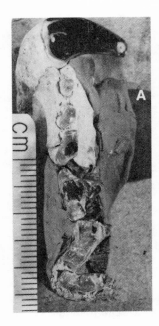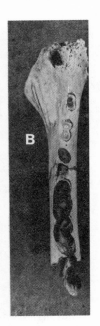

Figure 1. *Machaeroides simpsoni* (stereophotographs in occlusal view). *A*, CM 45115, C-M^2, holotype; and *B*, CM 36397, lower jaw with P_1, P_4-M_1.

TABLE 1. MEASUREMENTS (IN MILLIMETERS) OF
Machaeroides simpsoni AND *M. eothen*.

	CM 45115	CM 36397	USNM 17059
C length at alveolus	8.2		8.5
P^1 length : width	2.4 : 2.9		3.2 : 2.1
P^2 length : width	4.0 : 2.2		5.4 : 2.0
P^3 length : width	7.2 : 3.6		8.0 : 5.0
P^4 length : width	9.3 : 7.6		10.0 : 8.2
M^1 length : width	9.4 : 11.2		9.4 : 7.5
M^2 length : width	3.7 : 10.5		3.5 : 8.1
depth jaw at flange		27.0	22.0
depth jaw below P_4		14.7	
P_1 length : width		2.3 : 1.0	3.6 : 1.6
P_4 length : width		8.0 : 3.5	9.1 : 4.0

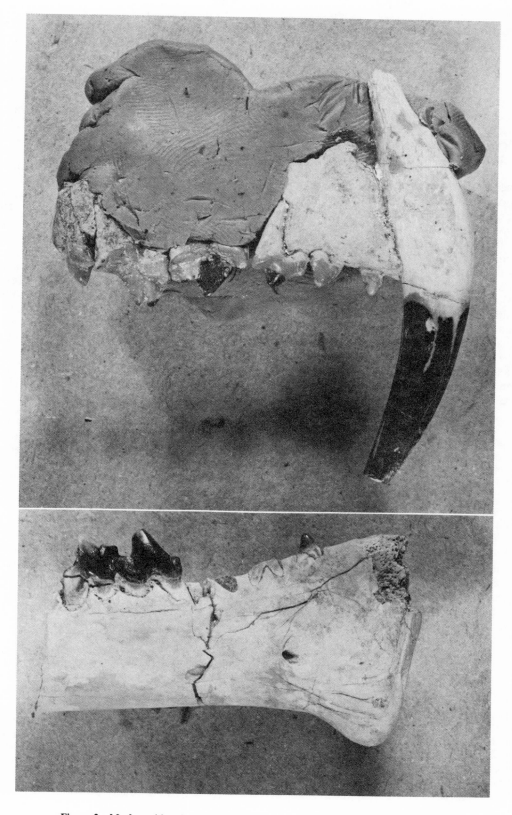

Figure 2. *Machaeroides simpsoni*, lateral views of same specimens as in Figure 1.

distinct anterior and posterior projecting rims as in *Machaeroides eothen*. A piece of the right maxilla (Figs. 1, 2) preserves the posterior border of the canine alveolus and has P^1-P^3 in place.

Each of the three upper incisors is compressed transversely, has a sharp, piercing crown, and a heavy, somewhat bulbous root. I^3 is the largest, I^1, the smallest. Wear facets occur posterolaterally on I^3, and both medially and laterally on I^1 (I^2 is too incomplete to determine). The upper canine is a long (extending at least 25 mm below the alveolus), transversely compressed saber with its greatest anteroposterior thickness at about the midpoint. A finely serrated enamel crest forms the anterior and posterior edges of the tooth. The tip is broken from both canines, but both anterior and posterior crests extend at least as far as the break. The curvature of the canine is such that the axis of curvature occurs just posterior to the maxillary tooth row. As in *Machaeroides eothen* the maxilla must have been high and fairly steep anteriorly to accommodate the heavy root of the canine (Fig. 2).

There is a short diastema both anterior and posterior to P^1, which is a conical, single rooted tooth. The two-rooted P^2 is oriented obliquely (anterolingual to posterobuccally) and slightly overlaps P^3. P^2 is longer than P^1. The enamel on the anterior and posterior crests that extend from its central cusp is serrated. P^3 is two rooted; the posterior root is wide transversly, supporting both the metastyle and a small protocone lobe. The enamel on the crest is serrated. In *Machaeroides eothen* P^3 is three rooted, and has a more prominent protocone lobe and lingual cingulum. P^4 is robust, three rooted, and wide buccally, with a large paracone, distinct parastyle, larger, blade-like metastyle, and protocone supported by a distinct root. The postmetacrista is more prominent than in *M. eothen*. As on the other premolars, the enamel on the crests is finely serrated.

The largest upper cheek tooth, M^1, is a well developed carnassial. The paracone and metacone are closely spaced, joined at their bases and form a relatively tall buccal side of the trigon. The parastyle is well developed and the metastyle is elongate. The posterolingual side of the tooth (postvallum) is worn to a long, smooth blade, which extends from the base of the metacone posteriorly to the metastyle. The protocone is smaller than that on P^4 and a shallow, rounded wear facet. The buccal wall is concave, with the metastyle protruding outward. This contrasts with a straighter buccal wall in *Machaeroides eothen*. The enamel of the anteroposterior crest is serrated where it is not worn through. M^2, supported by two roots, is a robust tooth, distinctly larger than in *M. eothen*. It has a rounded buccal shelf and a single, central cusp. The small protocone shelf is worn.

The upper left and right dentitions show different patterns of wear. The protocone of right P^4 has two distinct, anteroposteriorly oriented wear facets; that of left P^4 has one large posterior wear facet. The anterior face (prevallum) of right M^1 is unworn, that of left M^1, worn. The protocone of right M^2 is less worn than the deeply notched protocone of left M^2.

The lower jaw, CM 36397 (Figs. 1B, 2) is from a different locality, but approximately the same horizon (Lostcabinian) as the type. The jaw is wide anteriorly and has a well developed, relatively vertically oriented symphysis (measuring 12.3 mm anteroposteriorly, 22.0 mm dorsoventrally) that occupies the entire depth of the jaw anteriorly. A distinct flange projects ventrally. Two foramina occur on the anterior surface of the jaw, and the mental foramen is on the lateral surface below P_2. The specimen does not show clearly the number of incisors; there appears to be one alveolus anteromedial to that of the lower canine. It is probable that, as in *Machaeroides eothen*, there were two lower incisors in each jaw. The alveolus of the lower canine shows that this tooth has a transversely narrow root. The two rooted, peg-like P_1 is bordered by a diastema anteriorly and a shorter diastema posteriorly. P_2 and P_3, both missing in this specimen, were two rooted and were separated by a short diastema. The posterior alveolus of P_3 is more buccally situated than the anterior alveolus, suggesting that P_3, like P^3, was somewhat obliquely oriented. P_4 has serrated enamel on the crests, a small paraconid, a prominent protoconid, the anterior surface of which tilts posteriorly more than in *M. eothen,* and a talonid that is crested buccally and shows very slight development of a lingual basin. M_1, broken posteriorly, has a slender, sharply crested paraconid. The paracristid and protocristid are worn; the angle at which these crests intersect is wider than in *M. eothen*. A small metaconid is present on the posterolingual slope of the protoconid. Broken remnants suggest that the talonid was basined. M_2, though badly broken, was clearly larger than M_1. The paraconid is a sharp blade, the protoconid is high, and a metaconid appears to have been present. The tooth is too incomplete to provide any clues to the nature of the talonid.

Discussion: Comparisons with Bridgerian *Machaeroides eothen*, especially with USNM 17059, the excellent specimen described by Gazin (1946), show general similarity but interesting differences between the two species. The differences seem to be of two sorts: those suggestive of a more primitive condition in *M. simpsoni* and others suggestive of a more highly evolved saber tooth specialization in *M. simpsoni*. In the former category are: in *M. simpsoni* P^3 is two rooted and has a less distinct protocone; M^1 is less sectorial, the postvallum is less blade-like and the protocone more distinct; and M^2 is larger, with a well developed lingual portion. In the latter category are: in *M. simpsoni* P^2 is more oblique in position and the upper premolars are generally more crowded; the metastyle of P^4 is sharper and more elongate; P_4 has a more inclined anterior face; and the ridges formed by the paraconid and protoconid of M_1 intersect at a larger angle, forming a straighter shearing surface.

CM 37342, the Gardnerbuttean specimen, is referred to *M. simpsoni* largely on the basis of the structure of P^3, two rooted and lacking a well developed protocone. In some other characters, especially in the shape of M^1, this specimen is intermediate between *M. simpsoni* and *M. eothen*.

RELATIONSHIPS OF THE MACHAEROIDINAE

Machaeroides was described by Matthew (1909), with one species, *M. eothen*. The taxon was based on two jaws from the Lower Bridger and one tooth from the Upper Bridger. The presence of four two-rooted premolars and two molars of equal size, as well as the form of the symphysis, led Matthew to refer this genus to the Oxyaenidae. He noted (1909, p. 461), "With a better knowledge of the genus it would probably be necessary to place it in a distinct sub-family of equal rank with Oxyaeninae and Limnocyoninae, but for the present I refer it to the latter group." Elsewhere in the same paper (1909, p. 330), Matthew named the subfamily Machairoidinae (sic).

The next sabertooth creodont to be described (Scott, 1938) was *Apataelurus kayi*, coming from the Uinta B. Scott considered *Apataelurus* to be related either to the Hyaenodontidae or, "more likely" to the Oxyaenidae. Like *Machaeroides* at the time of its description, *Apataelurus* was known only from lower jaws.

Denison (1938), in his thorough review of the "broad-skulled Pseudocreodi" (that is, subfamilies Oxyaeninae, Palaeonictinae, Limnocyoninae, and Machaeroidinae), united *Apataelurus* with *Machaeroides* in the Machaeroidinae, considered, as were the Limnocyoninae, to be a subfamily of the Hyaenodontidae. He held the Machaeroidinae to be "an offshoot of the Limnocyoninae showing a remarkable parallelism to the true sabre-toothed cats" (1938, p. 181). Like Matthew, Denison did not discuss characters in which the machaeroidines resembled the limnocyonids. Gazin (1946) had the advantage of working with a superb specimen USNM 17059 of *Machaeroides eothen*, a skull, jaws, and postcranial parts. He accepted the association of machaeroidines with limnocyonines, but recommended that these subfamilies be united to form the family Limnocyonidae, which was separated from the Hyaenodontidae. Presence of $\frac{M_1}{M_2}$ as the carnassial pair was given as the most important character differentiating the limnocyonids from the hyaenodontids. Affinities between the machaeroidines and the limnocyonines have been accepted as well by more recent workers (*e. g.*, Van Valen, 1969).

Revision of the oxyaenids, hyaenodontids, and limnocyonids is clearly outside the scope of this paper. However, some characters of *Machaeroides* that might have a bearing on the familial relationships of the Machaeroidinae must be considered. (1) *Upper incisors decreasing in size from I^3 to I^1*. In this character *Machaeroides* resembles the oxyaenids, in which I^3 is usually enlarged, I^1, reduced. In the limnocyonids *Oxyaenodon* and *Thereutherium* the upper incisors are subequal, but in other members of the family I^3 is reduced or lost and I^2 is frequently enlarged. (2) *Depth of jaw and strength of symphysis*. (3) *Deeply cupped glenoid with distinct pre-glenoid crest*. Characters 2 and 3 are also shared with the oxyaenids. Although they may be adaptations related to the total sabertooth morphology, their acquisition from oxyaenid ancestors seems more probable than from the limnocyonids. Limnocyonids are characterized by a shallow jaw with an anteriorly projecting canine, and a relatively open glenoid cavity lacking a preglenoid crest. (4) *Presence of a distinct paraconid on P_4*. This is shared with the primitive oxyaenid *Dipsalidictides amplus* as well as with most other oxyaenids. The paraconid of P_4 is weak and variable in limnocyonids. (5) *In M. simpsoni absence of a protocone and third root on P^3*. In this character, the two known species of *Machaeroides* differ. In *M. simpsoni*, as in *D. amplus*, P^3 has two roots and a rudimentary lingual lobe; the younger *M. eothen* has a three rooted P^3 and better development of the protocone lobe. This trend in the machaeroidines is seen in oxyaenids. In *Dipsalidictides amplus* and some other primitive oxyaenids, P^3 is two rooted, and the more advanced oxyaenids have three roots and a distinct protocone. It should be noted that the oldest known oxyaenid, *Tytthaena*, has been described as having a three-rooted P^3 (Gingerich, 1980), which raises questions about variability and reliability of this character. (6) *Two-rooted P_1*. This is shared with the limnocyonids; in oxyaenids P_1 is single rooted or absent.

Although evaluation of these few characters does not constitute a family level revision, the evidence from them does at least throw open the question of familial relationships of the Machaeroidinae, and the validity and composition of the Oxyaenidae and Limnocyonidae. Most of the characters discussed above suggest that the machaeroidines are closer to the oxyaenids than to the limnocyonids. The Oxyaenidae can now be traced into the middle Tiffanian of North America, following which they began a considerable radiation (Gingerich, 1980). The Machaeroidinae may have resulted from this early diversification among primitive oxyaenids.

The Machaeroidinae are not well enough known to follow patterns of development within the group. The large temporal and especially morphologic gap between *Machaeroides* and *Apataelurus* provides little on which to base estimates of phylogenetic development. *Apataelurus* differs from both species of *Machaeroides* in its much lower coronoid process, long diastema from the canine to P_1, and M_2 distinctly larger than M_1. Some indications that *M. simpsoni* may be more closely related to *Apataelurus* than was *M. eothen* come from the inclined protoconid on P_4, the slightly greater opening of the paracristid and protocristid on M_{1-2}, and the crowded premolars in the Lostcabinian and Gardnerbuttean species.

THE SABERTOOTH ADAPTATION IN MAMMALIAN CARNIVORES

The sabertooth adaptation has been the subect of much speculation as to just how the animals with this feature made a living. Simpson's careful analysis (1941) of the morphology of a variety of sabertooths showed clearly that the sabers were used largely for stabbing. Relatively thick-skinned ungulates are usually cited as the prey for the sabertooths, and the extinction of many of these ungulates at the end of the Pleistocene is often given as being responsible for the demise of sabertooths. The extinction of sabertooths does not imply that their adaptation was not successful. The adaptation is now known to have occurred over a time span of at least 50 million years (late early Eocene to late Pleistocene). Furthermore, the sabertooth complex arose independently at least three times (in the machaeroidine creodonts, thylacosmilid marsupials, and felids) and perhaps as many as six times, depending on one's interpretation of the phylogeny of the Feloidea (compare, for example, Thenius, 1967, and Martin, 1980).

The suite of characters that occurs in all of the sabertooth carnivores includes: enlarged, transversely flattened upper canine, and reduced lower canine; crowding and/or reduction of the premolars; emphasis on the shearing function of a carnassial pair; flanged lower jaw, with a low condyle, reduced coronoid process and sturdy symphysis; and skull with a low, strongly cupped, glenoid cavity, large mastoid process, and high occiput with well developed ridges for muscle attachment (Matthew, 1910). The co-occurrence of these characters in at least three different lineages suggests that the sabertooth morphological complex is under strong pleiotropic control.

Machaeroides is the smallest known sabertooth but can, nonetheless, be visualized as having been a voracious carnivore. Its strong canines, shearing molars, and serrations on the enamel are dental features indicating a well developed predatory nature. Herbivores such as *Phenacodus* and *Esthonyx* may have been its regular prey, and even young *Coryphodon* may have been vulnerable to its predation.

REFERENCES CITED

Denison, R. H., 1938, The broad-skulled Pseudocreodi: New York Academy of Sciences, Annals, v. 37, p. 163-256.

Gazin, C. L., 1946, *Machaeroides eothen* Matthew, the saber-tooth creodont of the Bridger Eocene: United States National Museum, Proceedings, v. 96, p. 335-347.

Gingerich, P. D., 1980, *Tytthaena parrisi,* oldest known oxyaenid (Mammalia, Creodonta) from the Paleocene of western North America: Journal of Paleontology, v. 54, p. 570-576.

Martin, L. D., 1980, Functional morphology and the evolution of cats: Nebraska Academy of Sciences, Transactions, v. 8, p. 141-154.

Matthew, W. D., 1909, The Carnivora and Insectivora of the Bridger Basin, middle Eocene: American Museum of Natural History, Memoir, v. 9, p. 289-567.

_____ 1910, The phylogeny of the Felidae. American Museum of Natural History, Bulletin, v. 28, p. 289-316.

Scott, W. B., 1938, A problematical cat-like mandible from the Uinta Eocene, *Apataelurus kayi,* Scott: Annals of Carnegie Museum, v. 27, p. 113-120.

Simpson, G. G., 1941, The function of the saber-like canines in carnivorous mammals: American Museum Novitates, no. 1130, 12 p.

_____ 1944, Tempo and mode in evolution: New York, Columbia University Press, 237 p.

_____ 1978, Concession to the improbable: New Haven, Yale University Press, 281 p.

Thenius, E., 1967, Zur Phylogenie der Feliden (Carnivora, Mamm.): Zeitschrift für zoologische Systematik und Evolutionsforschung, v. 5, p. 129-143.

Van Valen, L., 1969, Evolution of dental growth and adaptation in mammalian carnivores: Evolution, v. 23, p. 96-117.

MANUSCRIPT RECEIVED JANUARY 29, 1986
REVISED MANUSCRIPT RECEIVED APRIL 25, 1986
MANUSCRIPT ACCEPTED JUNE 16, 1986

Early Eocene artiodactyls from the San Juan Basin, New Mexico, and the Piceance Basin, Colorado

LEONARD KRISHTALKA
RICHARD K. STUCKY

Section of Vertebrate Fossils, Carnegie Museum of Natural History, Pittsburgh, Pennsylvania 15213

ABSTRACT

Three species of artiodactyls (*Diacodexis* sp. cf. *D. secans*, *Simpsonodus chacensis* [new genus and species], and *Wasatchia grangeri*) are reported from the San Jose Formation, New Mexico; these three and an additional four taxa (*Simpsonodus* sp., *W. pattersoni*, *Bunophorus sinclairi*, and *Hexacodus pelodes*) occur in the Debeque Formation, Colorado. The new diacodexeid genus, *Simpsonodus*, includes *D. chacensis* and *W. lysitensis*. *Wasatchia pattersoni*, new species, the most derived known species of *Wasatchia*, is morphologically suitable as an ancestor to *Bunophorus*, with which it overlaps stratigraphically. *Diacodexis* sp. cf. *D. secans* is morphologically and apparently phylogenetically intermediate between *D. secans*, on the one hand, and *Hexacodus* and *Antiacodon* on the other. The early Eocene record of diacodexeids is robust enough to indicate: (1) the gradual and continuous divergence of *D. secans* and *Diacodexis* sp. cf. *D. secans* in the mosaic distribution of derived character states; and (2) from these taxa, respectively, the gradational origin of other diacodexeids/leptochoerids and antiacodontids/homacodontids. In these instances, microevolutionary processes appear to account for the macroevolutionary pattern.

INTRODUCTION

In 1946, George Gaylord Simpson, then at The American Museum of Natural History, began an extensive program of paleontological exploration of Tertiary deposits of the San Juan Basin, New Mexico. Among other achievements, his eight years of field work there (1946-1951, 1953, 1958) resulted in recovery of an outstanding collection of early Eocene vertebrates from the San Jose Formation, a rock unit he described and named (Simpson, 1948). Simpson's plan to ". . . use the San Jose collections . . . for a full revision and discussion of the mammalian faunas in the formation" (Simpson, 1981, p. 21) was dashed by his departure from the American Museum in 1959. Our studies of the artiodactyls from the San Jose Formation (Krishtalka and Stucky, 1985; this paper; Stucky and Krishtalka, *in press*) are dedicated to Simpson as a collector of fossil vertebrates and one of the founders of the Modern Synthesis.

The first North American early Eocene artiodactyls were recovered from the San Jose Formation and described by E. D. Cope (1875) as *Pantolestes chacensis* and *Sarcolemur crassus*. The type specimens of both have been lost, but *S. crassus* is now included in *Wasatchia grangeri* and *P. chacensis* is the type species of *Simpsonodus*, a new genus of artiodactyl, with Simpson's superb specimen from Quarry 88 (AMNH 48694) as the designated neotype (see below).

Other artiodactyls from the San Jose Formation described here were recovered by field parties from the American Museum (1912-1914, W. Granger), the Webb School of California (1949, M. C. McKenna), Carnegie Museum of Natural History (1963, M. R. Dawson; 1978-1981, L. Krishtalka), and the University of Arizona (1973, 1977, L. L. Jacobs). Apart from Sinclair's (1914) description of some of the American Museum material, this paper is the first systematic treatment of artiodactyls from the San Jose Formation. The Piceance Basin artiodactyls, collected by field parties from the Carnegie Museum, the Field Museum of Natural History, and the University of Colorado, were described by Kihm (1984), but are systematically revised here (except for his *Wasatchia* sp. and ?Dichobunidae, genus unknown, original material of which was unavailable for study).

Specimens of *Diacodexis* from both basins were included in a recent revision of that genus (Krishtalka and Stucky, 1985), which indicated that artiodactyls from the San Jose and Debeque formations are pivotal in reconstructing the origin and early radiation of diacodexeids, homacodontids, and antiacodontids.

Granger (1914) divided the San Jose record into Almagre and Largo faunas, which he considered diachronic and Graybullian and Lostcabinian, respectively; Van Houten (1945) and Simpson (1948) thought they were Graybullian and Lysitean, respectively. According to later workers (Lucas, 1977; Gingerich and Simons, 1977; Froelich and Reser, 1981) the two faunas were penecontemporaneous and Lostcabinian. Most recently, however, studies by Lucas and others (1981) and Stucky (1984) indicate that both the Almagre and Largo faunas are late Graybullian and/or early Lysitean.

Mammalian faunas with artiodactyls from the Piceance Basin are Wasatchian, ranging from the late Graybullian through the Lostcabinian. Kihm's (1984) biostratigraphic terminology for the Piceance Basin differs from the current and well established zonation of the early Eocene (West and others, *in press*) and can be misleading. As emended accordingly, his "early Wasatchian" is equivalent to much of the Graybullian (= Schankler's, 1980, *Haplomylus-Ectocion* Range Zone), his "lower middle Wasatchian" is late Graybullian (= *Bunophorus* Interval Zone), his "upper middle Wasatchian" is Lysitean, and his "late Wasatchian" is Lostcabinian.

Abbreviations are as follows: AMNH, American Museum of Natural History, New York; CM, Carnegie Museum of Natural History, Pittsburgh; FMNH, Field Museum of Natural History, Chicago; GGS, George Gaylord Simpson; PU, Princeton University, Princeton; UA, University of Arizona, Tucson; UCM, University of Colorado Museum, Boulder; USNM, U.S. National Museum, Washington, D.C.; and WCM, Western Colorado Museum, Grand Junction.

SYSTEMATIC PALEONTOLOGY

Order ARTIODACTYLA Owen, 1848
Family DIACODEXEIDAE Gazin, 1955
Diacodexis Cope, 1882

A recent review of the early and middle Eocene record of *Diacodexis* (see Krishtalka and Stucky, 1985) recognized four species: (1) *D. secans*, a Sandcouleean (earliest Wasatchian) through Blacksforkian (early Bridgerian) species-lineage, which most commonly occurs in the Bighorn and Wind River basins of Wyoming (it was divided into four informal lineage segments to reflect significant anagenetic change in frequency and degree of expression of derived features); (2) *D. gracilis*, from the late Graybullian of the Bighorn Basin; (3) *D. woltonensis*, restricted to the Lostcabinian of the Wind River Basin; and (4) *D. minutus*, from the Lostcabinian and Gardnerbuttean of the Wind River and Huerfano basins.

Four previously named species of *Diacodexis* were synonymized with *D. secans* (*D. olseni, D. metsiacus, D. laticuneus,* and *D. brachystomus*), and two (*D. robustus,* and *D. chacensis*) were removed from the genus. *Diacodexis robustus* is referred elsewhere to *Wasatchia* (see Stucky and Krishtalka, *in press*); *D. chacensis* is assigned below to *Simpsonodus*, a new genus of early Eocene artiodactyl.

Two additional *Diacodexis*-like taxa, dubbed Artiodactyla sp. A and sp. B, were recognized on limited material from the San Juan and Piceance basins. The former is assigned below to *Diacodexis* sp. cf. *D. secans*, which now includes a larger hypodigm, especially from the Piceance Basin; the latter is referred below to *Simpsonodus chacensis*.

Diacodexis sp. cf. *D. secans*
Figure 1; Table 1

Referred specimens: CM 43129-43132, 43134-43136; AMNH 16296; UA 11186 (all from the San Juan Basin). CM 10475, 43702; UCM 40831, 47052-47054, 48517, 48526, 47889, 47891; FMNH-P 15605, 15615, 15730, 26455, 26475, 26572, 26894, 26919, 37259 (all from the Piceance Basin).

Localities: CM locs. 941, 945, 1017, 1025; AMNH loc. "Ojo San Jose, East Bluff"; UA loc. 7747 (all late Graybullian to Lysitean, San Jose Fm., San Juan Basin). CM locs. 754, "Indian Valley"; UCM locs. 78033, 78049, 78016; FMNH locs. 28-37, 37-37, 473-41, 500-41 (all late Graybullian to Lysitean, Debeque Fm., Piceance Basin).

Known distribution: Early Wasatchian (late Graybullian to Lysitean), San Juan and Piceance basins.

Discussion: Material from the San Juan Basin is, on average, slightly smaller than that from the Piceance Basin. As described elsewhere (Krishtalka and Stucky, 1985), lower molars in this sample closely resemble those of the *Diacodexis secans-metsiacus* lineage segment of *D. secans*. Some specimens (CM 10475; UCM 47053; FMNH-P 26894, 26455; all CM material from the San Jose Fm.), however, differ from the latter in having one

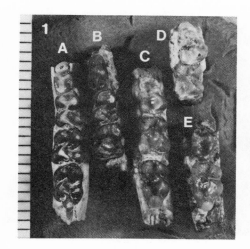

Figure 1. Stereophotographs (occlusal views) of *Diacodexis* sp. cf. *D. secans*. *A*, FMNH-P 26455, LP$_4$-M$_3$; *B*, UCM 48526, LP$_4$-M$_2$; *C*, CM 10475, RM$_{1-3}$; *D*, CM 43132, RM$_{1-2}$; and *E*, CM 43130, RM$_{2-3}$. Scale in millimeters.

TABLE 1. DIMENSIONS (IN MILLIMETERS) OF LOWER TEETH OF *Diacodexis* sp. cf. *D. secans*.

Specimen Number	Locality	L P/3	W P/3	L P/4	W P/4	L M/1	W M/1	L M/2	W M/2	L M/3	W M/3
FMNH P15730	UCM 78016	4.7	1.8	4.1	2.2						
CM 43135	CM 941		1.5	4.9	2.1						
UCM 48526	UCM 78049			4.1	2.3	3.7	2.8	4.0	3.4		
FMNH PM 293	FMNH 98-C-47				2.6	4.3	3.2	4.3	3.7		
FMNH P26475	UCM 78033			3.8	2.2	3.7	3.0				
FMNH P26919	FMNH 500-41			4.2	2.3	3.7	2.9				
CM 43135	CM 945			4.5	2.1						
CM 10475	Indian					4.2	2.8	4.2	3.3		
CM 10475	Valley						2.9	4.2	3.5	5.1	3.1
FMNH P26683	UCM 78018					3.9	2.8	3.8	3.1	5.1	3.0
FMNH P26572	UCM 78016					4.0	3.0	4.1	3.5		3.1
UCM 47053	UCM 78049						2.8	4.0	3.3	4.8	2.8
CM 43132	CM 941						2.7	4.1	3.2		
FMNH P26894	FMNH 473-41							4.2	3.4	4.9	3.1
UCM 47052	UCM 78049							4.2	3.6	4.9	3.3
AMNH 16296	Ojo San Jose E. Bluff							4.3	3.4	5.2	3.1
FMNH P26455	UCM 78016							3.9		5.5	3.6
CM 43702	Scenery Gulch							3.4		4.8	3.2
FMNH P15615	FMNH 37-37							3.8		5.3	3.5
FMNH P15605	FMNH 28-37							4.0	3.3		
UA 11186	UA 7747							4.1	3.1		
CM 43130	CM 1017									5.2	3.2
CM 43134	CM 1017									4.7	2.9
UCM 47891	UCM 78049									5.0	3.1
UCM 40831	UCM 78033									5.0	3.1
MEAN		4.7	1.7	4.3	2.3	3.9	2.9	4.1	3.4	5.0	3.2
N		1	2	6	7	7	10	13	16	13	14
SD				.38	.17	.25	.14	.14	.23	.23	.21
CV				9	8	6	5	3	7	4	7

or more of the following characteristics: a slightly to moderately more elevated trigonid; a non-inflated metaconid; a small paraconid more closely appressed and anterior or barely anteromedial to the metaconid; a more elongate talonid; a higher, larger, more conical entoconid; a hypocristid directed toward (but not reaching) the posterolabial aspect of the entoconid; and a hypoconulid on a broader postcingulid. These features are part of the suite of specializations consistently expressed and shared by antiacodontids and homacodontids.

The entire sample, however, shows a continuous range of variation from specimens (*e. g.,* AMNH 16296, UCM 47052) that are virtually indistinguishable from *Diacodexis secans-metsiacus* (especially AMNH 4696, cotype of *D. metsiacus;* UCM 44126) to those (*e. g.,* CM 10475, FMNH-P 26455, 26894) in which most exhibit the derived features listed above and seem morphologically distinct from *D. secans*. Moreover, the primitive and derived features show a mosaic rather than consistent pattern of distribution, so that, for example, some specimens with an elevated trigonid (*e. g.,* UCM 47053) also bear a primitive, non-conical entoconid, and a hypocristid that extends directly to the hypoconulid. Other specimens with a hypocristid directed toward the posterolabial base of the entoconid lack a conical entoconid (*e. g.,* FMNH-P 26455), and vice versa (*e. g.,* CM 10475). Many other permutations and combinations of primitive and derived features occur in this sample.

This material may represent one or more taxa. The implication of one taxon is supported by a continuous range of variation that is no larger than that found in penecontemporaneous samples of single species of

Diacodexis, including those from one locality (Krishtalka and Stucky, 1985). Given this conclusion, which we favor, this taxon may be a geographic variant of *D. secans* (lineage segment *D. s.-metsiacus*), or a discrete species. The apparently continuous morphologic overlap between the San Juan/Piceance basin material and *D. secans* implies the former (*i. e.*, geographic variant); the mosaic array of derived features in many of the specimens implies the latter (*i. e.*, a discrete species), despite the degree of morphologic overlap with *D. secans*. Such a species would have fuzzy boundaries indeed.

Crucial here is the interpretation of patterns of morphological variation in intra- and interbasinal samples, especially when that variation is apparently continuous— the dilemma of a good fossil record. At one end of the continuum, the San Juan/Piceance basin material is indistinguishable from *D. secans;* at the other end it is morphologically derived and discrete. Such a pattern cannot be reflected adequately by Linnaean taxonomy, and seems to defy objective resolution; yet taxonomic decisions (*i. e.*, species in the fossil record) are interpreted as real biological entities, and often provide theoretical grist for the macroevolutionary mill.

Alternatively, the material could be divided into two or more species in a variety of permutations, all of which, given the mosaic distribution of primitive and derived features, would involve both the arbitrary assignment of some specimens and a plethora of species (perhaps as many as there are specimens). Here the different species would reflect the degree of difference along a continuum in derived morphological features from *Diacodexis secans,* and from one another. Such a solution, although perhaps cladistically heuristic, seems biologically unrealistic: specimens from the same locality differing in morphological minutiae would have to be placed in different species despite the integrity of the sample.

No matter the taxonomic solution adopted, this taxon of *Diacodexis* from the San Juan and Piceance basins is morphologically intermediate between *D. secans* (lineage segment *D. s.-metsiacus*) and the most primitive species of antiacodontids and homacodontids. It seems basal to the origin of the latter in that the lower molars bear a mosaic of derived features that are fixed and further derived in *Hexacodus* and *Antiacodon*.

Simpsonodus, new genus

Etymology: In honor of George Gaylord Simpson's unparalleled contributions to paleobiology; *odous*, G., tooth.

Type species: Diacodexis chacensis Cope, 1875.

Included species: Type species and *Wasatchia lysitensis* Sinclair, 1914 (Stucky and Krishtalka, in press).

Known distribution: Late Graybullian to Lysitean, western North America.

Diagnosis: Differs from penecontemporaneous *Diacodexis* in its larger size, slightly elevated molar trigonid, and proportionately less expanded metaconid (except *Diacodexis* sp. cf. *D. secans*), relatively longer talonid, and more quadrate upper molars. Differs from *Wasatchia* and *Bunophorus* in having more gracile P_4, larger P_4 talonid basin, non-bulbous molar cusps and crowns, more elevated molar trigonids, and deeper talonid basins.

Discussion: Recognition of *Simpsonodus* clarifies the confusion surrounding the systematics of early Eocene species *Diacodexis chacensis* Cope, 1875 and *D. robustus* Sinclair, 1914, neither of which belong in *Diacodexis sensu stricto* (see Krishtalka and Stucky, 1985). This confusion began with Cope's (1884, p. 719, Pl. 24e, Fig. 5; AMNH 4691) referral of Bighorn Basin material to *Pantolestes chacensis* and the loss of the type specimen of *D. chacensis* (Cope, 1877, Pl. 45, Fig. 17), which was collected in the San Juan Basin. Sinclair (1914) did not see the holotype of *D. chacensis,* but based his concept of this species exclusively on Cope's and his own material from the Bighorn Basin, including AMNH 4691, which he designated as the paratype of *D. chacensis*. However, AMNH 4691 is significantly smaller than and differs morphologically from the lost holotype of *D. chacensis* (see Cope's, 1877, figure and measurements), but closely resembles the holotype of *D. brachystomus* (AMNH 4700). As a result, Sinclair (1914) confused *D. chacensis* with *D. brachystomus* and *D. metsiacus* (Gazin, 1952), species that since have been included in *D. secans* along with *D. laticuneus* and *D. olseni* (see Krishtalka and Stucky, 1985).

In addition, Sinclair's (1914) hypodigm of *Diacodexis robustus* was a composite of two taxa: the holotype (AMNH 15514; Sinclair, 1914, Fig. 27A) and one paratype (AMNH 15512, Fig. 27C) belong in *Wasatchia* (see below), whereas one of the other paratypes (AMNH 15513, Fig. 27B) is similar to Cope's lost holotype of *Pantolestes chacensis*.

Pantolestes chacensis and *Sarcolemur crassus* were the only two artiodactyls present in Cope's (1875, 1877, 1884) comprehensive accounts of the mammalian faunas of the San Juan Basin. His figured specimen of *S. crassus* (now a *nomen oblitum*) is referred below to *Wasatchia grangeri*. His figured specimen (and holotype; Cope, 1877, Pl. 45, Fig. 17) of *P. chacensis* conforms in size and morphology to a large sample from the San Juan, Bighorn, and Piceance basins that is generically distinct from *Diacodexis* (*sensu stricto,* Krishtalka and Stucky, 1985).

Simpsonodus chacensis (Cope, 1875)
Figure 2; Tables 2 and 3

Synonyms: Pantolestes chacensis Cope, 1875, p. 15; Cope, 1877, p. 146, Pl. 45, Fig. 17. *Trigonolestes chacensis* (Cope, 1875), Matthew, 1899, p. 34, in part. *Diacodexis chacensis* (Cope, 1875), Sinclair, 1914, p. 291, in part (AMNH 15520). *Diacodexis robustus* Sinclair, 1914, in part, p. 293-295, Figs. 27B, 28.

Holotype: Original holotype lost (USNM collections, *fide* Matthew, 1899).

Designated neotype: AMNH 48694, RP_3-M_3, LP_4-M_3, LM^{1-3}, collected by G. G. Simpson from Quarry 88, Regina Member, San Jose Formation, San Juan Basin, New Mexico.

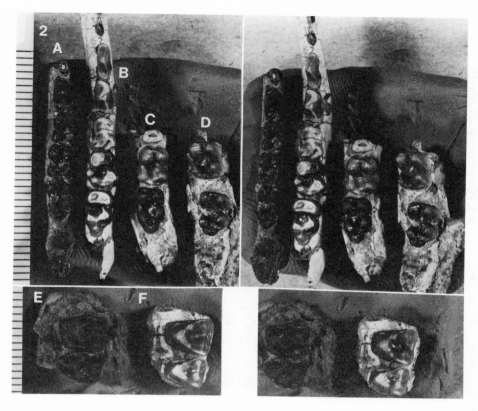

Figure 2. Stereophotographs (occlusal views) of *Simpsonodus chacensis*. *A*, AMNH 48694, neotype, RP_4-M_3; *B*, AMNH 15520, RP_3-M_3; *C*, USNM 19224, RM_{2-3}; *D*, UCM 53926, RM_{2-3}; *E*, AMNH 48694, neotype, LM^{2-3}; and *F*, CM 43139, RM^{1-2}. Scales in millimeters.

Referred specimens: CM 43128, 43133, 43137, 43139; UA 7000; USNM 19224; AMNH 237, 16297, 48694, 55516, 86289 (all from the San Juan Basin). FMNH-P 1222, PM 316 (both from the Piceance Basin). AMNH 15510, 15511, 15513, 15520, 92843; CM 58133, 12348, 12232; UCM 44124, 44129, 44130, 53920-53927 (all from the Bighorn Basin).

Localities: AMNH locs. Quarry 88, "7 mi. W. Ojo San Jose," "center of mosaic 45"; USNM loc. "1-1/2 mi. W. Regina"; UA loc. 7400; CM loc. 941 (all San Jose Fm., San Juan Basin, New Mexico). FMNH-P loc. 328-41; UCM loc. 78049 (both Debeque Fm., Piceance Basin, Colorado). AMNH locs. "lower Graybull Valley," "5 mi. S. Otto," "2 mi. S. St. Joe"; UCM locs. 80066, 84167, 85231, 85341-85345; CM locs. 144, 147, 680 (all Willwood Fm., Bighorn Basin, Wyoming).

Emended diagnosis: 50 percent smaller than *Simpsonodus lysitensis*; P_{3-4} more gracile than in *Simpsonodus* sp. and P_4 without metaconid.

Description: *Simpsonodus chacensis* is known from P_3-M_3 and M^{1-3}. P_3 is shorter than, or as long as, P_4 and has a small paraconid, predominant protoconid, basined heel, and a cristid obliqua on the posterior face of the protoconid that reaches the postcingulid.

P_4, slightly longer than M_1 and as wide as the M_1 trigonid, lacks a metaconid and is gracile and *Diacodexis*-like, unlike the more robust condition of P_4 (and P_3) in *Simpsonodus* sp. (see below). P_4, like the P_3, is dominated by the protoconid, and bears a strong, low paraconid, a cristid obliqua (with a small bulge at its anterior end), and a short, basined talonid with two cusps, an entoconid, and ?hypoconid at the posterior end of the cristid obliqua.

The molars are not bulbous, in contrast to *Wasatchia, Bunophorus, Diacodexis woltonensis,* and most *D. secans*. M_{1-3} have a strong metaconid and anterior or slightly anteromedial paraconid; their bases are appressed, but their apices are distinct and subequal, as in some specimens of *D. secans,* lineage segment *D. s.-primus* (see Krishtalka and Stucky, 1985). In posterior view, the paraconid-metaconid wall is higher than the protoconid. On most specimens, the metaconid is proportionately not expanded lingually, the M_{1-2} talonid is elongate, and the hypoconulid is always strong and occurs on a broad postcingulid at the end of the hypocristid. A postmetacristid usually is absent, as is a true talonid notch, although some specimens have a gap between the bases of the metaconid and entoconid. M_3 bears a large hypoconulid lobe.

On the lost holotype of *Simpsonodus chacensis* from the San Juan Basin (Cope, 1877, Pl. 45, Fig. 17), a large accessory cuspule occurs on M_2 just anterolingual to the entoconid. Among the specimens referred here, only AMNH 15520 from the Bighorn Basin bears the same

TABLE 2. DIMENSIONS (IN MILLIMETERS) OF LOWER TEETH OF Simpsonodus chacensis and Simpsonodus sp.

Simpsonodus chacensis

Specimen Number	Locality	L P/3	W P/3	L P/4	W P/4	L M/1	W M/1	L M/2	W M/2	L M/3	W M/3
AMNH 15520	SW St. Joe	5.5	2.1	5.4	2.9	4.9	3.4	5.4	4.5	6.6	4.3
AMNH 15520				5.3	2.8	4.8	3.5	5.5	4.6	6.6	4.2
AMNH 48694*	Quarry 88			5.7	2.9	4.9	3.7	5.2	4.3	6.5	3.8
AMNH 48694*	GGS 1947			5.6	3.0	4.8	3.8	5.3	4.4	6.2	4.0
UCM 53920	UCM 84167			5.5	2.8	5.0		5.2	4.3		
CM 12232	CM 144			4.7	2.7						
UCM 53925	UCM 85342					5.1	3.9	5.3	4.4	6.5	4.1
UCM 53926	UCM 85344					4.6	3.7	5.9	4.6	6.8	4.1
UCM 53927	UCM 85345					5.2	3.8	6.1	4.6	6.9	4.6
CM 43133	CM 941					4.4	3.2	5.0	3.7	6.4	3.6
AMNH 55516	Mosaic 45, NM						3.4	5.1	4.1		
CM 58133	CM 680					5.3	3.9	5.8	4.5		
AMNH 92843	SW St. Joe					5.4	3.8	5.9	4.7		
AMNH 15513	Lower Graybull Valley					5.5	3.8	5.9	4.4		
UCM 44124	UCM 80066					5.0	3.5				
FMNH PM 316	FMNH 328-41					5.5	4.0				
AMNH 86289	GGS 1947							4.7	3.7	5.9	3.7
USNM 19224	Regina, NM							5.8	4.7	7.2	4.3
UCM 44130	UCM 80066							5.4	4.3	6.4	4.3
UCM 44129	UCM 80066							5.7	4.7	7.5	4.7
AMNH 15511	Near St. Joe							5.6	4.5	7.2	4.5
CM 43137	CM 941							5.5	4.3	6.3	4.0
UCM 53921	UCM 85231							6.0	4.4	6.5	4.3
UCM 53924	UCM 85342							5.5	4.4	6.7	4.2
CM 12348	CM 147							5.4	4.5		
UA 7000	UA 7400							5.4	4.0		
UCM 53923	UCM 85341							5.5	4.9		
AMNH 16297	Ojo San Jose, NM								4.2		
CM 43128	CM 941							4.8	3.7		
AMNH 237	GGS 1949									6.2	3.6
MEAN		5.5	2.1	5.4	2.9	5.0	3.7	5.5	3.4	6.6	4.1
N		1	1	6	6	14	14	25	26	17	17
SD				.36	.10	.33	.23	.36	.31	.41	.30
CV				7	4	7	6	7	7	6	8

* Designated Neotype of Simpsonodus chacensis.

Simpsonodus sp.

Specimen Number	Locality	L P/3	W P/3	L P/4	W P/4	L M/1	W M/1	L M/2	W M/2	L M/3	W M/3
WCMD 755	Kladder E	6.2	2.7	6.2	3.2	4.4	5.2				
AMNH 16778	Head, 5 mi Creek			6.0	3.3	3.6	5.6	4.3			

TABLE 3. DIMENSIONS (IN MILLIMETERS) OF UPPER MOLARS OF *Simpsonodus chacensis*.

Specimen Number	Locality	L M1/	W M1/	L M2/	W M2/	L M3/	W M3/
AMNH 48694*	Quarry 88	4.5	5.7	5.1	6.3	4.8	6.3
CM 43139	CM 941	4.4	5.6	5.1	6.5		
AMNH 15510	S. of Otto			6.0	7.2	5.2	6.7
UCM 53926	UCM 85345			5.1	6.7	5.0	5.9
FMNH PM1222	UCM 78049			5.2	6.2		
UCM 53923	UCM 85341			5.6	7.1		
MEAN		4.5	5.7	5.4	6.7	5.0	6.3
N		2	2	6	6	3	3
SD				.37	.41		
CV				7	6		

* Designated Neotype of *S. chacensis*.

anomalous cuspule supported by an extra root, and curiously, is remarkably similar (but not identical) to Cope's figure of the lost holotype. Apart from the accessory cuspule, AMNH 15520 is indistinguishable from the hypodigm of *S. chacensis*.

The crowns of M^{1-2} are slightly longer labially than lingually, and have squared corners. The pre- and postcingulae are strong and bear, respectively, a weak periconal bulge and a larger hypoconal shelf. The paracone and metacone are conical and high, the conules low, well-developed and pyramidal with strong external wings, and the protocone is central with broad protocristae. M^3 resembles M^{1-2} except for its triangular outline.

Discussion: *Simpsonodus chacensis* is larger than penecontemporaneous (late Graybullian to Lysitean) species of *Diacodexis* (*D. secans*, *D. gracillis*, and *Diacodexis* sp. cf. *D. secans*), smaller than those of *Wasatchia* (*W. grangeri*) and *Bunophorus* (*B. etsagicus*), and overlaps in size with the early Graybullian *W. robustus* and the Lysitean to Gardnerbuttean *W. pattersoni* (see below).

As in the case in *Diacodexis secans* (see Krishtalka and Stucky, 1985), *Simpsonodus chacensis* shows a moderate amount of variation in size and morphology. The smallest specimens overlap non-lithosympatric samples of *D. secans*, but are larger than penecontemporaneous *D. s.-metsiacus* and *Diacodexis* sp. cf. *D. secans*; the largest specimens of *S. chacensis* overlap in size with *Wasatchia robustus* and *W. pattersoni*. Although the variation in size is continuous, specimens from the San Juan Basin are, on average, smaller than those from the Bighorn Basin. Variable morphological features include: placement and degree of separation of the paraconid (anterior or slightly anteromedial to the metaconid); inflation of the metaconid (not inflated to slightly so); a postmetacristid (absent to weak); protocristid (absent to low and strong); postcingulid (usually broad); enamel thickness; and presence of a talonid gap lingually. As in *D. secans* and *Diacodexis* sp. cf. *D. secans* (see above), morphologic features vary in a mosaic pattern in the sample of *S. chacensis*.

Simpsonodus chacensis most closely resembles the *Diacodexis secans-primus* and *D. s.-metsiacus* lineage segments of *D. secans* (see Krishtalka and Stucky, 1985) and *Diacodexis* sp. cf. *D. secans* in structure of the molar paraconid and non-bulbous nature of the molar cusps. *Simpsonodus chacensis* lacks the derived molar features of later lineage segments of *D. secans*, such as a reduced, anteromedial paraconid, highly inflated metaconid, strong talonid notch and postmetacristid, and an elongate P_3. *Simpsonodus chacensis* also differs from *D. s.-primus* and *D. s.-metsiacus* in its larger size and in that most specimens have a less inflated molar metaconid, a more elongate talonid, a broader postcingulid, and a higher trigonid. These features also occur in *Diacodexis* sp. cf. *D. secans* from the Debeque and San Jose formations (see above), but this species is smaller than *S. chacensis* and has a more reduced molar paraconid.

Simpsonodus sp.
Figure 3; Table 2

Referred specimens: AMNH 16778 (Bighorn Basin); WCM D-755 (Piceance Basin).

Localities: AMNH loc. "Head of 5 mi. Cr." (Willwood Fm., Bighorn Basin, Wyoming). WCM loc. "Kladder E" (Debeque Fm., Piceance Basin, Colorado).

Known distribution: Early Wasatchian (late Graybullian and/or early Lysitean), Piceance and Bighorn basins.

Figure 3. Stereophotographs (occlusal views) of *Simpsonodus* sp. AMNH 16778. LP$_4$ (left), RP$_4$-M$_2$ (right). Scale in millimeters.

Discussion: The two specimens of *Simpsonodus* sp. preserve P$_3$-M$_3$ (P$_2$ on AMNH 16778 is shattered), and differ from *S. chacensis* in having a more robust protoconid on the premolars and a metaconid bulge on P$_4$. Based on these differences, we tentatively recognize two taxa, although the molars of *Simpsonodus* sp. are indistinguishable from those on the larger specimens of *S. chacensis,* and only the smaller specimens of the latter have associated premolars. A greater sample of *Simpsonodus* may support presence of two species, one of which (*Simpsonodus* sp.) has larger molars associated with more robust and molariform premolars, in which case some of the larger molars referred to *S. chacensis* may belong to *Simpsonodus* sp. Alternatively, a greater sample may indicate that more robust premolars are allometric in larger individuals of *S. chacensis,* in which case the material referred to *Simpsonodus* sp. would belong to *S. chacensis.*

Wasatchia Sinclair, 1914

The systematics of North American species of *Wasatchia (W. grangeri, W. dorseyana,* and *W. lysitensis)* and *Bunophorus (B. etsagicus, B. macropternus, B. sinclairi,* and *B. gazini)* will be treated elsewhere (Stucky and Krishtalka, *in press*). Briefly, in this paper we recognize *W. grangeri* (including *W. dorseyana*) and all species of *Bunophorus* except *B. gazini,* as junior synonyms of *B. sinclairi.* Also included in *Wasatchia* are "*Diacodexis*" *robustus* and *W. pattersoni,* a new species described below. *Wasatchia lysitensis* was transferred above to *Simpsonodus.*

Wasatchia differs from *Diacodexis* in its larger size and pronounced robusticity and bunodonty. As currently defined, *Wasatchia,* unlike *Bunophorus,* retains a well-developed paraconid on M$_{1-3}$. Other distinguishing characters cited by Sinclair (1914) are variable in species of both genera and, as Van Valen (1971) suggested, generic distinction of *Wasatchia* and *Bunophorus* may not be warranted (Stucky and Krishtalka, *in press*).

Wasatchia grangeri Sinclair, 1914
Figure 4; Table 4

Referred specimens: AMNH 16295, 48003; CM 14933, 43140; UA 7154, 7155 (all San Juan Basin). UCM 40974 (Piceance Basin).

Localities: AMNH locs. "10 mi. W. Laguna Colorado," "2 mi. E. Lindrith"; CM locs. "Almagre Rincon (Red Lake)," 945; UA loc. 7411 (all San Jose Fm., San Juan Basin, New Mexico). UCM loc. 78013 (Debeque Fm., Piceance Basin, Colorado).

Known distribution: Early Wasatchian (late Graybullian and/or early Lysitean), San Juan Basin, New Mexico, Piceance Basin, Colorado, and Bighorn Basin (Willwood Fm.), Wyoming.

Discussion: Wasatchia grangeri is larger than *W. robustus* (= *Diacodexis robustus,* Stucky and Krishtalka, *in press*), is more derived in having a vestigial paraconid on P$_4$, but is more primitive than all other species of *Wasatchia* and *Bunophorus* in retaining a prominent paraconid on M$_{1-3}$ and in having less inflated P4/4-M3/3. The material from the San Juan and Piceance basins does not differ significantly from the type

Figure 4. Stereophotographs (occlusal views) of *Wasatchia grangeri*. *A*, UCM 40974, RM$_3$; and *B*, AMNH 48003, LM$_{1-3}$. Scale in millimeters.

TABLE 4. DIMENSIONS (IN MILLIMETERS) OF LOWER TEETH OF *Wasatchia pattersoni*, sp. nov. AND *Wasatchia grangeri*.

Wasatchia pattersoni

Specimen Number	Locality	L P/3	W P/3	L P/4	W P/4	L M/1	W M/1	L M/2	W M/2	L M/3	W M/3
FMNH P26593	FMNH 147-41	6.7	3.0	6.7	4.0						
FMNH P26590*	FMNH 199-41			6.5	4.1	5.7	5.4	5.6		7.2	5.6
WCM A634	Kladder, J			6.6	3.7	5.9	5.0	6.3	5.8		
FMNH PM248	FMNH 54-C-47			7.0	3.9	5.8	5.3	6.3	5.8		
FMNH P26536	FMNH 109-41					5.7	5.0	6.3	5.7	8.2	5.7
FMNH P26636	FMNH 177-41					4.6		5.4	5.0	8.0	5.2
FMNH P26807	FMNH 372-41					4.4	4.1	4.3	4.6		
AMNH 17561	Farasita Fm.							6.4	5.6	7.7	5.5
MEAN		6.7	3.0	6.7	3.9	5.4	5.0	5.8	5.4	7.8	5.5
N		1	1	4	4	6	5	7	6	4	4
SD				.22	.17	.67	.51	.77	.50	.43	.22
CV				3	4	12	10	13	9	6	4

* Holotype of *W. pattersoni*.

Wasatchia grangeri

Specimen Number	Locality	L P/3	W P/3	L P/4	W P/4	L M/1	W M/1	L M/2	W M/2	L M/3	W M/3
AMNH 48003	Lindrith, NM					7.5	6.0	8.0	7.2	8.6	6.3
AMNH 48003								7.9	7.4	9.1	6.2
CM 14933	Red Lake, NM					7.0	5.6	7.7	6.8		
UA 7154	UA 7411					7.3	5.7	7.5	6.6		
CM 43140	CM 945							8.0	6.9		
UCM 40974	UCM 78013									8.5	6.5
UA 7155	UA 7411									9.4	6.3
MEAN						7.3	5.8	7.8	7.0	8.9	6.3
N						3	3	5	5	4	4
SD								.22	.32	.42	.13
CV								3	5	5	2

(AMNH 15516), and large collections of *W. grangeri* (= *W. dorseyana*, Stucky and Krishtalka, *in press*) from the late Graybullian of the Bighorn Basin.

Wasatchia grangeri was first recovered from the San Juan Basin in 1877 by Cope, who attributed the specimen to a new species of *Sarcolemur*, *S. crassus* (see Cope, 1877, p. 149, Pl. 45, Fig. 16), which he later (1884, p. 233) removed from *Sarcolemur* and relegated to paleontological obscurity. "*Sarcolemur crassus*" was last mentioned by Trouessart (1904, *fide* Hay, 1930), and is thus a *nomen oblitum*. Cope's specimen, apparently deposited in the USNM collections (Osborn, 1902), is now lost. The only other occurrences of *W. grangeri* in the San Jose Formation were recorded by Sinclair (1914; as *W. dorseyana*) and Lucas and others (1981; as *Bunophorus dorseyanus* and *B. grangeri*).

Kihm (1984) noted the presence of *Wasatchia* sp. in the Debeque Formation, but we have not examined his referred material. The specimen assigned here to *W. grangeri* (UCM 40974) was part of the hypodigm of Kihm's (1984) *Bunophorus* sp. cf. *B. macropternus*, the remainder of which is referred below either to *W. pattersoni*, sp. nov. or *B. sinclairi*.

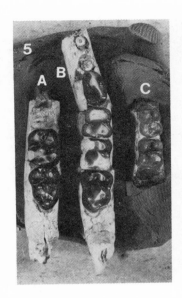

Figure 5. Stereophotographs (occlusal views) of *Wasatchia pattersoni*. A, AMNH 17561, LM_{2-3}; B, FMNH-P 26590, holotype, RP_4-M_3; and C, FMNH-P 26807, RM_{1-2}. Scale in millimeters.

Wasatchia pattersoni, new species
Figure 5; Table 4

Etymology: In honor of Bryan Patterson for his contributions to paleontology of the Piceance Basin, Colorado.

Holotype: FMNH-P 26590, RP_4-M_3.

Type locality: FMNH 163-41 (= FMNH 199-41 in Kihm, 1984, p. 285), Debeque Formation, Roan Cliffs, Garfield County, Colorado.

Referred specimens: FMNH-P 26636, 26536, 26593, 26807, PM 360, PM 248; WCM A-634 (all from the Piceance Basin). AMNH 17561 (Farisita Formation, Huerfano Basin, Colorado).

Localities: Lysitean — FMNH locs. 372-41, 109-41; Lostcabinian — FMNH locs. 54-C47, 166-41, 209-41; WCM loc. "Kladder J" (all Piceance Basin). Gardnerbuttean — AMNH loc. "5 mi. N. Gardner Butte, Colorado" (Huerfano Basin).

Known distribution: Middle to late Wasatchian (Lysitean to Lostcabinian), Piceance Basin (Debeque Fm.), Colorado; earliest Bridgerian (Gardnerbuttean), Huerfano Basin (Farisita Fm.), Colorado.

Diagnosis: Unlike *Wasatchia robustus*, P_4 with vestigial paraconid; unlike *W. robustus* and *W. grangeri*, M_{1-3} with reduced paraconid and P_4-M_3 inflated; similar in size to *W. robustus*, smaller than *W. grangeri* and all species of *Bunophorus*.

Description: The material referred here preserves P_3-M_3. P_3 (FMNH-P 26593 only), slightly longer than P_4, bears a vestigial, basal paraconid, a robust protoconid, and a short, unicuspid talonid. On P_4 the paraconid is slightly larger (although still vestigial), the protoconid more bulbous, and the talonid larger, with three or four barely discernable cuspules. All of the molar cusps are bunodont and inflated at the base. M_{1-2} are nearly square, with a slightly lower and wider talonid than trigonid, a tiny paraconid directly anterior or anteromedial to the metaconid, a well developed medial or barely labial cristid obliqua, and a slightly lingual hypoconulid. On M_3 the talonid is narrower than the trigonid, and tapers posteriorly.

Discussion: P_3-M_3 in *Wasatchia pattersoni*, as in all species of *Wasatchia* and *Bunophorus*, are much more robust and inflated than in *Diacodexis*. The talonid and paraconid on P_4, unlike those of *W. robustus*, are reduced; compared to *W. robustus* and *W. grangeri*, the molars are proportionately more inflated and have a reduced paraconid and a slightly larger metaconid. *Wasatchia pattersoni* is smaller than species of *Bunophorus*, and has a less robust P_4 and a somewhat larger molar paraconid on most specimens. As such, *W. pattersoni* is the most derived species of *Wasatchia*, but is more primitive than known species of *Bunophorus*, all of which are further derived in a number of dental features (Stucky and Krishtalka, *in press*). Known dental morphology of *W. pattersoni* is structurally suitable to characterize the common ancestor of species of *Bunophorus*.

Wasatchia pattersoni is penecontemporaneous with *Bunophorus macropternus* and *B. etsagicus* in the Lysitean, and with *B. etsagicus* and *B. sinclairi* in the Lostcabinian and Gardnerbuttean. Throughout its biostratigraphic range (Lysitean to Gardnerbuttean), *W. pattersoni* shows morphologic stasis.

Bunophorus Sinclair, 1914

The systematics of *Bunophorus* will be treated elsewhere (Stucky and Krishtalka, *in press*). Briefly, in this paper we recognize three species: *B. etsagicus*, the type species; *B. macropternus*; and *B. sinclairi* (= *B. gazini* Guthrie, 1971).

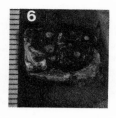 

Figure 6. Stereophotographs (occlusal view) of *Bunophorus sinclairi.* FMNH-P 26792. LM$^{2-3}$. Scale in millimeters.

Bunophorus sinclairi Guthrie, 1966
Figure 6

Referred specimen: FMNH-P 26792.

Locality: FMNH 369-41 (Lostcabinian), Debeque Fm., Piceance Basin, Colorado.

Known distribution: Late Wasatchian to early Bridgerian (Lostcabinian to Gardnerbuttean), Wind River Basin (Wind River Fm.), Wyoming; late Wasatchian (Lostcabinian), Green River Basin (Wasatch Fm.), Wyoming and Piceance Basin (Debeque Fm.), Colorado.

Discussion: *Bunophorus sinclairi* is most derived among known species of *Bunophorus* in having a bulbous, well developed hypocone on M$^{1-2}$ and a vestigial (or absent) cristid obliqua on M$_{1-3}$. This species was named and then synonymized to *B. etsagicus* by Guthrie (1966, 1971). Our studies (Stucky and Krishtalka, *in press*) indicate that the two species shared an immediate common ancestry and, although similar in size, are morphologically distinct. *Bunophorus etsagicus* is more primitive in having a better developed cristid obliqua on the lower molars and a weaker hypocone on M$^{1-2}$.

Kihm (1984) assigned this Piceance Basin specimen to *Bunophorus* sp. cf. *B. macropternus,* but FMNH-P 26792 (M^2: L, 7.1 mm; W, 10.0 mm. M^3: L, 7.0 mm; W, 9.6 mm) is larger than *B. macropternus,* has a bulbous hypocone on M^2, and lacks the molar hyperinflation of the latter. Other specimens identified by Kihm (1984) as *Bunophorus* sp. cf. *B. macropternus* were referred above to *Wasatchia grangeri* (UCM 40974) and *W. pattersoni* (FMNH-P 26590, 26593, PM 248, WCM A-634).

Family *Homacodontidae* Gazin, 1955
Hexacodus Gazin, 1952

The first unequivocal occurrence of *Hexacodus,* the most primitive known homacodontid, is in Lysitean horizons in southwestern Wyoming (Gazin, 1962) and in the Wind River Basin (Krishtalka and Stucky, *unpublished information*). Gazin's (1962) citation of Graybullian *Hexacodus (H. uintensis)* in the Bighorn Basin is unwarranted; the specimen (AMNH 4140) was collected for Cope by Wortman in the 1880s, and lacks specific locality data.

The two recognized species of *Hexacodus* (*i. e.,* the type species, *H. pelodes* and *H. uintensis*) were synonymized and placed in *Protodichobune* by Van Valen (1971), actions with which we disagree. A study of a larger sample of *Hexacodus* (according to Krishtalka and Stucky, *unpublished information*) indicates that the type specimen of *Protodichobune oweni* is generically distinct from *Hexacodus* in the bulbous nature of the molar cusps, quadrate molar crowns, crennulate enamel, reduced hypoconulid on M$_3$, large accessory cuspule in the hypoflexid on M$_{1-3}$, and unique pattern of the hypocristid complex (a neomorphic ridge extends between the entoconid and hypoconid) on M$_{1-2}$.

The two species of *Hexacodus* also appear to be distinct. *Hexacodus pelodes* is more derived in reduction of the paraconid on M$_2$ and its more medial occurrence on M$_{1-2}$. On the type of *H. uintensis,* the paraconid on M$_{1-2}$ is well developed and directly anterior to the metaconid.

Hexacodus pelodes Gazin, 1952
Figure 7

Referred specimen: FMNH PM 289.

Locality: FMNH loc. 98C-47, Gulch north of Rifle Golf Course, Debeque Fm., Piceance Basin, Colorado.

Known distribution: Middle to late Wasatchian (Lysitean to Lostcabinian)—Wind River Basin (Wind River Fm.), Green River Basin (Wasatch Fm.), Wyoming; Wasatchian (subage uncertain)—Piceance Basin (Debeque Fm.), Colorado, Bighorn Basin (Willwood Fm.), Wyoming.

Discussion: PM 289 (an M$_2$: L, 4.6 mm; W, 3.4 mm), originally identified as *Diacodexis metsiacus* (see Kihm, 1984), is the first record of *Hexacodus* from the Debeque Formation, Piceance Basin, Colorado, and bears the characters noted above for *H. pelodes.* Unlike most diacodexeids, the molar entoconid is high and conical, metaconid is unexpanded, hypoconulid is lingual to the midline on a broad postcingulum, hypocristid is strong and directed toward the posterolingual part of the base of the entoconid before deflecting to the hypoconulid, and paraconid/metaconid wall is elevated. On some specimens of *H. pelodes* (including PM 289) and on the type of *H. uintensis* (PU 16175) a neomorphic ridge or bulge connects the hypocristid to the base of the entoconid. These features define the origin of the homacodontid clade, and are further developed in more advanced middle Eocene (*Homacodon, Microsus*) and late Eocene descendants (Gazin, 1955; Krishtalka and Stucky, *unpublished information*). A number of these derived features also occur (albeit to a lesser degree) in some specimens of *Diacodexis* sp. cf. *D. secans* described above, and the latter appears to be structurally (and perhaps phylogenetically) intermediate between *D. secans* and the homacodontid-antiacodontid clade.

The radiation of early homacodontids, beginning with *Hexacodus,* may have occurred on the western front of the central Rocky Mountains. Evidence here is the disproportionate frequency of occurrence of late Graybullian to Lostcabinian *Hexacurrence of late Graybullian to Lostcabinian Hexacodus* and *Hexacodus*-like *Diacodexis* sp. cf. *D. secans* on the western side of the Rocky Mountains compared to that of orthodox *Diacodexis* (*i. e., D. secans* and relatives; Krishtalka and Stucky, 1985) along the eastern front in Wyoming and

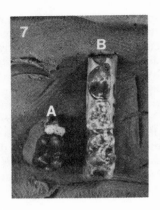 

Figure 7. Stereophotographs (occlusal views) of *Hexacodus pelodes*. *A*, FMNH PM 289, LM$_2$; and *B*, USNM 19215, holotype, LP$_4$-M$_2$. Scale in millimeters.

Colorado. The fossil record of these taxa seems rich enough spatially and stratigraphically to negate biases of preservation and collecting.

The age of PM 289 cannot as yet be pinpointed within the Wasatchian. Its provenance ("Gulch north of the Rifle Golf Course") cannot be relocated precisely (Kihm, 1984), and the fauna from that locality (cf. *Tetonius, Cantius* sp. cf. *C. trigonodus, Meniscotherium chamense, Homogalax protapirinus, Homogalax* small sp., and *Hyracotherium* sp. A and sp. B) can range biostratigraphically from the middle Graybullian through the Lysitean, and could occur in the Lostcabinian. From Kihm's (1984) discussion of these taxa it is evident that his assignment of mid Graybullian (= his "upper early Wasatchian") age to PM 289 is premature.

CONCLUSIONS

Three species of diacodexeids (*Diacodexis* sp. cf. *D. secans, Simpsonodus chacensis,* and *Wasatchia grangeri*) occur in the San Jose Formation, San Juan Basin, New Mexico; these three species, as well as *Simpsonodus* sp., *Wasatchia pattersoni, Bunophorus sinclairi,* and *Hexacodus pelodes* are recorded from the Debeque Formation, Piceance Basin, Colorado.

Simpsonodus, a new genus, includes *Diacodexis chacensis* and *Wasatchia lysitensis. Simpsonodus* sp., from the Piceance and Bighorn basins, is tentatively separated from *S. chacensis* until a larger sample is available to indicate whether the two taxa are distinct or parts of the variation of *S. chacensis. Wasatchia pattersoni* is a new species that is morphologically intermediate between *W. grangeri* and all species of *Bunophorus. Diacodexis* sp. cf. *D. secans* is either a geographic variant of *D. secans* in the San Juan and Piceance basins, or a discrete species. The mosaic distribution of primitive and derived features in *Diacodexis* sp. cf. *D. secans* and its degree of overlap with *D. secans* do not allow a finer taxonomic resolution—the welcome paradox of a fossil record good enough to lack appreciable gaps and preserve apparently continuous variation (and change or stasis) across time and space. The nature of the derived characters in *Diacodexis* sp. cf. *D. secans* implies that this taxon is near the origin of homacodontids and antiacodontids.

Significantly, the fossil record of diacodexeids from New Mexico, Colorado, and the western interior is sufficiently robust to test evolutionary models and to indicate ancestor-descendant relationships at the species level, which in some instances represent the origin of higher taxa. For example, the artiodactyl record implies the origin of the Diacodexeidae and Leptochoeridae from *Diacodexis secans* (lineage segment *D. s.-metsiacus*), and Antiacodontidae, Homacodontidae, and selenodont artiodactyls from *Diacodexis* sp. cf. *D. secans*. Whether these two ancestral taxa are considered separate species or geographic variants of one species (see above), they show continuous morphologic overlap at one end of their respective ranges of variation and divergence at the other end in the frequency and degree of expression of derived (and different) features; here, these two taxa approach the morphology of two different monophyletic clades (other diacodexeids/leptochoerids; antiacodontids/homacodontids). As such, the known record of *D. secans* (lineage segment *D. s.-metsiacus;* see Krishtalka and Stucky, 1985) and *Diacodexis* sp. cf. *D. secans* appears to preserve both their initial divergence as well as the subsequent origin of descendant clades. Both events involved gradual change, and both branches show equal and gradational morphological divergence from a single species. This pattern, as well as the fossil record of apparently continuous spatial and temporal variation in *D. secans* (see Krishtalka and Stucky, 1985), belie notions of punctuation and stasis in these taxa. The evidence here reaffirms Simpson's view of the primacy of microevolutionary processes in producing macroevolutionary events.

ACKNOWLEDGMENTS

We thank M. C. McKenna and J. P. Alexander (AMNH), R. Emry and R. Purdy (USNM), E. Lindsay (UA), P. Robinson, and R. Bakker (UCM), D. Baird (PU) and W. Turnbull (FMNH) for the loan of specimens in their care. This study was supported in part by grants from M. Graham Netting Research Fund (CM) and NSF grant BSR-8402051.

REFERENCES CITED

Cope, E. D., 1875, Systematic catalogue of Vertebrata of the Eocene of New Mexico, collected in 1874: Report of the Engineers Department, U.S. Army, in charge of Lieut. Geo. M. Wheeler, Washington, p. 5-37.

—— 1877, Report upon the extinct Vertebrata obtained in New Mexico by parties of the expedition of 1874: Geographical Survey West of the 100th Meridian, U.S. Corps of Engineers, p. 1-370

—— 1882, Two new genera of Mammalia from the Wasatch Eocene: American Naturalist, v. 16, p. 1029.

—— 1884, The Vertebrata of the Tertiary formations of the West: Report of the U.S. Geological Survey of the Territories, v. 3, 1009 p.

Froelich, J. W., and Reser, P. K., 1981, First occurrence of *Homogalax* (Mammalia; Perissodactyla; Tapiroidea) in the Regina Member of the San Jose Formation, San Juan Basin, New Mexico, *in* Lucas, S. G., Rigby, J. K., and Kues, B. S., eds., Advances in San Juan Basin paleontology: Albuquerque, University of New Mexico Press, p. 293-303.

Gazin, C. L., 1952, The Lower Eocene Knight Formation of western Wyoming and its mammalian faunas: Smithsonian Miscellaneous Collections, v. 117, no. 18, 82 p.

—— 1955, A review of the Upper Eocene Artiodactyla of North America: *ibid.*, v. 128, no. 8, *iii* + 96 p.

—— 1962, A further study of the Lower Eocene mammalian faunas of southwestern Wyoming: *ibid.*, v. 144, no. 1, *v* + 98 p.

Gingerich, P. D., and Simons, E. L., 1977, Systematics, phylogeny, and evolution of early Eocene Adapidae (Mammalia, Primates) in North America: University of Michigan, Museum of Paleontology, Contributions, v. 24, p. 245-279.

Granger, W., 1914, On the names of Lower Eocene faunal horizons of Wyoming and New Mexico: American Museum of Natural History, Bulletin, v. 33, p. 201-207.

Guthrie, D. A., 1966, A new species of dichobunid artiodactyl from the early Eocene of Wyoming: Journal of Mammalogy, v. 47, p. 487-490.

—— 1971, The mammalian fauna of the Lost Cabin Member, Wind River Formation (Lower Eocene) of Wyoming: Annals of Carnegie Museum, v. 43, p. 47-113.

Hay, O. P., 1930, Second bibliography and catalogue of the fossil Vertebrata of North America: Carnegie Institution of Washington, Publication No. 390, 916 p. (volume 1), 1074 p. (volume 2).

Kihm, A. J., 1984, Early Eocene mammalian faunas of the Piceance Creek Basin, northwestern Colorado [Ph.D. thesis]: Boulder, Colorado, University of Colorado, 381 p.

Krishtalka, L., and Stucky, R. K., 1985, Revision of the Wind River faunas, early Eocene of central Wyoming. Part 7. Revision of *Diacodexis* (Mammalia, Artiodactyla): Annals of Carnegie Museum, v. 54, p. 413-486.

Lucas, S. G., 1977, Vertebrate paleontology of the San Jose Formation, east-central San Juan Basin, New Mexico: New Mexico Geological Society, 28th Field Conference Guidebook, p. 221-225.

Lucas, S. G., Schoch, R. M., Manning, E., and Tsentas, C., 1981, The Eocene biostratigraphy of New Mexico: Geological Society of America, Bulletin, v. 92, p. 951-967.

Matthew, W. D., 1899, A provisional classification of the fresh-water Tertiary of the West: American Museum of Natural History, Bulletin, v. 12, p. 19-75.

Osborn, H. F., 1902, American Eocene primates and the supposed rodent Family Mixodectidae: *ibid.*, v. 16, p. 169-214.

Schankler, D. M., 1980, Faunal zonation of the Willwood Formation in the central Bighorn Basin, Wyoming, *in* Gingerich, P. D., ed., Early Cenozoic paleontology and stratigraphy of the Bighorn Basin, Wyoming: University of Michigan, Papers in Paleontology, v. 24, p. 99-114.

Simpson, G. G., 1948, The Eocene of the San Juan Basin, New Mexico: American Journal of Science, v. 246, p. 257-282, 363-385.

—— 1981, History of vertebrate paleontology in the San Juan Basin, *in* Lucas, S. G., Rigby, J. K., and Kues, B. S., eds., Advances in San Juan Basin paleontology: Albuquerque, University of New Mexico Press, p. 3-25.

Sinclair, W. J., 1914, A revision of the bunodont Artiodactyla of the middle and lower Eocene of North America: American Museum of Natural History, Bulletin, v. 23, p. 267-295.

Stucky, R. K., 1984, The Wasatchian-Bridgerian land mammal age boundary (early to middle Eocene) in western North America: Annals of Carnegie Museum, v. 53, p. 347-382.

Stucky, R. K., and Krishtalka, L., *in press*, Revision of the Wind River faunas, early Eocene of central Wyoming. Part 8. Revision of *Simpsonodus* and *Bunophorus* (Mammalia, Artiodactyla): *ibid.*

Trouessart, E. L., 1904, Catalogus Mammalium tam vivendum quam fossilium: Berolini, Quinquincale supplementum, 546 p.

Van Houten, F., 1945, Review of latest Paleocene and early Eocene mammalian faunas: Journal of Paleontology, v. 19, p. 421-461.

Van Valen, L., 1971, Toward the origin of artiodactyls: Evolution, v. 25, p. 523-529.

West, R. M., McKenna, M. C., Krishtalka, L., Stucky, R. K., Black, C. C., Bown, T. M., Dawson, M. R., Golz, D. J., Lillegraven, J. A., and Turnbull, W. D., *in press,* Eocene biochronology of North America, *in* Woodburne, M. O., ed., Cenozoic biochronology of North America: Berkeley, University of California Press.

MANUSCRIPT RECEIVED DECEMBER 16, 1985
REVISED MANUSCRIPT RECEIVED MAY 2, 1986
MANUSCRIPT ACCEPTED JUNE 9, 1986

Early Eocene rodents from the San Jose Formation, San Juan Basin, New Mexico

KATHRYN M. FLANAGAN *Department of Geology and Geophysics, The University of Wyoming, Laramie, Wyoming 82071*

ABSTRACT

A rodent fauna was collected from three localities in the Almagre facies of the San Jose Formation, San Juan Basin, New Mexico. The fauna contains approximately 350 isolated rodent teeth, predominantly from the Sciuravidae. The taxa include *Knightomys depressus, Pauromys sp., Lophiparamys debequensis, Paramys copei, Paramys excavatus, Mattimys kalicola,* and two new species *(Apatosciuravus jacobsi* and *Knightomys* cf. *K. minor). Microparamys reginensis* has been placed in the genus *Knightomys* based on discovery of the upper dentition and additional specimens of lower dentition. This is the first record of *Mattimys kalicola, Lophiparamys debequensis,* and *Pauromys* in the San Jose Formation.

This rodent fauna is early Eocene (Wasatchian Land Mammal Age), and correlates closely to the Lysitean "subage" based on presence of *Lophiparamys debequensis* and *Knightomys* cf. *K. minor.*

INTRODUCTION

The small mammal fauna of the San Jose Formation yields a sizeable collection of rodents, insectivores, primates, and a multituberculate. This paper discusses the rodent fauna of three localities of the Almagre facies of Simpson (1948) or Regina Member of Baltz (1967), San Jose Formation, San Juan Basin, New Mexico.

All three localities appear to be on the same stratigraphic level within the Almagre facies (Haskin, 1980). UALP locality 7747 is found in a fairly indurated, buff sandstone lens within brightly colored varigated clays; the locality has been tied stratigraphically, following tracer beds, to UALP Locality 7745 (across the highway on an adjoining rincon). UALP 7745 also is located in a buff sandstone lens, though the lenticular nature of the sands of the Almagre facies does not display clear-cut stratigraphic relationships. By following distinct tracer or marker beds, however, correlation is possible. UALP Locality 7746 appears to be approximately 31 m (100 ft) higher than 7745 and 7747, on another rincon, and in the variegated clay layers (Fig. 1).

The San Jose faunas are of the Wasatchian Land Mammal Age, but their position within the Wasatchian remains a question. The perissodactyls which characterize the Graybullian, Lysitean, and Lostcabinian subages are not found in the San Jose Formation. By comparing early Eocene rodent faunas from New Mexico to rodent faunas from the Wasatchian of Wyoming, it may be possible to assign a subage to the San Jose faunas.

Three families of rodents were collected from the localities studied, the Ischyromyidae, ?Eutypomyidae, and Sciuravidae. Of the Ischyromyidae, three genera *(Paramys, Lophiparamys,* and *Apatosciuravus)* and four species were collected. Of the Sciuravidae, two genera *(Knightomys* and *Pauromys)* and four species were found. A new species of *Apatosciuravus* and a new species of *Knightomys* are described. *Mattimys kalicola* is the only taxon of the ?Eutypomyidae.

The record of evolution and radiation of early Cenozoic rodents is fragmentary. Rodents evolved in Asia and subsequently dispersed to North America during the Paleocene (Korth, 1984). By early Eocene, the rodents found throughout deposits in the Rocky Mountain region had radiated and diversified. Although large rodents both from the northern basins (Bighorn and Wind River Basins) and the San Juan Basin are similiar, comparative material has been lacking for the smaller rodents, especially sciuravids and small ischyromyids. The present addition to the rodent fauna of the San Jose Formation may shed some light on stage of evolution of the small rodents, and have interesting biogeographical implications.

MATERIALS AND METHODS

UALP 7745 and 7747 are located in T. 23 N., R. 1 W., of the Regina Quadrangle, Sandoval County, New Mexico. They are found within a buff colored sandstone lens of the Almagre facies (Simpson, 1948) or the Regina Member of Baltz (1967). The region was prospected in 1973 by L. L. Jacobs and J. H. Honey; subsequently, it was collected in 1977 by the UALP crew, under direction of E. H. Lindsay. The collection was part of the field work of Richard Haskin (1980), which included mapping and magnetostratigraphy.

The matrix was collected and brought back to the University of Arizona Laboratory of Paleontology in Tucson, Arizona. It was then soaked in water and screen washed. Heavy liquid separation with tetrabromomethane was used to separate fossil material from the

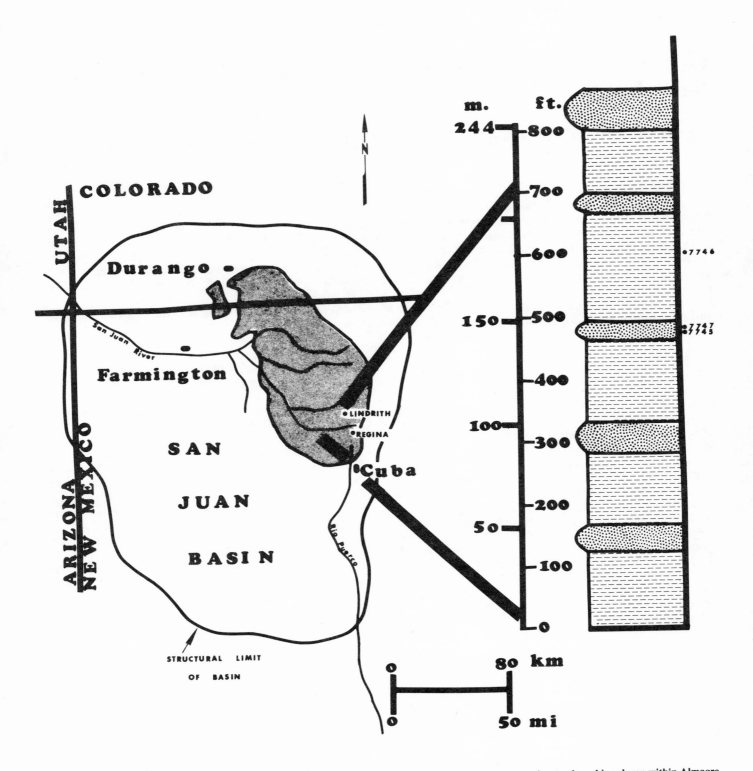

Figure 1. Regional map of San Juan Basin, showing San Jose Formation in shaded area and a composite stratigraphic column within Almagre facies of San Jose Formation. Stratigraphic occurrences of fossil localities are shown to right of column (modified from Haskin, 1980).

predominantly quartz concentrate, then hand picked under a microscope.

The fossil material included isolated teeth and skeletal fragments of rodents, multituberculates, and insectivores. Specimens of isolated teeth were mounted on pins and measured with a binocular microscope at The University of Arizona and again with an EPOI Shopscope at The University of Wyoming (Lillegraven and Bieber, 1986).

No jaw material was found, thus indirect means were used to associate composite dentitions. Clustering of common morphs and size of teeth proved to be the most trustworthy diagnostic methods.

The classification of the early Cenozoic rodents by Korth (1984) is followed in this paper. Tooth measurements and terminology follow those of A. E. Wood (1962, p. 6-9). All dental measurements are in millimeters, to 3 decimal places.

Abbreviations used in this paper are: ACM, Pratt Museum, Amherst College; AMNH, American Museum of Natural History; CM, Carnegie Museum; FMNH, Field Museum of Natural History; UALP, University of Arizona Laboratory of Paleontology (if specimen numbers are unspecified, they are UALP); N, number; OR, observed range; $\overline{X}$, mean; S, standard deviation; V, coefficient of variation; A-P, anterior-posterior length; TRA, transverse anterior width; TRP, transverse posterior width.

SYSTEMATIC PALEONTOLOGY

Order RODENTIA Bowdich, 1821
Suborder SCIUROMORPHA Brandt, 1855
Family ISCHYROMYIDAE Alston, 1876
Subfamily PARAMYINAE (Simpson, 1945)
Paramys Leidy, 1871
Paramys copei Loomis, 1907
Table 1

Type specimen: AMNH 4755, skull with lower jaws and postcranial material.

Referred specimens: UALP Locality 7745-10234, left M2/; UALP Locality 7746-11223, right M/1; 11229, right M/2; UALP Locality 7747-16338, right dP4/; 11212, left P4/; 13359, 13377, left M1/; 16334, right M2/; 16330, right P/4; 16329, left P/4; 16335, right M/1; 16339, left M/1; 11214, 15289, left M/3.

Location: UALP Localities 7745, 7746, 7747,

TABLE 1. MEASUREMENTS (IN MILLIMETERS) OF TEETH OF *Paramys copei*.

Specimen number	Element	A/P	TRA	TRP
16338	Rt. dP4/	?	3.015	?
11212	Lt. P4/	3.236	3.678	3.634
13359	Lt. M1/	3.390	3.825	3.436
13377	Lt. M1/	3.091	3.500	3.300
16334	Rt. M2/	3.295	3.626	3.300
10234	Lt. M2/	3.157	3.700	3.406
16330	Rt. P/4	3.136	2.229	2.628
16329	Lt. P/4	3.170	2.366	2.845
11223	Rt. M/1	3.122	2.674	3.005
16339	Lt. M/1	3.205	2.540	2.836
11229	Rt. M/2	3.654	3.475	3.574
11214	Lt. M/3	3.942	2.950	3.069
15289	Lt. M/3	4.000	3.043	2.935

MEASUREMENTS (IN MILLIMETERS) OF TEETH OF *Paramys excavatus*.

16341	Rt. dP4/	2.000	2.064	2.193
11333	Lt. P/4	2.386	?	2.144

Almagre facies, San Jose Formation, Sandoval County, New Mexico.

Age: Early Eocene (Wasatchian Land Mammal Age).

Description: Fourteen isolated teeth from three localities in the San Jose Formation were collected. Many of the specimens are badly worn, but identification is still possible. The teeth are placed in *Paramys copei* on the basis of size and cusp morphology.

dP4/—The dP4/ is broken but a small paracone, larger protocone, and minute hypocone are still present. The anterior cingulum joins the anterolabial edge of the paracone to anterior apex of the protocone. There is a small cuspule adjacent to the paracone, in the valley between the anterior cingulum and the talon basin. The protoconule is small but distinct, resting midway between the paracone and protocone.

M1/—Two isolated first upper molars have worn protocones and hypocones. There is no separation between the minute hypocone and large protocone. The anterior cingulum joins the anterolabial edge of the paracone to the anterolingual edge of the protocone, forming a small valley. The paracone is large relative to a smaller metacone. A tiny mesostyle is present. There is a small cuspule on the labial margin of the anterior cingulum. A large metaconule is present midway between the metacone and protocone, but it has been heavily worn on 1 of 2 specimens. A tiny double protoconule is slightly anterior and midway between the paracone and protocone on 1 of 2 teeth. Faint crenulations are present in the basin on 1 of 2 teeth.

M2/—The M2/ is similar to the M1/ in the size of the anterior cingulum and well developed posterior cingulum. The M2/ differs from M1/ in the: (1) position of the hypocone; (2) development of the mesostyle; and (3) more pronounced conules. The hypocone on M1/ is located directly adjacent to the protocone, whereas the hypocone on the M2/ is positioned slightly posterolabially to the protocone. The M2/ has a less quadrate shape than M1/. The mesostyle is well developed, and there is a double mesostyle on 1 (10234) of 2 specimens. There is evidence for a large metaconule on all specimens, though the metaconule on 10234 is indistinct due to wear. The metaconule rests midway between the metacone and protocone in the talon basin, and there is no protoconule on either teeth. Weak lophs converge on the protocone from the metacone and paracone.

P/4—The two heavily worn isolated premolars have a narrow trigonid relative to the much wider talonid. No anterior cingulid is present on these specimens. They possess a large metaconid and hypoconid relative to the smaller protoconid and entoconid. The hypoconulid is indistinct due to wear on 16330, and it is minute on 16329. A tiny mesoconid is present.

M/1—The trigonid of M/1 is small and elevated above the talonid. An anterior cingulum, connecting the anterolabial edge of the metaconid to anterior the apex of the protoconid, closes the anterior trigonid. A small ridge extending labially from the protoconid and ascending the metaconid closes the posterior trigonid on 1 (16339) of 2 specimens. One (16339) of 2 teeth shows faint crenulations in the talonid basin. The metaconid is the highest cusp on the tooth. The entoconid and protoconid are small relative to a larger hypoconid. The mesoconid is minute and an ectolophid is present. There is a continuous posterolophid with an indistinct hypoconulid.

M/2—The metaconid is the largest cusp on the tooth, positioned anterolingually to the protoconid, giving the tooth a rhomboidal shape. The protoconid and hypoconid are large relative to a smaller entoconid. The hypoconulid is indistinct. The posterolophid joins the entoconid to the hypoconid with no separation. There is a small valley on the trigonid between the anterior cingulid and the protoconid. The anterior cingulid runs from the base of the metaconid to the base of the protoconid. A small mesoconid is joined to the entoconid and hypoconid by an ectolophid.

M/3—As on the other lower molars, the metaconid is the largest cusp, and the protoconid and hypoconid are large relative to the smaller entoconid. The hypoconulid is indistinct and a small mesoconid is present. The elongate, noncuspate posterolophid joins the entoconid to the hypoconid. The talonid basin is wide and shallow and the trigonid basin is minute.

Discussion: Paramys copei was described previously from the San Jose Formation (Cope, 1877; Wood, 1962; Korth, 1984). It has been found in both the Largo and Almagre facies of the San Jose, and throughout Wasatchian deposits of North America.

Unworn teeth show crenulations in the basins, the mesostyles are large, and the M/2 is the largest tooth in area. These characters, including their size range, are the basis for their placement in *P. copei*.

Paramys excavatus Loomis, 1907
Table 1

Type specimen: ACM 327, right jaw with P/4 - M/1.

Referred specimens: Locality 7745—11333, left P/4. Locality 7747—16341, right dP4/.

Location: Almagre facies, San Jose Formation, Sandoval County, New Mexico.

Age: Early Eocene (Wasatchian Land Mammal Age).

Description: dP4/—The dP4/ is very small. The anterior cingulum is wide at the anterolabial edge of the tooth where it joins the paracone. It thins to a small ridge at the anterolingual edge of the protocone, terminating in a cuspule. There are three distinct cusps. The hypocone is just a minor swelling on the posteroloph. The paracone and metacone are equal in size, and small relative to the large protocone. The metaconule and protoconule are distinct, both converging on the protocone. The presence of a mesostyle is unknown due to breakage.

P/4—The trigonid of the P/4 is broken, but the protoconid is large relative to the entoconid and is joined by the continuous posterolophid. There is no evidence of a hypoconulid. An ectolophid connects a tiny mesoconid to the hypoconid.

Discussion: Paramys excavatus is a small paramyid, previously known from the San Jose Formation (Cope,

1877; Wood, 1962, Korth, 1984). Teeth used in the present study are badly worn, and identification is tentative based on size, small trigonid, and relative lack of crenulations.

Subfamily REITHROPARAMYINAE
Lophiparamys Wood, 1962
Lophiparamys debequensis Wood, 1962
Figure 2, Table 2

Type specimen: FMNH PM 1217, LM/1.

Referred specimens: UALP Locality 7745—15725, right M1/; 11304, 11311, 11319, right M2/; 11305, left M2/; 11302, left M3/; 11330, 16343, left M/2; 11350, 11584, 15776, left M/3.

UALP Locality 7747—15847, right P4/; 15877, left P4/; 13358, 15768, left M1/; 15772, 15773, 15840, 15876, left M3/; 13366, 15766, 15767, 15878, right P/4; 15769, 15770, right M/1; 15879, 17031, left M/1; 11213, left M/2; 13367, 15774, 15775, right M/3; 15771, 15807, left M/3.

Location: UALP Localities 7745 and 7747, Almagre facies, San Jose Formation, Sandoval County, New Mexico.

Age: Early Eocene (Wasatchian Land Mammal Age).

Emended diagnosis: A small rodent with highly crenulated upper and lower teeth; minute, distinct hypocone on P4/-M2/; upper teeth with small accessory lophs; cuspule on anterior cingulum; metaloph, protoloph, and loph from anterior cingulum converge on the protocone. The anterior cingulum is continuous with the protocone on P4/-M2/, but separated on M3/. The lower teeth have: an indistinct hypoconulid; no separation between the entoconid and posterolophid; small, incomplete hypolophid; and separation of the anterior cingulid from the protoconid.

Description: P4/—The P4/ is anteroposteriorly shorter than the molars, but highly crenulated. The paracone is large relative to a minute metacone. The hypocone, located on the lingual posteroloph, is small but distinct and separated from the larger protocone. The metaloph is low and discontinuous, converging on the protocone. The anterior cingulum bifurcates midway, at the base of the paracone, with the cingulum continuing anterolingually while the other loph joins the protocone. An arm of the paracone is directed lingually into the talon basin, but terminates before joining the loph uniting the anterior cingulum and protocone. There is a minute mesostyle.

M1/—The M1/ is crenulated, with non-parallel ridges. A weak metaloph converges on the protocone. As on the P4/, an arm of the paracone is directed lingually into the talon basin, but remains separate from a small loph directed anterolabially from the protocone. The protoloph terminates just before joining the anterior cingulum midway. The anterior cingulum forms a low ridge from the labial edge of the paracone to the anterolingual edge of the protocone, terminating in a small cuspule. The hypocone is small relative to the large, round protocone. The protocone and hypocone are connected by a thin ridge. A mesostyle is present.

M2/—The crenulated M2/ closely resembles M1/ except for: (1) the mesostyle is larger on the M2/ and, on 1 of 4 specimens, joins a short loph directed lingually into the talon basin; and (2) the hypocone, though still connected to the protocone on 2 of 4 teeth as on the M1/, has shifted to a slightly more posterolabial position on the posteroloph. The lingually directed arm of the paracone forms a weak connection with the protoloph on 3 of 4 specimens. The metaloph joins the protocone, but a small weak loph extends from the metaloph and connects with the hypocone, on unworn specimens.

M3/—The M3/ is smaller than the first two molars, and has a more triangular shape due to loss of the hypocone, increase in size of the protocone, and posterior expansion of the posteroloph. The paracone is large, and the metacone is absent as a distinct cusp forming a large posterior extension of the posteroloph. The anterior cingulum forms a low shelf from the anterolabial edge of the paracone to the anterolingual margin of the protocone, terminating in a small lingual cuspule. The protoloph joins the paracone with the protocone in 3 of 5 specimens, with only a weak connection to the anterior cingulum. The arm of the paracone forms a weak connection with the protoloph on specimens 11302, 15772, and 15773. Specimen 15840 is too heavily worn to see loph development, and 15773 shows moderate wear. An indistinct metaloph extends from the protocone to the posterolabial metacone swelling on 15876, the least worn specimen of the five. The crenulations form small, parallel ridges in the talon basin. There is a minute mesostyle adjacent to the paracone.

P/4—The P/4 is molariform. The trigonid is considerably narrower relative to the molars and elevated above the talonid basin. A small, low anterior cingulid closes the anterior margin of the trigonid. The posterior arm of the protoconid joins the metaconid, cutting off the back of the trigonid basin. There are distinct crenulations within the talonid basin, but no mesoconid. A continuous posterolophid connects a well developed entoconid and hypoconid, but the hypoconulid is absent.

M/1—The M/1 is subrectangular in occlusal outline. The trigonid is narrower than the talonid, and slightly elevated above the talonid basin. The metaconid and entoconid are high and large relative to the large, low, and rounded protoconid and hypoconid. The hypoconulid is indistinct on all the molars. Crenulations are present in both the talonid and trigonid basins, forming minute ridges on all lower molars. The anterior cingulid is separated from the protoconid by a small valley on both unworn specimens (15769, 15770). The posterolophid forms a complete connection with the entoconid. A small hypolophid extends labially into the talonid basin from the entoconid, but teminates before reaching the posterolophid in 1 of 3 specimens. In 15770, the hypolophid is directed labially into the talonid basin, and terminates at the posterolingual base of the mesoconid. A small mesoconid is joined to both the protoconid and hypoconid by an ectolophid, on all 3 specimens.

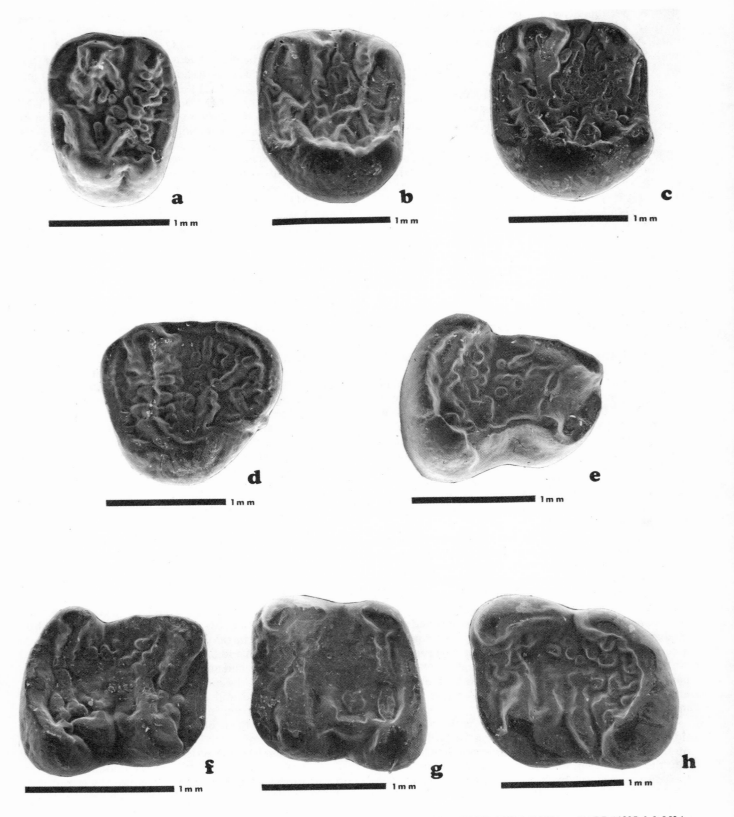

Figure 2. Teeth of *Lophiparamys debequensis* (occlusal view). *a*, UALP 15877, left P4/; *b*, UALP 1568, left M1/; *c*, UALP 11305, left M2/; *d*, UALP 15876, left M3/; *e*, UALP 15766, right P/4; *f*, UALP 15770, right M/1; *g*, UALP 11330, left M/2; and *h*, UALP 15776, left M/3.

TABLE 2. MEASUREMENTS (IN MILLIMETERS) OF TEETH OF Lophiparamys debequensis.

ELEMENT	N	OR	$\bar{X}$	S	V
P4/					
a/p	2	1.037-1.153	1.095	0.082	-----
tra	2	1.270-1.563	1.417	0.207	-----
trp	2	1.244-1.466	1.355	0.157	-----
M1/					
a/p	3	1.237-1.423	1.323	0.094	7.105
tra	3	1.350-1.635	1.503	0.144	9.558
trp	3	1.290-1.500	1.380	0.108	7.826
M2/					
a/p	4	1.320-1.480	1.399	0.066	4.718
tra	4	1.419-1.677	1.583	0.114	7.202
trp	4	1.237-1.481	1.396	0.111	7.951
M3/					
a/p	5	1.401-1.595	1.503	0.074	4.924
tra	5	1.305-1.482	1.408	0.073	5.185
trp	5	1.037-1.300	1.203	0.102	8.479
P/4					
a/p	4	1.284-1.440	1.373	0.065	4.734
tra	4	0.770-0.973	0.919	0.099	10.773
trp	4	1.100-1.192	1.150	0.038	3.304
M/1					
a/p	3	1.342-1.477	1.412	0.068	4.816
tra	3	1.150-1.184	1.163	0.018	1.548
trp	3	1.253-1.324	1.279	0.039	3.049
M/2					
a/p	3	1.404-1.487	1.437	0.044	3.062
tra	3	1.365-1.533	1.421	0.097	6.826
trp	2	1.395-1.440	1.418	0.032	-----
M/3					
a/p	7	1.413-1.820	1.654	0.129	7.799
tra	7	1.195-1.562	1.373	0.109	7.939
trp	7	1.200-1.511	1.362	0.123	9.031

M/2—The M/2 is slightly rectangular in occlusal outline. The teeth are similar to the M/1 except M/2 tends to be more rectangular, with the trigonid and talonid approximately the same width. The trigonid is not elevated above the talonid basin. The hypolophid is directed toward the posterolophid on 1 (11330) of 3 teeth and is worn away on 11213 and 16343.

M/3—The lower third molar has a rounded, posteriorly expanded posterolophid. There is no valley separating the entoconid from the posterolophid. The talonid basin is large relative to the trigonid basin. Both the trigonid and talonid are of equal width, giving the tooth a more rectangular shape. The trigonid is closed posteriorly on 5 of 8 specimens by the posterolingual arm of the protoconid, which extends to but terminates at the base of the metaconid in 4 of the 5 specimens. A hypolophid is directed toward the hypoconid from the entoconid on 1 of 8 specimens. The mesoconid is shortened anteroposteriorly and lengthened transversely relative to the other molars in 5 of 8 specimens.

Discussion: *Lophiparamys* has been collected previously from the Almagre facies of the San Jose Formation (Korth, 1984). This collection of 33 isolated teeth greatly increases the number of *Lophiparamys* specimens found in the southern basins. It also strengthens the temporal assignment of the San Jose UALP localities to Lysitean, in agreement with other San Jose taxa in the UALP collections. The *Lophiparamys* described above, agrees in morphology and size to *Lophiparamys* from the Lysite equivalent of the Wind River Basin, Wyoming and Debeque Formation, Colorado (Guthrie, 1967; Wood, 1962).

Upper molars of *Lophiparamys debequensis* have not been described, although lower molars are well known (Wood, 1962, 1965; Guthrie, 1967; Korth, 1984). The upper molars of *Lophiparamys* closely resemble *Microparamys* in: (1) shape; (2) the valley separating the anterior cingulum from the protoconid; and (3) the weak hypocone. The P4/ closely resembles that of *Microparamys,* and is similar to *Knightomys;* but P4/ of *Lophiparamys* has a small hypocone relative to those of sciuravids. Loph development on the upper molars is relatively weak, with both lophs converging on the protocone. Conules are indistinct or absent. Crenulations on all teeth are well developed, forming subparallel ridges.

Subfamily REITHROPARAMYINAE Wood, 1962
Subfamily ?REITHROPARAMYINAE
Apatosciuravus Korth, 1984
Apatosciuravus jacobsi, new species
Figure 3, Table 3

Holotype: UALP 15729, left M1/.

Hypodigm: UALP Locality 7745—16777, 16780, right dP4/; 11286, left dP4/; 16772, left P4/; 11578, 15724, right M1/; 11309, 15726, 15728, 15729, left M1/; 11317, 11318, 15727, right M2/; 11303, 11310, 15730, 15731, left M2/; 11228, 11299, 15740, right M3/; 11298, 15785, left M3/; 11322, 11323, right dP/4; 11355, 15741, left P/4; 11327, 11336, 15751, right M/1; 11328, 11341, 11354, left M/1; 11339, 15748, left M/2; 11347, 11349, left M/3.

UALP Locality 7746—16783, right M1/; 15739, left M3/.

UALP Locality 7747—16814, right dP4/; 16812, 16813, left dP4/; 15809, 16795, right P4/; 15719, 15844, 15846, 16791, left P4/; 11208, 16809, right M1/; 15780, 15829, 15839, left M1/; 15736, 15781, 15832, 15833, right M2/; 15722, 15723, 15834, 15836, left M2/; 15738, right M3/; 13371, 15743, 15745, 15747, 15853, right M/1; 15744, 15746, 15854, 16793, left M/1; 13360, 13363, 16804, right M/2; 11207, 13362, 13364, 13365, 15752, 15753, 15754, 15755, 15758, 16798, 16808, left M/2; 11203, 15763, 15792, 15857, 16318, 16805, 16806, right M/3; 11209, 15760, 15761, 16316, 16317, 16807, left M/3.

Location: UALP Localities 7745 and 7747, Almagre facies, San Jose Formation, Sandoval County, New Mexico.

Age: Early Eocene (Wasatchian Land Mammal Age).

Diagnosis: The smallest species of *Apatosciuravus;* lacking small cuspule on the anterior cingulum; P4/ with well developed metaloph, but lacking a mesostyle; a short hypolophid on the M/1 and M/2 directed toward but rarely reaching the hypoconid or hypoconulid; both metaloph and the protoloph converge on protocone on M1/ and M2/.

Etymology: Named for Louis L. Jacobs, who found the fossil localities.

Description: dP4/—The anterior cingulum is low and shelf-like, connecting midline to the protocone. The protoloph joins a distinct paracone and protocone. No protoconules are present on any teeth. The metacone is small relative to the paracone. The hypocone is subequal to the large protocone, forming a distinct cusp on the lingual end of the posteroloph. The hypocone is separated from the protocone by a shallow valley. The tooth forms a somewhat triangular shape due to the posterior and lingual position of the hypocone relative to the protocone. A metaloph is well developed, connecting the metacone and protocone. An indistinct metaconule is present on 2 of 6 teeth. A minute mesostyle is present on all teeth.

P4/—The P4/ is small and much shorter relative to the molars. The hypocone is a distinctive swelling on the lingual posteroloph, joining the posterior protocone. A shallow lingual fold exists between the protocone and hypocone in slightly worn specimens. The metacone is reduced relative to the paracone. The metaloph and protoloph converge on the large protocone, and one or the other is constricted before reaching the protocone. A weak and short loph arising from the protocone and directed towards the paracone, joins (in 6 of 7 specimens) the middle of the protoloph with the anterior cingulum. The anterior cingulum forms a low shelf anterolingually where it joins the anterior protocone. The posteroloph is higher than the anterior cingulum, and forms a wider shelf, especially lingually where it ascends the hypocone. No mesostyle is present.

M1/—The first upper molar is subquadrate in occlusal outline. The anterior cingulum forms a low,

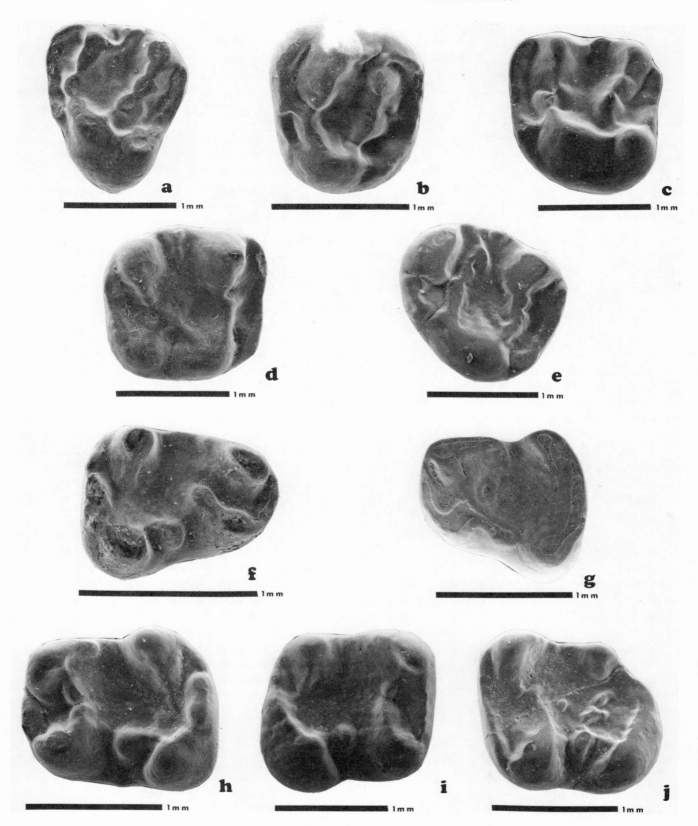

Figure 3. Teeth of *Apatosciuravus jacobsi* (occlusal view). *a*, UALP 16812, left dP4/; *b*, UALP 15844, left P4/; *c*, UALP 15729, left M1/; *d*, UALP 11317, right M2/; *e*, UALP 15740, right M3/; *f*, UALP 11323, right dP/4; *g*, UALP 11355, left P/4; *h*, UALP 11341, left M/1; *i*, UALP 13360, right M/2; and *j*, UALP 11209, left M/3.

TABLE 3. MEASUREMENTS (IN MILLIMETERS) OF TEETH OF *Apatosciuravus jacobsi*.

ELEMENT	N	OR	$\bar{X}$	S	V
dP4/					
a/p	6	1.122–1.238	1.162	0.054	4.647
tra	6	1.063–1.340	1.188	0.122	10.269
trp	6	1.126–1.465	1.281	0.123	9.602
P4/					
a/p	7	1.050–1.172	1.090	0.043	3.945
tra	7	1.305–1.437	1.360	0.053	3.897
trp	7	1.245–1.464	1.336	0.073	5.464
M1/					
a/p	11	1.276–1.475	1.379	0.072	5.221
tra	12	1.370–1.598	1.503	0.079	5.256
trp	10	1.309–1.540	1.468	0.082	5.586
M2/					
a/p	15	1.274–1.490	1.394	0.063	4.519
tra	15	1.344–1.602	1.488	0.066	4.436
trp	15	1.250–1.450	1.347	0.057	4.232
M3/					
a/p	7	1.340–1.513	1.447	0.061	4.216
tra	7	1.197–1.436	1.335	0.074	5.543
trp	7	0.965–1.282	1.110	0.117	10.541
dP/4					
a/p	2	1.303–1.418	1.360	0.081	-----
tra	2	0.737–0.844	0.791	0.076	-----
trp	2	1.050–1.070	1.060	0.014	-----
P/4					
a/p	2	1.220–1.282	1.251	0.044	-----
tra	1	0.912	-----	-----	-----
trp	2	1.101–1.109	1.105	-----	-----
M/1					
a/p	15	1.316–1.485	1.412	0.049	3.470
tra	15	0.964–1.213	1.094	0.065	5.942
trp	15	1.109–1.382	1.259	0.063	5.004
M/2					
a/p	15	1.343–1.576	1.488	0.069	4.637
tra	15	1.204–1.406	1.333	0.066	4.951
trp	15	1.303–1.491	1.406	0.053	3.770
M/3					
a/p	14	1.414–1.693	1.571	0.082	5.220
tra	13	1.250–1.502	1.354	0.068	5.022
trp	13	1.164–1.351	1.279	0.051	3.988

narrow shelf, joining the protocone anterolingually as in the P4/. There are four well developed cusps. The metacone is slightly smaller than the paracone. The protocone and hypocone are large and subequal (the hypocone smaller). A small valley separates the hypocone from the protocone on unworn specimens. The metaloph and protoloph, which converge on the protocone, are well developed. A metaconule is present on all teeth. The protoconule is distinct on only 5 of 12 specimens, and on 6 of 12 specimens the protoloph is discontinuous, with the lingual side slightly displaced anteriorly. There is an incipient loph connecting the metaloph and hypocone on 4 of 12 specimens (15724, 15726, 15780, 16783). A minute mesostyle is present between the paracone and metacone. The posteroloph ascends the hypocone, forming a narrow posterior shelf.

M2/—The M2/ closely resembles M1/, but the M2/ length/width ratio is closer to 1, giving it a more quadrate appearance relative to the other molars. The M2/ differs from M1/ in: (1) the metacone is more reduced relative to the paracone; (2) the hypocone is more reduced relative to the protocone; (3) the valley between the hypocone and protocone tends to be shallower; and (4) the posterior end of the tooth is slightly narrower. The metaloph and protoloph converge on the protocone. The small loph connecting the hypocone and metaloph is distinct and better developed on 7 specimens (11310, 11317, 15722, 15723, 15781, 15834, 15836) but absent on 6 (11318, 15730, 15731, 15736, 15832, 15833). A minute mesostyle is present.

M3/—The M3/ is smaller than the anterior molars and the hypocone is greatly reduced in size, giving the tooth a triangular shape. The metacone is a less distinct cusp and appears as a swelling on the posteroloph. The paracone is the highest cusp, with the protocone forming a large, broad anterolingual cusp. The metaloph and protoloph converge on the protocone, and the metaloph is reduced posterolabially. The mesostyle is minute or indistinct. The anterior cingulum is long and low, joining the anterior paracone and protocone, with no distinct connection to the protoloph. The central basin is broad and shallow, with short low lophs, suggesting remains of a metaloph.

dP/4—The anterior cingulid, when present on unworn specimens, forms a small cuspule slightly anterior to and between the metaconid and protoconid. The metaconid and protoconid are distinct cusps forming the labial and lingual edge of a narrow trigonid. The small compressed trigonid is elevated above the talonid basin. The entoconid and hypoconid form sharp cusps, and a minute hypoconulid is present on both specimens. The posterolophid is separated from the entoconid by a small notch. One (11323) of 2 teeth has small cuspules on the posterolophid between the hypoconulid and entoconid. A small mesoconid is present on both specimens. A hypolophid extends from the entoconid toward the posterolophid on 1 of 2 teeth.

P/4—The P/4 is molariform and small relative to molars. The trigonid is narrow and elevated slightly above a much wider talonid. The metaconid is prominent and situated anterolingually to the small protoconid. The hypoconid and entoconid are large, joined posteriorly by the posterolophid. The hypoconulid is small but distinct, centrally located on the posterolophid on unworn specimens. A shallow notch separates the entoconid from the posterolophid. A small mesoconid is present, and tends to join the protoconid and hypoconid by a low labially located ectolophid. A short hypolophid is directed toward the hypoconid, but terminates toward the midline. A short mesolophid is present on one of two specimens.

M/1—The trigonid on the M/1 is narrower than the talonid, and like the P/4 the trigonid is slightly elevated above the talonid basin. The metaconid is the highest cusp; the protoconid and hypoconid are large and inflated. The entoconid is the smallest prominent cusp. The hypoconulid is a distinct swelling on the posterolophid. A small mesoconid tends to join the hypoconid in late wear. The posterior arm of the protoconid is directed toward the base of the metaconid, terminating at its base (except on 11336, 15743, 15746, 15751, and 15854, on which it ascends the metaconid) to form an incipient metalophid. The hypolophid is short, directed toward the hypoconid in 5 specimens, and directed toward the hypoconulid in 9 specimens (but terminating before reaching that cusp). The posterior arm of the protoconid separates the trigonid from the larger and lower talonid basin. The anterior cingulid is low in 7 of 15 (high in 11327, 11336, 11341, 15746, 15751, 15854, and 16793) and connects to the anterior protoconid and metaconid, forming a low anterior wall of the trigonid basin. A double mesoconid is present on 16793.

M/2—The M/2 is subrectangular in shape, being longer than wide, with the width of the talonid slightly less than the trigonid. The M/2 closely resembles the M/1, but the entoconid is much smaller than the metaconid on the M/2; the protoconid is large, almost equal to the hypoconid. The mesoconid is slightly larger than in the M/1, and the notch between the posterolophid and entoconid is narrower. The trigonid and talonid basin are more equal in height on M/2, and the hypolophid is directed toward the less distinct hypoconulid in 8 of 16 specimens. On two specimens (15752, 15756) the hypolophid extends labially into the talonid basin, toward the hypoconid.

M/3—The M/3 is large, rounded posteriorly with the expansion of an arcuate, noncuspate posterolophid. The trigonid valley is separated from the talonid basin by the posterior arm of the protoconid, but approximates the level of the talonid basin. The mesoconid is slightly more prominent, and joins the protoconid and hypoconid after moderate wear. The hypolophid is indistinct in 11 of 15 specimens. On 4 of 15 specimens it terminates prior to the mesoconid.

Discussion: Apatosciuravus jacobsi is similar in cusp morphology to *A. bifax* Korth, 1984, but the characters that warrant recognition of *A. jacobsi* as a separate species are: (1) smaller size; (2) better development of the metaloph on P4/; (3) absence of a mesostyle on P4/; and (4) increased lophate development on the lower molars. *Apatosciuravus jacobsi* appears to be a more advanced species than *A. bifax*, showing increased loph development on the molars and trending toward a more molariform premolar. *Apatosciuravus jacobsi* may be close to the ancestry of *Knightomys*.

Apatosciuravus differs from *Knightomys* in development of the anterior cingulum, which is a low, broad shelf in *Knightomys* and a narrow shelf extending to the top-midline of the protocone in *Apatosciuravus*. *Knightomys* has wider, more quadrate upper molars. The size of the hypocone, relative to the protocone, on M2/ of *Apatosciuravus* is larger than *Knightomys*. There is a strong connection between the metaloph and protocone in both species of *Apatosciuravus;* in *Knightomys* the connection is weak or lost altogether, and a loph between the metacone and hypocone is developed.

Family SCIURAVIDAE Miller and Gidley, 1918
Knightomys Gazin, 1961
Knightomys depressus (Loomis, 1907)
Figure 4, Table 4

Type specimen: ACM 432, partial skull with right M1/-M3/.

Referred specimens: UALP Locality 7745—11222, 11288, 11289, left dP4/; 11292, 11293, 11582, left P4/; 11312, 11313, 11315, right M1/; 11316, left M1/; 11306, 11308, right M2/; 11320, 11321, left M2/; 11337, left P/4; 11325, right M/1; 15787, left M/1; 11331, 11576, right M/2; 11345, 11583, 15791, left M/3.

UALP Locality 7746—11224, left M/1.

UALP Locality 7747—15850, 16811, left dP4/; 13374, left P4/; 11215, 15861, 15862, 16790, right M1/; 11217, 13376, 15779, 15859, left M1/; 13373, 15737, right M2/; 13379, 15783, left M2/; 15784, right M3/; 15875, left M3/; 15764, left dP/4; 15786, left P/4; 13372, 15868, right M/1; 13361, left M/1; 13368, 13370, 15788, 15869, 15870, 15871, right M/2; 13369, 15789, 15872, left M/2; 15873, 15874, right M/3; 11199, left M/3.

Location: UALP Localities 7745 and 7747, Almagre facies, San Jose Formation, Sandoval County, New Mexico.

Age: Early Eocene (Wasatchian Land Mammal Age).

Description: Knightomys depressus is known only from isolated teeth in the present collection. The lower molars appear to have a slightly more lophate trigonid relative to molars described from other collections. The incisors and P3s are not identifiable, and several of the teeth show moderate wear. The upper molars tend to have a less developed lophate pattern than molars of *Knightomys depressus* from the northern basins, but the lower molars possess a distinct loph which closes the posterior trigonid basin.

dP4/—The anterior cingulum is broad, forming a shelf connecting the anterolobial edge of the paracone to the anterior apex of the protocone. There are four distinct cusps. The metacone and paracone are equal in size. The protocone is the largest cusp, and a small valley separates it from the slightly smaller hypocone. The posterolingual position of the hypocone relative to the protocone gives the tooth a somewhat triangular shape. A posteroloph closes off the back of the talon basin. Both the metaloph and protoloph converge on the protocone. There is a weak connection between the metaloph and hypocone on 2 (11289 and 16811) of 5 specimens. No conules are present on either lophs. There is a minute mesostyle on all specimens.

P4/—The anterior cingulum joins with the protocone, forming a broad low shelf. An arm from the protocone extends toward the paracone, but midway it is directed anterolabially, connecting to the anterior cingulum. Another arm extends lingually from the paracone towards the protocone, but terminates halfway into the trigon basin. The protocone is large relative to a robust paracone. The metacone is small and indistinct, continuous with the metaloph. The metaloph extends from the small metacone to the protocone, and a protoloph extends from the paracone to the protocone. The hypocone is small but distinct, positioned on the lingual base of the posteroloph. There is no sharp valley separating the protocone from the hypocone. An indistinct metaconule is present low on the metaloph and disappears with moderate wear. A mesostyle is absent.

M1/—The M1/ is quadrate in shape. The anterior cingulum forms a low, broad shelf which connects to the protocone. The paracone and metacone are approximately equal in size. The protocone is large and the hypocone is well developed, separated from the protocone by a small valley which disappears with wear. A protoloph joins the paracone to the protocone with an indistinct and small protoconule present on 3 of the 12 specimens. A metaloph connects the metacone to the hypocone, with no evidence of a connection to the protocone on 10 of the 12 isolated teeth. A large metaconule is present on the metaloph. A mesostyle is distinct and well developed. The posteroloph extends from the metacone to the hypocone, closing off the back of the tooth.

M2/—The M2/ is similar to the M1/ except: (1) the mesostyle is larger; (2) the metacone is slightly smaller relative to the paracone; (3) the hypocone is smaller than the hypocone on the M1/, but still distinct and separated from the protocone by a small valley; and (4) the hypocone on M2/ is positioned more posterolabially than on M1/, giving the tooth a less quadrate shape. A protoloph connects the paracone to the protocone, and a metaloph joins the metacone and hypocone, with a metaconule midway. There appears to be a poorly developed loph between the metaconule and protocone on 3 of 8 specimens. The mesostyle is joined to the paracone by a small ridge in 6 of 8 specimens.

M3/—There are only two M3/s, both badly worn. The teeth are triangular in shape, smaller than the first two anterior molars. The protocone and paracone are large. The metacone is a swelling on the posteroloph and not a distinct cusp. The metaloph appears to run towards the protocone, but is too worn to determine whether it terminates at the metaconule or the protocone.

dP/4—The trigonid is narrow, anteriorly compressed, and elevated above the talonid. The talonid is elongate anteroposteriorly, and forms a wide basin. The anterior cingulid forms a ridge, joining the base of the

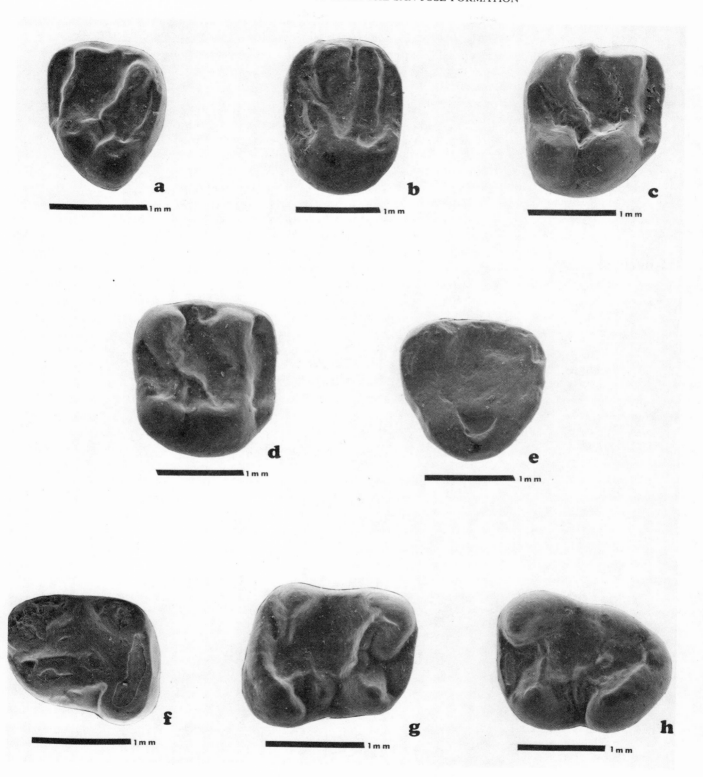

Figure 4. Teeth of *Knightomys depressus* (occlusal view). *a*, UALP 16811, left dP4/; *b*, UALP 11582, left P4/; *c*, UALP 11312, right M1/; *d*, UALP 11308, right M2/; *e*, UALP 15875, left M3/; *f*, UALP 15786, left P/4; *g*, UALP 11325, right M/1; and *h*, UALP 11583, left M/3.

TABLE 4. MEASUREMENTS (IN MILLIMETERS) OF TEETH OF Knightomys depressus.

ELEMENT	N	OR	$\bar{X}$	S	V
dP4/					
a/p	5	1.224–1.312	1.260	0.033	2.216
tra	5	1.144–1.375	1.291	0.096	7.436
trp	5	1.265–1.440	1.368	0.069	5.044
P4/					
a/p	4	1.292–1.400	1.355	0.047	3.469
tra	4	1.715–1.818	1.780	0.048	2.697
trp	4	1.633–1.745	1.687	0.047	2.786
M1/					
a/p	11	1.530–1.690	1.624	0.048	2.956
tra	10	1.753–1.950	1.851	0.054	2.917
trp	10	1.623–1.858	1.728	0.077	4.456
M2/					
a/p	7	1.620–1.735	1.677	0.048	2.862
tra	7	1.775–1.960	1.853	0.064	3.454
trp	7	1.535–1.717	1.627	0.061	3.749
M3/					
a/p	2	1.640–1.677	1.659	0.026	-----
tra	2	1.678–1.695	1.687	0.012	-----
trp	2	1.284–1.323	1.304	0.028	-----
dP/4					
a/p	2	1.307–1.484	1.396	0.125	-----
tra	2	0.900–0.931	0.916	0.022	-----
trp	2	1.181–1.236	1.209	0.039	-----
P/4					
a/p	2	1.590–1.635	1.613	0.032	-----
tra	2	1.190–1.194	1.192	-----	-----
trp	2	1.410–1.485	1.448	0.053	-----
M/1					
a/p	6	1.780–1.920	1.815	0.053	2.920
tra	5	1.356–1.537	1.430	0.069	4.825
trp	6	1.564–1.678	1.619	0.048	2.965
M/2					
a/p	10	1.655–1.850	1.760	0.052	2.955
tra	10	1.566–1.745	1.640	0.057	3.476
trp	8	1.633–1.841	1.716	0.065	3.788
M/3					
a/p	6	1.765–1.988	1.906	0.082	4.302
tra	6	1.510–1.674	1.587	0.063	3.970
trp	6	1.450–1.633	1.551	0.066	4.255

metaconid to the protoconid. A posterior arm of the protoconid extends lingually to the base of the metaconid, and ascends the metaconid on both specimens. The arm closes off the posterior edge of the trigonid. A large hypoconid and hypoconulid are joined by the posterolophid. There is a small valley between the hypoconulid and entoconid on all specimens, with minute cuspules located in the valley. The hypolophid extends a short distance from the entoconid toward the posterolophid in 1 of 2 specimens. A small mesoconid is present, but is not elongated as on the molars. Both teeth have a well developed posterior ectolophid, connecting the mesoconid to the hypoconid and a poorly developed or absent anterior ectolophid, connecting the mesoconid to the protoconid.

P/4—There are only two badly worn specimens of the lower P/4. The anterior cingulid is small, and worn away on 1 of the 2 specimens. The trigonid is narrow and elevated above the talonid basin. There is a large metaconid relative to a fairly small protoconid. The P/4 has no hypoconulid present on these teeth, but the hypoconid and entoconid are large. The mesoconid is small and indistinct, due to wear.

M/1—The anterior cingulid is separated from the metaconid and protoconid by shallow valleys. The trigonid is elevated above the talonid on the molars, and the trigonid is slightly narrower than the talonid on M/1. A posterior arm of the protoconid extends to and ascends the metaconid on 5 of 6 specimens, forming the posterior wall of the trigonid. The metaconid is a larger cusp relative to the protoconid. The entoconid is separated from the posterolophid by a small valley. A short hypolophid extends from the entoconid towards the posterolophid on 5 of 6 specimens, and terminates before reaching posterolophid on all 5 specimens. The hypoconid is large, and the hypoconulid is a poorly developed cusp on the posterolophid. The mesoconid is well developed and slightly elongate, extending a short distance into the talonid basin on 2 of 6 specimens. The mesoconid joins the hypoconid with wear.

M/2—Some teeth are worn, and the anterior cingulid (separated from the protoconid and metaconid by shallow valleys) is lost with wear. The M/2 is subrectangular in shape, with the trigonid as wide as the talonid. As on the M/1, the metaconid is the largest cusp on the trigonid. The entoconid is separated from the posterolophid on both the M/1 and M/2 by a shallow valley. The hypoconid is large and connected to a small, indistinct hypoconulid by the posterolophid. A hypolophid extends from the entoconid toward the posterolophid on 3 of 11 teeth and, as on M/1, terminates before reaching it. The hypolophid extends to the hypoconid on 3 of 11 specimens, and is indistinct on 5 specimens. The mesoconid is elongate and extends far into the talonid basin.

M/3—The M/3 resembles M/1 and M/2 except the posterolophid is large and round, extending posteriorly toward the back of the tooth row. Again, the talonid and trigonid are equal in width, as on M/2. The mesoconid extends into the talonid basin (posterolabially directed), and almost connects with the midline of the posterolophid. With moderate wear, an ectolophid tends to join the mesoconid with the protoconid and hypoconid. The hypoconulid is indistinct on 3 of the 6 specimens.

Discussion: The present collection of 57 isolated teeth increases the number of teeth of *Knightomys depressus* known from the San Jose Formation. *Knightomys depressus* was mentioned from the Largo facies of the San Jose Formation by Korth (1984). The present sample extends the range into the Almagre facies.

The size range of these teeth falls within the lower average range for *Knightomys depressus* reported by Korth (1984, p. 53). The teeth are identified as *K. depressus* on the basis of: (1) elongated mesoconid on lower molars; (2) the anterior cingulid separated from the metaconid and protoconid; and (3) a broad mesoconid of the M/3 extending to the posterolophid on unworn specimens. The hypocone on P4/ is small relative to the hypocone of the molars. These teeth differ from *K. minor, K. reginensis,* and *K. huerfanensis* in size. They fall within the size range of *K. senior,* but differ from this species in elongation of the mesoconid and lack of connection between the anterior cingulid and protoconid.

The isolated dP4s tentatively are placed in *Knightomys depressus* on the basis of size. They differ from dP4s of *Apatosciuravus jacobsi* in the lesser development of conules, wider trigonids, and larger size. There appears to be an incipient connection between the hypocone and metaloph on unworn specimens of dP4/s, which disappears with moderate wear.

Knightomys cf. *K. minor* (Wood, 1965; Korth, 1984)
Figure 5, Table 5

Referred specimens: UALP Locality 7745—11360, left P/4; 11326, right M/1; 11332, left M/1; 11338, 11340, right M/2.

UALP Locality 7747—15845, left P4/; 15830, 15831, 15838, 16794, 17033, left M1/; 13378, 17032, right M2/; 15734, 15735, left M2/; 11219, right M3/; 15863, 16801, right M/1; 15864, left M/1; 15866, left M/2; 15762, left M/3.

Location: UALP Localities 7745 and 7747, Almagre facies, San Jose Formation, Sandoval County, New Mexico.

Age: Early Eocene (Wasatchian Land Mammal Age).

Description: P4/—The anterior cingulum forms a continuous connection between the paracone and protocone. The protocone is large relative to other cusps. An arm arising from the anterior edge of the protocone runs labially toward, and joins with, the anterolabial edge of the paracone. The metacone is an indistinct swelling on the labial edge of the metaloph. A well developed metaloph converges on the protocone. The hypocone is small, and there is no mesostyle.

M1/—The anterior cingulum forms a low, broad shelf, joining the anterolabial edge of the paracone to the anterolabial edge of the protocone. A protoloph arises from the protocone and connects to the paracone. The metacone is small relative to the protocone but equal in

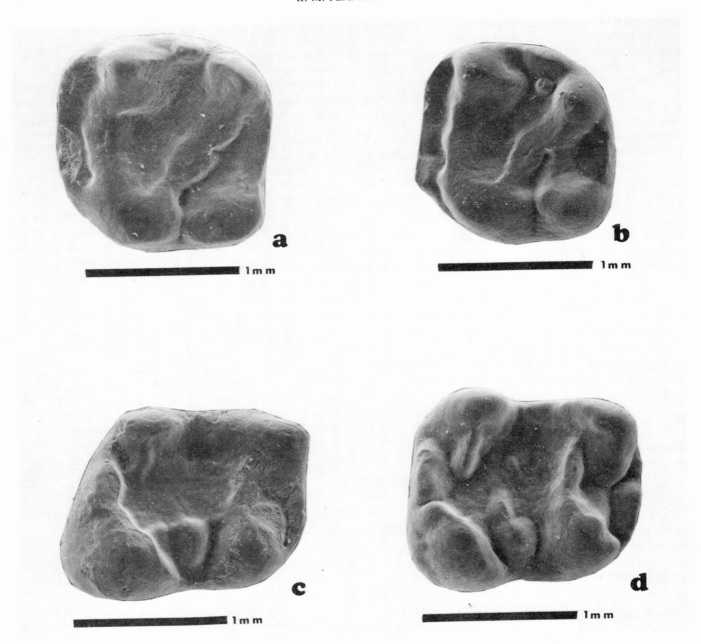

Figure 5. Teeth of *Knightomys cf. K. minor* (occlusal view). *a*, UALP 15831, left M1/; *b*, UALP 15734, left M2/; *c*, UALP 15863, right M/1; and *d*, UALP 11338, right M/2.

size to the paracone. A well developed metaloph converges on the protocone in all teeth, but weakly joins the protocone on 3 (15831, 16794, 17033) of 5 specimens. On 15838, there is a weak connection to the hypocone. The hypocone is large, giving the tooth a quadrate shape. It is separated from the protocone by a valley on unworn teeth. A mesostyle is present.

M2/—The M2/ differs from M1/ in: (1) the hypocone is smaller in size and located more posterolabially on the tooth; and (2) the metacone is subequal to the paracone. A metaloph and protoloph converges on the protocone in 2 of 4 specimens. There is a weak connection from the metaloph to the hypocone in 3 of 4 specimens and a strong connection in 1 (15735) of 4 specimens.

M3/—The M3/ is smaller than the two anterior molars, and triangular in shape. The anterior cingulum joins the paracone to the midline of the protocone. The protocone is large relative to other cusps. The metacone appears as a swelling on the posteroloph. Both the metaloph and protoloph converge with, and join to, the protocone. The hypocone is minute and centrally located on the posteroloph. The mesostyle is indistinct.

P/4—The P/4 is worn, but the trigonid is much nar-

TABLE 5. MEASUREMENTS (IN MILLIMETERS) OF TEETH OF *Knightomys* cf. *K. minor*.

NUMBER	ELEMENT	A/P	TRA	TRP
15845	Lt. P4/	1.240	1.490	1.455
15830	Lt. M1/	1.387	1.525	1.360
15831	Lt. M1/	1.400	1.521	1.407
15838	Lt. M1/	1.250	1.425	1.355
16794	Lt. M1/	1.263	1.536	1.350
17033	Lt. M1/	1.356	1.432	1.340
13378	Rt. M2/	1.350	1.448	1.272
17032	Rt. M2/	1.402	1.491	1.418
15734	Lt. M2/	1.363	1.419	1.255
15735	Lt. M2/	1.480	1.545	1.321
11219	Rt. M3/	1.447	1.323	1.120
11360	Lt. P/4	1.377	1.077	1.248
11326	Rt. M/1	1.678	1.260	1.490
15863	Rt. M/1	1.602	1.284	1.420
16801	Rt. M/1	1.666	1.313	1.506
11332	Lt. M/1	1.680	1.375	1.537
15864	Lt. M/1	1.715	1.235	1.500
11338	Rt. M/2	1.638	1.235	1.414
11340	Rt. M/2	-----	1.157	-----
15866	Lt. M/2	1.610	1.470	1.503
15762	Lt. M/3	1.632	1.355	1.375

rower than the talonid and compressed anteroposteriorly. The anterior cingulid is not present. There is no mesoconid or hypoconulid.

M/1—The M/1 is subrectangular in occlusal outline, with the trigonid narrower than the talonid. The anterior cingulid is high, joining the anterolabial base of the metaconid to the anterolingual edge of the protoconid in unworn specimens. An arm of the protoconid closes off the posterior trigonid from the talonid basin and ascends the metaconid. The trigonid is slightly compressed anteriorly, and not elevated above the talonid. The posterolophid is separated from the entoconid by a small valley. The hypoconulid is a distinct swelling on the posterolophid in 2 (11332, 15863) of 5 specimens. The mesoconid is small, round, and connected to the large hypoconid by a small ectolophid. The hypolophid is directed toward the posterolophid in 2 specimens (11326, 15864), and towards the hypoconid in 3 teeth (11332, 15863, 16804).

M/2—The M/2 closely resembles M/1, but is: (1) more rectangular in shape, with the trigonid and talonid being equal in length; and (2) the mesoconid is larger, extending into the talonid basin. The trigonid is narrower than the talonid as on M/1, and the anterior cingulid is high and small. The hypoconulid is distinct on 1 (11338) of 3 specimens. The hypolophid is directed toward the posterolophid on 2 of 3 teeth and towards the hypoconid on 1 of 3 teeth.

M/3—The M/3 closely resembles the two anterior molars, but the posterolophid is expanded toward the back of the tooth row. The mesoconid is small, compressed, and extends into the talonid basin toward the posterolophid. The hypolophid joins the entoconid with the posterolophid.

Discussion: In size and morphology, these teeth resemble *Knightomys minor*, described by Korth (1984) from the Lysite Member of the Wind River Formation. They differ in: (1) the lophs on the upper molars converge on the protocone and form only a weak connection to the hypocone; (2) the arm of the protoconid ascends the metaconid on *Knightomys* cf. *K. minor*, completely closing the posterior trigonid; and (3) the hypoconulid is less distinct on *Knightomys* cf. *K. minor* (but this may be due to wear).

This small collection of isolated teeth shows minor differences with *Knightomys minor* from the Wind River Basin, but it supports the Lysitean subage of these localities in the San Juan Basin.

Knightomys reginensis (Korth, 1984)
Figure 6, Table 6

Type specimen: CM 38762, right mandible with M/2-M/3.

Referred specimens: UALP Locality 7745—11296, right P4/; 15793, right M1/; 11307, left M1/; 11359, right M2/; 11324, right dP/4; 15798, 15800, 15801, 16779, 17034, right M/1; 11329, 11585, 15802, 15803, right M/2; 11342, left M/2.

UALP Locality 7747—15795, left M1/; 15837, right M2/; 15794, left M2/; 15805, left M3/; 15810, right dP/4; 11205, right M/2; 11204, left M/2; 15880, right M/3; 15806, left M/3.

Location: The type is from CM locality 2, Almagre Member, San Jose Formation, Sandoval County, New Mexico. Referred specimens come from CM Locality Almagre 4. UALP Localities, Almagre facies, 7745 and 7747, San Jose Formation, Sandoval County, New Mexico.

Age: Early Eocene (Wasatchian Land Mammal Age).

Emended diagnosis: Smallest known species of *Knightomys;* hypocone on the upper molars is large and well developed, but subequal relative to the protocone; metaloph and protoloph converge on the protocone, although the connection of metaconule to protocone is weak; a strong connection exists between the metaconule and hypocone; protoconule and metaconule are reduced but distinct; anterior cingulid on the lower molars connects midline to protoconid; small hypoconulid is separated from entoconid by a valley; hypolophid is short, terminating before joining the posterolophid; mesoconid is well developed; a small metastylid is present on the posterolingual edge of the metaconid on unworn specimens.

Descriptions: P4/—The anterior cingulum is low, connecting the apex of the paracone to the midline of the protocone. The hypocone is a distinct swelling on the ridge of the posteroloph. The protocone is the largest cusp. The metacone is small relative to the paracone, with a metaloph and protoloph extending to and joining with the protocone. An indistinct metaconule is present on the metaloph. There is a minute mesostyle.

M1/—The upper first molar is quadrate in occlusal outline, with a well developed but subequal hypocone relative to the other cusps. The paracone and metacone are equal in size. The anterior cingulum forms a low shelf from the anterolabial corner of the tooth to its termination as a small cuspule anterior to, and at the midline of, the protocone. The metaloph and protoloph are low and discontinuous. The protoloph connects with the protocone, whereas the metaloph forms a weak connection with the protocone. There is a short hypoloph from the hypocone to the metaconule. A metaconule is present on all teeth, located in the center of the metaloph. A poorly developed protoconule can be seen on unworn teeth. There is a minute mesostyle.

M2/—The M2/ is similar to M1/. The hypocone on M2/ is distinct, but has shifted to a more posterolabial position than on the M1/, and is reduced in size. This gives the tooth a less quadrate shape than M1/. The metacone on M2/ is smaller relative to the paracone. The metaloph and protoloph converge on the protocone, but the connection between the metaconule and protocone is weak and the hypoloph connection between metaconule and hypocone is stronger than seen on specimens of M1/. The protoconule is absent, and a minute mesostyle is present.

M3/—The third upper molar is triangular in shape and smaller than the anterior molars. The paracone is the highest cusp. The protocone is large, and the metacone is indistinct. The hypocone is reduced, and appears as a slight swelling on the lingual end of the posteroloph. The anterior cingulum joins the midline of the protocone, terminating in a small cuspule. The protoloph is low and connects to the protocone. There is a short, low, discontinuous loph directed labially from the posterior protocone terminating near the center of the talon basin. A mesostyle, protoconule, and metaconule are absent.

dP/4—The trigonid is narrow and compressed transversely, whereas the talonid basin is elongate anteroposteriorly. The anterior cingulid is a minute cuspule, closing off the anterior trigonid on 1 of 2 specimens. The metaconid is the highest cusp of the tooth, and the protoconid is reduced. A small arm of the protoconid extends posterolingually into the talonid basin, partially closing the trigonid. The entoconid is large, separated from the hypoconulid by a small notch. The hypoconid and hypoconulid are located on the posterolophid. A small mesoconid is present on all teeth. A minute hypolophid extends lingually from the entoconid on 1 of 2 specimens.

M/1—The lower first molar has a narrower trigonid than talonid. On unworn specimens (15801, 16799), the trigonid is elevated slightly above the talonid basin and the metaconid is the highest (but not largest) cusp on the tooth. The protoconid and hypoconid are large and more inflated than the metaconid and entoconid. The hypoconulid is indistinct or absent. A low but distinct metalophid and hypolophid tend to join the posterior arm of the protoconid and posterolophid, respectively. The metalophid-posterior arm of the protoconid tend to separate a higher and smaller trigonid basin from a lower and larger talonid basin. The anterior cingulid is high, and connects to the protoconid, forming the anterior wall

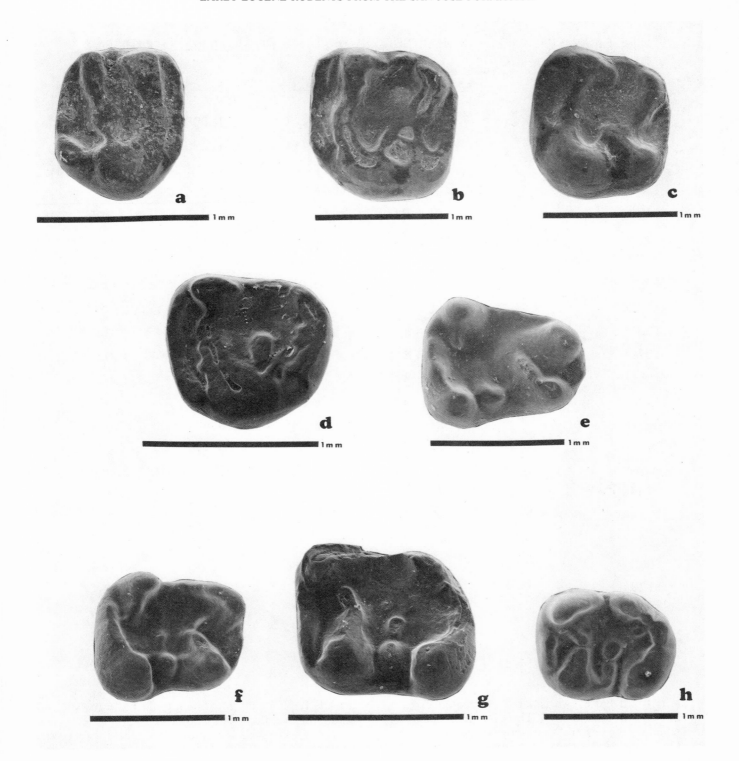

Figure 6. Teeth of *Knightomys reginensis* (occlusal view). *a*, UALP 11296, right P4/; *b*, UALP 15795, left M1/; *c*, UALP 15794, left M2/; *d*, UALP 15805, left M3/; *e*, UALP 15810, right P/4; *f*, UALP 16779, right M/1; *g*, UALP 11342, left M/2; and *h*, UALP 15806, left M/3.

TABLE 6. MEASUREMENTS (IN MILLIMETERS) OF TEETH OF Knightomys reginensis.

Specimen number	Element	A/P	TRA	TRP
11296	Rt. P4/	0.850	1.008	0.888
15793	Rt. M1/	1.000	1.090	1.055
11307	Lt. M1/	1.070	1.101	1.055
15795	Lt. M1/	1.075	1.136	1.085
11359	Rt. M2/	?	0.945	0.895
15837	Rt. M2/	1.072	1.185	1.100
15794	Lt. M2/	1.005	1.120	1.000
15805	Lt. M3/	0.977	0.934	0.790
11324	Rt. dP/4	1.000	0.720	0.920
15810	Rt. dP/4	1.175	0.717	1.000
15798	Rt. M/1	1.030	0.885	1.020
15800	Rt. M/1	1.151	0.878	1.017
15801	Rt. M/1	1.070	0.995	1.058
16779	Rt. M/1	1.060	0.802	0.974
17034	Rt. M/1	1.043	0.777	0.910
11205	Rt. M/2	1.117	1.005	1.020
11329	Rt. M/2	1.112	0.974	?
11585	Rt. M/2	1.120	0.935	0.985
15802	Rt. M/2	1.065	0.940	0.979
15803	Rt. M/2	1.084	0.936	0.990
11204	Lt. M/2	?	0.985	?
11342	Lt. M/2	1.075	0.936	0.946
15880	Rt. M/3	1.165	0.925	0.920
15806	Lt. M/3	1.035	0.895	0.867

of the trigonid basin. An isolated mesoconid extends linguinally into the talonid basin. There is a small, indistinct metastylid on the posterolingual edge of the metaconid.

M/2—The M/2 is rectangular in shape, longer than wide, with the width of the talonid and trigonid being almost equal. In cusp morphology, the M/2 closely resembles M/1, but the entoconid is distinctly smaller than the metaconid on M/2. Both molars have a small valley which separates the entoconid from the posterolophid. The trigonid and talonid basins are nearly equal in height on M/2. The metalophid and hypolophid are small but distinct on unworn specimens. The mesoconid extends far linguinally into the talonid basin toward the minute metastylid.

M/3—The M/3 is rounded posteriorly with an expansion of the noncuspate posterolophid. The trigonid valley is separated from the talonid basin by the posterior arm of the protoconid, and approximates the level of the talonid. The mesoconid expands linguinally into the talonid basin. The hypoconid is continuous with the posterolophid, the latter of which terminates at the posterior base of the entoconid. A small cuspule is present in the talonid basin, anterior to the posterolophid.

Discussion: With collection of more specimens from

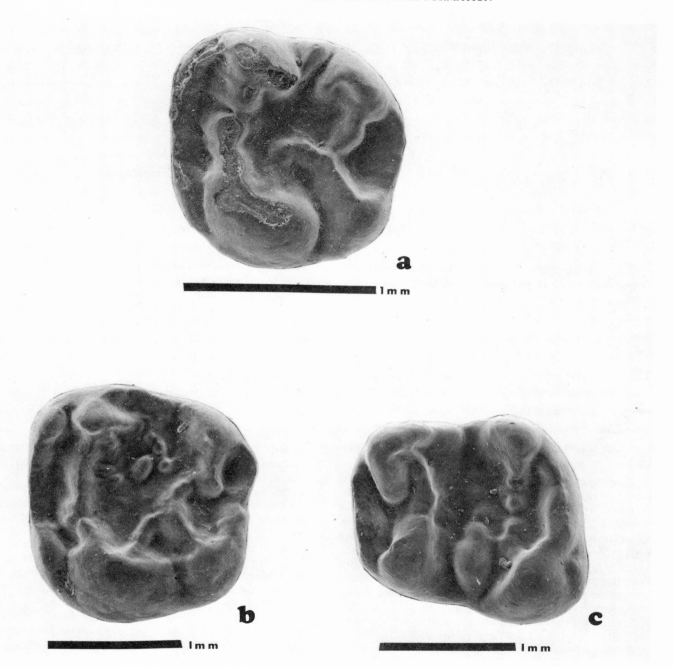

Figure 7. Teeth of *Pauromys* sp. and *Mattimys kalicola* (occlusal view). *a*, *Pauromys* sp.: UALP 16810, left M1/ or M2/; and *Mattimys kalicola*: *b*, UALP 11581, left M2/; and *c*, UALP 11334, left M/2.

the San Jose Formation, this small rodent is removed from *Microparamys* and placed in the genus *Knightomys*. The generic characters of *Knightomys* are: (1) cheek teeth with rudimentary loph development; (2) both the metaloph and protoloph are complete and separate on the upper molars; (3) the entoconid is isolated with an incomplete hypolophid extending into the talonid basin; (4) the mesoconid on the lower molars is distinct and extends lingually into the talonid basin; and (5) a minute metastylid is present on the lower cheek teeth (Korth, 1984).

All of these characters are present in the newly collected specimens. The upper molars have a well developed, fairly large hypocone, in contrast to the small, reduced hypocone of *Microparamys*. The anterior cingulid joins the protoconid in these specimens, while there is a small valley separating the anterior cingulid from the protoconid in *Microparamys*. *Knightomys reginensis* appears to be a transitional form between *Apatosciuravus* and later forms of *Knightomys*, as seen in the loph development. These forms of small sciuravids maintain the weak loph con-

TABLE 7. MEASUREMENTS (IN MILLIMETERS) OF TEETH OF *Mattimys kalicola*.

Specimen number	Element	A/P	TRA	TRP
11581	Rt. M2/	1.722	1.962	1.645
11334	Lt. M/2	1.847	1.420	1.652

MEASUREMENTS (IN MILLIMETERS) OF TEETH OF *Pauromys sp*.

16810	Lt. M^x	1.190	1.269	1.110

nection with the protocone, while the larger species of *Knightomys* lose all connection between the metaloph and protocone.

Pauromys Troxell, 1923
Pauromys sp.
Figure 7, Table 7

Referred specimen: UALP 16810, left M1/ or 2/.

Location: UALP locality 7747, Almagre facies, San Jose Formation, Sandoval County, New Mexico.

Age: Early Eocene (Wasatchian Land Mammal Age).

Description: M1/ or 2/—The anterior cingulum is wide and low, joining the anterolabial edge of the paracone to the anterior base of the protocone. The hypocone and metacone are slightly smaller than the paracone, and the protocone is the largest cusp. A small mesostyle is present. The protoloph runs parallel to the anterior cingulum. The metaloph connects with the hypocone, but midway between the metacone and hypocone another loph extends anterolingually into the talon basin. No distinct conule occurs at the junction, and the small loph terminates before reaching the protocone. A strong posteroloph closes off the back of the tooth, forming a small valley between it and the metaloph.

Discussion: This is the first recorded occurrence of *Pauromys* in the San Jose Formation, although it is not the first record in the Wasatchian (see Korth, 1984).

The upper molar of *Pauromys* sp. from the San Jose Formation closely resembles the upper molar of *Pauromys* sp. from the Wind River Formation of Wyoming (Korth, 1984), but is a larger tooth. Since the specimens are so few, the Wasatchian size range of *Pauromys* sp. is unknown. The loph development is weaker than in the Bridgerian species of *Pauromys* sp. from Powder Wash (Dawson, 1968) and *Pauromys* sp. from the Fowkes Formation (Nelson, 1974). Upper molars have not been described from *Pauromys perditus* and *Pauromys schaubi*.

Family EUTYPOMYIDAE Miller and Gidley, 1918
Family ?EUTYPOMYIDAE
Mattimys Korth, 1984
Mattimys kalicola (Matthew, 1918; Korth, 1984)
Figure 7, Table 7

Referred specimen: UALP 11581, left M2/ and UALP 11334, left M/2.

Location: UALP Locality 7745, Almagre facies, San Jose Formation, Sandoval County, New Mexico.

Age: Early Eocene (Wasatchian Land Mammal Age).

Description: M2/—The anterior cingulum diverges labially from the paracone, forming a slim notch; the cingulum is directed lingually to the protocone, terminating in a cuspule. There are four well developed cusps, with the metacone and paracone small relative to the hypocone and protocone. A minute double mesostyle is present on the labial margin of the tooth. A protoloph joins the paracone with the protocone. The metaloph is directed slightly anterior from the metacone toward the hypocone, with a minute metaconule in midlength. From the metaconule, a weak loph extends anterolingually to the protocone. Another small loph is directed posterolingually from the metacone at the base of the metaloph to join the posteroloph. The paracone also possesses a weak loph, directed posterolabially into the talon basin. The talon basin contains a series of irregular, low, broad ridges and tiny cuspules.

M/2—The anterior cingulid is directed from the lingual edge of the metaconid to the base of the protoconid, forming a high ridge and closing the anterior trigonid basin. The metaconid is the highest cusp. The protoconid and entoconid are small relative to a broad hypoconid. The small hypoconulid is joined by the posterolophid to the entoconid and hypoconid. The mesoconid is well developed and slightly shortened anteroposteriorly, extending a short distance into the talonid basin. The talonid basin contains low, broad, irregular ridges. An arm is directed posterolingually from the protoconid and ascends the metaconid, closing the back of the trigonid. A hypolophid, composed of cuspules, joins the entoconid and hypoconid.

Discussion: Both molars are tentatively identified as *Mattimys kalicola* on the basis of: (1) low, broad, irregular ridges within the talon and talonid basins; (2) well developed lophs; (3) lower molar lacking an ectolophid; and (4) size. Although the upper molars of *Mattimys kalicola* never have been described, the ridge development, multiple lophs, and cuspule on the anterior cingulum justify the identification. The lower molar of the sample from the San Jose Formation falls within the size range of *Mattimys kalicola* from Wyoming, but M2/ is large. The number of specimens of *M. kalicola* from New Mexico is low and, until a larger collection is available (including upper dentition), the extent of size variation remains uncertain. The present sample extends the geographic range of *Mattimys kalicola* to southern basins.

CONCLUSIONS

The three UALP localities (7745, 7746, 7747) are in the interbedded, variegated clays and sands of the Almagre facies of the San Jose Formation, New Mexico. The Almagre facies appears to represent a floodplain deposit controlled by fluvial sedimentation experiencing periodic flooding. UALP 7745 and 7747 occur within lenticular sand channels, possibly of a meandering stream system. UALP 7746 occurs in an overbank deposit. All the rodent specimens collected are isolated teeth, although jaws of other taxa and postcranial material are present in these UALP localities. An abundance of sciuravids characterizes the rodent fauna. This is probably not due to fluvial sorting, as various elements of large mammals *(Coryphodon* and *Hyracotherium)* are common in all three localities, along with several medium sized taxa.

The rodent fauna of 7745, 7746, and 7747 is diverse; it includes five taxa known from mid-Wasatchian deposits in the more northerly basins of Wyoming and Colorado. These taxa are *Lophiparamys debequensis, Paramys excavatus, Paramys copei, Knightomys depressus,* and *Mattimys kalicola.* The remaining taxa are morphologically close to species of the same genera from northern Rocky Mountain localities.

The dominant rodent taxon of the UALP collections is *Apatosciuravus jacobsi,* contributing 46 percent of the total rodent fauna in terms of available specimens. *Apatosciuravus jacobsi* is similar in cusp morphology to *Knightomys,* and may have been an ecological equivalent to medium sized sciuravids such as *Knightomys senior* or the smaller *Knightomys minor. Apatosciuravus* may have been close to the ancestry of *Pauromys,* a small rodent common in the middle Eocene (Bridgerian). Both genera have a small loph connecting the protoloph and anterior cingulum on P4/ and dP4/. This character is present, but less developed, in some specimens of *Knightomys;* it may be a primitive character of sciuravids. The metaloph pattern of *Pauromys* could be derived from that of *Apatosciuravus jacobsi* by strengthening the crest development, following replacement of the distinct metaconule with elongation to a loph. Both *Pauromys* and *Apatosciuravus* from the San Jose Formation are similar to *Sciuravus* sp. and *Pauromys* sp. from Powder Wash (Dawson, 1968), but are less lophate. *Knightomys reginensis* is the smallest rodent in the present fauna, and the smallest known species of *Knightomys. Knightomys depressus* from the San Jose Formation is similar in tooth morphology in upper and lower molars to *K. huerfanensis,* differing only in size. *Knightomys depressus* from the San Jose Formation may have been close to the ancestry of *K. huerfanensis,* with both species continuing through the late Wasatchian.

Lophiparamys debequensis is dentally identical to the same taxon from Wyoming and Colorado, and strongly supports a Lysitean subage for the New Mexican fauna. *Mattimys kalicola* and *Pauromys* are found elsewhere in Lostcabinian and younger deposits, but the evidence is scanty; thus they are not useful for time correlation. All rodent species from the three UALP localities are common to a Lysitean subage within the Wasatchian, and close to an early Lostcabinian subage. No rodent taxa restricted to the Graybullian were identified, although *P. copei* exists throughout the entire Wasatchian.

The two species of *Paramys (P. copei* and *P. excavatus)* are the largest rodents collected from the New Mexican sites, but represent only eight percent of the total sample on the basis of identified specimens. Many of the larger rodents described by Korth (1984) were not found. There is no obvious sedimentological explanation for this size restriction; thus the fauna may represent an ecological grouping. The sciuravids in the fauna may have been common members of a riparian environment. The Almagre facies appears to be a low energy, fluvial deposit with small channels and thick floodplain sediments. The brightly colored, variegated clays possibly represent palesols, contributing to preservation of these small mammals.

The present rodent fauna is interesting paleobiogeographically in suggesting the continuity of small mammalian species both in the northern and southern Rocky Mountain region.

ACKNOWLEDGMENTS

I express my thanks to my committee, E. H. Lindsay, K. Flessa, and E. L. Cockrum for their helpful suggestions and comments. I am especially grateful to R. F. Butler for providing a job during my Master's program. I benefitted from helpful discussions with Brent Breithaupt, Jeanne Davidson, Mary Dawson, Miao Desui, Dick Haskin, W. W. Korth, Jason Lillegraven, Bob McCord, Charlotte Otts, Louis Taylor, and Yuki Tomida. Will Downs and Kevin Moodie helped with preparation of the fossils, and Mark Bierei with the SEM. Lawrence Flynn and Louis L. Jacobs gave invaluable help throughout the project. I am indebted to Mary Dawson, P. D. Gingerich, and Peter Robinson for generous use of museum specimens. I am especially grateful to G. G. Simpson and L. M. Gould for their encouragement and friendship.

REFERENCES CITED

Baltz, E. H., 1967, Stratigraphic and regional tectonic implications of part of Upper Cretaceous and Tertiary rocks, east-central San Juan Basin, New Mexico: U.S. Geological Survey, Professional Paper 552, 101 p.

Cope, E. D., 1877, Report upon the extinct Vertebrata obtained in New Mexico by parties of the expedition of 1874: Report of Geographical Survey West of 100th Meridian, v. 4, part 2, p. 1-370.

Dawson, M. R., 1968, Middle Eocene rodents (Mammalia) from northeastern Utah: Annals of Carnegie Museum, v. 39, p. 327-370.

Gazin, C. L., 1961, New sciuravid rodents from the Lower Eocene Knight Formation of western Wyoming: Biological Society of Washington, Proceedings, v. 74, p. 193-194.

Guthrie, D. A., 1967, The mammalian fauna of the Lysite Member, Wind River Formation (early Eocene) of Wyoming: Southern California Academy of Science, Memoirs, v. 5, p. 1-53.

Haskin, R. A., 1980, Magnetic polarity stratigraphy and fossil Mammalia of the San Jose Formation, Eocene, New Mexico [M.S. thesis]: Tucson, Arizona, University of Arizona, 75 p.

Korth, W. W., 1984, Earliest Tertiary evolution and radiation of rodents in North America: Carnegie Museum of Natural History, Bulletin, v. 24, p. 1-71.

Leidy, J., 1871, Notice of some extinct rodents: Academy of Natural Sciences of Philadelphia, Proceedings, v. 22, p. 230-232.

Lillegraven, J. A., and Bieber, S. L., 1986, Repeatability of measurements of small mammalian fossils with an industrial measuring microscope: Journal of Vertebrate Paleontology, v. 6, p. 96-100.

Loomis, F. B., 1907, Wasatch and Wind River rodents: American Journal of Science, ser. 4, v. 23, p. 123-130.

Matthew, W. D., 1918, A revision of the Lower Eocene Wasatch and Wind River faunas. Part 5. Insectivora (continued), Glires, edentates: American Museum of Natural History, Bulletin, v. 38, p. 565-657.

Miller, G. S., Jr., and Gidley, J. W., 1918, Synopsis of the supergeneric groups of rodents: Washington Academy of Science, Journal, v. 8, p. 431-448.

Nelson, M. E., 1974, Middle Eocene rodents (Mammalia) from southwestern Wyoming: Contributions to Geology, University of Wyoming, v. 13, p. 1-10.

Simpson, G. G., 1948, The Eocene of the San Juan Basin: American Journal of Science, v. 246, p. 257-282 (part 1) and 363-385 (part 2).

Troxell, E. L., 1923, *Pauromys perditus,* a small rodent: *ibid.,* v. 5, p. 155-156.

Wood, A. E., 1962, The early Tertiary rodents of the Family Paramyidae: American Philosophical Society, Transactions, v. 52, p. 1-261.

―――― 1965, Small rodents from the early Eocene Lysite Member, Wind River Formation of Wyoming: Journal of Paleontology, v. 39, p. 124-134.

MANUSCRIPT RECEIVED DECEMBER 12, 1985
REVISED MANUSCRIPT RECEIVED JULY 15, 1986
MANUSCRIPT ACCEPTED JULY 18, 1986

Fossil vertebrates from the latest Eocene, Skyline channels, Trans-Pecos Texas

JOHN A. WILSON — *Department of Geological Sciences, The University of Texas at Austin, Austin, Texas 78758*

MARGARET S. STEVENS — *Department of Geology, Lamar University of Orange County, Orange, Texas 77710*

ABSTRACT

The fauna of the Skyline channels, West Texas, of early Duchesnean age, contains the following taxa: *Simidectes magnus, Harpagolestes* sp., *Hyaenodon* cf. *H. vetus, Mahgarita stevensi,* ?*Leptotomus coelumensis* (new species), *Duchesneodus* cf. *uintensis,* ?*Hyracodon* sp. indet., *Amynodontopsis bodei, Protoreodon pumilus, Agriochoerus* sp. indet., *Leptoreodon* sp., and *Hendryomeryx* sp. *Simidectes* is the most common form; this apparently is its last appearance, along with *Harpagolestes* and *Leptoreodon.* It is the first appearance of *Hyaenodon,* ?*Duchesneodus,* and *Hyracodon.*

INTRODUCTION

We have now passed the fortieth anniversary of publication of an article by G. G. Simpson (1946) entitled "The Duchesnean Fauna and the Eocene-Oligocene Boundary." The position of a boundary was somewhat of a problem then, and it is still somewhat of a problem. Simpson (1946, p. 52) summarized that the Duchesne River Formation lay conformably upon the Eocene Uinta Formation and was formerly "considered as a barren part of the Uinta Formation and as latest Eocene in age." Somewhat later field parties from the Carnegie Museum found identifiable mammals in the so-called barren Uinta. Kay (1934) established the Duchesne River Formation as separate from the Uinta Formation, and interpreted the faunas as Oligocene. Wood and others (1941) "placed the Duchesne River in the Eocene, and proposed Duchesnean as a provincial time term for latest Eocene, although they did note that there was no general agreement as to whether it was latest Eocene or earliest Oligocene" (Simpson, 1946, p. 53). The previous year Scott (1945) had revised the Duchesne River fauna and returned the Duchesne River Formation to the Oligocene.

Simpson (1946, p. 53) then went on to say "Such a boundary is largely arbitrary and is a matter of convenience rather than of right or wrong. The most important thing is that all students should follow the same usage, whatever it is. To promote such agreement, it should be demonstrated that some particular boundary is more convenient than the alternatives, or at least that it is not less convenient." One of us (Wilson) can look back over the last forty years and see that Simpson's hope that "all students should follow the same usage" has not been realized, nor does it seem likely to be in the near future.

Simpson (1946) did not follow Kay (1934) in subdividing the Duchesne River Formation into three "horizons" (Randlett, Halfway, and Lapoint, in ascending order), but treated the formation with its fauna as a single unit. Most other authors dealing with the Duchesne River have subdivided both the formation and the faunas following Kay (1934). The latest additions and revision of the fauna of the Duchesne River Formation are by Emry (1981). Only the fauna of the Lapoint is considered as the type of the Duchesnean land mammal age; the earlier ones are generally considered Uintan.

Forty years ago when Simpson discussed the problem it was relatively simple. Identification of the age of sediments was based on the age of the contained fauna. For practical purposes, that was all that was available. Unfortunately the meager fauna of the Duchesne River Formation made correlation difficult. A new method of dating by the potassium-argon content of volcanic rocks evolved, and became well established with publication of the classic paper by Evernden and others (1964). Four years later, potassium/argon dates for the volcanic units of the Vieja area were given by Wilson and others (1968), and associated faunal units were summarized later by Wilson (1977b).

The fossil fauna of the Vieja area, first described by Stovall (1948) and later enhanced by Bryan Patterson and J. H. Quinn for the Field Museum of Natural History, was recognized as early Oligocene. Samples of volcanic rocks taken in the early 1960s were submitted for dating to J. F. Evernden, and were incorporated in Evernden and others (1964). Two of their dates, KA1274 at 42.7 Ma from the Alamo Creek Basalt of Big Bend National Park and KA1010 at 36.8 Ma from the Bracks Rhyolite of the Vieja area, confirmed that we were working within the general time span of the Duchesnean. Additional samples from the Vieja area were dated and published by Wilson and others (1968) and for Big Bend National Park by Maxwell and others (1967). Meanwhile through the courtesy of Dr. John Clark, then of the Field Museum of Natural History, a sample of a tuff or ashy stiltstone was obtained and dated by Dr. Fred McDowell of The University of Texas at Austin. The sample came from the contact between the Halfway and Lapoint Members of the Duchesne River Formation and gave an age (McDowell and others, 1973) of 39.3 ± 0.8 Ma (40.3 corrected). In this way an entirely different means of correlation, not available to Simpson, permitted a direct tie to the type section. Since then, approximately 250 samples have been dated radioisotopically from the Trans-Pecos volcanic field of West Texas.

The K-Ar method has not been without problems. Some dates were from samples for which there was poor locality data, a problem also found in paleontology. Some dates were from samples that could not be associated with vertebrate faunas, and some dates seem to have had problems of sample freshness or problems of analysis. Nonetheless, the method has been firmly established as a highly valuable tool, although neither the data from Utah nor the dates from West Texas establish an Eocene-Oligocene boundary.

The potassium-argon dates from the continental interior of North America have been based on biotite, feldspars, or whole rock analyses of igneous rocks. Another source of potassium and argon has been the mineral glauconite. Glauconite occurs in marine and nearshore sediments, and it was hoped this would give a key to correlations to the classic continental sediments, particularly in Europe, and, through that correlation, to a date on their Eocene-Oligocene boundary. Unfortunately, there has been difficulty with dates on glauconite, because they rather consistently turn out to be younger than those based on volcanic rocks. One has to know what has been dated, as well as where it came from.

Still another technique for correlation, also developed since Simpson's (1946) work, uses the pattern of paleomagnetic normals and reversals. Some early data using magnetic direction and K/Ar data were given by Evernden and others (1964) for the late Neogene. Since then, magnetic polarity timescales have expanded to include the Paleogene and older rocks. A succession of these has appeared; the following list is not intended to be complete, but only some we have seen cited for correlation by workers investigating continental sediments: La Brecque and others (1977); Berggren and others (1978); Hardenbol and Berggren (1978); Lowrie and Alvarez (1981); and Harlan and others (1982). A polarity timescale by itself is repetitive, and must be tied to an independent non-repetitive scale such as a potassium-argon dated scale or a biostratigraphic scale, or preferably to both. A recent paper by Flynn and others (1984) uses the acronym GMPTS, to refer to "*the* Geomagnetic Polarity Time Scale" (italics ours), but nowhere do they tell the reader whose GMPTS they use. No GMPTS was available in 1946. The paleomagnetic pattern for the Vieja section was published by Testarmata and Gose (1979), and has been used, with modification, by Prothero and others (1982) in their calibration of the Oligocene.

Because of uncertainty of correlations between stratotypes of the Eocene and Oligocene and the composite marine sections identified as Eocene and Oligocene, a boundary between them has shown far more fluctuation than Simpson (1946) might have expected. It has been proposed to be as young as middle Arikareean (24 to 26 Ma) by Lander (1983) and as old as 38 Ma by Wilson (1977b). A boundary proposed by Wolfe (1981) at 32 Ma would be above the Chadronian faunas as Simpson (1946) would have identified them.

With the present state of the art, a currently favored method of correlation uses first appearance of a migrant taxon. A good example is the first appearance in North America of *Hyaenodon*. According to Mellett (1977), *H. vetus* and *H. venturae* first appear in the Sespe Formation of California in CIT locality 150 (Stock 1933a). *Hyaenodon* cf. *H. vetus* also was found in the Lapoint Member of the Duchesne River Formation. *Hyaenodon (Protohyaenodon)* sp. was identified by Schrodt (1980) and Lucas (1983) from the Baca Formation of New Mexico. The Baca Formation underlies the Spears Formation in west-central New Mexico, and dates from the Spears Formation "are consistent with a Duchesnean Age" (Lucas, 1983, p. 190). *Hyaenodon* cf. *H. vetus* (Gustafson, *in press*) is found in the Skyline channels, a marker bed at the base of the Bandera Mesa Member of the Devil's Graveyard Formation in West Texas (Stevens and others, 1984). Fortunately, two of these localities, Utah and Texas, are associated with potassium-argon dates. The Utah date in McDowell and others (1973) of 40.3 ± 0.8 Ma and the Texas date of 42.7 ± 1.6 Ma (Wilson and Schiebout, 1981) give the approximate time of arrival of *Hyaenodon* in North America.

There very well may be a *Hyaenodon* datum at 40-42 Ma. This does not help in placing an Eocene-Oligocene boundary, but it does help in placing a lower boundary for the Duchesnean land mammal age.

A summary of the faunas found in superposition in the Agua Fria area has been given by one of us (Wilson, *in press*). These faunas and the associated K/Ar dates overlap with those from the Vieja area, which is not surprising because both occur in the same volcanic province. A critical fauna in the Agua Fria area is that from the Skyline and Cotter channels. This fauna is small like that from the type Duchesnean, but it rests on a biotite bearing ash dated at 42.7 ± 1.6 Ma (Wilson, 1984). Parts of this fauna have been described previously, but except for the list in Wilson *(in press)*, it has not been summarized. Gustafson (1979) described *Simidectes;* Gustafson *(in press)* described *Harpagolestes* and *Hyaenodon;* Wilson and Szalay (1976) described *Mahgarita;* and Wilson and Schiebout (1981) described *Amynodontopsis bodei*. All of the specimens were found and stratigraphically documented by Margaret S. Stevens except the lower jaw of *Protoreodon*, which was discovered by James B. Stevens. J. A. Wilson is responsible for the identifications. The drawings are by Margaret S. Stevens.

ABBREVIATIONS

CM	Carnegie Museum of Natural History, Pittsburgh
CIT	California Institute of Technology, Pasadena
FMNH	Field Museum of Natural History, Chicago
KU	Natural History Museum, University of Kansas, Lawrence
LACM	Natural History Museum of Los Angeles County
NMC	National Museum of Canada, Ottawa
OU	Oklahoma University, Norman
PU	Princeton University, Philadelphia
SDSM	Museum of Geology, South Dakota School of Mines and Technology, Rapid City
TMM	Texas Memorial Museum, Austin
USNM	United States National Museum of Natural History
UUVP	University of Utah, Salt Lake City
YPM	Peabody Museum of Natural History, Yale University

This publication is a contribution of the Vertebrate Paleontology Laboratory, Texas Memorial Museum, The University of Texas at Austin. We are grateful to Dr. Mary Dawson for the loan of materials from the Carnegie Museum and also to Dr. R. M. West, formerly of the Geology Section, Milwaukee Public Museum, and presently at the Carnegie Museum. Mr. and Mrs. Billy Pat McKinney, the lessors and Messers H. J. Burton, Sid Burton, and Macon Richmond, owners of the Agua Fria Ranch, and Dr. Walter W. Dalquest generously permitted access to their lands. We are also grateful for access to the San Jacinto Ranch, M-Ranch, Coffee Cup Ranch, and the Montgomery ranches in the Jordan Gap and Crystal Creek area, owned or leased by Clegg and D'Ette Fowlkes. We thank Dr. Robert J. Emry for his critical reading of the manuscript and most helpful suggestions.

All potassium-argon dates used in this report have been recalculated using the tables of Dalrymple (1979).

SYSTEMATIC PALEONTOLOGY

Order INSECTIVORA Bowdich, 1821
Superfamily PANTOLESTOIDEA Cope, 1887
incertae sedis
Simidectes Stock, 1933b
Simidectes magnus (Peterson, 1919)
Figured in Gustafson (1979), Table 1

Synonyms: Pleurocyon Peterson, 1919, *Pleurocyon (Simidectes)* Stock, 1933b, *Petersonella* Kraglievich, 1948.

Type: CM 2928, mandible with P_3-M_2, Uinta Formation (Uinta C), Utah.

Referred material: TMM 41580-11, mandible with part of P_3, P_4, M_1, and part of M_2; 41578-21, partial mandible with alveoli for I_{1-3}, partial canine, alveoli for P_1 and P_2, badly worn M_{1-3}; 41578-24, mandible with partial canine, alveoli for P_1 and P_2, roots of P_3, P_4, M_1, talonid of M_2, roots of M_3; 41578-14 mandible with most of I_1, most of I_2, alveolus for I_3, C, alveoli for P_1 and 2, P_3, P_4, part of M_1, M_2; 42254-4 upper molar; all from Skyline channels. The other material referred by Gustafson (1979) is from the Serendipity local fauna of Uinta C age.

Stratigraphic position: Skyline local fauna, Skyline channels, base of Bandera Mesa Member, Devil's Graveyard Formation, Stevens and others (1984); Serendipity local fauna (listed as Whistler Squat local fauna in Gustafson, 1979).

Age: Early Duchesnean *sensu* Wilson (1984).

Description: Simidectes material from Utah and California was thoroughly described by Coombs (1971). Gustafson (1979) described the Texas material as it was known at that time. Since 1979 three additional lower jaws and a single upper molar have been collected from the Skyline channels. Some of the lower dentition (I_2, C, P_3-M_2) for the Texas *Simidectes* is now known. Nothing can be added to the description of the cheek teeth. In 41580-11, 41578-14, and -21 the symphysis is at least partly preserved to show the incisors. They are arranged almost vertically directly below the canine. Alveolus for I_1 is the smallest, and is slightly medial and ventral to I_2; I_2 is slightly larger and directly beneath the canine. All alveoli and the preserved I_2 on 41578-14 show that the incisors were procumbant and compressed. *Simidectes* must have had a very pointed, fox-like chin. The alveolus for P_2 in the Texas specimens is single rooted, whereas Coombs (1971, p. 3) said that the P_2 is "double rooted but with the roots close together." Until a larger sample of *Simidectes* is known both from Utah and Texas, we follow Gustafson (1979) and refer the Texas sample to *S. magnus*.

Order CONDYLARTHRA Cope, 1881
Family MESONYCHIDAE Cope, 1875
Subfamily MESONYCHINAE Wortman, 1901
Harpagolestes Wortman, 1901
Harpagolestes sp.

Figures and measurements given in Gustafson *(in press)*

Referred material: TMM 41715-8, right maxilla with alveoli for C^1 and P^1, P^2-M^2.

TABLE 1. MEASUREMENTS OF LOWER TEETH OF Simidectes.

		TMM 41580-11	TMM 41578-14	TMM 41578-21	TMM 41578-24
I_2	L		4.4		
C	L	9.2	7.7		
	W	4.8	4.8		
P_3	L		6.7		
	W		3.7		
P_4	L	8.5	8.9		
	W	5.3	5.0		
M_1	L	8.8			8.5
	W	5.5			
M_2	L	8.5	7.6	7.7	8.7
	W		3.9	4.6	
M_3	L	6.0e			
	W				

Stratigraphic position: Skyline local fauna, Skyline channels, base of Bandera Mesa Member, Devil's Graveyard Formation, Stevens and others (1984); Candelaria local fauna.

Age: Early Duchesnean *sensu* Wilson (1984).

Discussion: A right maxilla with unworn P^2-M^2, collected by Margaret S. Stevens in June of 1980, was prepared and sent to Eric Gustafson for study. Gustafson had reviewed the carnivorous mammals of West Texas, but the manuscript has been delayed for publication. He added a description of this specimen to the manuscript, which is scheduled to appear in 1986. *Harpagolestes* also is present in the Candelaria local fauna of Uintan age in the Vieja area (Wilson 1977b, Table 1). This may be the youngest known occurrence of *Harpagolestes,* and the only North American locality where *Harpagolestes* and *Hyaenodon* are found together.

Family HYAENODONTIDAE Leidy, 1869
Subfamily HYAENODONTINAE Leidy, 1869
Hyaenodon Laizer and de Parieu, 1838
Hyaenodon cf. *H. vetus,* Stock, 1933a

Figures and measurements given in Gustafson *(in press)*

Type: CIT 1243, skull from Sespe upper Eocene, north of Simi Valley, Ventura Co., California.

Referred material: TMM 41715-4, skull with left M^{1-2} lacking snout; 41535-1, mandible with P_3 and M_3.

Stratigraphic position: Skyline local fauna, Skyline channels, base of Bandera Mesa Member, Devil's Graveyard Formation (Stevens and others, 1984).

Age: Early Duchesnean *sensu* (Wilson, 1984).

Discussion: These specimens, as well as others, are described by Gustafson *(in press).* This is the earliest appearance of *Hyaenodon* in the Texas section, and is one of the reasons for placing the Skyline in the Duchesnean Age.

Order PRIMATES Linnaeus, 1758
Suborder STREPSIRHINI Geoffroy, 1812
Infraorder LEMURIFORMES Gregory, 1915
Superfamily ADAPOIDEA Trouessart, 1879
Family ADAPIDAE Trouessart, 1879
Mahgarita Wilson and Szalay, 1976
Mahgarita stevensi Wilson and Szalay, 1976

Type: TMM 41578-9, a crushed skull with C, P^2-M^3.

Referred material: 41578-8, uncrushed lower jaws with C, P_2-M_3; 41578-20, a second crushed skull with C, P^3-M^3 (it is doubtful whether there is room for the very small P^2, although this area is not fully prepared).

Stratigraphic position: Skyline channels, base of Bandera Mesa Member, Devil's Graveyard Formation, Stevens and others (1984).

Age: Early Duchesnean *sensu* Wilson (1984).

Discussion: Nothing new can be added to the description of *Mahgarita* at this time. The referred skull has even better preserved upper dentition than the type.

Order RODENTIA Bowdich, 1821
Family PARAMYIDAE Miller and Gidley, 1918
Subfamily PARAMYINAE Simpson, 1945
?*Leptotomus* Matthew, 1910

A comparatively well preserved right mandible with P_4-M_3 was found in the summer of 1983. Unfortunately, the incisor is missing and the jaw is somewhat flattened. The shape of the lower incisor is apparently a most important character in recognizing the genus. In the Skyline specimen the flattening may obscure the robustness of the jaw beneath the molar teeth. The teeth are well preserved and have smooth enamel. Mesoconids and mesostylids are present on M_2 and M_3. The wear on M_1 is such that a mesostylid, if originally present, was worn off, but a mesoconid is present. According to Black (1971, p. 181) mesoconids and mesostylids are "usually absent" in the subfamily, and "entoconids generally [are] continuous with posterolophids" except in *Mytonomys* where the entoconid is separate from the posterolophid. *Mytonomys* is another possible taxon to which this individual might belong. The genotype of *Mytonomys, M. robustus,* is a lower jaw fragment with only M_3. A referred specimen of *M. robustus,* PU 14658, has a prominent chin and a ridge along the lateral margin above the alveolus (Wood, 1956). In the Texas specimen, the chin area is missing and there is no strong ridge.

The lower jaw from the Skyline, although flattened, appears to have been more delicate than *Mytonomys robustus.* It is a deeper jaw than both specimens of *Leptotomus leptodus* from the Candelaria l.f. of the Vieja area described by Wood (1974), and is fully as deep as *Mytonomys robustus,* PU 14658.

Wood (1974) identified two specimens, TMM 40630-5 and 40498-6, from the Colmena Tuff of the Vieja area as *Leptotomus leptodus.* In his discussion of these specimens he does not compare them with *L. mytonensis.* Because *L. leptodus* is the genotypic species, it was hoped that the Colmena specimens might aid in the generic identification of the Skyline specimen. However, lack of the incisor frustrated this comparison. In all the taxa mentioned above, the pattern of the lower molars is very similar. This is especially true if one tries to account for smoothness of enamel because of wear. *Leptotomus kayi* is known from upper teeth only. We therefore assign this specimen to ?*Leptotomus,* with the hope that more material will be found to help solve the problem.

The proposal by Wood (1976) that *Mytonomys* should probably be allied with the European genus *Ailuravus* was questioned later by Wood (1977). However, Wood (1976) was followed by Korth (1984), who placed *Mytonomys* in the Ailuravinae Michaux, 1968. Wood (1976, p. 125-126) in describing *Ailuravus* stated: "The most striking feature of the lower incisor is its shortness. I know of no other rodent in which this tooth is so short. . . . The closest approach that I have seen in this respect is in the late Eocene *Mytonomys* of North America (Ferrusquia and Wood, 1969, Fig. 1*B*)." In the genoholotype (*Mytonomys robustus,* CM 2925, a mandibular fragment with M_3), the incisor continues for

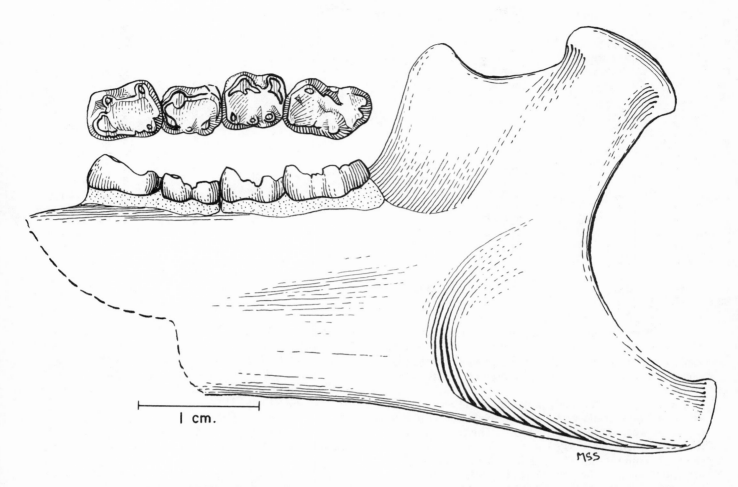

Figure 1. *?Leptotomus coelumensis* new species. TMM 41580-31. Lateral (below) and occlusal (above) views of right mandible, with P_4-M_3. Skyline channels, early Duchesnean, Trans-Pecos Texas.

a considerable distance, at least 12 mm beyond the posterior end of M_3. This seems to be a normal length. However, the subfamily relationship of *Mytonomys* is beyond the scope of this paper.

?Leptotomus coelumensis new species
Figure 1; Table 2

Type: TMM 41580-31, a right mandible with P_4-M_3.
Material: Type only.
Stratigraphic position: Skyline channels, base of Bandera Mesa Member, Devil's Graveyard Formation, Stevens and others (1984).
Age: Early Duchesnean *sensu* Wilson (1984).
Etymology: Leptotomus coelum, sky; *ensis,* resident of, in reference to the difficult climb to localities in the Skyline channels.
Diagnosis: Medium-sized paramyine; trigonid basin moderately large on P_4, very small on molars; mesostylid and mesoconid not present on P_4, both prominent on M_{1-3}; broad open talonid basin (but this may be due to wear); entoconid crest short and directed toward hypoconid; entoconid separate from posterolophid; M_3 elongate; and jaw slender but deep.

Description: The lower cheek teeth are moderately worn. The enamel is smooth on both the wear surfaces and the lower crown surfaces. However, we have found that in a larger sample from one quarry lower in the section, teeth with less wear show more surface pattern on the triturating surfaces. The lower jaw is delicate and deep (16.1 mm) beneath M_2. This is almost as deep as in *Leptotomus gigans* (17 mm estimate from cast of FMNH PM 47) of the Porvenir l.f., but the mandible is not as thick.

The P_4 is much narrower anteriorly than posteriorly (Table 2), making this tooth more triangular. In *Mytonomys* the same measurements are more nearly equal. The Texas specimen appears to have lacked the accessory ridge from the metaconid between the anterolophid and the metalophid which, according to Black (1968), evidently is unique in *Mytonomys*. The anterolophid is prominent, and the trigonid basin large. The latter is closed posteriorly by the meeting of metalophids from the protoconid and the metaconid. Although the wear surface on the metaconid is large, we do not believe it to have been as complex as the same area of the P_4 on *Mytonomys robustus*, KU 11830, as described by Black (1968). The talonid basin is broad and

TABLE 2. MEASUREMENTS OF LOWER TEETH OF SOME LARGE LATE EOCENE RODENTS.

		TYPE L. coelumensis TMM 41580-31	L. leptodus TMM 40498-6	L. leptodus TMM 40630-5	TYPE L. mytonensis CM 9253	M. robustus KU 11838	M. robustus after N	M. robustus Wood (1962) M
P_4	L	6.1			4.8	6.0	2	5.49
	W M	4.4				4.2	3	4.16
	W H	5.2				4.5	3	4.48
M_1	L	5.0	4.6	5.5	3.8	5.2	4	4.92
	W M	4.4	4.0	5.0	3.9	3.9	4	3.99
	W H		4.9	5.4	4.1		4	4.46
M_2	L	5.5	4.6	5.5		5.2	6	5.05
	W M	4.7	4.6	5.4			6	4.49
	W H	5.1	4.9	5.5		4.4	4	4.63
M_3	L	6.9	5.6		5.6		8	6.64
	W M	5.1	4.5		4.2		7	4.96
	W H	4.5	4.2		4.0	4.2	9	4.60
P_4–M_3		23.6	22.0	24.3	19.7		2	23.2

smooth. There is no mesoconid nor mesostylid on P_4. The hypoconid is prominent like the condition in CM 9387, a lower jaw referred to *Leptotomus mytonensis*. The posterolophid is not as prominent on P_4 as on the molars.

The hypoconid on M_1 is damaged. There is a well developed mesoconid and a low mesostylid, too low to reach the wear surface of the tooth. The trigonid basin is almost worn away. M_2 has a small remnant of a V-shaped trigonid basin. The metalophid and anterolophid are rounded, and the former slopes posterior into the large smooth talonid basin. Both M_2 and M_3 have prominent mesoconids and mesostylids. M_3 has a small trigonid basin. The posterolophid is separated from the entoconid as in *Mytonomys robustus*; in fact the M_3s of 41580-31 and that of PU 14658 are very similar. In the Texas specimen, however, there is no accessory ridge from the metaconid between the anterolophid and the metalophid.

Discussion: ?*Leptotomus coelumensis* is a late Eocene paramyine closely related to *Leptotomus mytonensis, Mytonomys robustus,* and probably *Leptotomus kayi*, if lower teeth were known for the last. Other paramyines are known from the Purple Bunch and Serendipity localities lower in the section, so it seems best to await their description before speculating further on relationships of the specimen from the Skyline.

Order PERISSODACTYLA Owen, 1848
Family BRONTOTHERIIDAE Marsh, 1873
Duchesneodus Lucas and Schoch, 1982
Duchesneodus uintensis (Peterson, 1931)
?*Duchesneodus* cf. *uintensis*
Figures 2 and 3; Table 3 and 4

Material: TMM 41715-6, left mandible with posteroexternal part of P_2, external half of P_3, P_4-M_3; TMM 41580-30, premaxillary symphysis with left and right I^{2-3}, left C; TMM 41578-10, partial maxilla with external portions of C, P^2-M^3; TMM 41715-7, fragmentary right M_2-M_3; probable reference 41853-17, palate with left and right I^{2-3}, C, right P^1-M^3, left P^2-M^3.

Stratigraphic position: Skyline channels, base of Bandera Mesa Member, Devil's Graveyard Formation, Stevens and others (1984).

Age: Early Duchesnean *sensu* Wilson (1984).

Discussion: Lucas and Schoch (1982) provided a valuable contribution in showing that the holotype of "*Teleodus*" *avus*, YPM 10321, had only two lower incisors instead of three. They proposed the genus *Ducheneodus*, with *D. uintensis* as the type species, based on material in the collection of the Carnegie Museum from the titanothere quarry eleven miles west of Vernal, in the Lapoint Member of the Duchesne River Formation. Lucas kindly pointed out to Wilson that a well preserved skull in the Field Museum of Natural History, FMNH PM 136, from the Blue Cliff horizon in the Chambers Tuff at Rifle Range Hollow, has a prominent convexity on the top of the skull. Lucas (*correspondence* to Wilson, 1981) stated "This specimen appears to be *Duchesneodus uintensis* and that taxon would become a part of the Porvenir local fauna." He went on to say: "The greatly exaggerated cranial convexity and massive zygomatic arches of the Field Museum specimen suggest that it is a male (this sort of sexual dimorphism can be seen in the Carnegie quarry sample)."

Lucas compared the palate of TMM 41853-17 from the Stone Corral locality in the Cotter channels to one from the Porvenir, and tentatively concluded that because of certain differences and smaller size 41853-17 might be a female, and perhaps a different species of "*Teleodus*."

We agree with Lucas' comment that FMNH PM 136 is *Duchesneodus*, but Lucas and Schoch (1982) did not give statistical data on which to properly judge specific assignments. They gave measurements for the cheek teeth of the holotype CM 11809 and the paratype CM 11761, both of which were identified as females by Peterson

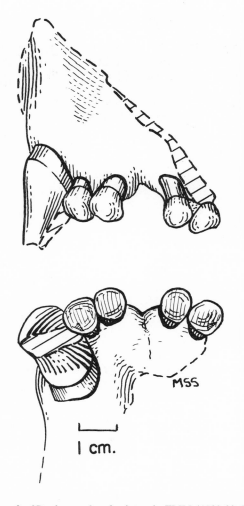

Figure 2. *?Duchesneodus* cf. *uintensis*. TMM 41580-30. Premaxillary symphysis. Top anterior view; bottom posterior view. Left and right I^{2-3}, left C^1. Skyline channels, early Duchesnean, Trans-Pecos Texas.

(1931). A lower jaw (OU 18-4-5-17), that Wilson (1977a) identified as *Menodus bakeri* from the Porvenir local fauna, is much larger (Table 3) than either of the Carnegie Museum specimens, and the lower canines are procumbent and long rooted.

Unfortunately, no statistical data for sexual dimorphism in titanotheres is available, and samples of the Skyline and Porvenir are not adequate to come to any conclusions as to species. We can only say that the Skyline mandible, 41715-6, is a smaller, long, shallow jaw, like *Dolichorhinus* and different from those in the Porvenir local fauna. We hesitate to claim that three Porvenir jaws are male and that the single Skyline jaw is female.

There are some similarities between *Notiotitanops mississippiensis* and *?Duchesneodus* from Texas. Both have the blunt, rounded incisors and a diastema between C and P^1. However, the tetartocone is a distinct round cusp on the Texas specimen 41853-17, and better developed than it is on *Notiotitanops*. The hypocones on the upper molars are about equally developed.

Family HYRACODONTIDAE Cope, 1879
?Hyracodon Leidy, 1856
?Hyracodon sp. indet.
Figure 4; Table 5

Material: TMM 41580-28, left mandibular ramus with P_2-M_3 badly worn.

Stratigraphic position: Skyline channels, base of Bandera Mesa Member, Devil's Graveyard Formation, Stevens and others (1984).

Age: Early Duchesnean *sensu* Wilson (1984).

Description: A small hyracodontid with no P_1. The occlusal pattern of P_2 is completely worn off. On P_3 and P_4 the buccal portion of the groove anterior to the hypolophid is all of the pattern that remains. M_1 and M_2 are completely worn. Both the meta- and hypolophid are preserved on M_3, and no hypoconulid is present.

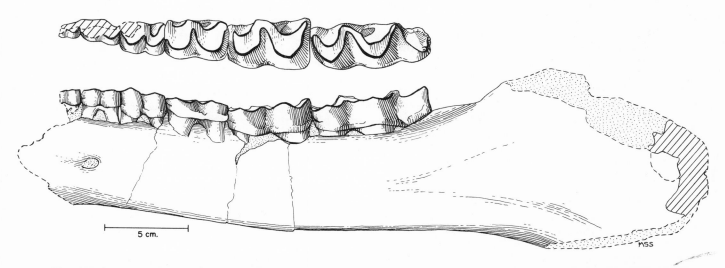

Figure 3. *?Duchesneodus* cf. *uintensis*. TMM 41715-6 left mandible in occlusal (above) and lateral (below) views, with P_2-M_3. Skyline channels, early Duchesnean, Trans-Pecos Texas.

TABLE 3. MEASUREMENTS OF UPPER TEETH OF Duchesneodus.

		Duchesneodus sp. Texas TMM 41853-17	TMM 41578-10	D. uintensis Peterson (1931) Paratype CM 11759	Notiotitanops Gazin and Sullivan (1942) USNM 16646	D. thyboi South Dakota Bjork (1967) SDSM 63690	SDSM 63689	D. uintensis Utah Nelson et al. (1980) UUVP 9501
I^2	L	9.0						
	W	8.3						
I^3	L	9.6			10.5			
	W	8.4			8.8			
C	L	17.8	26.7@		27			
	W	14.3			24			
P^1	L	14.0@		14	17.2			
	W	11.9		14	13.2			
P^2	L	25.8	22@	20	20.7	22.5		24
	W	23.5		24	23.3	24.6		
P^3	L	26.9	27.8	27	25.2	26.8	26.5	28
	W	30.8		29	30.7	29.5	31.4	
P^4	L	30.0	32.8	30	29.3		33.8	32
	W	38.0		32	37.4		39.9	
M^1	L	45.0		36	43.6		46.3	47
	W	44.2		39	42.3		45.8	
M^2	L	57.5	64.0	50	53.5		57.6	57
	W	54.6		49	51.4		57.4	54
M^3	L	65.3	68.6	55	58.0		60.4	67
	W	57.0		53	56.5		57.1	61
$C-M^3$		290	350@		282		291	
P^1-M^3		250		230	229			
P^1-P^4		91		88	90.7			
M^1-M^3		166		140	147			

Discussion: This small hyracodontid is an unexpected occurrence. It cannot belong with *Triplopus* because it has lost P_1. Teeth of the Texas specimen fall within the range of measurements of *Triplopus obliquidens* given by Radinsky (1967, Table 3) and, although no measurements were given for a P_1, presumably it was present. Wood (1927, p. 33) stated that in the holotype of *"Eotrigonias" rhinocerinus,* YPM 13331, "P_1 is represented by part of the alveolus. It was probably of fair size." "*E.*" *rhinocerinus* was placed in *Triplopus* by Radinsky (1967).

The second premolar of the Skyline specimen is smaller than that of *Hyracodon primus* of the Porvenir local fauna of the Vieja area. The Skyline specimen may have been an immigrant from Asia, having been a larger form belonging a *Rhodopagus*-like group. Like *Rhodopagus,* the Texas specimen has no P_1 and no hypoconulid, but the Texas specimen is about twice the size of *R. minimus.* The Skyline specimen does not seem to have short enough premolars to be a small amynodontid. *Toxotherium,* another small ceratomorph, has been shown by Emry (1979) to have lost both P_1 and P_2. In addition, *Toxotherium* has a procumbent tusk, an alveolus for which is lacking in TMM 41580-28.

Amynodontopsis Stock, 1933c
Amynodontopsis bodei Stock, 1933c

Material: TMM 41715-2, crushed skull with LP^1, dP^{2-4}, erupting M^1, RdP^{3-4}, erupting M^{1-2}; TMM 41668-8, uncrushed lower jaw with LI_1, 2, alv. I_3, RI_1, alv. I_{2-3}, RLC, P_3-M_3; TMM 41580-6, RM_1.

Stratigraphic position: Skyline channels, base of Bandera Mesa Member, Devil's Graveyard Formation, Stevens and others (1984).

Age: Early Duchesnean *sensu* Wilson (1984).

Discussion: Stock (1933c) described *Amynodontopsis bodei* from the Sespe Formation, CIT locality 150. Nothing can be added to the description or discussion given by Wilson and Schiebout (1981).

Order ARTIODACTYLA Owen, 1848
Family AGRIOCHOERIDAE Leidy, 1869
Protoreodon Scott and Osborn, 1887
Protoreodon pumilus (Marsh, 1875)

Material: TMM 41580-1, left mandible with I_3, C, P_1-M_3.

Stratigraphic position: Skyline channels, base of Bandera Mesa Member, Devil's Graveyard Formation, Stevens and others (1984).

TABLE 4. MEASUREMENTS OF LOWER TEETH OF Duchesneodus.

		Lucas and Schoch 1982			?Duchesneodus sp. Texas		
		Duchesneodus uintensis		Duchesneodus californicus			
		CM 11809 Holotype	CM 11761 Paratype	LACM 1398 Holotype	TMM 41715-6 Skyline	TMM 40840-38 Porvenir	OU 18-4-5-17 Porvenir
P_2	L	21.9	19.2	21.7			26.8
	AW	13.0	11.9	12.6			16.3
	PW	15.6	12.2@	13.4			20.2
P_3	L	26.3	23.8	23.8	24.2		32.3
	AW	17.8	14.3	15.6			21.0
	PW	20.2	16.2@	18.8			25.1
P_4	L	32.2	27.2	27.8	28.0		
	AW	21.9	17.2	21.0			
	PW	23.6	19.2	23.1	20.5		
M_1	L	36.4	34.7	38.6	38.5		
	AW	23.4	20.4	24.9	21.8		
	PW	24.1	23.4	27.4	24.7		
M_2	L	47.2	48.0	48.5	49.5	57.9	62.8
	AW	27.4	26.2	28.5	27.0		36.3
	PW	31.1	28.1	32.2	30.8		38.0@
M_3	L	77.7	66.4		74.5	79.3	90.2
	AW	31.2	28.4	31.3@	28.1	33.5	36.8
	PW	30.6	29.0		29.6	34.4	38.0

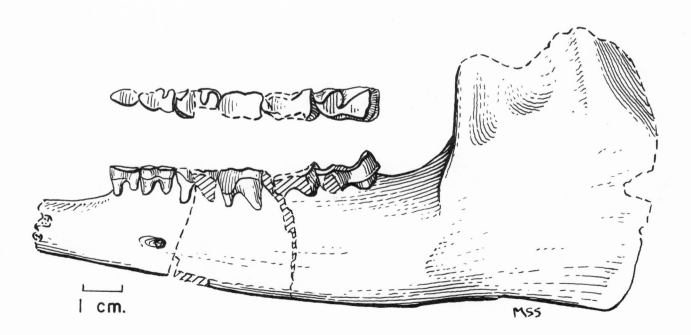

Figure 4. *?Hyracodon* sp. indet. TMM 41580-28. Left mandible with occlusal (above) and lateral (below) views, with P_2-M_3. Skyline channels, early Duchesnean, Trans-Pecos Texas.

TABLE 5. MEASUREMENTS OF LOWER TEETH OF SMALL LATE EOCENE CERATOMORPHS.

		?Hyracodon sp. indet. TMM 41580-28	Hyracodon primus TMM 40203-21	Toxotherium hunteri Emry (1979) NMC 8918 (type)
P_2	L	7.8	10.8	
	W	4.6	7.4	
P_3	L	9.7	12.5	
	W	7.2	8.9	
P_4	L		12.6	8.3
	W		9.9	5.6
M_1	L	12@	13.7	10.3
	W	8@	10.0	6.1
M_2	L		15.8	14.1
	W		12.2@	8.0
M_3	L	15.1		
	W	8.6		
P_2-M_3		72.0		
P_2-P_4		28.2		
M_1-M_3		42.5		

Age: Early Duchesnean *sensu* Wilson (1984).

Description: Protoreodon pumilus was described adequately by Wilson (1971), and nothing new can be added. The length of M_1 is 9.3 mm, the length of M_3 is 17.5 mm, and, if plotted on Figure 5 of Wilson (1971), the specimen falls within the size range of the sample of *P. pumilus* from Myton Pocket.

Agriochoerus Leidy, 1850
Agriochoerus sp. indet.
Figure 5; Table 6

Material: TMM 42254-6, palate with right C, P^1, partial P^3, broken P^4, M^{1-3}, left P^3-M^3, TMM 41580-14, fragmentary M^{1-2}, M^3.

Stratigraphic position: Skyline Channels, base of Bandera Mesa Member, Devil's Graveyard Formation, Stevens and others (1984).

Age: Early Duchesnean *sensu* Wilson (1984).

Description: The skull of TMM 42254-6 was removed by erosion, and only the crowns of the teeth are preserved, in their approximate natural position. A diastema separates C and P^1, and P^1 and P^2. P^4 has a single external cusp. On the molars there is only a slight indication of a swelling where a protoconule would be expected.

Discussion: Black (1978) reviewed the Agriochoeridae of the Badwater fauna. He synonymyzed *Protoreodon* near *P. pumilus* Gazin 1956, *Protoreodon pearcei* Gazin 1956, and *Diplobunops* cf. *matthewi* Gazin 1956 with *Diplobunops matthewi*. As Black (1978, p. 234) stated: "From its inception *Diplobunops* is distinguished from *Protoreodon* by characters of the rostrum and anterior premolar dentition." The broad snout seems to be the most consistent character in *Diplobunops*. In addition, *Diplobunops* has well-developed protoconules as in *Protoreodon*. None of the larger agriochoerids from the Skyline channels, Montgomery bonebed (Wilson, 1984, Fig. 1), nor Porvenir local fauna has protoconules, nor do they seem to have a particularly broad snout. A single external cusp on P^4 is not characteristic of *Agriochoerus antiquus* of the Oligocene, but, as pointed out in Wilson (1971) and Black (1978), the P^4 during the Uintan and Duchesnean is highly variable in this group. The loss of the protoconule, however, is not variable in the Texas population. The diasetemas between C and P^1 and between P^1 and P^2 were considered by Gazin (1955) as characteristic of *Diplobunops,* but, as Black (1978) pointed out, there is no diastema between P^1 and P^2 in *D. matthewi*. Peterson (1931) compared *Diplobunops* with *Agriochoerus*, particularly the limbs and forefoot. Unfortunately, no postcranial material is known from the Skyline channels that is associated with an agriochoerid dentition.

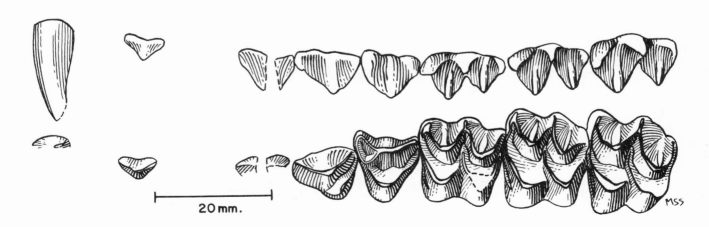

Figure 5. *Agriochoerus* cf. *antiquus*. TMM 42254-6. Lateral (above) and occlusal (below) views, with C^1-M^3. C^1 and P^{1-2} reversed, missing parts of P^2-M^3 restored from opposite side. Skyline channels, early Duchesnean, Trans-Pecos Texas.

TABLE 6. MEASUREMENTS OF UPPER TEETH OF ARTIODACTYLS FROM SKYLINE CHANNELS.

		Agriochoerus cf. antiquus TMM 42254-6	Hendryomeryx sp. TMM 41580-2	Leptoreodon sp. TMM 41580-27
C	L	7.6@		
P^3	L	11.5		
	W	10.1		
P^4	L	11.5@		6.2
	W	13.0		3.5
M^1	L	14.6		6.0
	W	16.3		4.8
M^2	L	15.1		6.1@
	W	18.7		
M^3	L	17.0	9.8	9.8@
	W	17.8	4.6	
$C-M^3$		102@		
P^1-M^3		96@		
M^1-M^3		45.0		

Family PROTOCERATIDAE Marsh, 1891
Leptoreodon Wortman, 1898
Leptoreodon sp.
Figure 6; Table 6

Material: TMM 71580-27, right mandible with partial P3, worn P4-M3.

Stratigraphic position: Skyline channels, base of Bandera Mesa Member, Devil's Graveyard Formation, Stevens and others (1984).

Age: Early Duchesnean *sensu* Wilson (1984).

Discussion: Leptoreodon has been described by Gazin (1956), Golz (1976), Black (1978), and Wilson (1984). All agree that the differences between *Leptoreodon* and *Leptotragulus* are minor and involve slight differences in the lower premolars, and that the two genera possibly are synonyms. The Skyline specimen is badly worn and does not help solve the problem. M2 has enough of the pattern remaining to show that the metaconid overlaps the entoconid lingually, and the anterior cristid of the protoconid is cut off from the lingual edge of the tooth by the metaconid, as is the posterior cristid of the hypoconid. This is the typical pattern in *Leptoreodon* and *Leptotragulus*, as described by Black (1978). P4 is short, has a paraconid, and seems to lack a prominent metaconid. The last character is known to be variable in the larger samples from the Whistler Squat quarry and the Purple Bench locality, both of which are lower in the section.

Family LEPTOMERYCIDAE Scott, 1899
Hendryomeryx Black, 1978
Hendryomeryx sp.
Figure 7; Table 6

Material: TMM 41580-2, isolated M3.

Stratigraphic position: Skyline Channels, base of Bandera Mesa Member, Devil's Graveyard Formation, Stevens and others (1984).

Age: Early Duchesnean *sensu* Wilson (1984).

Description: A worn isolated M3 (Fig. 7), with the metaconid not quite fused to the entoconid (as in *Hendryomeryx wilsoni*). The anterior and posterior crests of the protoconid connect with the metaconid. The hypoconid is isolated. The heel of M3 is open lingually, with only a buccal crest connected to the worn base of a cuspule that lies posterior to the entoconid.

Discussion: Hendryomeryx was named and described by Black (1978, p. 254), who pointed out the distinctive pattern on the lower molars in which the "entoconid and metaconid are connected through a straight lingual crest." Black named *H. wilsoni* based on specimens from Badwater Locality 20, and referred *Leptomeryx defordi* Wilson (1974) to *H. defordi*. The latter taxon is part of the Porvenir local fauna. Although Black (1978, p. 254) stated that "selenes [are] moderately developed on M1-M3", he did not list a specimen with an M3 in the hypodigm, nor did he figure one. Identification of the isolated M3 from the Skyline is based on comparison with material collected since Wilson (1974) was published.

I agree with Black that *Hendryomeryx* is a distinct genus. There is, however, considerable variability within the sample of *Hendryomeryx* from the Vieja area. For example, connection of the entoconid-metaconid is present only in worn specimens. The heel on M3 may be open or closed, but the lingual crest is always stronger, and the buccal closure is formed by cuspules posterior to the entoconid.

CONCLUSIONS

The known fauna of the Skyline Channels is small, but growing. It contains a mixture of taxa, some of which are, for Texas, last appearances or first appearances, and others that pass on through. It has the last known appearance of *Simidectes, Harpagolestes,* and *Leptoreodon*. It has the first appearance for *Hyaenodon, ?Duchesneodus, ?Hyracodon, Agriochoerus,* and *Hendryomeryx*. If our queried *Leptotomus* is correctly placed, it is a "pass through," along with *Protoreodon pumilus*. *Mahgarita* is, so far at least, endemic. The most common taxon, *Simidectes*, is not known in the Porvenir local fauna, nor are *Harpagolestes* and *Leptoreodon*, and this is the basis for Wilson (1984) recognizing an earlier and a later Duchesnean. The appearance of *Hyaenodon* seems to be a useful stratigraphic datum. *Mahgarita* and *?Hyracodon* are not widely enough known to be useful for correlation and, therefore, to use Simpson's (1946) criteria, are not convenient for recognizing a boundary. In 1946 the Duchesnean controversy was confined to Utah. Now the Duchesnean has been recognized at Badwater, Wyoming, southern California, New Mexico, and Texas. Little wonder the disagreement has increased since 1946. As for agreement on an Eocene-Oligocene boundary, we wonder whose terminal event will prove to be most convenient.

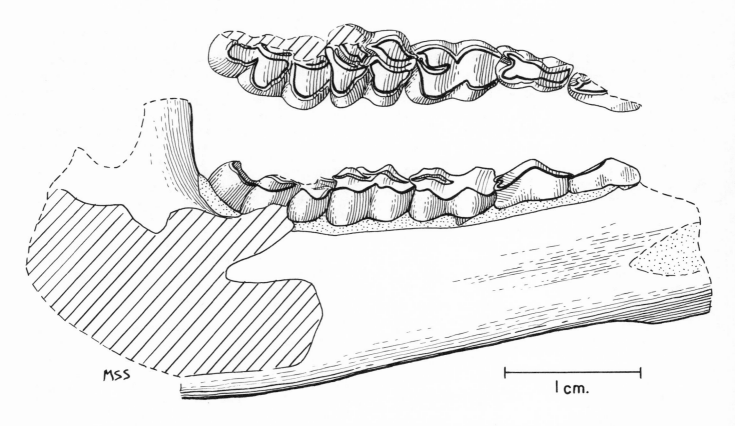

Figure 6. *Leptoreodon* sp. TMM 41580-27. Occlusal (above) and lateral (below) views of right mandible, with P_3-M_3. Skyline channels, early Duchesnean, Trans-Pecos Texas.

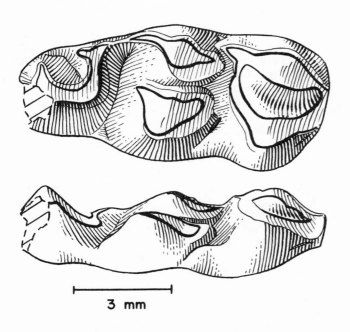

Figure 7. *Hendryomeryx* sp. TMM 41580-2. Occlusal (above) and lateral (below) views of M_3. Skyline channels, early Duchesnean, Trans-Pecos Texas.

REFERENCES CITED

Berggren, W. A., McKenna, M. C., Hardenbol, J., and Obradovich, J. D., 1978, Revised Paleogene polarity time scale: Journal of Geology, v. 86, p. 67-81.

Bjork, P. R., 1967, Latest Eocene vertebrates from northwestern South Dakota: Journal of Paleontology, v. 41, p. 227-236.

Black, C. C., 1968, The Uintan rodent *Mytonomys: ibid.*, v. 42, p. 853-856.

_____ 1971, Paleontology and geology of the Badwater Creek area, central Wyoming Part 7. Rodents of the family Ischyromyidae: Annals of Carnegie Museum, v. 43, p. 179-217.

_____ 1978, Paleontology and Geology of the Badwater Creek area, Central Wyoming Part 14. The artiodactyls: *ibid.*, v. 47, p. 223-259.

Coombs, M. C., 1971, Status of *Simidectes* (Insectivora, Pantolestoidea) of the late Eocene of North America: American Museum Novitates, no. 2455, 41 p.

Dalrymple, G. B., 1979, Critical tables for conversion of K-Ar ages from old to new constants: Geology, v. 7, p. 558-560.

Emry, R. J., 1979, Review of *Toxotherium* (Perissodactyla: Rhinocerotoidea) with new material from the Oligocene of Wyoming: Biological Society of Washington, Proceedings, v. 92, p. 28-41.

―――― 1981, Additions to the mammalian fauna of the type Duchesnean, with comments on the status of the Duchesnean "Age": Journal of Paleontology, v. 55, p. 563-570.

Evernden, J. F., Savage, D. E., Curtis, G. H., and James, G. T., 1964, Potassium-argon dates and the Cenozoic mammalian chronology of North America: American Journal of Science, v. 262, p. 145-198.

Ferrusquia-Villafranca, I., and Wood, A. E., 1969, New fossil rodents from the early Oligocene Rancho Gaitan local fauna, northeastern Chihuahua, Mexico: Texas Memorial Museum, Pearce-Sellards Series no. 16, 11 p.

Flynn, J. J., MacFadden, B. J., and McKenna, M. C., 1984, Land mammal ages, faunal heterochrony, and temporal resolution in Cenozoic terrestrial sequence: Journal of Geology, v. 92, p. 687-705.

Gazin, C. L., 1955, A review of the upper Eocene artiodactyla of North America: Smithsonian Miscellaneous Collections, v. 128, no. 8, iii + 96 p.

―――― 1956, The geology and vertebrate paleontology of upper Eocene strata in the northeastern part of the Wind River Basin, Wyoming Part 2. The mammalian fauna of the Badwater area: *ibid.*, v. 131, no. 8, 35 p.

Gazin, C. L., and Sullivan, J. M., 1942, A new titanothere from the Eocene of Mississippi with notes on the correlation of the Gulf Coastal Plain and the continental Eocene of the Rocky Mountain Region: *ibid.*, v. 101, no. 13, 13 p.

Golz, D. J., 1976, Eocene Artiodactyla of southern California: Natural History Museum of Los Angeles County, Science Bulletin 26, 85 p.

Gustafson, E. P., 1979, Early Tertiary vertebrate faunas Big Bend area, Trans-Pecos Texas: *Simidectes* (Mammalia, Insectivora): Texas Memorial Museum, Pearce-Sellards Series no. 31, 9 p.

―――― in press, Carnivorous mammals of the late Eocene and early Oligocene of Trans-Pecos Texas: Texas Memorial Museum, Bulletin 33.

Hardenbol, J., and Berggren, W. A., 1978, A new Paleogene numerical time scale, *in* Cohee, G. V., Glaessner, M. F., and Hedberg, H. D., eds., Contributions to the geologic time scale: American Association of Petroleum Geologists, Studies in Geology no. 6, p. 213-234.

Harland, W. B., Cox, A. V., Llewellyn, P. G., Pickton, C. A. G., Smith, A. G., and Walters, R., 1982, A geologic time scale: Cambridge, Cambridge University Press, 128 p.

Kay, J. L., 1934, The Tertiary formation of the Uinta Basin, Utah: Annals of Carnegie Museum, v. 23, p. 357-371.

Korth, W. W., 1984, Earliest Tertiary evolution and radiation of rodents in North America: Carnegie Museum of Natural History, Bulletin, no. 24, 71 p.

Kraglievich, L. J., 1948, Substitucion de un nombre generico: Anales de la Sociedad de Ciencias de Argentina, v. 146, p. 161-162.

LaBrecque, J. L., Kent, D. V., and Cande, S. C., 1977, Revised magnetic polarity time scale for late Cretaceous and Cenozoic time: Geology, v. 5, p. 330-335.

Laizer, L. de, and Parieu, de, 1838, Description et détermination d'une mâchoire fossile appartenant à un mammifère jusqu'à présent inconnu, *Hyaenodon leptorhynchus:* Comptes-rendus hebdomadaires des séances de l'Académie des sciences, Paris, v. 7, p. 442.

Lander, E. B., 1983, Continental vertebrate faunas from the upper member of the Sespe Formation, Simi Valley California, and the terminal Eocene event, *in* Squires, R. R., and Filewicz, M. V., eds., Cenozoic geology of the Simi Vally area, Southern California: Pacific Section, Society of Economic Paleontologists and Mineralogists, Fall Field Trip Volume and Guidebook, p. 142-144.

Leidy, J., 1850, Descriptions of *Rhinoceros nebrascensis, Agriochoerus antiquus, Palaeotherium proutii,* and *P. bairdii:* Academy of Natural Sciences of Philadelphia, Proceedings, v. 5, p. 121-122.

―――― 1856, Notices of several genera of extinct Mammalia, previously less perfectly characterized: *ibid.*, v. 8, p. 91-92.

Lowrie, W., and Alvarez, W., 1981, One hundred million years of geomagnetic polarity history: Geology, v. 9, p. 392-397.

Lucas, S. G., 1983, The Baca Formation and the Eocene-Oligocene boundary in New Mexico: New Mexico Geological Society Guidebook, 34th Field Conference, Socorro Region II, p. 187-192.

Lucas, S. G., and Schoch, R. M., 1982, *Duchesneodus,* a new name for some titanotheres (Perissodactyla, Brontotheriidae) from the late Eocene of western North America: Journal of Paleontology, v. 56, p. 1018-1023.

McDowell, F. W., Wilson, J. A., and Clark, J., 1973, K-Ar dates for biotite from two paleontologically significant localities: Duchesne River Formation, Utah and Chadron Formation, South Dakota: Isochron West no. 7, p. 11-12.

Marsh, O. C., 1875, Notice of new Tertiary mammals, IV: American Journal of Science and Arts, v. 9, p. 239-250.

Matthew, W. D., 1910, On the osteology relationships of *Paramys,* and the affinities of the Ischyromyidae: American Museum of Natural History, Bulletin, v. 28, p. 43-72.

Maxwell, R. A., Lonsdale, J. T., Hazzard, R. T., and Wilson, J. A., 1967, Geology of Big Bend National Park, Brewster County Texas: Bureau of Economic Geology, The University of Texas, Publication no. 6711, 320 p.

Mellett, J. S., 1977, Paleobiology of North American *Hyaenodon* (Mammalia, Creodonta), *in* Hecht, M. K., and Szalay, F. S., eds., Contributions to evolution: Basel, S. Karger, 134 p.

Michaux, J., 1968, Les Paramyidae (Rodentia) de l'Eocene inferieur du Basin de Paris: Paleovertebrata, v. 1, p. 135-193.

Nelson, M. E., Madsen, J. H., Jr., and Stokes, W. L., 1980, A titanothere from the Green River Formation, central Utah: *Teleodus uintensis* (Perissodactyla: Brontotheriidae): Contributions to Geology, University of Wyoming, v. 18, p. 127-134.

Peterson, O. A., 1919, Report on the material discovered in the upper Eocene of the Uinta Basin by Earl Douglass in the years 1908-1909 and by O. A. Peterson in 1912: Annals of Carnegie Museum, v. 12, p. 40-178.

―――― 1931, New species of the genus *Teleodus* from the upper Uinta of northeastern Utah: *ibid.*, v. 20, p. 307-312.

Prothero, D. R., Denham, C. R., and Farmer, H. G., 1982, Oligocene calibration of the magnetic-polarity timescale: Geology, v. 10, p. 650-653.

Radinsky, L. B., 1967, A review of the rhinocerotoid family Hyracodontidae (Perissodactyla): American Museum of Natural History, Bulletin, v. 136, p. 1-46.

Schrodt, A. K., 1980, Depositional environments, provenance, and vertebrate paleontology of the Eocene-Oligocene Baca Formation, Catron County, New Mexico [M.S. thesis]: Baton Rouge, Louisiana, Louisiana State University, 174 p.

Scott, W. B., 1945, The Mammalia of the Duchesne River Oligocene: American Philosophical Society, Transactions, new series, v. 34, p. 209-253.

Scott, W. B., and Osborn, H. F., 1887, Preliminary report on the vertebrate fossils of the Uinta Formation, collected by the Princeton expedition of 1886: American Philosophical Society, Proceedings, v. 24, p. 255-264.

Simpson, G. G., 1946, The Duchesnean fauna and the Eocene-Oligocene boundary: American Journal of Science, v. 244, p. 52-57.

Stevens, J. B., Stevens, M. S., and Wilson, J. A., 1984, Devil's Graveyard Formation (new) Eocene and Oligocene Age Trans-Pecos Texas: Texas Memorial Museum, Bulletin 32, 21 p.

Stock, C., 1933a, Hyaenodontidae of the upper Eocene of California: National Academy of Sciences, USA, Proceedings, v. 19, p. 434-440.

―――― 1933b, A miacid from the Sespe upper Eocene, California: *ibid.*, v. 19, p. 481-486.

―――― 1933c, An amynodont skull from the Sespe deposits, California: *ibid.*, v. 19, p. 762-767.

Stovall, J. W., 1948, Chadron vertebrate fossils from below the Rim Rock of Presidio County, Texas: American Journal of Science, v. 246, p. 78-95.

Testarmata, M. M., and W. A. Gose, 1979, Magnetostratigraphy of the Eocene-Oligocene Vieja Group, Trans-Pecos Texas: Bureau of Economic Geology, University of Texas at Austin, Guidebook, 19, p. 55-66.

Wilson, J. A., 1971, Early Tertiary vertebrate faunas, Vieja Group Trans-Pecos Texas: Agriochoeridae and Merycoidodontidae: Texas Memorial Museum, Bulletin 18, 83 p.

―――― 1974, Early Tertiary vertebrate faunas, Vieja Group and Buck Hill Group, Trans-Pecos Texas; Protoceratidae, Camelidae, Hypertragulidae: *ibid.*, no. 23, 34 p.

―――― 1977a, Early Tertiary vertebrate faunas, Big Bend area, Trans-Pecos Texas: Brontotheriidae: Texas Memorial Museum, Pearce-Sellards Series no. 25, 17 p.

―――― 1977b, Stratigraphic occurrence and correlation of early Tertiary vertebrate faunas, Trans-Pecos Texas, Part 1: Vieja area: Texas Memorial Museum, Bulletin 25, 42 p.

―――― 1984, Vertebrate faunas 49 to 36 million years ago and additions to the species of *Leptoreodon* (Mammalia: Artiodactyla) found in Texas: Journal of Vertebrate Paleontology, v. 4, p. 199-207.

―――― *in press,* Stratigraphic occurrence and correlation of early Tertiary vertebrate faunas, Trans-Pecos Texas, Part 2. Agua Fria-Green Valley areas: Journal of Vertebrate Paleontology.

Wilson, J. A., and Schiebout, J. A., 1981, Early Tertiary vertebrate faunas Trans-Pecos Texas: Amynodontidae: Texas Memorial Museum, Pearce-Sellards Series no. 33, 62 p.

Wilson, J. A., and Szalay, F. S., 1976, New adapid primate of European affinities from Texas: Folia Primatologica, v. 25, p. 294-312.

Wilson, J. A., Twiss, P. C., De Ford, R. K., and Clabaugh, S. E., 1968, Stratigraphic succession, potassium-argon dates and vertebrate faunas, Vieja Group, Rim Rock Country, Trans-Pecos Texas: American Journal of Science, v. 266, p. 590-604.

Wolfe, J. A., 1981, A chronologic framework for Cenozoic megafossil floras of northwestern North America and its relation to marine geochronology: Geological Society of America, Special Paper 184, p. 39-47.

Wood, A. E., 1956, *Mytonomys,* a new genus of paramyid rodent from the upper Eocene: Journal of Paleontology, v. 30, p. 753-755.

―――― 1974, Early Tertiary vertebrate faunas Vieja Group Trans-Pecos Texas: Rodentia: Texas Memorial Museum, Bulletin 21, 112 p.

―――― 1976, The paramyid rodent *Ailuravus* from the middle and late Eocene of Europe, and its relationships: Palaeovertebrata, v. 7, p. 117-149.

―――― 1977, The Rodentia as clues to Cenozoic migrations between the Americas and Europe and Africa, *in* West, R. M., ed., Paleontology and plate tectonics with special reference to the history of the Atlantic Ocean: Proceedings of a symposium presented at the North American Paleontological Convention II, Lawrence, Kansas, Milwaukee Public Museum, Special Publications in Biology and Geology no. 2, p. 95-109.

Wood, H. E., 2nd, 1927, Some early Tertiary rhinoceroses and hyracodonts: Bulletins of American Paleontology, v. 13, no. 50, 105 p.

Wood, H. E., 2nd, Chaney, R. W., Clark, J., Colbert, E. H., Jepsen, G. L., Reeside, J. B., Jr., and Stock, C., 1941, Nomenclature and correlation of the North American continental Tertiary: Geological Society of America, Bulletin, v. 52, p. 1-48.

Wortman, J. L., 1898, The extinct Camelidae of North America, and some associated forms: American Museum of Natural History, Bulletin, v. 10, p. 93-142.

⸺ 1901, Studies of Eocene Mammalia in the Marsh collection, Peabody Museum. Part I: Carnivora: American Journal of Science, Ser. 4, v. 11, p. 333-348, 437-450; v. 12, p. 143-154, 193-206, 281-296, 377-382, 421-432.

MANUSCRIPT RECEIVED NOVEMBER 18, 1985
REVISED MANUSCRIPT RECEIVED MAY 27, 1986
MANUSCRIPT ACCEPTED JUNE 26, 1986

Systematics and evolution of *Pseudhipparion* (Mammalia, Equidae) from the late Neogene of the Gulf Coastal Plain and the Great Plains

S. DAVID WEBB   *Florida State Museum, University of Florida, Gainesville, Florida 32611*

RICHARD C. HULBERT, JR.   *Department of Zoology, University of Florida, Gainesville, Florida 32611*

ABSTRACT

The seven species of *Pseudhipparion* are, from oldest to youngest, as follows: unnamed species; *P. retrusum*; *P. curtivallum*, new combination; *P. hessei*, new species; *P. gratum*; *P. skinneri*, new species; and *P. simpsoni*, new species. They range from late Barstovian through latest Hemphillian in the Great Plains and the Gulf Coastal Plain. Both the phylogenetic relationships among these species and the phylogenetic position of this genus among other hipparionine horses are discussed.

A notable feature of *Pseudhipparion* evolution is its prevailing tendency toward dwarfing. All crown dimensions except unworn height decrease through time, although analysis is complicated by the fact that sample means are larger in the Great Plains than in contemporaneous Gulf Coast samples, in keeping with Bergmann's Rule. Rates of change in several dental measurements between various *Pseudhipparion* species pairs, calculated over intervals of about one million years, have a mean value of 0.11 darwins, which is equal to or greater than in other hipparionine species pairs. In *P. simpsoni*, the final late Hemphillian species, root formation was delayed ontogenetically, producing extremely high-crowned (incipiently hypsodont) cheek teeth and incisors. Potential crown heights are 85 mm in upper premolars and 110 mm in upper molars; and the enamel patterns are greatly simplified. Such extreme hypsodonty was attained during no more than 1.5 million years within the late Hemphillian at a rate of at least 0.58 darwins, roughly six times the normal rate of crown height increase in hipparionine horses.

INTRODUCTION

The genus *Pseudhipparion* stands out as an enigma among the generally well-studied hypsodont horses of North America. One of the earliest named hipparionine species, *P. gratum* (Leidy, 1869) was regarded by Matthew and Stirton (1930) as central to the origin of Old World *Hipparion* and yet by others (e. g., Quinn, 1955) as a pliohippine horse closely related to *Calippus*. Appropriately, the new generic name proposed by Quinn for Leidy's species was *Griphippus*, Greek for "puzzling horse." Apparently Quinn did not know of Ameghino's (1904) prior publication of another generic name, *Pseudhipparion*, with similar connotations of phylogenetic ambiguity, typified by a closely related species, Cope's (1889) *Hippotherium retrusum*. Skinner and others (1968) and Webb (1969) showed that the generic name *Pseudhipparion* is valid, and that *G. gratum* occurs in rocks overlying those producing *P. retrusum*. The latter paper advocated synonymizing *Griphippus* with *Pseudhipparion*, and presented evidence that the phylogenetic affinities of *Pseudhipparion* lie with the hipparionines rather than with *Calippus*.

The impetus for this contribution was our discovery that, among the late Neogene faunas of the Gulf Coastal Plain, *Pseudhipparion* occurred both earlier and later than the two classic species from the northern Great Plains. We now recognize a total of seven species, although the earliest is unnamed for lack of adequate samples. New evidence extends the chronologic range of the genus over a span of about ten million years. These new records clarify phylogenetic relationships of *Pseudhipparion* to other hipparionine genera, and amplify our knowledge of evolutionary patterns within the genus.

We dedicate this work to George Gaylord Simpson. His penetrating brilliance and broad vision have enriched our geological and biological perspectives beyond measure. In the 1930s when Simpson inherited his interest in fossil Equidae from W. D. Matthew it formed but a small fraction of his total work; yet in the 1940s it came to play a vital part in his extraordinary contributions to the evolutionary synthesis. We are pleased in the following account to name the most hypsodont of all horses for the most influential of all recent vertebrate paleontologists.

ABBREVIATIONS

AMNH - Department of Vertebrate Paleontology, The American Museum of Natural History, New York
F:AM - Frick American Mammals, AMNH
KUVP - University of Kansas Museum of Natural History, Lawrence, Kansas

PPM - Panhandle-Plains Historical Museum, Canyon, Texas
TMM - Texas Memorial Museum, The University of Texas, Austin
TRO - Timberlane Research Organization, private collection of J. S. Waldrop, Lake Wales, Florida
UCMP - Museum of Paleontology, The University of California, Berkeley
UF - Vertebrate Paleontology Collection, Florida State Museum, The University of Florida, Gainesville
UNSM - University of Nebraska State Museum, Lincoln
USNM - National Museum of Natural History, Smithsonian Institution, Washington, D.C.
WT - West Texas State University collection, housed in PPM
R, L - right, left
D - deciduous
I/i - upper/lower incisor
C/c - upper/lower canine
P/p - upper/lower premolar (e. g., P4 is an upper fourth premolar)
M/m - upper/lower molar (e. g., m2 is a lower second molar)
DPOF - dorsal preorbital fossa (= lacrimal or nasomaxillary fossa)
mybp - million years before present
$\bar{X}$ - sample mean
S - sample standard deviation
O.R. - observed range of a sample
V - coefficient of variation
N - sample size
Abbreviations for dental measurements are defined in the Materials and Methods section.

BIOSTRATIGRAPHIC CONTEXT

Pseudhipparion is present at numerous localities of late Barstovian and Clarendonian age (about 13.5 to 9.0 mybp) from the Great Plains, and is frequently one of the most abundant equid taxa encountered during that interval. We have not attempted to document all known occurrences of *Pseudhipparion* from the Great Plains. Rather, we have concentrated on specimens from five principal faunas (Fig. 1) that are well known in the literature, reasonably well dated, and contain large samples of *Pseudhipparion*. These are the Devil's Gulch Fauna (Devil's Gulch Horse Quarry and stratigraphic equivalents; specimens housed in the F:AM collection), the Burge Fauna (Burge, Midway, and June Quarries; F:AM and UCMP), the Minnechaduza Fauna (numerous sites from the Cap Rock and lower Merritt Dam Members of the Ash Hollow Formation, including the Little Beaver B, Clayton, East Clayton, Mensinger, Gallup Gulch, and Bear Creek Quarries in Nebraska and the stratigraphically equivalent Hollow Horn Bear Quarry in South Dakota; F:AM and UCMP), the Clarendon Fauna (MacAdams Quarry, Grant Lease Quarry, Noble Ranch Site, and Charles Risley Ranch Site; F:AM, PPM, TMM, and UCMP), and the Xmas-Kat Fauna (Xmas Quarry, Kat Quarry, and stratigraphic equivalents in the upper part of the Merritt Dam Member; F:AM and UCMP). For detailed stratigraphic and locality information see Schultz (1977) for the Clarendon Fauna, and Webb (1969) plus Skinner and Johnson (1984) for faunas from the northern Great Plains.

In the Gulf Coastal Plain, *Pseudhipparion* ranges from late Barstovian through late Hemphillian, an interval of about ten million years. We have studied specimens of *Pseudhipparion* from the Gulf Coastal Plain in a similar manner to those from the Great Plains. In particular, reasonably large sample sizes were obtained from three faunas: the Lapara Creek Fauna (localities listed in Quinn, 1955 and Forsten, 1975; TMM collection); the Archer Fauna (combined sample from the Love, Cofrin Creek, McGehee and Haile 19A localities of north-central Florida and temporal equivalents from the Bone Valley Region of south-central Florida; UF collection); and the Palmetto Fauna (Upper Bone Valley Formation; UF, TRO, and F:AM collections). Additionally, we include in the present study other samples represented by fewer individuals, but which document *Pseudhipparion* in otherwise unrecorded faunas (the Agricola and Bradley Faunas) in the southeastern United States (Fig. 1). The faunal terms introduced here for Florida will be fully described and documented by Waldrop and Webb *(in preparation)*. Their work demonstrates that all four faunas (i. e., the Bradley, Agricola, Archer, and Palmetto Faunas) can be recognized in the Bone Valley Phosphate Mining Region in stratigraphic superposition. Biostratigraphic correlation between faunas from the Gulf Coastal Plain and the Great Plains (Fig. 1) is based on revised definitions of the North American Land Mammal "Ages" by Tedford and others *(in press)* and personal observations of both authors.

MATERIALS AND METHODS

Equid samples from the Gulf Coastal Plain consist primarily of isolated teeth, with a limited number of mandibles, post-cranial elements, and maxillae recovered from the more productive localities. Skulls of equids are exceedingly rare (the few found are almost always crushed), and to our knowledge none is referable to *Pseudhipparion*. In appropriately aged deposits of the Great Plains, however, skulls of *Pseudhipparion*, and even associated skeletons, are not uncommon, and the cranial morphology of *P. retrusum* has been described in detail (Webb, 1969). In the present study, dental characters are emphasized in the diagnoses and descriptions, as teeth are the common, identifiable element represented from both faunal provinces. At present, most isolated post-cranial elements of *Pseudhipparion* from Gulf Coastal Plain faunas are not yet distinguishable from those of other contemporaneous small equids, and thus cannot be used in systematic studies. For each of our major faunal samples, we have surveyed cranial (when available) and dental characters in order to place each population into a phylogenetic framework. Measurements (described below) were taken on individual cheekteeth and

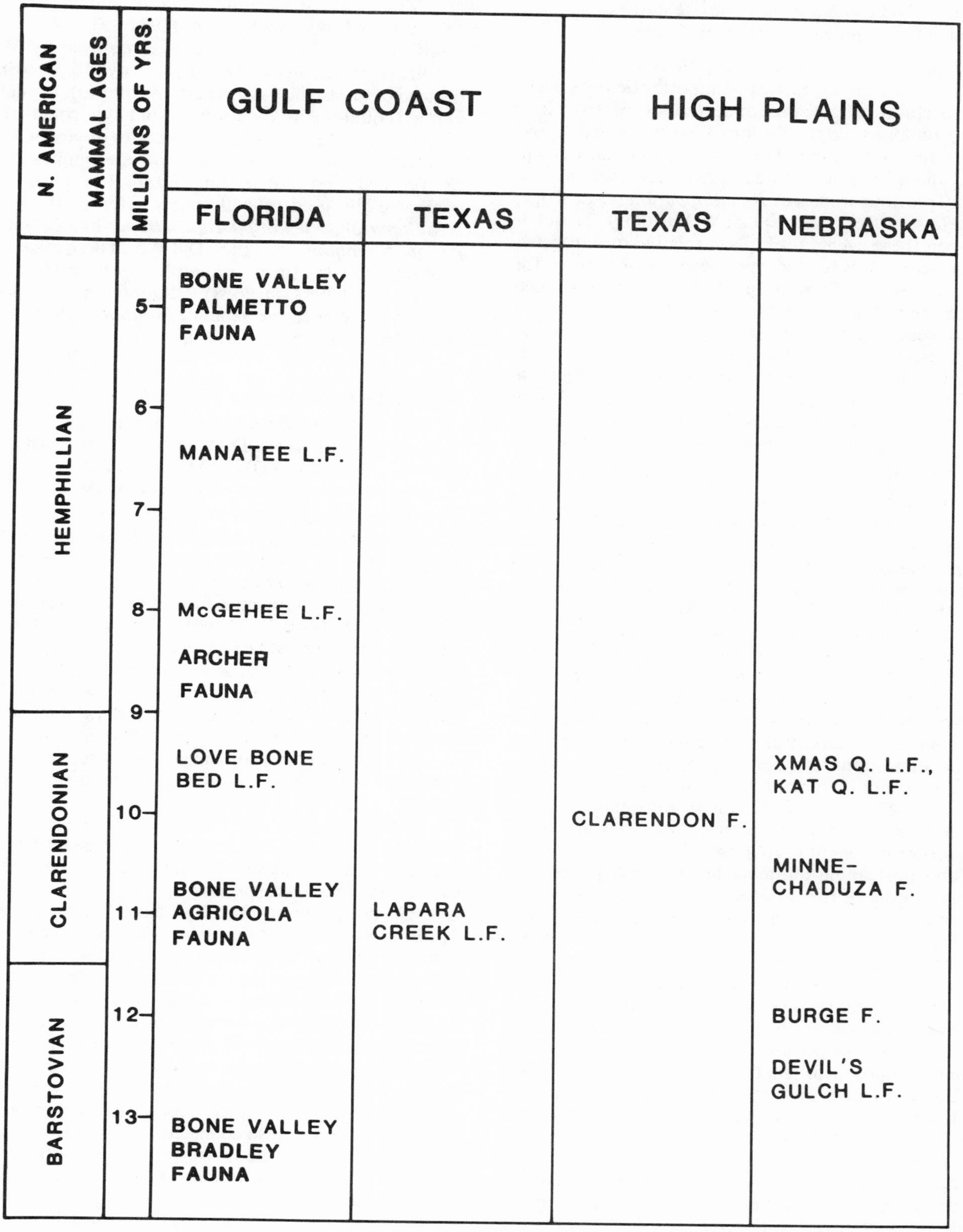

N. AMERICAN MAMMAL AGES	MILLIONS OF YRS.	GULF COAST		HIGH PLAINS	
		FLORIDA	TEXAS	TEXAS	NEBRASKA
HEMPHILLIAN	5–	BONE VALLEY PALMETTO FAUNA			
	6–				
		MANATEE L.F.			
	7–				
	8–	McGEHEE L.F.			
		ARCHER FAUNA			
	9–				
CLARENDONIAN		LOVE BONE BED L.F.			XMAS Q. L.F., KAT Q. L.F.
	10–			CLARENDON F.	
					MINNE-CHADUZA F.
	11–	BONE VALLEY AGRICOLA FAUNA	LAPARA CREEK L.F.		
BARSTOVIAN	12–				BURGE F.
					DEVIL'S GULCH L.F.
	13–	BONE VALLEY BRADLEY FAUNA			

Figure 1. Stratigraphic and geographic distribution of *Pseudhipparion;* principal samples organized by land mammal ages (Barstovian, oldest through Hemphillian, youngest), in High (= Great) Plains and Gulf Coastal Plain. Vertical position of each local fauna is based upon a biostratigraphic estimate of its age, and may be accurate within about half a million years.

toothrows, and comparative univariate statistics were calculated when sample sizes were sufficiently large.

During the attritional wear of a cheektooth in *Pseudhipparion*, two phylogenetically important events occur on the occlusal surface: the connection of the protocone to the protoloph; and the closure of the hypoconal groove to form an isolated enamel lake, which eventually disappears. These have been noted, but rather qualitatively, in many previous studies of *Pseudhipparion* (*e. g.,* Cope, 1892; Hesse, 1936; Quinn, 1955; Webb, 1969). To quantify when these events took place relative to ontogeny, the character states for these two features were noted for each individual cheektooth for which crown height could be measured or accurately estimated. Crown height is a good indicator of relative age in equids, and can even be used to estimate absolute age at death for fossil horses (Hulbert, 1982 and references listed therein). Using these data, timing of connection of the protocone and closure of the hypoconal groove can be contrasted among various populations and between molars and premolars of the same species.

For quantitative analyses, cheekteeth were grouped into six categories: P2, P3 and P4, M1 and M2, p2, p3 and p4, and m1 and m2. Upper and lower third molars were not measured often enough to produce sample sizes sufficient for statistical analysis. Measurements made on teeth from associated toothrows were combined with those taken on isolated teeth to increase sample sizes. Six measurements were taken on individual upper cheekteeth: APL, anteroposterior length exclusive of ectoloph; BAPL, anteroposterior length at base of crown; TRW, transverse width from mesostyle to lingual-most part of protocone; PRL, maximum protocone length, excluding spur or connections to either the protoloph or hypocone; PRW, maximum width of protocone perpendicular to PRL; and MSCH, crown height from occlusal surface to base of crown along mesostyle. On skulls and maxillae, toothrow length (TRL) was measured from P2 to M3, and postcanine diastema length (UDL) from the alveolar border of the C to the anteriormost part of P2. In species diagnoses, mean TRL refers only to mature individuals, *i. e.,* those with erupted and worn M3s. Younger individuals can have substantially longer toothrows (up to 15% greater than the mean), and very old individuals generally have much shorter toothrows due to ontogenetic variation of tooth length. For lower cheekteeth, seven measurements were taken: apl, maximum anteroposterior length from paralophid to hypoconulid, excluding protostylid; bapl, anteroposterior length at base of crown; atw, transverse width from protoconid to metaconid; ptw, transverse width from hypoconid to metastylid; mml, anteroposterior length of metaconid-metastylid complex; entl, anteroposterior length of entoflexid; and mcch, crown height measured at the metaconid. All were taken to the nearest 0.1 mm with dial calipers and, except for basal lengths and crown heights, measured on the occlusal surface (exclusive of cement). To enhance consistency, all dental measurements were made by one person (RCH).

Modern standard equid dental terminology is used, as illustrated by MacFadden (1984, Fig. 4). Terms significantly different from those employed by Webb (1969) are ectostylid (instead of protostylid), protostylid (instead of parastylid), ectoflexid (hypoflexid), linguaflexid (metaconid-metastylid groove), and hypoconal groove (hypoglyph). Durations and boundaries of North American Land Mammal "Ages" follow the recommendations of Tedford and others *(in press)*. The term "hipparionine" is used to collectively refer to the genera *Neohipparion, Pseudhipparion, Hipparion, Cormohipparion, Nannippus,* and their Old World derivatives, that together probably form a monophyletic group (see Forsten, 1984; Hulbert, *in press*). "Pliohippine" is used to collectively refer to the genera *Pliohippus, Dinohippus, Protohippus, Calippus,* and their derivatives.

SYSTEMATIC PALEONTOLOGY

Class MAMMALIA Linnaeus, 1758
Order PERISSODACTYLA Owen, 1848
Family EQUIDAE Gray, 1821
Pseudhipparion Ameghino, 1904

Synonym: Griphippus Quinn, 1955.

Type species and locality: Pseudhipparion retrusum (Cope), 1889. In his review of the genus *Hippotherium,* Cope (1889, p. 447) named this species, and noted that the type series was collected from "Phillips County, Kansas, from the Loup Fork bed." The fluvial deposits in this area would now be referred to the Ogallala Group or Formation, but it is also now known that several channel-cutting cycles are represented in that area. Webb (1969) showed that large samples from the Burge Fauna in northcentral Nebraska very probably represent Cope's species. The generic synonymy and early references to *P. retrusum* and *gratum* are more fully discussed in Webb (1969).

Included species: Pseudhipparion curtivallum (Quinn, 1955), *P. hessei,* new species, *P. gratum* (Leidy, 1869), *P. skinneri,* new species, *P. simpsoni,* new species, and a new unnamed species.

Generic diagnosis: Small, tridactyl horses, moderately to extremely hypsodont. Short rostrum with deep nasal notch, retracted to point above P2; DPOF anteriorly placed and lacking anterior and ventral rims as well as posterior pocket; malar fossa vestigial or absent; zygomatic arch lacking dorsal "buckle"; postglenoid process reduced and medially placed. Incisors form convex arcade, and develop progressively elongate to incipiently hypselodont crowns. Protocone large and elliptical, isolated in early wear stages, connected with protoloph in advanced stages, also connected posteriorly with hypocone in molars of some species; hypoconal groove tends to form fossette; styles prominent with grooved parastyle in premolars; enamel plications moderate to simple. Metastylid subequal to metaconid and widely separated from it by U-shaped linguaflexid; ectoflexid shallow on premolars; protostylid prominent in most species; ectostylid present on deciduous lower premolars in all but one species.

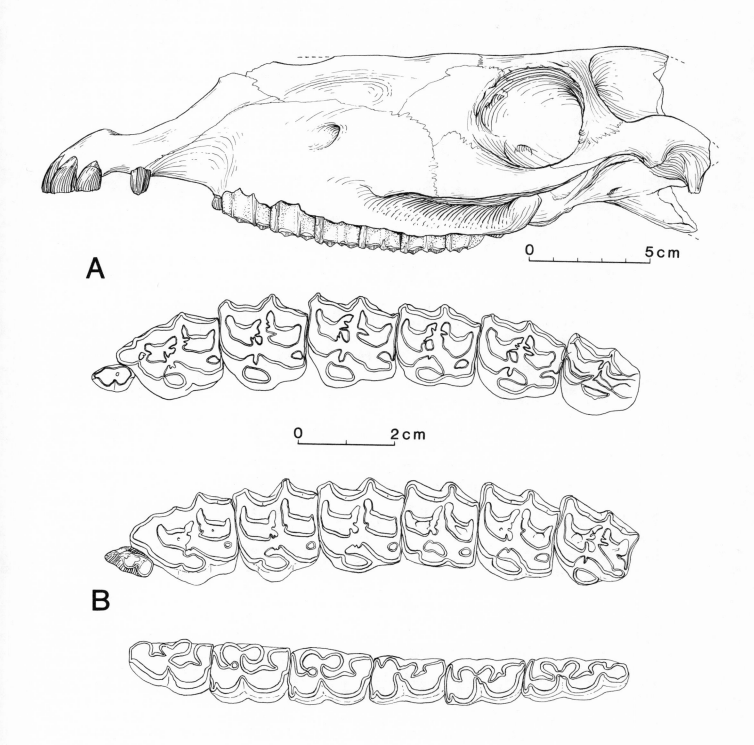

Figure 2. *Pseudhipparion retrusum* from Burge Member of Valentine Formation, north-central Nebraska, late Barstovian. *A*, F:AM 117073, June Quarry, Brown Co., Nebraska, left lateral view of skull and occlusal view of LD P1-M3. *B*, F:AM 70000, Lucht Quarry, Brown County, Nebraska, occlusal views of LD P1-M3 and L p2-m3.

TABLE 1. STANDARD UNIVARIATE STATISTICS FOR UPPER CHEEKTEETH OF *Pseudhipparion retrusum*, *P. curtivallum*, *P. hessei* n. sp., *P. gratum*, *P. skinneri* n. sp., AND *P. simpsoni* n. sp. THE FIRST LINE OF EACH ENTRY GIVES $\bar{X}$, S AND N. THE SECOND LINE GIVES O.R. AND V. ALL MEASUREMENTS IN MILLIMETERS.

SPECIES	RETRUSUM	RETRUSUM	CURTIVALLUM	HESSEI
FAUNA	DEVIL'S GULCH	BURGE	LAPARA CREEK	CLARENDON
P2				
APL	23.4,0.75,4 22.3-23.9,3.23	23.0,0.96,36 20.5-24.8,4.17	19.0,0.90,9 17.4-19.9,4.75	20.9,0.83,43 19.1-22.9,4.00
BAPL	———	19.3,0.54,12 18.4-20.3,2.81	15.9,0.42,8 15.1-16.5,2.69	17.0,1.09,36 14.8-19.3,6.40
TRW	18.0,0.65,4 17.3-18.6,3.60	17.5,1.00,34 14.2-19.5,5.73	14.7,1.05,9 13.1-16.5,7.16	16.2,0.79,42 13.2-17.9,4.90
PRL	5.8,0.17,4 5.6-6.0,2.96	5.7,0.60,35 4.9-7.8,10.56	4.8,0.55,9 3.7-5.7,11.41	5.2,0.45,43 4.3-6.2,8.49
PRW	4.6,0.69,4 3.8-5.4,14.97	3.9,0.59,35 2.4-5.2,15.09	3.4,0.35,9 3.1-4.1,10.09	3.6,0.28,43 3.1-4.2,7.59
P3 AND P4 (COMBINED)				
APL	19.1,1.34,11 17.2-21.0,7.03	18.8,1.36,75 16.0-22.0,7.23	16.2,0.75,22 14.4-17.7,4.65	17.6,0.98,70 14.6-19.3,5.58
BAPL	———	14.8,0.50,24 13.6-15.5,3.39	12.7,0.63,17 11.8-13.9,5.00	13.6,0.68,56 12.2-15.0,5.03
TRW	20.3,0.84,10 18.9-21.4,4.13	19.9,1.00,75 17.5-22.0,5.05	16.5,1.02,23 14.2-18.0,6.16	17.8,0.96,71 13.7-19.9,5.35
PRL	6.8,0.53,11 5.8-7.8,7.73	7.1,0.69,75 5.6-9.1,9.74	6.0,0.68,23 4.8-7.0,11.44	6.8,0.58,71 5.8-8.1,8.20
PRW	4.3,0.54,11 3.6-5.4,12.43	3.9,0.38,75 3.1-4.8,9.69	3.4,0.31,23 2.7-4.0,9.17	3.7,0.31,72 2.6-4,5,8.31
M1 AND M2 (COMBINED)				
APL	18.5,1.12,14 16.8-20.8,6.06	17.9,1.57,67 14.8-20.9,8.78	15.9,1.46,26 13.4-18.2,9.21	16.9,1.18,71 14.9-20.5,6.97
BAPL	14.6,0.42,2 14.3-14.9,2.91	14.0,0.49,19 13.4-15.1,3.54	12.3,0.72,19 11.3-13.7,5.81	13.1,0.87,53 11.6-15.1,6.64
TRW	18.4,1.32,14 14.9-19.9,7.17	18.0,1.03,69 15.0-19.9,5.73	15.1,0.93,26 13.1-16.8,6.19	16.5,0.92,70 14.1-18.0,5.55
PRL	6.6,0.57,14 5.7-7.8,8.71	6.9,0.66,69 5.4-8.7,9.58	5.7,0.54,26 4.8-7.0,9.43	6.4,0.65,71 5.0-7.8,10.19
PRW	3.8,0.56,14 2.9-5.0,14.83	3.6,0.40,68 2.9-4.7,11.09	3.1,0.33,27 2.4-3.8,10.59	3.4,0.31,71 2.7-4.2,9.12

TABLE 1 (CONTINUED)

SPECIES FAUNA	GRATUM MINNECHADUZA	SKINNERI XMAS-KAT	SKINNERI ALACHUA	SIMPSONI PALMETTO
		P2		
APL	21.6,1.04,23 19.4-23.1,4.82	18.6,0.89,8 17.7-20.3,4.80	16.9,0.21,2 16.7-17.0,1.26	13.0,1.21,5 11.6-14.4,9.29
BAPL	17.5,1.18,14 15.6-19.2,6.73	14.2,0.66,5 13.6-15.1,4.67	13.7,0.46,3 13.2-14.0,3.36	11.7,0.51,3 11.1-12.1,4.40
TRW	16.8,0.80,23 15.2-18.1,4.76	14.4,0.89,8 12.5-15.1,6.21	13.3,0.64,3 12.6-13.8,4.82	11.5,0.53,5 10.9-12.0,4.59
PRL	5.6,0.48,22 4.6-6.6,8.57	5.0,0.68,8 4.2-6.1,13.63	4.5,0.21,4 4.2-4.7,4.61	4.0,0.34,5 3.4-4.3,8.59
PRW	3.6,0.24,22 3.1-4.0,6.59	3.2,0.23,8 2.8-3.5,7.09	3.0,0.21,4 2.8-3.3,6.82	2.9,0.24,5 2.7-3.3,8.18
		P3 AND P4 (COMBINED)		
APL	18.5,1.08,47 16.2-20.9,5.84	15.4,1.19,11 13.0-16.8,7.67	13.9,0.80,23 12.7-15.5,5.76	12.6,1.60,18 9.6-15.6,12.77
BAPL	13.8,0.59,25 12.6-15.0,4.26	11.4,0.37,5 10.9-11.9,3.21	9.9,0.52,17 9.2-10.9,5.26	8.8,0.53,12 7.7-9.5,6.05
TRW	18.8,1.33,46 15.0-21.2,7.06	16.2,0.73,11 15.0-17.1,4.48	14.3,0.56,23 13.3-15.4,3.92	13.8,1.31,17 11.7-16.0,9.47
PRL	7.2,0.68,46 6.0-9.2,9.42	6.2,0.61,11 5.2-7.0,9.90	5.6,0.57,25 4.6-7.3,10.14	6.2,0.48,19 5.5-6.9,7.70
PRW	3.7,0.27,46 3.2-4.2,7.43	3.4,0.30,11 2.8-3.8,8.65	3.1,0.17,24 2.8-3.4,5.61	3.2,0.36,19 2.5-3.8,11.22
		M1 AND M2 (COMBINED)		
APL	17.6,1.05,53 15.1-20.0,5.99	15.5,1.64,21 13.1-18.4,10.57	13.2,0.78,23 11.9-14.8,5.92	12.6,1.47,24 10.1-15.4,11.65
BAPL	13.3,0.49,30 12.3-14.0,3.67	11.3,0.61,13 9.9-12.2,5.43	10.0,0.50,18 9.0-10.7,5.03	8.1,0.57,8 7.4-9.0,6.99
TRW	17.1,1.32,53 11.5-19.3,7.71	14.8,0.56,19 13.6-15.9,3.77	12.7,0.51,22 11.9-13.8,3.98	12.3,0.94,22 10.9-14.3,7.59
PRL	6.7,0.57,53 5.4-8.2,8.63	5.8,0.72,20 4.8-7.3,12.29	5.0,0.45,22 4.2-5.9,8.92	6.0,0.42,23 5.2-6.7,6.92
PRW	3.4,0.27,53 2.7-3.9,8.11	3.2,0.15,18 2.9-3.5,4.52	2.8,0.18,21 2.5-3.1,6.49	2.9,0.24,23 2.2-3.2,8.38

TABLE 2. STANDARD UNIVARIATE STATISTICS FOR LOWER CHEEKTEETH OF
Pseudhipparion retrusum, *P. curtivallum*, *P. hessei* n. sp., *P. gratum*,
P. skinneri n. sp., AND *P. simpsoni* n.sp. FORMAT AS IN TABLE 1.

SPECIES	RETRUSUM	RETRUSUM	CURTIVALLUM	HESSEI
FAUNA	DEVIL'S GULCH	BURGE	LAPARA CREEK	CLARENDON

p2

apl	20.1,0.85,6 19.2-21.1,4.21	19.2,1.30,22 16.4-21.5,6.78	15.8,0.60,10 14.9-16.8,3.78	17.3,0.94,26 15.6-19.1,5.40
bapl	————	16.2,0.58,5 15.7-17.2,3.58	13.6,0.87,7 12.4-14.7,6.36	14.1,0.83,11 12.9-15.6,5.88
atw	7.9,0.33,6 7.5-8.4,4.17	7.8,0.58,20 7.0-9.1,7.42	6.4,0.65,10 4.7-7.1,10.07	7.1,0.56,25 5.9-8.0,7.88
ptw	9.9,0.38,5 9.6-10.5,3.87	9.6,0.81,21 7.3-11.3,8.46	8.2,0.79,10 6.1-9.0,9.66	9.1,0.61,26 7.8-10.3,6.71
mml	6.0,0.76,6 5.0-6.7,12.71	6.6,0.82,20 4.8-7.7,12.53	5.7,0.97,10 3.3-6.9,17.09	6.6,0.84,25 4.9-7.9,12.70
entl	6.4,1.61,6 3.2-7.5,25.21	7.5,0.80,21 6.1-8.6,10.64	6.8,0.67,10 5.6-7.7,9.98	7.5,0.72,26 5.2-8.4,9.64

p3 and p4 (COMBINED)

apl	19.4,0.86,5 18.4-20.7,4.40	18.6,1.33,33 16.8-21.8,7.19	16.1,1.06,29 13.7-17.9,6.56	17.6,0.95,42 15.2-19.8,5.39
bapl	————	15.3,0.84,10 13.5-16.4,5.49	13.0,0.99,18 11.4-15.3,7.62	14.0,0.61,29 13.0-15.4,4.36
atw	10.8,0.97,5 9.5-12.2,9.03	10.1,0.89,33 8.0-12.3,8.80	8.6,0.84,29 6.9-10.2,9.77	9.4,0.86,40 7.6-10.9,9.12
ptw	10.8,0.55,5 10.1-11.5,5.11	10.6,0.76,33 8.9-12.6,7.14	8.8,0.77,29 6.5-9.7,8.81	9.8,0.69,42 7.8-11.2,7.08
mml	9.5,0.68,5 8.9-10.5,7.19	9.6,0.44,32 8.7-10.4,4.56	8.6,0.68,29 7.3-9.9,7.82	9.5,0.70,42 8.2-11.1,7.35
entl	7.5,0.86,5 6.1-8.4,11.46	7.9,1.19,33 5.2-10.1,15.14	7.3,0.71,29 6.0-8.5,9.74	8.0,0.65,42 5.6-9.2,8.14

m1 and m2 (COMBINED)

apl	17.2,1.80,7 13.7-19.7,10.46	17.8,2.09,35 14.8-21.6,11.73	16.4,1.77,44 12.4-19.8,10.72	17.7,1.47,48 14.7-20.7,8.33
bapl	————	14.3,0.49,11 13.6-15.0,3.42	12.4,0.77,21 11.3-14.4,6.21	13.4,0.45,34 12.4-14.6,3.33
atw	9.0,0.82,7 7.7-9.8,9.03	8.1,0.69,35 6.7-9.7,8.59	7.1,0.81,44 5.4-8.8,11.37	7.8,0.72,48 5.9-9.4,9.17
ptw	7.4,0.44,7 6.7-8.1,5.95	7.2,0.53,35 6.3-8.6,7.37	6.4,0.59,43 5.2-8.5,9.20	7.1,0.52,48 5.4-8.1,7.37
mml	8.8,0.69,6 8.1-10.1,7.84	8.5,0.74,35 7.5-10.6,8.67	7.7,0.72,44 6.0-9.9,9.41	8.7,0.73,48 7.0-10.1,8.39
entl	3.8,2.02,7 1.3-7.3,53.4	4.7,1.56,35 1.6-7.2,33.07	5.1,1.36,44 2.1-8.4,26.59	5.3,0.75,48 3.1-6.9,14.04

TABLE 2 (CONTINUED)

SPECIES	GRATUM	SKINNERI	SKINNERI	SIMPSONI
FAUNA	MINNECHADUZA	XMAS-KAT	ALACHUA	PALMETTO
p2				
apl	18.1, -- ,1	14.8,0.97,14 13.2-16.7,6.54	14.5,0.21,3 14.3-14.7,1.44	11.9,1.13,4 10.9-12.9,9.45
bapl	————	11.7, -- ,1	10.0,0.07,2 9.9-10.0,0.41	————
atw	7.1, -- ,1	6.8,0.53,14 5.7-7.6,7.72	6.7,0.21,3 6.5-6.9,3.12	6.3,0.35,4 5.9-6.7,5.58
ptw	9.6, -- ,1	8.7,0.70,14 7.2-9.8,8.07	8.5,0.20,3 8.3-8.7,2.35	7.1,0.35,4 6.9-7.6,4.95
mml	5.7, -- ,1	5.9,0.57,13 4.7-6.6,9.70	6.4,0.42,3 6.1-6.9,6.44	5.8,0.90,4 4.6-6.7,15.64
entl	7.3, -- ,1	6.1,0.76,14 4.9-7.2,12.48	6.5,0.72,3 6.0-7.3,11.19	5.2,2.07,4 2.9-7.1,39.88
p3 and p4 (COMBINED)				
apl	17.5,0.42,2 17.2-17.8,2.42	14.8,1.02,26 13.1-16.8,6.85	14.1,0.66,10 13.0-14.8,4.68	12.4,1.08,9 11.3-14.4,8.73
bapl	————	11.6,0.10,3 11.5-11.7,0.86	10.6,0.54,7 9.9-11.5,5.10	————
atw	10.2,0.21,2 10.0-10.3,2.09	9.5,0.63,26 8.3-10.5,6.66	8.1,0.36,10 7.5-8.6,4.42	8.2,0.72,8 6.9-9.3,8.78
ptw	10.0,0.35,2 9.7-10.2,3.55	9.2,0.46,26 8.5-10.1,5.05	8.1,0.30,10 7.8-8.8,3.65	8.2,0.42,9 7.3-8.6,5.10
mml	9.5,0.21,2 9.3-9.6,2.25	8.9,0.55,26 8.0-10.5,6.18	8.7,0.60,10 7.7-9.4,6.92	7.4,0.44,9 6.7-8.2,6.01
entl	7.7,0.35,2 7.4-7.9,4.62	6.3,0.98,26 4.7-8.3,15.53	6.6,0.45,10 6.1-7.7,6.81	5.8,1.11,9 3.7-7.3,19.31
m1 and m2 (COMBINED)				
apl	19.3,0.07,2 19.2-19.3,0.37	14.5,1.32,30 12.3-17.2,9.17	14.8,1.38,7 13.4-17.7,9.26	13.2,1.69,13 10.9-16.2,12.82
bapl	13.3, -- ,1	11.4,0.42,9 10.8-12.1,3.65	11.0,0.49,6 10.0-11.3,4.50	9.4,1.02,3 8.7-10.6,10.83
atw	7.9,0.50,2 7.5-8.2,6.31	7.8,0.44,30 6.6-8.5,5.69	7.1,0.47,6 6.2-7.5,6.56	7.8,0.88,14 6.2-9.1,11.38
ptw	7.8,0.92,2 7.1-8.4,11.86	7.1,0.34,30 6.5-7.7,4.82	6.9,0.45,6 6.0-7.2,6.58	7.3,0.82,14 5.7-8.3,11.26
mml	10.1,1.13,2 9.3-10.9,11.20	8.4,0.56,30 7.4-9.8,6.68	9.0,0.23,7 8.7-9.3,2.55	7.1,1.11,14 5.0-8.6,15.59
entl	5.7,0.42,2 5.4-6.0,7.44	4.5,1.09,30 2.6-7.0,24.26	5.5,1.35,7 4.3-8.1,24.67	5.9,1.10,14 3.4-7.3,18.68

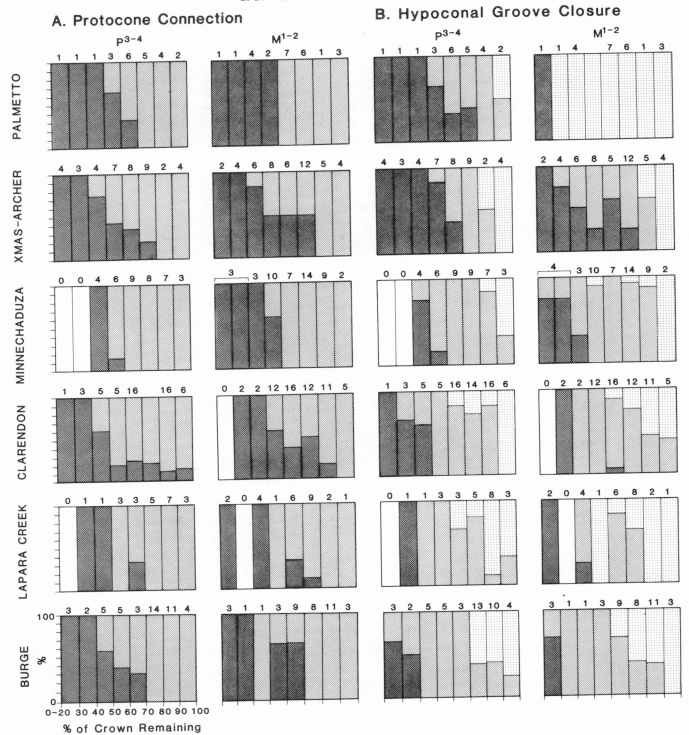

Figure 3. Histograms depicting differences in timing of protocone connection and hypoconal groove closure relative to upper cheektooth crown height for six species of *Pseudhipparion*. Percentage of crown remaining is calculated by dividing measured mesostyle height by unworn maximum value for fauna. These values are (in mm): Burge *(P. retrusum)*, P3/P4 = 42, M1/M2 = 45; Lapara Creek *(P. curtivallum)*, P3/P4 = 39, M1/M2 = 42; Clarendon *(P. hessei)*, P3/P4 = 45, M1/M2 = 48; Minnechaduza *(P. gratum)*, P3/P4 = 50, M1/M2 = 54; Xmas-Kat and Archer samples combined *(P. skinneri)*, P3/P4 = 43, M1/M2 = 47; and Palmetto *(P. simpsoni)*, P3/P4 = 85, M1/M2 = 110. Crown heights for *P. simpsoni* specimens which had not initiated root formation are estimated from regression equations based on specimens of known crown height and anteroposterior length. Sample sizes for each cohort are at top of each bar. *A*, Protocone connection for combined samples of upper third and fourth premolars (left) and first and second molars (right). Lighter shading indicates percentage of each wear class with isolated protocones; darker shading, connected protocones; unshaded bars represent unsampled wear classes (N = 0). *B*, Hypoconal groove closure of upper third and fourth premolars (left) and first and second molars (right). Lightest shading indicates percentage of each wear class with open hypoconal grooves; medium shading, those with isolated hypoconal lakes; darkest shading represents percentage lacking any trace of hypoconal groove.

Pseudhipparion retrusum (Cope), 1889
Figure 2; Tables 1-2

Type specimen and locality: AMNH 8350, R M2. Cope (1889, p. 446) stated that two associated right upper molars, about half worn, ". . . present the characters of the species best." Six associated upper cheekteeth (AMNH 8349) from the same locality as AMNH 8350 and discussed by Cope in the same paper represent *Protohippus profectus,* and need not be linked with the typology of *Pseudhipparion.* The two type specimens have been well figured by Osborn (1918, p. 142), who designated the M2 as lectotype.

Diagnosis: Unworn M1 and M2 crown height about 45 mm; cheekteeth larger than those of any other species, mean upper molar BAPL = 14.0 mm, mean TRL = 110 mm; protocone tends to connect with protoloph in moderate wear stages in molars, also posteriorly with hypocone in molars; and hypoconal groove closes to fossette in early to moderate wear stages, persists into very late wear stages. Upper cheekteeth markedly curved. Metaflexid enclosed by connection between metaconid and protolophid in later wear stages.

Distribution: Large samples are from the Devil's Gulch and Burge Faunas of northcentral Nebraska; the type sample is from western Kansas; and Cassiliano (1980) has described a third diagnostic sample from southeastern Wyoming. Probably this species was widely distributed in the northern Great Plains during the latest Barstovian, about 11.5 to 12.5 mybp.

Referred material: Numerous quarries in the Burge Member of the Valentine Formation in Cherry and Brown counties, Nebraska (reviewed by Skinner and Johnson, 1984, p. 284-292) have yielded large samples of this species, including several skeletons, housed at F:AM, UNSM, and UCMP. These last collections are described and partly figured in Webb (1969). Some of the F:AM material examined during this study, including a sample of skulls and a few of the better maxillae and mandibles, are listed here. Skulls: F:AM 70003, 70004, 70005, 70010, 70015, 70016, 70020, 70024, 70025, 70026, 70028, 70031, 70033-36, 70038, 70039, 70052, 70060, and 117073. Maxillae: 70059, 70063, and 70461. Mandibles: 70154, 70156, 70159, 70163, 70168, 70178, 70179, 70196, 70235, and 70237.

A stratigraphically subjacent sample from the Devil's Gulch Horse Quarry in the Devil's Gulch Member of the Valentine Formation in Brown County, Nebraska (Skinner and others, 1968, p. 406; Skinner and Johnson, 1984, p. 282-283) also yielded a large sample in the F:AM collection. We note a partial selection of the best material. Skulls: F:AM 60442, 69607, 69609, 69611, 70090, and 70091. Mandibles: 70367-70, 70372, and 70373.

Discussion: Complete synonymy and description of this species are presented in Webb (1969). Typical representatives of skull and dental morphology are illustrated here in Figure 2. It is fortunate that this species has been recognized by Skinner and others (1968) and by Webb (1969) in the Burge Fauna of northcentral Nebraska where excellent samples are available and where the chronostratigraphic and biostratigraphic contexts have been thoroughly worked out (Skinner and Johnson, 1984). We also assign the Devil's Gulch sample to *Pseudhipparion retrusum,* although teeth from that sample are somewhat larger (Tables 1 and 2) and in other more subtle features represent a slightly more primitive stage of evolution.

Pseudhipparion curtivallum (Quinn), 1955
Figures 4 and 5; Tables 1-3

Synonyms: Astrohippus curtivallis Quinn, 1955 (in part); *Griphippus* species, Quinn, 1955.

Type specimen and locality: TMM 30896-196, L mandibular fragment with p4-m2, from Buckner Ranch, Bee County, Texas; Lapara Creek Fauna, figured by Quinn (1955, Pl. 7, Fig. 9).

Revised diagnosis: Unworn M1 and M2 crown height about 42 mm; cheekteeth smaller than in *Pseudhipparion retrusum, P. hessei,* and *P. gratum;* upper molar mean BAPL = 12.3 mm, mean TRL = 95 mm; protocone rarely connected with protoloph until late wear stages; hypoconal groove closes to a fossette in early wear stages and persists into latest wear stages; upper cheekteeth not markedly curved. Metaconid and metastylid long but rounded; metaflexid deep and persistently open; linguaflexid relatively shallow; strong protostylid.

Distribution: The principal sample comes from the Lapara Creek Fauna, early Clarendonian of the Gulf Coastal Plain of Texas. Additional referred material comes from the Agricola Fauna in the Bone Valley District, Polk County, Florida, also of early Clarendonian age (c. 10.5-11.5 mybp).

Referred material: Other specimens from Lapara Creek sites: maxillae, TMM 31081-1017, RP2-3, RM1-3, LP3, LM1-2; 31081-1182, LP3-M1; 31132-236, LP4-M3; P2s, 31170-38, -69, -130, -131; upper premolars, 30896-80, -351, 31081-371, -1161, -597, 31132-106, -126, -133, -265, -368, -438, -538, -617-622, 31170-36, -132-136; upper molars, 30936-74, -146, -258, -363, 31132-96, -138, -205, -226, -248, -277, -286, -493, -535, -623-627, 31170-137, -138; M3s, 31132-500, -628-631, 31081-1139, -1438, -1501; mandibles and associated lower teeth, TMM 30896-581, p4-m3; 30936-421, m1-2; 30936-134, p2-3; 31081-479, p2-m3; 31132-168, m1-2; 31132-250, p2-m3; isolated p2, 30936-422; lower premolars, 30936-141, -423-426, 31132-405; lower molars, 30936-427-430, 31018-1138, 31132-20, -119, -336, -377, -632-634; m3s, 30936-431, 31132-635, -636.

Gray Zone, Phosphoria Mine, Polk County, Florida, Agricola Fauna: upper premolars, UF 28546, 55885, 61301, 61305; upper molars, 28547, 28550, 61302; lower molars, 28575, 50754, 57375, 61303; m3s, 23978 and 28581-2.

Description: No cranial or facial material of *Pseudhipparion curtivallum* is known from these samples. Upper and lower dentitions and isolated cheekteeth are well represented at several sites within the Lapara Creek Fauna (Figs. 4 and 5; Tables 1 and 2), and a smaller but indistinguishable sample occurs in the Agricola Fauna in Florida (Fig. 4; Table 3).

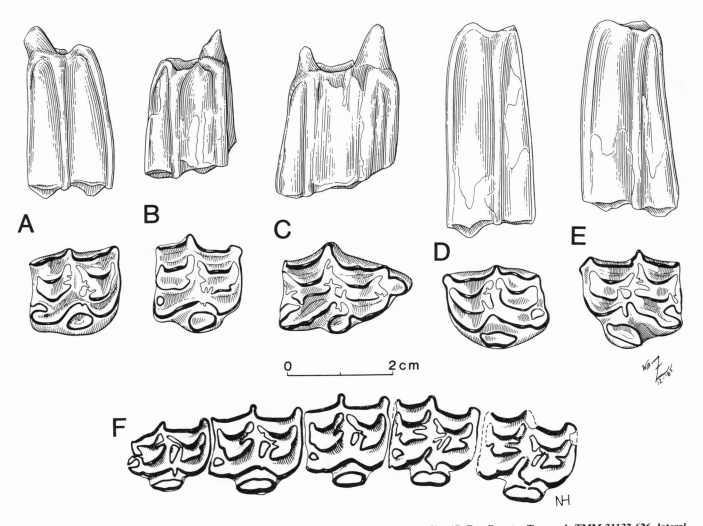

Figure 4. *Pseudhipparion curtivallum* upper cheekteeth. *A-C*, Lapara Creek Fauna, Site 17, Bee County, Texas. *A*, TMM 31132-626, lateral and occlusal views of R M1. *B*, TMM 31132-617, lateral and occlusal views of R P4. *C*, TMM 31132-375, lateral and occlusal views of R P2. *D-E*, Agricola Fauna, Gray Zone, Phosphoria Mine, Polk County, Florida. *D*, UF 61302, lateral and occlusal views of L M2. *E*, UF 61301, lateral and occlusal views of L P4. *F*, TMM 30896-177, from Buckner Ranch, Site 1, Bee County, Texas, occlusal view of R P3-M3.

Upper cheekteeth of *Pseudhipparion curtivallum* are readily distinguished from those of *P. retrusum* by their smaller size (Tables 1 and 2) and straighter crowns. As in *P. retrusum* the protocone is much longer than wide, and tends to bear a "spur" directed toward the protoloph (Fig. 4). Length of protocone relative to mean BAPL is less than in all more progressive species of *Pseudhipparion*. The lingual wall of the protocone appears flat in most specimens, but occasionally (as in TMM 30896-177) molars and premolars in early wear stages have concave lingual borders. The protocone usually connects with the protoloph at crown heights less than 20 mm, although a few cheekteeth have connected protocones above that height, for example TMM 31170-137 which is 31 mm tall (Fig. 3). In *P. retrusum*, by contrast, most upper cheekteeth have connected protocones at crown heights of 30 mm or less and some as high as 36 mm. In *P. curtivallum* a distinct preprotoconal valley appears even in well-worn upper cheekteeth, and this often provides a useful distinction from worn specimens of *Calippus*. The protocone connects posteriorly with the hypocone only rarely, for example in UF 28550, and never prior to the connection with the protoloph.

The hypocone is posteriorly directed and set off labially by a deep groove in early wear stages. This hypoconal groove closes to form a persistent lake by the time crown height reaches about 32 mm (Fig. 3). Only two teeth in these samples have been observed to retain an open hypoconal groove as late as 26 mm crown height, and the earliest loss of the lake in any intermediate molar occurs at a height of 12.8 mm.

The mandible is unusual among *Pseudhipparion* samples for its weak premasseteric notch (Fig. 5C); in mature mandibles of *P. retrusum* the ventral border deepens behind that notch below m2 by about 10 mm, whereas in both TMM 31081-479 and 31132-250 there is a gentle convexity that deepens posteriorly by only a few

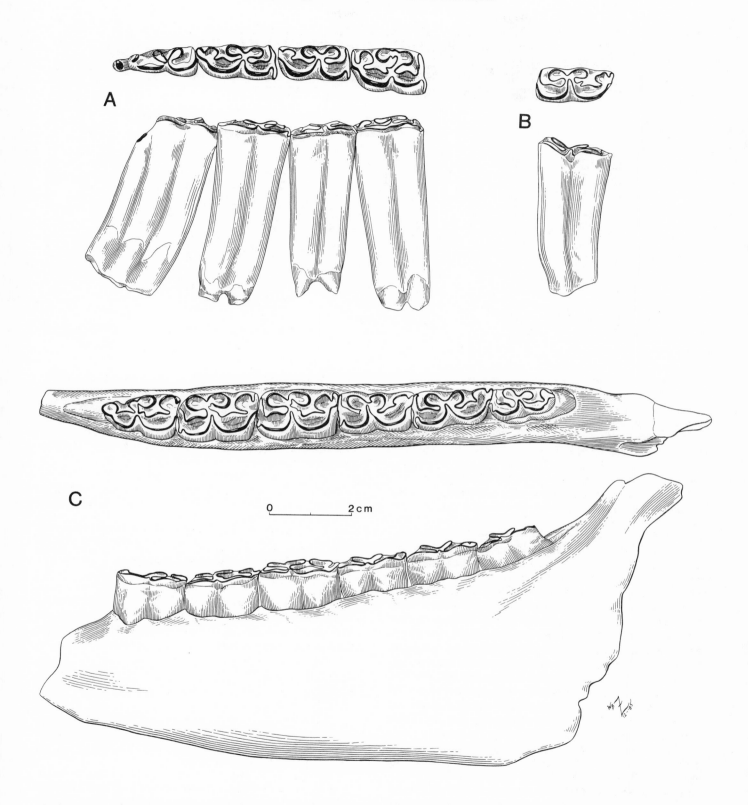

Figure 5. *Pseudhipparion curtivallum* mandible and lower cheekteeth. *A,* TMM 30896-581, Lapara Creek Fauna, Buckner Ranch, Site 1, Bee County, Texas, lateral and occlusal views of L p4-m3. *B,* UF 61303, lateral and occlusal views of L m2, Agricola Fauna, Gray Zone, Phosphoria Mine, Polk County, Florida. *C,* TMM 31132-250, Lapara Creek Fauna, Site 17, Bee County, Texas, lateral and occlusal views of L mandible with p2-m3.

TABLE 3. MEASUREMENTS (IN MM) OF *Pseudhipparion curtivallum* FROM THE GRAY ZONE, PHOSPHORIA MINE, POLK COUNTY, FLORIDA (AGRICOLA FAUNA, EARLY CLARENDONIAN). SPECIMENS THAT ARE EITHER THIRD OR FOURTH PREMOLARS ARE CODED AS P34; LIKEWISE, FIRST OR SECOND MOLARS ARE M12. a = APPROXIMATE.

	TOOTH	SIDE	APL	TRW	PRL	PRW	BAPL	MSCH
UF 28546	P34	L	–	–	5.0	3.2	–	a25
UF 55885	P34	R	15.5	15.7	5.4	3.1	11.7	26.6
UF 61301	P34	L	16.6	16.1	6.6	3.4	12.3	31.9
UF 61302	M12	L	16.0	14.1	6.2	2.8	11.7	33.9
UF 28547	M12	R	a14	–	5.5	3.0	–	–
			apl	ptw	mml	entl	bapl	mcch
UF 28575	m12	L	16.4	5.7	6.8	6.8	12.1	a36
UF 50754	m12	L	16.0	6.8	7.6	5.3	11.5	28.1
UF 61303	m12	L	16.9	7.2	8.7	6.4	–	–
UF 28582	m3	R	16.2	5.5	6.7	4.4	16.1	28.6

mm. The depth below m1 in the latter specimen is 43.5 mm. No Dp1 is present in this specimen.

Maximum crown heights observed in the large Lapara Creek sample of lower molars range from 40 to 43 mm, thus nearly equalling those found in the Burge sample. The metaflexid is deep and persistent in the type and other lower cheekteeth. The metaconid and metastylid are considerably longer than in *Calippus*, for instance, but they are generally shorter and more rounded lingually than in other species of *Pseudhipparion*. The linguaflexid is correspondingly shallower on many lower cheekteeth, for example in TMM 30896-581 (Fig. 5A). The protostylid is strong on all lower cheekteeth except p2, as in *P. retrusum*. The ectoflexid is quite variable in depth, but in premolars nearly always it is shallower than in other species of *Pseudhipparion*. Such a shallow ectoflexid is illustrated on p4 in Fig. 5A, and also occurs on the molars in the type specimen. Quinn (1955, p. 40) described this feature as "median valley of molars shortened," and it formed the basis for the species name *P. curtivallum*. Simplicity of enamel pattern is exemplified by rarity of the pli-caballinid which occurs in only two out of 90 specimens, both premolars in early wear. In this near absence of a pli-caballinid and in the general simplicity of its enamel pattern *P. curtivallum* resembles *P. retrusum* more than *P. gratum* or *P. hessei*.

Discussion: We do not recognize any *Pseudhipparion* upper cheekteeth in the equid material described and figured by Quinn (1955). Of the lower cheekteeth, only the type specimen of Quinn's "*Astrohippus*" *curtivallis* represents *Pseudhipparion*. We thus do not support Forsten's (1975, p. 46) recognition of ". . . all specimens described and figured in Quinn (1955, Plates 6 and 7) as *Calippus optimus*, *C.* cf. *placidus* and *Astrohippus curtivallis* as belonging to *Pseudhipparion*." The majority of these specimens are referable to *C. placidus* (Hulbert, *in preparation*). We accept Forsten's suggestion only with respect to the type lower jaw fragment. The rest of the Lapara Creek sample of *Pseudhipparion* was cited, without specimen numbers, by Quinn (1955, p. 42-43) as "*Griphippus* species," and the cheekteeth were stated to compare closely with "*G. gratus* . . . except for their smaller size and shorter crown height." We dispute Quinn's (1955, p. 42) assignment of an associated skull and mandible (TMM 31183-41), and a number of isolated upper teeth from the Cold Spring Fauna of Barstovian age to *Pseudhipparion*. Another study (Hulbert, *in preparation*) will show that *Pseudhipparion* is absent from the Cold Spring Fauna and that Quinn's sample belongs to *Calippus*.

Pseudhipparion hessei, new species
Figures 6-7; Tables 1-2, 4

Synonyms: Nannippus sp., Hesse (1936, p. 64-65). *Nannippus gratus*, Johnston (1938, p. 245-248, in part).

Type specimen and locality: F:AM 116906, male skull and mandible lacking symphysial region, from MacAdams Quarry, Donley County, Texas, collected in 1936, mid Clarendonian in age.

Etymology: Named in honor of Curtis J. Hesse, former curator and professor at Texas A&M University, in recognition of his extensive work on late Miocene faunas of the Ogallala Formation.

Diagnosis: Unworn M1 and M2 crown height about 48 mm, thus shorter than in *Pseudhipparion gratum*; cheekteeth smaller than those of *P. retrusum*, about equal in size to those of *P. gratum* and larger than those of all other species; M1 and M2 mean BAPL = 13.1 mm, mean TRL = 100 mm; long protocone connects with protoloph earlier than in *P. gratum* and about as in *P. retrusum*;

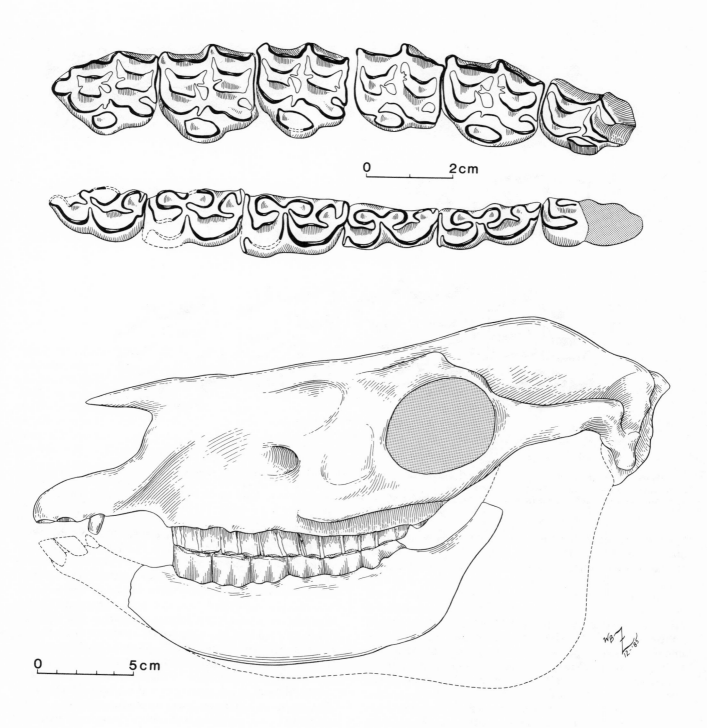

Figure 6. *Pseudhipparion hessei* type skull and mandible, F:AM 116906, Clarendon Fauna, MacAdams Quarry, Locality 17, Donley County, Texas, left lateral view of skull and mandible, and occlusal views of L P2-M3 and L p2-m3.

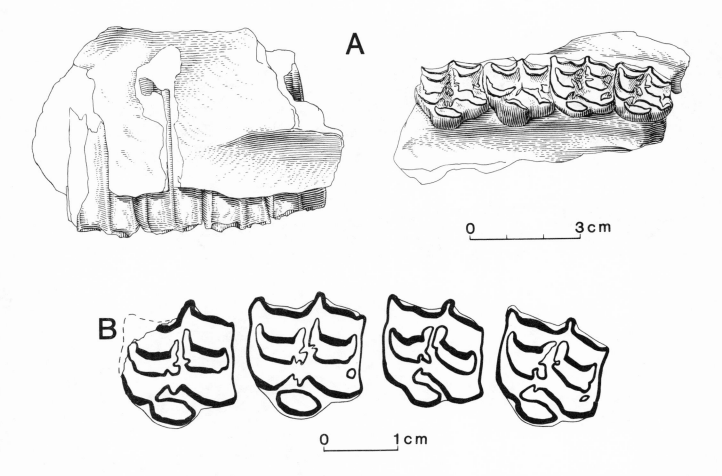

Figure 7. *Pseudhipparion hessei*, UCMP 31251, L maxillary with P3-M2 from Clarendon Fauna, Donley County, Texas. *A*, Left lateral and occlusal views of upper dentition in early wear stage. *B*, Occlusal view of same dentition sectioned 10 mm below crown surface. Note connection of protocones on three of these four cheekteeth, and closure of two hypoconal grooves to lakes and loss of two others (on P3 and M1).

hypoconal groove closes to a fossette in early wear stages and persists into moderately late wear stages; upper cheekteeth not markedly curved, but flared at the crown almost as much as in *P. retrusum*.

Referred material: The large F:AM series from MacAdams Quarry is the topotypic sample; a number of other sites also is well represented at F:AM; other major collections of the Clarendon Fauna that contain specimens of *Pseudhipparion hessei* are housed at the PPM, UCMP, and the TMM, including the original collections reported by Cope (1893). Another excellent specimen is USNM 1406, L P2-M2.

Description: The type cranium (F:AM 116906) represents a male with its third molars just beginning to wear (Fig. 6, Table 4). Its most noticeable feature is its short rostrum, with a postcanine diastema of only 28.3 mm. The nasals are correspondingly short, with a notch only 32 mm deep and a transverse width above the premaxillaries of just over 30 mm. The interorbital region, near the anterior end of the frontals, is broadly concave and contrasts with the highly domed region near the posterior end of the frontals. Similar frontal contours are noted in crania of other *Pseudhipparion* species (Webb, 1969, p. 116). The postorbital width is 117 mm. The DPOF is centered on the nasomaxillary suture, at the same level as the top of the orbit, about 65 mm rostral to the anterior margin of the orbit. The DPOF is long and shallow with no anterior or ventral rim, and no posterior pocket as is present in various other hypsodont horses. The infraorbital foramen lies 34.5 mm dorsal to the anterior end of P4 and well below the DPOF. There is a slight indication of a malar depression on the maxillo-jugal suture, but it is partly covered with plaster and if present was surely vestigial as in several other *Pseudhipparion* crania. The malar crest extends from under the orbit to a point above M1, and spreads as a nearly horizontal plate 25 mm lateral to the M3 mesostyle. As in most *Pseudhipparion* crania, the widest part of the specimen is across the malar crests, where it measures 124 mm, rather than across the postorbital bar as in most other equid skulls. The parietal region of the skull is well-rounded, more nearly resembling juvenile crania of other equid genera. The zygomatic arch is relatively shallow, and tapers gradually at its posterior end above the glenoid fossa; in

TABLE 4. MEASUREMENTS (IN MM) OF F:AM 116906, HOLOTYPE OF *Pseudhipparion hessei*, F:AM 70112, HOLOTYPE OF *P. skinneri*, AND UF 12943, HOLOTYPE OF *P. simpsoni*.

SPECIMEN	TOOTH	SIDE	APL	TRW	PRL	PRW	BAPL	MSCH
F:AM 116906	P2	R	22.9	16.5	5.6	3.5	-	-
"	P3	R	19.3	18.5	7.7	3.6	-	-
"	P4	R	17.1	17.3	7.9	3.8	-	-
"	M1	R	17.6	17.2	6.5	3.4	-	-
"	M2	R	18.5	16.9	7.2	3.2	-	-
F:AM 70112	P2	R	19.6	14.9	5.1	3.3	-	-
"	P3	R	16.7	17.0	6.4	3.8	-	-
"	P4	R	16.4	16.2	6.4	3.6	-	-
"	M1	R	14.5	14.5	5.2	3.3	-	-
"	M2	R	15.8	14.8	5.7	3.2	-	-
UF 12943	P3	L	13.2	-	5.7	3.0	-	-
"	P4	L	13.9	14.0	6.7	3.1	-	-
"	M1	L	13.2	12.9	6.4	3.0	9.0	42.4

			apl	atw	ptw	mml	entl
F:AM 116906	p2	L	a18	-	a9	7.9	7.9
"	p3	L	18.0	-	9.9	11.1	9.2
"	p4	L	19.8	-	9.7	11.0	7.8
"	m1	L	17.7	8.0	7.2	10.1	5.5
"	m2	L	19.8	7.7	6.9	10.1	6.5

			TRL	UDL
F:AM 70112		R	94.0	33.6
F:AM 116906		R	a113	a31

this respect *Pseudhipparion* skulls contrast markedly with many equid skulls in which the zygomatic process of the squamosal bears an elaborate "buckle" dorsal to the glenoid fossa. Few details of the basicranial region are preserved in the type cranium.

The upper cheekteeth fall in many respects between those of *Pseudhipparion retrusum* and *P. gratum*. In their size and enamel complexity they especially resemble the latter species (Tables 1 and 2). The pli caballin tends to be strong, persistent and rarely double, and in premolars the pli protoloph is also strong and persistent, as in *P. gratum*. The overall shape of unworn and little-worn upper cheekteeth is straighter than in *P. retrusum* and more nearly resembles the shape of cheekteeth in *P. gratum* and other more progressive species. On the other hand, there is a clear tendency for the crown of unworn upper cheekteeth to flare almost as fully as in *P. retrusum*. Two additional features shared by *P. hessei* and *P. retrusum* are their lower unworn crown heights and earlier connection of the protocone with the protoloph. The nearly unworn crown height of the P4 in USNM 1406, for example, is 41 mm. Several upper molars have the protocone fully connected at crown heights of 32 to 35 mm, and one (in the lot WT 925) attained such a connection at a height of 37 mm. The protocone is connected to the protoloph much less commonly on upper premolars, presumably because they are somewhat wider; even so, there are numerous examples of connected protocones on premolars in early wear stages, amounting to about 20 percent of the sample studied and thus exceeding the frequency of early-wear connection in *P. retrusum* (see Fig. 3). The protocone in upper molars also occasionally becomes connected with the hypocone in *P. hessei* as in *P. retrusum*. Presumably this does not occur in upper premolars because of their more obliquely oriented protocones. This feature, which was the principal basis for Ameghino's (1904) diagnosis of the genus *Pseudhipparion*, is characteristic of *P. retrusum*, occasionally found in *P. curtivallum* and *P. hessei*, rare in *P. gratum*, and wholly absent in later species. *P. hessei* dentitions are also notable for the long, anteriorly oriented hypoconal groove, observable in virtually all unworn upper cheekteeth. In early wear stages, however, it becomes enclosed to form a lake, and then persists until about the last 10 mm of crown height (Fig. 3). The presence of this lake is one of the most reliable criteria for discriminating between well worn upper cheekteeth of *Pseudhipparion* and *Calippus* (see Gregory, 1942).

The lower dentition of *Pseudhipparion hessei* closely resembles that of *P. gratum*. The linguaflexid tends to be wider and flatter than in *P. retrusum*, and a pli caballinid occurs occasionally in early to moderately worn lower premolars, whereas in *P. retrusum* that plication is rare. The sample of lower cheekteeth in the Clarendon Fauna can be distinguished from that in the Burge Fauna by its lower average unworn crown height, more elongate metaconid and metastylid, and its longer entoflexid.

Discussion: Curtis J. Hesse (1936) described as *Nannippus* sp. a sample of 27 cheekteeth from the Beaver Quarry in Beaver County in the panhandle of Oklahoma, but noted that they were "... distinctly smaller than those of Gidley's neotype of the latter species [*P. gratum*]" (Hesse, 1936, p. 64). He further suggested that the same "small *Nannippus*" occurred in the Clarendon Fauna, but that "The Clarendon specimens show fewer cases of the protocone connecting with the protoloph...". Hesse (1936, p. 65) concluded that the Clarendon population was specifically distinct from *P. gratum,* but chose not to formally describe it. Johnston (1938), however, did not suggest any important distinction between the Clarendon sample and the northern plains samples, and referred it to *Nannippus gratus*. With larger samples at hand, Webb (1969) noted that the lower crown height and earlier protocone connection of the Clarendon sample did indicate a distinct species, although he too did not formally recognize it. Unlike Hesse (1936), Webb (1969) referred the Beaver Quarry sample to *P. gratum*. This apparent species distinction between roughly contemporaneous faunas at Clarendon in the panhandle of Texas and at Beaver Quarry in the panhandle of Oklahoma led Webb (1969, p. 27) to postulate that "... the boundary between *P. gratum* and the more southerly species [*i. e., P. hessei*] lay near the Oklahoma-Texas border during the late Clarendonian." This hypothesis needs to be tested by studying other Clarendonian samples in southern Nebraska and Kansas.

Pseudhipparion gratum (Leidy), 1869
Tables 1 and 2

Type specimen and locality: USNM 587, L P2, from "the Niobrara collection" (Leidy, 1869, p. 287). Webb (1969) concluded that this species was represented in the Minnechaduza Fauna in the Niobrara River area from which Leidy's sample probably came. Gidley (1906) and several other authors have stated or implied that the type came from the Little White River area, but this is not the stated geographic location of the type.

Revised diagnosis: Unworn M1 and M2 crown height about 54 mm; cheekteeth smaller than those of *Pseudhipparion retrusum,* about equal to those of *P. hessei,* and larger than all other species; upper molar mean BAPL = 13.3 mm; mean TRL = 99 mm; long protocone does not connect with protoloph until mid to late wear stages, rarely connects posteriorly with hypocone; hypoconal groove closes early to form a fossette and persists into late wear stages; upper cheekteeth not markedly curved.

Distribution: Widely distributed in northern Great Plains, and southward into the panhandle of Oklahoma. Mid-Clarendonian in age.

Referred material: Gidley (1906) described and figured an excellent skull, AMNH 10863, from the Little White River area in South Dakota, and this "neotype" was refigured by Osborn (1918). Other important samples from South Dakota are the Big Spring Canyon Local Fauna (Gregory, 1942) and the Mission Local Fauna (Macdonald, 1960). Probable topotypic samples are from the Minnechaduza Fauna in the Cap Rock and lower Merritt Dam Members of the Ash Hollow Formation in Cherry County, Nebraska in both the F:AM and the UCMP collections (Webb, 1969; Skinner and Johnson, 1984). Another important sample is from the classic Snake Creek Fauna (Matthew, 1924; Skinner and others, 1977).

Discussion: A complete synonymy of this species is presented in Webb (1969), along with description of the Minnechaduza sample. Tabrum (1981) described a number of features in the *Pseudhipparion* sample from the Mission Local Fauna in South Dakota which appear slightly more primitive than in the Minnechaduza sample. We find these differences comparable to those distinguishing the older Devil's Gulch samples of *Pseudhipparion retrusum* from the younger Burge samples. Such minor temporal and geographic variation falls within the evolving species concept.

Pseudhipparion skinneri, new species
Figures 8-9; Tables 1-2, 4

Synonyms: Calippus placidus, Webb (1969, p. 79-80, in part). *Pseudhipparion* cf. *gratum,* Webb and others (1981, p. 526). *Griphippus,* Skinner and Johnson (1984, p. 314).

Type specimen and locality: F:AM 70112, partial male skull with entire upper dentition except first incisors, lacks skull roof and basicranium. From Jonas Wilson Quarry, Brown County, Nebraska; collected by Morris F. Skinner in 1934. Upper Ash Hollow Formation, Merritt Dam Member, Xmas-Kat Quarry zone (Skinner and Johnson, 1984).

Etymology: Named in honor of Morris F. Skinner of the Frick Laboratory, AMNH, for his extraordinary contributions to earth science in general and equid phylogeny in particular, both through his own work and through his generous encouragement of others.

Diagnosis: M1 and M2 crown height about 47 mm, less than in *Pseudhipparion gratum;* cheekteeth smaller than all other species except *P. simpsoni,* upper molar mean BAPL = 11.3 mm; mean TRL = 86 mm; relatively simple enamel patterns; relatively large protocone often connects with protoloph in middle wear stages and always by late wear stages, never connects with hypocone; hypo-

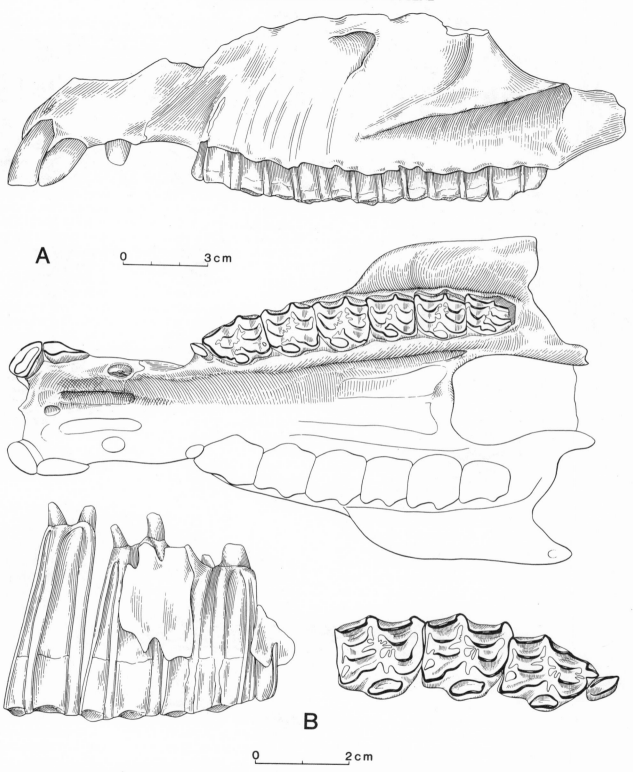

Figure 8. *Pseudhipparion skinneri* type skull and upper dentition from Xmas-Kat Quarry Zone, Merritt Dam Member, Upper Ash Hollow Formation, Nebraska. *A,* Left lateral and palatal views of type specimen, F:AM 70112, partial male skull with entire upper dentition except first incisors, from Jonas Wilson Quarry, Brown County, Nebraska. Note shallow malar fossa in lateral view. In palatal view left side is rendered in detail, right side in outline only. *B,* F:AM 70125, R maxillary fragment with DP1-P4, from Xmas Quarry, Cherry County, Nebraska, lateral and occlusal views.

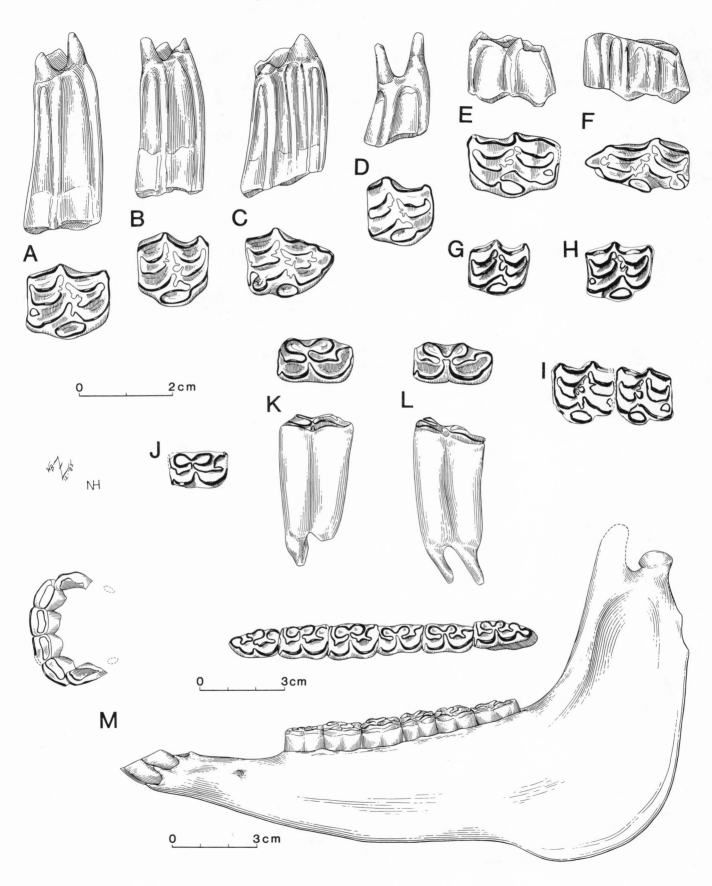

conal groove closes to form a fossette in early wear stages but is often lost in middle wear stages; upper cheekteeth not markedly curved. Protostylid weak.

Distribution: Known from the very late Clarendonian of northern Great Plains and late Clarendonian and early Hemphillian of Florida (*c.* 9.5-8.0 mybp).

Referred material: Jonas Wilson Quarry: P2, F:AM 70115; upper molars, 114184-5; M3, 114186-7; deciduous upper premolar, 114188.

Other Xmas-Kat Quarry Zone sites, Cherry County, Nebraska: maxillae, F:AM 70124, 70125, 70127; P2s, 114189-91, 114230; upper premolars, 114192-97, 114213, 114217, 114218, 114231; upper molars, 70129, 70132, 114198-203, 114227-29, 114214, 114232-34, UCMP 37128; M3s, F:AM 114215, 114216, 114225; mandibles, F:AM 114236-37, 114244-55, 114257-61, 114179-80; p2s, 114204, 114219; lower premolars, 114205, 114212, 114220-22; lower molars, 114206-11, 114223-24, 114226; m3, 114235.

Love Bone Bed, Alachua County, Florida: maxilla, UF 53621; P2, 53622-26; upper premolars; 53627-47, 59150-3; upper molars, 50353-54, 53647-50, 59140-48, 59155-60; M3s, 59154, 68841-44; deciduous upper premolars, 68845-48; mandibles, UF 32136; 32312; p2s, 32185, 64506-07; lower premolars, 59163-64, 64510-18; lower molars, 59168, 64519-26, 69801; m3, 64505.

McGehee Farm, Alachua County, Florida: upper premolars, UF 9508, 17109; upper molars, 17110, 17258, 19226, and 45637.

Haile 5B, Alachua County, Florida: upper molar, UF 17283.

Haile 19A, Alachua County, Florida: upper premolar, UF 47321.

Cofrin Creek, Alachua County, Florida: upper molars, UF 17108 and 61316; deciduous upper premolar, 61317; lower molars, 17298, 58554, and 61318.

Occidental Mine, Hamilton County, Florida: upper premolar, UF 36445.

Bone Valley Phosphate Mining District, Polk County, Florida: associated P4-M1, UF 23953; upper premolars, UF 23949, 57230, 69651; upper molars, UF 23951, 23952, 23958, 23959, 24649, 24654, 55825, 69652; M3, UF 90534; lower premolar, UF 23950.

Description: The type partial skull (Fig. 8; Table 4) indicates presence of the two facial fossae that often occur in *Pseudhipparion*. Position of the DPOF is suggested only by its indented ventral margin about 2 mm above the infraorbital foramen. The malar fossa is well preserved as a broad but shallow depression with a dorsoventral diameter of about 20 mm. The infraorbital foramen is unusually tall, measuring about 10 mm. The malar crest is broad, extending 30 mm lateral to M3, and reaches forward to a point above the posterior end of P4. The skull cap and basicranium are missing.

The upper incisors are relatively and absolutely longer than in comparable aged specimens of *Pseudhipparion gratum*. The estimated length of I1, based on the alveolar to occlusal distance, is about 26 mm; the measurable length of I2, excluding about 3 mm of embedded root, is 24 mm. Comparable lengths of upper incisors in a similar mature male skull of *P. gratum,* F:AM 70009, are 17 and 20 mm respectively. The lengths of I3s are not measured because they are deeply embedded in the premaxilla. The crown of I2 measures 17.5 mm in width by 7.3 mm in depth, and the infundibulae are transversely wide and persistent.

The upper cheekteeth are notable for their small occlusal dimensions (Table 1), large protocones relative to occlusal area, simple enamel patterns, and relatively ephemeral hypoconal fossettes. No unworn specimen attains the height observed in the tallest crowned specimens of *Pseudhipparion gratum;* the highest crowned upper cheekteeth range between 42 and 45 mm. The protocone is long in proportion to the crown APL, commonly bears a posterolateral spur, and tends to connect with the protoloph in early to middle wear stages. It is often connected by 30 mm crown height; for example, in UF 30354 it is broadly connected at a height of 30.4 mm, and it is already attached in UF 59148 which is 33.8 mm tall (Fig. 3). The protocone does not connect posteriorly to the hypocone in these samples as it does commonly in *P. retrusum* and occasionally in *P. gratum*. Simplicity of enamel pattern in this taxon is well illustrated by upper dentition of the type specimen (Fig. 8). There are only a few plications that persist after early wear stages. These include two on the posterior border of the prefossette, but they do not join to form a "subquadrate central loop" (Cope, 1889). The pli caballin tends to persist on premolars into middle wear stages, but never occurs on molars, in contrast to the condition in *P. gratum*. The hypoconal groove closes early to form a lake as is characteristic of *Pseudhipparion* species generally, but in this species it is frequently lost by middle wear stages and does not persist into late wear (Fig. 3). The two specimens cited above as having connected protocones at crown heights greater than 30 mm also already have lost their hypoconal lakes. About half the specimens with crown heights between 20 and 30 mm lack these lakes. We noted only one specimen, UF 24654, that was less than 20 mm in height and still retained a hypoconal fossette (see Fig. 3).

The mandible of this species is represented by F:AM 114236 from Xmas Quarry (Fig. 9*M*) and by UF 32312 from the Love Bone Bed. In both specimens the diastema between incisors and p2 (skipping the canine) is just over

Figure 9. *Pseudhipparion skinneri* (opposite page) cheekteeth from Florida and mandible from Nebraska. *A-H,* Upper cheekteeth from Love Bone Bed, Alachua Formation, Archer Fauna, Alachua County, Florida. *A,* UF 51951, R M1, lateral and occlusal views. *B,* UF 53631, R P4, lateral and occlusal views. *C,* UF 53623, R P2, lateral and occlusal views. *D,* UF 53633, R P3, lateral and occlusal views. *E,* UF 90264, L DP3, lateral and occlusal views. *F,* UF 90263, L DP2, lateral and occlusal views. *G,* UF 50354, R M1, occlusal view. *H,* UF 50353, R M1, occlusal view. *I,* UF 23953, from Nichols Mine, Bone Valley Formation, Polk County, Florida, associated L P4-M1. *J-L,* Lower cheekteeth from Love Bone Bed, Alachua Formation, Archer Fauna, Alachua County, Florida. *J,* UF 64511, L m1, occlusal view. *K,* UF 64513, L p4, occlusal and lateral views. *L,* UF 64525, L m2, occlusal and lateral views. *M,* F:AM 114236, occlusal and lateral views of L mandible and symphysis with complete dentition, from Xmas Quarry, Xmas-Kat Quarry Zone, Merritt Dam Member, Upper Ash Hollow Formation, Cherry County, Nebraska.

40 mm. The symphysis measures about 38 mm along the midline. The lower incisors form a smoothly convex arcade, between 30 and 35 mm wide. The mandible is 35 to 40 mm deep below m1, but deepens by about 10 mm behind the premasseteric notch.

The lower cheekteeth also are small (Table 2), with relatively simple patterns. The metaconid and metastylid are long. Often one or the other of these cuspids are curved lingually, but such pattern differences are extremely variable individually and with age. Size and shape of the entoconid also seem to be remarkably variable. Lower cheekteeth in these samples generally are wider in proportion to their length than in *Pseudhipparion gratum*. The metaflexid and linguaflexid are deep and persistent, as usual in *Pseudhipparion*. The ectoflexid is deeper on the premolars and wider and more persistent on the molars than in older species. The protostylid is present, but generally is much weaker than in *P. gratum* and earlier species of this genus.

Discussion: Skinner and Johnson (1984) thoroughly describe the stratigraphy and occurrence of the Xmas and Kat Channel Quarries. Among the 18 genera they cite from these highly fossiliferous beds is *"Griphippus"* (Skinner and Johnson, 1984, p. 314). The fauna is latest Clarendonian in age (about 9.5 mybp). Webb (1969) referred and figured an upper molar from the *Leptarctus* Quarry to *Calippus placidus* that is indistinguishable from *P. skinneri*. *C. placidus* is not recognized in the large F:AM sample from the Xmas-Kat Channel Quarries, which include the *Leptarctus* Quarry (Hulbert, *in preparation*). To judge from available stratigraphic records, this species had the longest chronological range of any in the genus (see below). No post-Clarendonian records of the species have been discovered in the Great Plains. In Florida, however, additional records at McGehee Farm, Haile 5B and Haile 19A indicate its continuation through at least the early Hemphillian. We realize that there are substantial size differences between the referred Florida sample of *P. skinneri* and that from the type region in Nebraska (Tables 1 and 2). However, with absence of significant morphological differences between the two samples, and because *Pseudhipparion* exhibits clinal variation in size (discussed below), we prefer to remain taxonomically conservative and recognize both samples as a single taxon.

Pseudhipparion cf. *skinneri*

Synonym: Griphippus species, Webb and Tessman (1968).

Referred material: Braden River Site, Manatee County, Florida: upper molars, UF 90544-5.

Manatee County Dam Site, Manatee County, Florida: upper premolar, UF 11923.

Discussion: A limited number of isolated teeth represents *Pseudhipparion* at a few low elevation coastal sites in Florida. In size and features of their crown pattern the teeth are indistinguishable from those of *P. skinneri* (see Webb and Tessman, 1968, Fig. 4). We do not include these among the Florida samples referred to *P. skinneri* because of their small sample size and because of the large stratigraphic gap (about two million years) separating them from the principal Florida samples. These sites probably are mid-Hemphillian in age, and probably represent estuarine sediments accumulated during the Messinian regression of sealevel (Webb and Tessman, 1968; Tedford and others, *in press*).

Pseudhipparion simpsoni, new species
Figures 10-11; Tables 1-2, 4

Type specimen and locality: Holotype, UF 12943, associated L P3-M1, moderately worn; from Fort Green Mine, Polk County, Florida; collected and donated by Larry Lawson in 1981; Upper Bone Valley Formation, Palmetto Fauna, late Hemphillian in age. Paratype: UF 24791, L mandibular fragment with p2-3, moderately worn; from Fort Green Mine; collected and donated by Frank Garcia in 1979.

Etymology: Named in honor of the late George Gaylord Simpson for his extraordinary contributions to human knowledge in general and to an understanding of horse evolution in particular.

Diagnosis: Incisors and cheekteeth incipiently hypselodont; total potential cheektooth crown height 85 to 110 mm; cheekteeth smaller than in any other species of *Pseudhipparion;* upper molar mean BAPL = 8.1 mm; relatively simple enamel patterns, fossettes absent in late wear stages; large protocone connects with protoloph in middle to late wear stages; in premolars the hypoconal groove closes to form a fossette in early wear stages and is lost in mid to late wear stages, in molars the hypoconal groove persists until latest wear stages. Metaconid and metastylid elongate and curved lingually; protostylid greatly reduced or lost; lower decidous premolars lack ectostylids and are very high-crowned.

Distribution: Upper Bone Valley Formation of central Florida and Delmore Formation of central Kansas; late Hemphillian about 4.5 mybp.

Referred material: Isolated teeth from the Bone Valley Phosphate Mining District of central Florida: P2s, UF 17104, 24407, 55841, 67977, and 68963; upper premolars, UF 17103, 17106, 32028, 43324, 52426, 53558, 53801, 53854, 55889, 57343, 58311, 58345, 60829, 61347, 68964, 68969, 68970, 68980, 68982, 68983, 68986, and F:AM 104875; upper molars, UF 12506, 17105, 17106, 18312, 21630, 22584, 24666, 43319, 43320, 43321, 43322, 43327, 47477, 53877, 55824, 55890, 55896, 57313, 57393,

Figure 10. *Pseudhipparion simpsoni* (opposite page) cheekteeth and median metacarpal from Palmetto Fauna, Upper Bone Valley Formation, Polk County, Florida. *A,* UF 12943, holotype, associated L P3-M1, from Fort Green Mine, Polk County, Florida, occlusal and lateral views. *B,* UF 18312, slightly worn L M2, occlusal and lateral views. *C,* UF 61347 (cast), moderately worn L P4, sectioned 11 mm above occlusal surface to show loss of fossettes (infundibula), lateral, occlusal and sectioned views. *D,* UF 24791, paratype, L mandibular fragment with L p2-p3, from Fort Green Mine, lateral and occlusal views. *E,* UF 23935, R m3, from Fort Green Mine, occlusal view. *F,* UF 24700, L p4 (reversed), from TRO Quarry, Payne Creek Mine, occlusal view. *G,* UF 69919, L m3 (reversed), occlusal view. *H,* UF 18313, L m1, from Palmetto Mine, occlusal and lateral views. *I,* UF 53960, median metacarpal, from Fort Green Mine, lateral and plantar views.

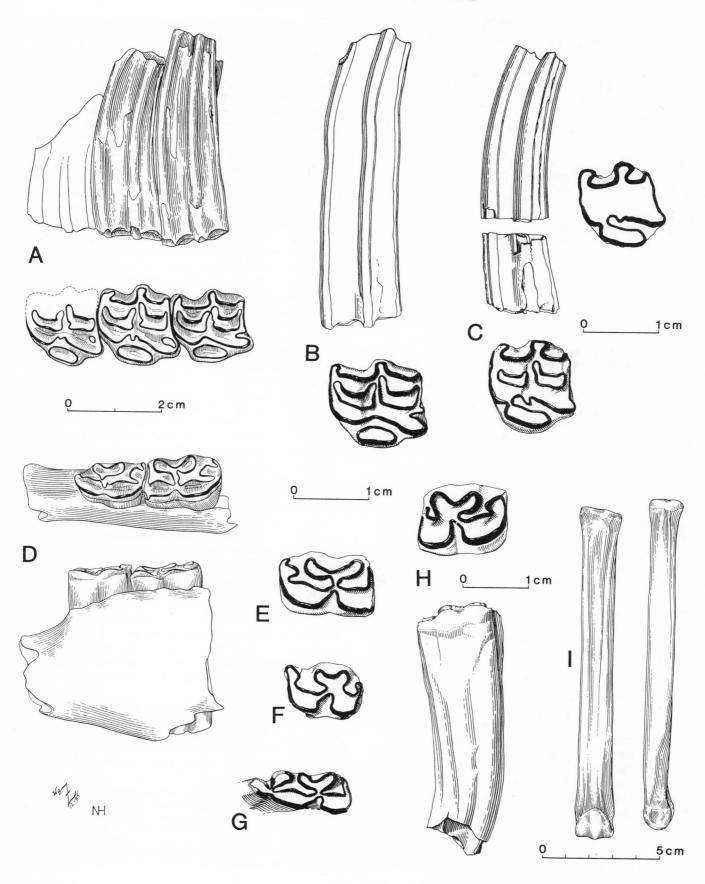

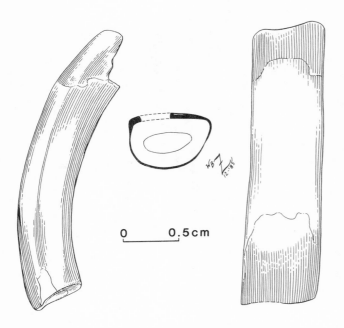

Figure 11. *Pseudhipparion simpsoni,* UF 69920, L upper incisor, lateral, occlusal and cranial views, from Palmetto Fauna, Upper Bone Valley Formation, Polk County, Florida.

58312, 58460, 60828, 64175, 65567, 65700, 65716, 68968, 68981, 68987, 68988; M3s, UF 18314, 43325, 55892, 67978, 68965-68967, 68984, 68985, 68989-68991; deciduous upper premolars, 58310, 61991, 65752, and 68999; and upper incisors, UF 57315 and 69920; lower cheekteeth including p2s, UF 17113, 18316, 52427, 58346, and 68979; lower premolars, UF 17111, 24700, 43323, 58347, 52428, 65680, 68971, 68975, and 69658; lower molars, UF 17112, 18313, 18315, 23935, 24667, 24707, 52429, 52430, 53574, 57314, 58313, 60830, 61498, 67979, 68972, 69654, 69655, and 69657; and m3s, UF 32030, 68973, 68974, and 68992; deciduous lower premolar, UF 65225; and median metapodials, UF 43565, 53960, 58376, 61315, 67986, 67992, 67993, and 91337. Precise locality information for most of these specimens is on file at UF.

Two isolated lower molars, KUVP 6872 and 7039, from Kinkerman's sand pit and Moundridge gravel pit respectively, Delmore Formation, central Kansas (Hibbard, 1952).

Description: Although this new species is known almost entirely from isolated teeth, it clearly represents an extremely advanced stage in the evolution of *Pseudhipparion*. In early to middle wear stages, upper cheekteeth of this species do not differ markedly from those of *P. skinneri*. The three teeth of the type specimen (Fig. 10A), for example, resemble upper cheekteeth of *P. skinneri* in their relatively long protocones and simple enamel patterns, although they are somewhat smaller (Tables 1, 2, and 4). Two further distinctions are that the protcone does not connect with the protoloph until late wear stages, and that on upper molars the hypoconal groove does not close to form a fossette. This latter distinction is not found in premolars which, as in the type specimen, are typical of *Pseudhipparion* with their persistent isolated hypoconal lakes (Figs. 3 and 10A).

An extraordinary transformation of features takes place in upper cheekteeth of *Pseudhipparion simpsoni* at the point at which about half the crown has worn away. A fuller discussion of ontogeny of these teeth and their attainment of incipient hypselodonty appears in a later section, but evidently the fundamental innovation is that the teeth approximately double their potential crown height by considerably delaying onset of root closure. At mid-height the fossettes disappear and the protocones connect with the protoloph. At about this point the crown dimensions diminish considerably, the hypocone becomes more oblique, and on molars the hypoconal groove becomes minute. The parastyle and mesostyle become more prominent, while the portion of the ectoloph posterior to the mesostyle narrows markedly and the metastyle disappears. The outer shell of enamel becomes thinner and is frequently warped and distorted. In addition to the ordinary dentine, two cores of softer "repair dentine" fill in the center of the teeth, about where the fossettes had disappeared. These bizarre late-wear teeth are readily distinguished from all other hypsodont horse teeth.

Depth of the paratype (UF 24791) measures 31 mm below p3, but rapidly shallows to a depth of only about 15 mm at a point 12 mm anterior to the p2. Lower cheekteeth of *Pseudhipparion simpsoni* most nearly resemble those of *P. skinneri,* but are considerably smaller and also can be distinguished by their longer and lingually curved metaconid and metastylid (Fig. 10D-H). In early wear stages the protostylid is small and sometimes isolated from the protoconid; it is lost in middle to late wear stages. The deciduous lower cheekteeth lack ectostylids, have small protostylids, and are extremely high-crowned.

Two upper incisors are referred to this species on the basis of their small size, extreme hypsodonty, and occurrence in the same deposits as *Pseudhipparion simpsoni* cheekteeth (Fig. 11). Both specimens are nearly 30 mm tall, but have open infundibula from wear surface to broken base. An exposed dentine tract about 2 mm wide extends the full length of the medial wall of each incisor. UF 69920 is 7.9 mm wide by 5.6 mm deep at the crown, and 7.3 by 5.9 mm at the broken base; UF 57315, an I3, measures 7.0 by 4.7 mm at the crown and 4.6 mm wide by 5.0 deep at the growing end. It appears probable that incisors of this species had attained an ever-growing condition.

One complete metacarpal III (Fig. 10I) and several partial medial metacarpals and metatarsals are closely comparable to but considerably smaller than those associated with the extensive skeletal samples of *Pseudhipparion retrusum* (see Webb, 1969). They are also smaller than metapodials in the Palmetto Fauna that seem to represent *Nannippus minor*. Large lateral appression surfaces running for much of the shafts' length indicate that these diminutive hipparionines remained functionally tridactyl.

Discussion: This species is known principally from the late Hemphillian of Florida. In a few sites, such as the TRO Quarry in Polk County, this species occurs in direct stratigraphic association with other characteristic taxa of the very late Hemphillian Palmetto Fauna (Tedford and others, *in press*). It is remarkable that it was not discovered in earlier studies of the Bone Valley Fauna (*e. g.*, Simpson, 1930), but indeed we have not found any examples in earlier collections.

Two additional specimens from the Delmore Formation in Kansas were referred by Hibbard (1952) to the genus *Nannippus*. They are indistinguishable from moderately worn lower molars of *Pseudhipparion simpsoni* from Florida. KUVP 7039, illustrated in Hibbard (1952), has the characteristic small size, simple pattern, and widely separated metaconid and metastylid. Although originally interpreted as age equivalent of the Higgins Fauna of probable earliest Hemphillian age, the two teeth here recognized as *P. simpsoni* along with specimens referable to *Neohipparion eurystyle* and *Nannippus lenticularis*, indicate presence of a very late Hemphillian horizon in central Kansas.

Pseudhipparion early species
Figure 12

Referred material: Kingsford Mine, Polk County, Florida: P2, UF 61311; upper premolars, UF 61301, 61306, 69660; upper molars, UF 61307-9; lower premolars, UF 61312, 69662; lower molar UF 61346; and deciduous lower premolar, UF 61310.

Nichols Mine, Polk County, Florida: upper premolar, UF 23964; upper molars, UF 23962, 23966, 25654; lower premolars, UF 23982, 24643; and lower molars, 23965, 23967, 23980.

Discussion: A few small samples of isolated cheekteeth clearly indicate presence of a small species of *Pseudhipparion* in the Bradley Fauna of late Barstovian age in the Bone Valley District of Florida. The age of this fauna is based on presence of such taxa as *Gomphotherium calvertensis*, *Megahippus* sp., *Protohippus perditus*, *Pliohippus mirabilis,* and *Procranioceras* cf. *skinneri*. These cheekteeth of a primitive *Pseudhipparion* species have occlusal dimensions slightly smaller even than those of *P. skinneri,* but they are far lower crowned, none exceeding 30 mm. The crown height of UF 23962, a complete but unworn M3, measures only 28.3 mm along the mesostyle; and UF 61307, a slightly worn upper molar, is only 24.9 mm tall. These cheekteeth clearly display characteristic features of an early species of the genus, for example early closure of the hypoconal groove, and a relatively short protocone that connects early with the protoselene (Fig. 12). The sample is only broadly comparable with *P. curtivallum*. We regard this as a distinct early species, but believe that it should remain unnamed until better samples become available.

PHYLOGENY AND RELATIONSHIPS OF *Pseudhipparion*

Introduction

Late Miocene Equinae traditionally have been separated into two basic groups, those with isolated pro-

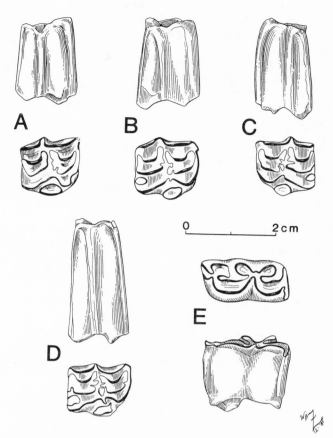

Figure 12. *Pseudhipparion*, early species. Cheekteeth from Bradley Fauna, Kingsford Mine, Lower Bone Valley Formation, Polk County, Florida. *A*, UF 61308, R M2, lateral and occlusal views. *B*, UF 61306, R P4, lateral and occlusal views. *C*, UF 61309, L M1, lateral and occlusal views. *D*, UF 61307, R M2, lateral and occlusal views. *E*, UF 61310, R Dp3, lateral and occlusal views.

tocones ("hipparionines") and those with connected protocones ("pliohippines"). *Pseudhipparion,* because it has isolated protocones in early wear stages but connected ones during later wear stages, has been placed into both of these groups by various authors. Phylogenetic relationships between *Pseudhipparion* and other equid genera can best be resolved through use of cladistic analysis of a diverse set of characters. Three phylogenetically important character-complexes are first discussed in general, to better demonstrate how we judged character state polarities.

Facial Fossae

Although occasionally perceived in the early 1900s (especially by Gidley, 1907), modern recognition of the phylogenetic significance of facial fossae in hypsodont Equidae can be directly traced to the Frick fossil horse collection, which includes many large, stratigraphically controlled samples of skulls amassed by Morris Skinner, Ted Galusha, and others (see also MacFadden, 1984, p. 4 and 6). The renaissance of facial fossae in equid taxonomy was initiated by Skinner and MacFadden's (1977) description of *Cormohipparion*. Subsequently, most workers have recognized the phylogenetic significance of facial fossae, although to varying degrees (see Eisenmann, 1981; Forsten, 1983; MacFadden, 1984; and Bernor and Hussain, 1985 for a variety of viewpoints). A survey of Oligocene and early Miocene equids reveals widespread presence of some type of facial fossa (or fossae), with a few exceptions (MacFadden, 1984). More particularly, within the Equinae, presence of a deep DPOF is primitive, while its absence is derived. The primitive position for the DPOF is near the orbit (i. e., a narrow pre-orbital bar) and for it to be comprised of three bones, the nasal, maxillary, and lacrimal. All known *Pseudhipparion* crania have a relatively shallow DPOF set well anterior to the orbit, and lacking anterior and ventral rims and a posterior pocket.

A shallow depression in the malar region is variably expressed in large quarry samples of *Pseudhipparion*. At most, it is a shallow fossa of limited area. A similar feature is present in *Merychippus insignis* (see Skinner and Taylor, 1967, p. 33), some specimens of *Neohipparion* (see MacFadden, 1984, p. 97), and variably in *Calippus*. We interpret the variable presence of a shallow malar fossa in *Pseudhipparion* to be a retained primitive character. The derived condition, notably developed in *Pliohippus*, is a large, deeply pocketed malar fossa.

Protocone Attachment

In lophodont Oligocene Equidae (e. g., *Mesohippus* and *Miohippus*) the protocone and protoconule are connected, even in extremely early wear stages, by a narrow ridge whose unworn height nearly equals that of the cusps. In the more hypsodont unworn teeth of *Parahippus*, the protocone and protoconule are separated from each other by a significant valley; the connecting ridge has failed to increase its height to keep pace with them. This, coupled with increased invagination by the preprotoconal groove and postprotoconal valley, isolates the protocone from the protoconule during early wear stages. The degree of isolation varies among the species of *Parahippus*. In *Merychippus*-grade horses, similar variation is observed (Stirton, 1941). However, they tend to segregate themselves into two groups: those in which the connection occurs relatively high on the crown (e. g., "*M.*" *primus*, "*M.*" *sejunctus*), and those in which the connection occurs about halfway down the crown (e. g., *M. insignis*). Presumably the former is the sister group of the pliohippines, the latter the sister group of the hipparionines. If this evolutionary scenario is correct: (1) the isolated hipparionine protocone-connection pattern is derived; (2) the symplesiomorphic condition for hipparionines is for the protocone to connect to the protoselene during midwear; and (3) the seemingly aberrant protocone-connection pattern of *Pseudhipparion* is, in fact, the retention of a primitive character state, but in a progressively more hypsodont tooth.

Metaconid-metastylid Complex

Primitively the metaconids and metastylids of equids are relatively small and unexpanded. With wear these two structures become widely confluent with each other and with the protoconid and hypoconid (e. g., the condition in *Parahippus leonensis* illustrated by Hulbert, 1984, Figs. 2 and 3). The latter occurs because the three major lingual flexids are: (1) shallow and poorly developed; and (2) rapidly reduced in depth by wear. This general pattern is observed in *Parahippus*, merychippines, and early pliohippines. In the latter two groups the metaconid often is expanded, but the metastylid remains relatively much smaller. Hipparionines share a derived metaconid-metastylid complex in which the metastylid is expanded along with the metaconid, so that the two are subequal or equal in size (Stirton, 1941; Forsten, 1982, 1984). Also, the lingual flexids are much deeper and are retained through a greater percentage of ontogeny; thus the metaconid and metastylid are well separated, both from each other and from the protoconid and hypoconid. The earliest hipparionines (e. g., *Cormohipparion goorisi*, MacFadden, 1984, Fig. 123) do exhibit these derived character states (contra Forsten, 1984, p. 170), especially in early wear stages, although not to the degree of later, more advanced taxa. In the Echo Quarry population referred to *Merychippus insignis* by Skinner and Taylor (1967), the lower cheekteeth are intermediate in their development of the metaconid-metastylid complex between that observed in primitive pliohippines and that of primitive hipparionines. The Echo Quarry *Merychippus* is thus useful as an outgroup for determining character state polarities within the hipparionines.

Relationships of Hipparionine Genera

As illustrated in Figure 13A, all hipparionine taxa share (at Point 1) the derived metaconid-metastylid complex, as well as increased crown height and heavily cemented deciduous and permanent cheekteeth (this discussion partially adapted from Hulbert, in press). The cheekteeth of *Pseudhipparion* possess these derived hipparionine character states. These, along with its "semi-isolated" protocone (symplesiomorphic for hipparionines, but derived relative to pliohippines), justify its inclusion among the hipparionines. The symplesiomorphic state of the DPOF in hipparionines is relatively deep, well rimmed, posteriorly pocketed, and located near the orbit, based on its configuration in the outgroup and its widespread appearance within merychippines and pliohippines. At Point 2, *Pseudhipparion*, *Neohipparion*, and two related species, "*Merychippus*" *coloradense* and "*Merychippus*" *republicanus*, can be united on the basis of having reduced the depth of the DPOF and decreased the distinctiveness of its ventral border. As reduction of the DPOF is a common trend observed in parallel in a number of equid lineages (Eisenmann, 1981; Bernor and Hussain, 1985), other characters were examined to corroborate the facial features. Dental evidence that the four taxa joined at Point 2 form a monophyletic group includes two shared derived character states: elongated protocone, and reduced depth of penetration of the ectoflexid into the isthmus on p2-p4. At Point 3, *Pseudhipparion*, *Neohipparion*, and "*M.*" *republicanus* possess the following shared derived character states relative to their sister group, "*M.*" *coloradense*: DPOF shallow, not

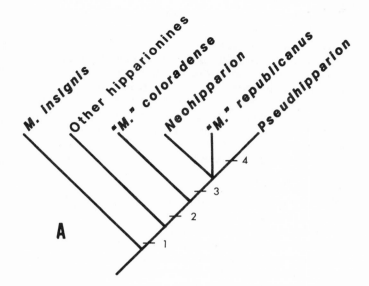

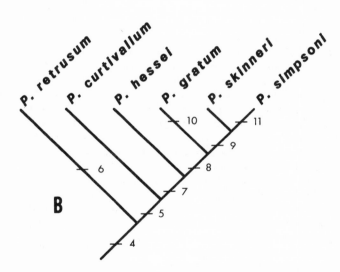

Figure 13. Phylogenetic relationships of *Pseudhipparion*. *A*, Cladogram illustrating hypothesis of relationships between *Pseudhipparion* and other hipparionines. *B*, Cladogram of proposed relationships among six named species of *Pseudhipparion*. See text for explanation of numbered character states and for discussion of phylogeny and relationships.

distinctly rimmed, lacks posterior pocket, and positioned more anteriorly relative to the orbit; long axis of the protocone of P2-P4 obliquely oriented (anterolabial-posterolingual); and protocone width less than about half its length.

The trichotomy at Point 3 results mainly from lack of information about *"M." republicanus*. The only figured specimen of this taxon is the holotype, AMNH 8347, a well preserved skull of a young male (Osborn, 1918, Fig. 99). When specimens from the Norden Bridge Quarry, tentatively referred to *"M." republicanus* (see Skinner and Johnson, 1984, p. 268), are eventually described, they may permit the resolution of this trichotomy. Two features of the holotype of *"M." republicanus* suggest a closer relationship with *Pseudhipparion* than with *Neohipparion*: (1) the hypoconal groove of the M2 is closed by a hypostylar plication to form a lake at an early wear stage (M3 just erupted); and (2) the nasal notch is retracted above the anterior portion of P2, the position usually observed in *P. retrusum* (see Webb, 1969, Fig. 30) and *P. hessei* (Fig. 6). In *M. insignis* (see Skinner and Taylor, 1967, Figs. 6 and 7), *"M." coloradense* (see MacFadden, 1984, Figs. 52-57), and *Neohipparion* (see MacFadden, 1984, Figs. 63 and 64) the nasal notch is less retracted, being located about half-way between the canine and the P2. Species of *Neohipparion* share a number of derived features relative to *Pseudhipparion* that unite them into a monophyletic group (Hulbert, *in press*).

Interrelationships of *Pseudhipparion* Species

At Point 4 (Fig. 13*B*), all species referred to *Pseudhipparion* share the following derived character states: deep hypoconal groove closed to form a lake relatively early in ontogeny; relatively small size; and a significant portion of each population with the protocone connected to the hypocone in advanced wear stages of the M1 and M2. The last character state is secondarily lost in *P. skinneri* and *P. simpsoni*. At Point 5 in Figure 13*B*, *P. curtivallum*, *P. hessei*, *P. gratum*, *P. skinneri*, and *P. simpsoni* are united on the basis of decreased size, less curved upper cheekteeth, earlier loss of the hypoconal groove, and more persistent lingual flexids (especially a better developed metaflexid). *P. retrusum* (Point 6) is uniquely derived with respect to these five species by its relatively early connection of the protocone to the hypocone on the M1 and M2, sometimes prior to its connection with the protoloph (Fig. 2*A*). At Point 7, *P. hessei*, *P. gratum*, *P. skinneri*, and *P. simpsoni* are united on the basis of increased unworn crown height, early closure of the hypoconal groove into a lake, and a relatively more elongated protocone. At Point 8, *P. gratum*, *P. skinneri*, and *P. simpsoni* are united on the basis of even earlier closure and subsequent loss of the hypoconal groove. At Point 9, *P. skinneri* and *P. simpsoni* share the derived character states of decreased basal cheektooth dimensions, reduced protostylid, decreased enamel complexity, relatively early loss of the hypoconal lake on the P2-P4, and the loss of connection between the protocone and the hypocone. *P. gratum* (Point 10) is derived relative to *P. skinneri* and *P. simpsoni* by its increased unworn crown height. *P. simpsoni* (Point 11) has numerous derived character states relative to its sister taxon, *P. skinneri*, including incipient hypselodonty, loss of fossettes about midway in its ontogeny, further reduction or loss of the protostylid, further decreases in enamel complexity and increase in relative protocone length, and an open hypoconal groove in the M1 and M2. The last is a reversal of a long term trend in *Pseudhipparion* evolution, that is best developed in *P. skinneri*. Upper cheekteeth of *P. skinneri* frequently lose any trace of the hypoconal groove after as little as 30 percent of the original crown height has worn off, ontogenetically early than other species of *Pseudhipparion*.

EVOLUTION OF HYPSELODONTY

Cheekteeth of *Pseudhipparion simpsoni* attain the most hypsodont condition known in any Equidae; indeed, that species even evolves incipiently hypselodont (evergrowing) incisors and cheekteeth. Crown heights in *P. simpsoni* regularly reach 45 mm, and the tallest specimen we observed is UF 17103, a left P4 with a MSCH of 50.1 mm. Hypsodonty is usually defined as a ratio of crown height to crown length or width (whichever is greater; e. g., Van Valen, 1960 and White, 1959). *Pseudhipparion* cheekteeth have smaller occlusal surfaces than those of most late Tertiary Equidae, especially in late wear stages when crown area reaches a minimum. The combination of tall crowns and small size in *Pseudhipparion simpsoni* gives measurable crown-height to crown-width (hypsodonty) ratios with maxima of about 3.8 to 4.0. These results approach those attained by *Nannippus peninsulatus (= N. phlegon)* of Blancan age, which is widely recognized as "extremely hypsodont" (MacFadden, 1984), and often is cited as the most hypsodont of all horses (Stirton, 1947; White, 1959).

The extraordinary degree of hypsodonty attained in *Pseudhipparion simpsoni,* however, cannot be measured directly from any single tooth, because in this species the total "life-span" of a given cheektooth greatly exceeds that represented by any one specimen. When the total potential crown height is estimated from the entire developmental sequence, a cheektooth of this species may be shown to have the potential of reaching at least twice the height that can be measured from a single specimen. Evidently this advanced species had evolved *incipient hypselodonty*. That is, tooth root formation was considerably delayed, although not permanently postponed, as would be the case in true hypselodonty. In order to demonstrate this, it is useful to compare development of cheekteeth in *Pseudhipparion simpsoni* with those in other hypsodont horses.

Figure 14 diagrammatically compares development of an upper premolar of *Pseudhipparion simpsoni* with that of *Equus caballus*. In *E. caballus,* as in *N. peninsulatus* and all other hypsodont equids, an unerupted but fully developed upper cheektooth has at its apex three sets of constrictions representing the three roots. Just below the roots are two closed enamel pockets (infundibula) that are connected to the inner enamel loops (fossettes) on the crown. As the tooth erupts and begins to wear, roots begin to form by constriction of the epithelial diaphragm at the tooth neck. At that point the tooth completes the long eruptive phase, and enters the short penetrative phase (terminology of Kovacs, 1971) in which the roots continue to elongate as dentin and cementum, but no enamel, are added. At this point maximum measurable crown height is attained. As the crown continues to wear, the roots become progressively more constricted, and by about the time that the cheektooth is half worn (about 40 mm tall in *E. caballus,* or about 30 mm in *N. peninsulatus*), the roots close except for a small apical foramen in each. Timing of root formation and root closure in lower cheekteeth resembles that in uppers, the essential differences being that there are no infundibula and that there are two roots instead of three.

In *Pseudhipparion simpsoni,* however, root formation is postponed considerably (Fig. 14). The eruptive phase of tooth development is, in most respects, indistinguishable from that in *Equus*, with the inner enamel fossettes closing at the apex of the upper cheekteeth at about the same time that the crown begins to wear. But whereas constriction for root formation would be expected to begin at the same time, no sign of it occurs. Instead, the outer enamel shell of the crown continues to grow upward. Apices of the infundibula provide useful markers to distinguish that part of the crown formed during the eruptive phase of development from that part formed during the penetrative phase in upper cheekteeth. A sample of eight upper cheekteeth in which the infundibula were nearly obliterated by wear but which had not yet initiated root formation had a mean crown height of 42.4 mm; thus the penetrative (post-infundibular) phase of development potentially contributes at least as much of the total crown height as the eruptive (infundibular) phase. Serial sections of two upper premolars illustrate (Fig. 15) how the two phases together can produce about 85 mm of effective crown height in individuals of *Pseudhipparion simpsoni* that live long enough. During the eruptive phase, tooth morphology is more or less "normal" for a hypsodont hipparionine horse. Subsequently, during the penetrative phase, the infundibula are lost, leaving only a hollow shell of enamel, the protocone becomes broadly connected to the protoloph, and the crown dimensions become smaller and smaller. The same extended crown elongation occurs in lower cheekteeth, but is more difficult to measure because there are no infundibula to serve as markers. During the penetrative phase of development in lowers, lingual and labial flexids are greatly retracted or lost, again resulting in a hollow enamel shell (e. g., UF 68979).

In *Pseudhipparion simpsoni* the effective crown heights of upper molars probably exceed those of premolars, attaining total heights of about 110 mm. We base this estimate on interpolation of all wear stages in the available sample of upper molars. Since the sample is quite limited, this estimate should be restudied when larger samples have been collected. Two appropriately worn upper molars demonstrate that a composite crown height exceeded 90 mm. UF 63634, in an early eruptive wear stage, has about 46 mm of crown from the occlusal surface to the base of the postfossette; UF 53877, in an early penetrative phase (with fossettes worn away) has a crown height of 45.0 mm. To this minimum estimate must be added about 10 mm of root development (not represented in UF 53877), a few millimeters of crown worn off UF 63634, and the amount of crown below the fossettes not represented in either UF 53877 or 63634. We estimate this missing middle section of the crown at about 10 mm on the following basis. In molars of hypsodont

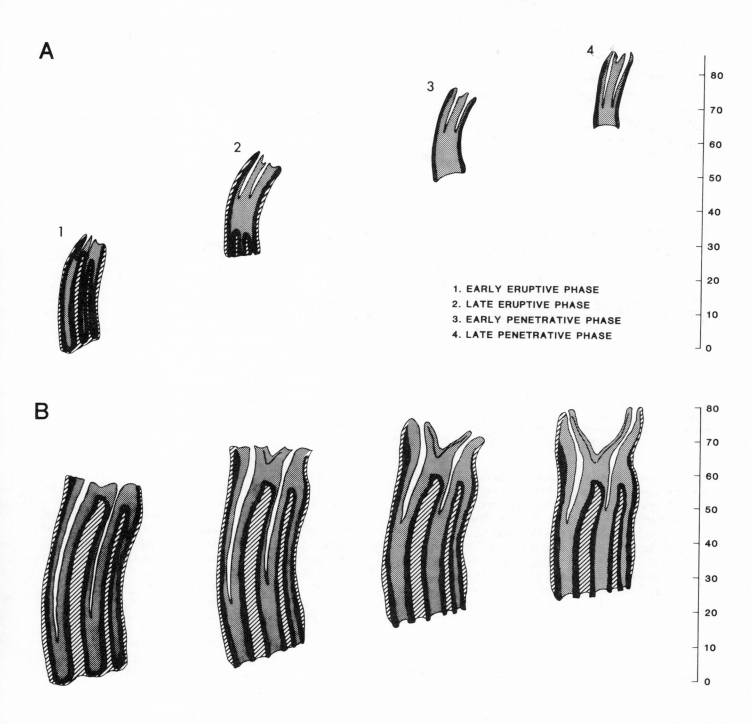

Figure 14. Diagrammatic anterior views of sectioned upper right premolars comparing development of *Pseudhipparion* and *Equus*. A, *Pseudhipparion simpsoni*. B, *Equus caballus* (after Getty, 1975). Thick black lines represent enamel; stippled areas indicate dentin; oblique lines are cement; and white area is pulp cavity. In Phase 1, early eruptive phase, roots have already begun to form in *Equus*, whereas their formation is postponed in *Pseudhipparion* until Phase 4, equivalent of late penetrative phase in *Equus*. Note also that infundibula in both genera cease to develop by Phase 1, so that in *Pseudhipparion* equivalents of Phase 3 and Phase 4 lack infundibula, and pulp cavities become filled with repair dentin. See text for further discussion.

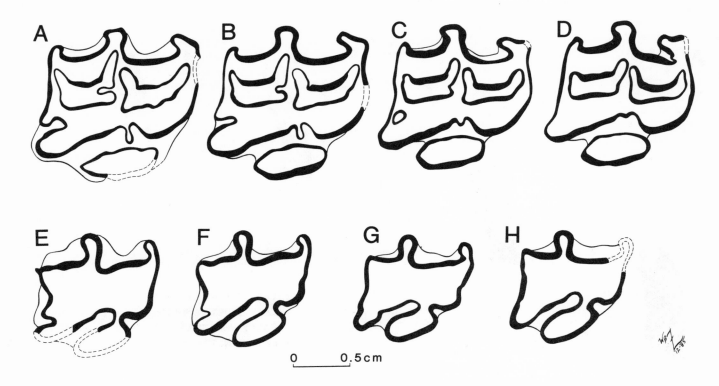

Figure 15. Wear stages in upper premolars of *Pseudhipparion simpsoni*. *A-D*, UF 58311, R P4 in early wear stage, crown height of 43 mm, measured along mesostyle. *A*, Original occlusal surface. *B*, Section 15 mm below *A*. *C*, Section 10 mm below *B*. *D*, Section 5 mm below *C*. Base of this tooth, 8 mm below section *D*, still retained fossettes. *E-H*, UF 57343, LP4 (reversed), in advanced wear stage without fossettes, crown height of 42 mm, measured along mesostyle. *E*, Original occlusal surface. *F*, Section 5 mm below *E*. *G*, Section 15 mm below *F*. *H*, Section 5 mm below *G*. Base of this tooth, 17 mm below section *H*, had initiated root formation.

Equidae there is a high positive correlation between the anteroposterior diameter of the crown and remaining crown height; this is simply the familiar observation that as cheekteeth wear they get smaller. The APL of UF 53877 is 10.6 mm and the APL measured at the point of postfossette loss is about 11.2 mm on UF 63634, resulting in a gap of crown height equivalent to an APL decrease of 0.6 mm. During the later three-fourths of crown wear, when the decrease in APL is nearly linear in *P. simpsoni*, MSCH and APL are related by the following regression equation:

$$APL = 0.06 \, MSCH + 7.5.$$

From this equation it can be calculated that a decrease of 0.6 mm in anteroposterior length results from attrition of about 10 mm of crown height. Using the estimated potential crown height of 110 mm and a middle wear molar APL of 11 mm, *Pseudhipparion simpsoni* attained a hypsodonty index of about ten, making it the most hypsodont of all known equids.

The crowns of *Pseudhipparion simpsoni* teeth continue to wear almost to the roots. Several upper cheekteeth, including UF 57313, 60829, and 43322, have completed root formation and have worn away the crown to within 10 to 15 mm of the root bases. The crown surfaces within the enamel shells are excavated to concavities between 1.1 and 2.5 mm deep. Increasingly in later wear stages a central infilling of repair dentin, presumably occupying space that was originally pulp cavity, can be distinguished by its lighter color form original dentin. Each of the rooted molars bears a heavy layer of cement extending 6 to 7 mm upward from the wear surface of the crown. Presumably, the cement helped hold the relatively short-crowned remnant of the tooth in the alveolar bone of the maxillary. An early phase of root formation is revealed in several upper cheekteeth, some with remnants of infundibula and others without. In UF 57393, for example, three dentinal processes have begun to constrict the epithelial diaphragm into three parts that would then each separately produce one of the three roots. In that specimen a large process extends posteriorly upward from the anterior groove of the tooth neck, meets a small process near the posterior groove and forms with it an interradicular ridge, while from the labial groove a third process can be seen in an early stage of development. The roots when fully formed are about 10 mm long, and each has a small apical foramen.

The relatively small enamel shell that forms the crown during the penetrative phase of its development is notable for its numerous wrinkles, contortions, and irregular thickenings and thinnings. Such features are evident in most upper and lower cheekteeth. Such undulations are expected in the penetrative phase of tooth development (Kovacs, 1971), but seem to be especially pronounced in *Pseudhipparion simpsoni,* in keeping with its exaggeration of this phase of development.

The question may be raised as to why *Pseudhipparion simpsoni* was the only equid that greatly extended its crown height by postponing root formation, thus attaining incipient hypselodonty or what White (1959) called "crown-base hypsodonty." Fortelius (1985) discussed a number of possible adaptive advantages of hypsodonty in ungulates, including indirect correlations with increased longevity and/or body size and direct function for improved mechanical processing of tougher food. He notes as a general rule that "increase in hypsodonty is often accompanied by a simplification of the occlusal morphology . . ." (Fortelius, 1985, p. 34). That trend does seem to apply in *Pseudhipparion* evolution. We are particularly impressed with the probable relationship between hypselodonty and small tooth size. First there must be stringent limits on the mechanical utility of a hollow shell of enamel when used in a transverse grinding mill like a horse's jaw. We suspect that the 10 mm by 12 mm dimensions attained by well-worn upper cheekteeth of *Pseudhipparion simpsoni* may lie near the upper limits of such a system. Even these small teeth develop deep dentine concavities between the enamel walls (over 2 mm) in some worn specimens. Secondly, it is at least suggestive that *Pseudhipparion* is the only equid lineage that clearly and persistently evolved toward decreasing body size (see discussion below). Among the other ungulate groups that achieve "crown-base hypsodonty," with hollow enamel crowns extending well beyond the level of fossette loss, are some Antilocapridae (e. g., *Antilocapra americana*) and some Bovidae (e. g., *Ovis* spp.) among the ruminants; and hegetotheres and interatheres among the notoungulates (Patterson and Pascual, 1972, p. 282). Each of these hypselodont ungulate groups has cheekteeth approximating those of *Pseudhipparion simpsoni* in size. Furthermore, most examples of crown-base hypsodonty and hollow enamel crowns occur within the Rodentia and Lagomorpha, and thus appear to be characteristic of even smaller sized mammals. This view is evidently too simplistic, however, since *Elasmotherium,* an Asiatic rhinocerotid, and *Toxodon,* a Neotropical ungulate, also attained "crown-base hypsodonty," but at very large body sizes. Evidently these two extinct exceptions avoided the mechanical problem of broad dentin exposures beyond the extent of the fossettes by elaborately infolding the outer shell of enamel. If small body size is indeed a general prerequisite for hypselodonty, then the limited number of hypselodont ungulates does not seem so surprising.

DWARFING TRENDS

Among the Equidae various dwarfing lineages have been noted, for example by Simpson (1953, p. 154). Three genera often cited are *Archaeohippus,* a smaller descendant of the genus *Miohippus,* and *Calippus* and *Nannippus,* each a smaller descendant of *Merychippus* s.l.

MacFadden (1985, p. 256), while accepting this latter example as a possibility at the generic level, finds on the contrary that ". . . within *Nannippus* a unidirectional trend toward decreased size is not observed in the ancestral-descendant species sequence." Ostensibly *Pseudhipparion* is another example of a small genus as compared to *Merychippus,* but the question of dwarfing in *Pseudhipparion* may be addressed more rigorously at the species level.

The following list of *Pseudhipparion* species proceeds downward from most recent to oldest and (in parentheses) indicates mean BAPL (in mm) for the largest faunal samples of M1 and M2:

P. simpsoni	(8.1)
P. skinneri	(11.3)
P. hessei	(13.1)
P. gratum	(13.3)
P. curtivallum	(12.3)
P. retrusum	(14.0)

These data indicate a strong general trend toward smaller size through time with respect to this measurement. The only striking anomaly is the small size of *P. curtivallum* at a much older stratigraphic level than might be expected. *P. gratum* and *P. hessei* are essentially contemporaneous species, and could just as well have been listed in reverse order; moreover, their mean values of 13.1 and 13.3 mm are not significantly different from each other based on Student's t-test.

A fuller view of size trends in *Pseudhipparion* emerges when geographic as well as chronologic patterns are considered. Webb (1969, p. 29) suggested that there were distinct northern and southern lineages of *Pseudhipparion,* at least during the late Clarendonian, and that there also were clinal size trends (Bergmann's Rule). If the samples from the Great Plains are compared only among themselves and, in a similar manner, the samples from the Gulf Coastal Plain, then the dwarfing progression may be viewed as two parallel reduction trends, with the southern series always smaller than their northern contemporaries. These separate northern (on the left) and southern series are as follows:

		P. simpsoni	(8.1)
P. skinneri (Xmas)	(11.3)	*P. skinneri* (Love)	(10.1)
P. hessei	(13.1)		
P. gratum	(13.3)		
P. retrusum	(14.0)	*P. curtivallum*	(12.3)

The cladistic analysis of relationships among *Pseudhipparion* species provides a more probable sequence of events. The primary exception to the size reduction series still remains the reversal from small *P. curtivallum* to its larger but otherwise more dervied northern sister taxa, *P. hessei* and *P. gratum* (but see below). Presence of an extremely small taxon in the late Barstovian of Florida also indicates that reduction in size of *Pseudhipparion* is not a simple orthogenetic trend.

TABLE 5. RATES OF MORPHOLOGICAL EVOLUTION (IN DARWINS) FOR SIX DENTAL CHARACTERS OF SEVEN SPECIES PAIRS OF *Pseudhipparion* AND AVERAGE VALUES FOR OTHER HIPPARIONINE GENERA (AFTER MACFADDEN, 1985).

	MSCH	APL	BAPL	TRW	PRL	mml	Δt (ma)
P. curtivallum to *P. hessei*	0.134	0.065	0.059	0.092	0.101	0.127	1.0
P. curtivallum to *P. gratum*	0.251	0.102	0.071	0.126	0.147	0.274	1.0
P. hessei to *P. skinneri*[1]	0.000	-0.086	-0.149	-0.109	-0.084	-0.030	1.0
P. curtivallum to *P. skinneri*[2]	0.045	-0.092	-0.105	-0.086	-0.068	0.078	2.0
P. skinneri[2] to *P. simpsoni*	0.218	-0.011	-0.052	-0.007	0.045	-0.057	4.0
P. skinneri[2] to *P. simpsoni*	0.581	-0.029	-0.139	-0.019	0.119	-0.152	1.5[3]
P. curtivallum to *P. simpsoni*	0.160	-0.038	-0.070	-0.033	0.007	-0.012	6.0
Mean hipparionine species pair rate	0.08	0.04	-	0.03	0.04	-	-
Mean hipparionine generic rate	0.10	0.04	-	0.02	0.04	-	-

[1] Great Plains sample only.
[2] Alachua Fauna (Florida) sample only.
[3] Assumes stasis in *P. skinneri* sample from 9.0 to 6.5 mybp. See text.

RATES OF EVOLUTION

Pioneering studies of Haldane (1949) and Simpson (1953) on quantifying evolution of morphological characters both included analyses of equids. MacFadden (1985), having benefits of greater chronological accuracy and a more detailed phylogeny, presented comparative evolutionary rates of four dental characters for eight pairs of "ancestral-descendant" species and four genera of hipparionines. MacFadden (1985) concluded that: (1) unworn crown height evolved at faster rates than other dental characters (a result also obtained and discussed by Simpson, 1953 and Haldane, 1949); (2) evolutionary rates of dental parameters exclusive of crown height for hypsodont equids were of average or normal magnitude for vertebrates; and (3) the four genera studied exhibited parallel evolution at similar (nonsignificantly different) rates.

We introduce similarly calculated evolutionary rates for *Pseudhipparion,* a fifth hipparionine genus. Table 5 presents evolutionary rates measured in darwins (d) for seven species pairs of *Pseudhipparion.* Only possible ancestral-descendant pairs were selected, based on our phylogenetic analysis of the genus. For example, no pairs include *P. retrusum* because derived features of its dentition prevent it from being ancestral to any other species. We wish to: (1) compare in magnitude and direction the evolutionary rates of *Pseudhipparion* with those of other hipparionines; (2) discuss effects of geographic size trends on evolutionary rates; and (3) determine if patterns of evolutionary change (*i. e.,* gradual or punctuated) can be detected in the fossil record of *Pseudhipparion.* In the following initial discussion, geographical differences among species are purposefully not considered.

The species pairs *Pseudhipparion curtivallum-P. hessei, P. curtivallum-P. gratum,* and *P. hessei-P. skinneri* each have intervals of about 1.0 million years (Fig. 1 and Table 5). The absolute value of rates of dental evolution in these three pairs ranges from 0.0 to 0.27 d, with a mean of 0.11 d (with or without MSCH). They are generally higher than those obtained by MacFadden (1985) for other hipparionine species pairs, but in his study time intervals ranged from 1.75 to 3.4 million years. Values of these three pairs of species would seem to be close to the average vertebrate rates for intervals of one million years calculated by Gingerich (1983), 0.12 d. If the latter is taken as a standard, then evolution of *P. hessei* from its putative ancestor *P. curtivallum* proceeded at about normal rates for MSCH, TRW, PRL, and mml, and about half the normal rate for APL and BAPL. In the lineage from *P. curtivallum* to *P. gratum,* however, crown height and possibly mml evolved relatively rapidly. The sample size for character mml in *P. gratum* is so small (N = 2) that its mean value may not truly reflect that of the population. The species pair *P. hessei-P. skinneri* displays average or slightly below average values, except for MSCH and mml, but of a negative sign (indicating size decrease). The low absolute value for mml is best interpreted as indicating an increase in mml relative to tooth size. Although the two species have equal crown heights, the smaller occlusal area in *P. skinneri* would produce a relatively more hypsodont tooth than that of *P. hessei*.

The time differential for two of the species pairs, *Pseudhipparion curtivallum-P. skinneri* and *P. skinneri-P. simpsoni,* are of comparable magnitude to those in MacFadden's (1985) study. The rates of these two pairs are in general higher than those of the other hipparionine species (Table 5), with mean rates (using absolute values) of 0.06 d (without MSCH) or 0.07 d (with MSCH).

The species pair *Pseudhipparion curtivallum-P. simpsoni* spans the history of the whole genus, and can be compared with the four generic lineages analyzed by MacFadden (1985). The rates for APL and TRW are within the observed range for the four other hipparionine genera, and are close to the generic means (Table 5). The rates of *Pseudhipparion* are the only ones with predominantly negative values, indicating that it is the only hipparionine genus with a consistent trend towards reduction in body size. The lower absolute values for PRL and mml can be interpreted as increases in their lengths relative to tooth size. The rate of increase in crown height for *Pseudhipparion,* 0.16 d, is the highest observed among hipparionines, not surprising considering the evolution of incipient hypselodonty in its youngest species. In conclusion, if geographic aspects are not considered, rates of evolution of dental characters of *Pseudhipparion* are somewhat greater than those of other hipparionines, but are within the same order of magnitude. A notable difference is the prevalence of size decreases within *Pseudhipparion,* the opposite of a general trend towards size increase observed in other hipparionines (MacFadden, 1985, p. 255).

Species of *Pseudhipparion* appear to have significant clinal variation in size (Bergmann's Rule) when populations from the Great Plains are contrasted with those of the Gulf Coastal Plain (Webb, 1969; this study). When ancestral and descendant species are from different geographic provinces, what proportion of the change in size is attributable to evolution, and what to geographic/climatic factors? Unfortunately, the only species with adequate samples in both geographic provinces is *P. skinneri*. *P. simpsoni* also spans both provinces, but is rare in the Great Plains. In *P. skinneri,* northern populations are significantly larger in occlusal dimensions than those from Florida (Tables 1 and 2). In fact, the differences are similar to or greater than those between southern *P. curtivallum* and northern *P. hessei* or *P. gratum*. If the clinal variation observed in *P. skinneri* is extrapolated to earlier Clarendonian populations, then differences in occlusal dimensions observed between *P. curtivallum* and both *P. gratum* and *P. hessei* do not necessarily reflect evolutionary increases in size. Based on the *P. skinneri* size differential, a hypothetical population of *P. curtivallum* from Nebraska would have greater mean values for APL, TRW, and BAPL than those actually observed in *P. gratum*. Thus, what appears to be an evolutionary trend for increased size, could have been a decrease. We therefore urge caution in interpretation of changes in body size through time for mammals such as equids that have wide geographic ranges and naturally vagile populations, unless the fossil record includes a sufficiently robust sample, both geographically and chronologically. For these reasons, the *P. skinneri* samples from Nebraska and Florida are analyzed separately in Table 5, to factor out geographic effects when possible and to limit comparisons to those between populations from the same geographic province.

Rates of evolution of dental characters between the early Clarendonian *Pseudhipparion curtivallum* and the late Hemphillian *P. simpsoni* are of average or slightly above average magnitude compared to other hipparionine genera (MacFadden, 1985) or to vertebrates in general (Gingerich, 1983). The fossil record of *Pseudhipparion,* excellent though it is, is not complete enough to clearly distinguish whether these rates were gradually and consistently in effect over the six million year period, or if there were alternating periods of relative stasis and of rapid change. For one character, crown height, there is some indication that its rate was not constant. Apparently there was a rapid increase in crown height from *P. curtivallum* to *P. hessei,* no increase from *P. hessei* to *P. skinneri,* and another period of rapid increase from *P. skinneri* to *P. simpsoni*.

There is additional evidence that the rate of crown height increase between *Pseudhipparion skinneri* and *P. simpsoni* was greater than these tabulations suggest. Although well documented in the late Clarendonian and earliest Hemphillian of Florida, *Pseudhipparion* is poorly represented in the later early Hemphillian between 8.0 to 6.5 mybp. Two coastal sites from central Florida, the Manatee Dam site (Webb and Tessman, 1968) and the Braden River site, produce three upper cheekteeth which are referred to *P.* cf. *skinneri* (see systematics above) and which have none of the diagnostic and highly derived features of *P. simpsoni*. Their occurrence with both *N. eurystyle* and *N. minor* and their low elevation, probably related to the Messinian low sealevel, indicate an age of about 6.5 mybp (Webb and Tessman, 1968; see also Fig. 1). The presence of *P.* cf. *skinneri* at these younger sites suggests relative morphologic stasis in *Pseudhipparion* evolution from about 9 mybp to at least about 6.5 mybp. The time interval between the last record of *P. skinneri* and the appearance of *P. simpsoni* did not exceed 1.5 million years. On this basis, a second set of rates is calculated in Table 5. The rate of crown height evolution between *P. skinneri* and *P. simpsoni* is thus 0.58 darwins, roughly six times the rate for other hipparionine species pairs calculated over similar short time intervals.

Accelerated crown height increase in *Pseudhipparion simpsoni* is correlated with delayed root formation and size decrease. Such retardation of ontogenetic events is associated frequently with rapid or radical evolutionary events (Gould, 1977).

CONCLUSIONS

The genus *Pseudhipparion* ranges from the late Barstovian (about 15 mybp) through the latest Hemphillian (about 5 mybp) in the Great Plains and in the

Gulf Coastal Plain. Seven species are recognized, from oldest to youngest: unnamed species, *P. retrusum, P. curtivallum, P. hessei, P. gratum, P. skinneri,* and *P. simpsoni. P. skinneri* of the late Clarendonian and *P. simpsoni* are the only species of this genus recorded from both the Gulf Coastal Plain and the Great Plains.

The relationships of *Pseudhipparion* are clearly with the hipparionine Equidae. Its enigmatic status and former alliance with *Calippus* were based only on its small size and on its primitive retention of a tendency to connect the protocone with the protoselene. The following features define a monophyletic group that allies *Pseudhipparion* with *"Merychippus" coloradense, Neohipparion,* and *"M." republicanus*: shallow DPOF with reduced ventral rim; elongate protocone; and reduced ectoflexid depth on lower premolars. *Pseudhipparion* and *"M." republicanus* further share the uniquely derived feature of early closure of a persistent hypoconal lake. Among the seven species of *Pseudhipparion* there are overall trends toward smaller size and, in the cheekteeth, simpler enamel patterns and taller crowns. The incisors and cheekteeth of *P. simpsoni* attained a condition of incipient hypselodonty; total potential crown height in the molars may have reached 110 mm.

Rates of evolution in various dental measurements among three species pairs over intervals of about one million years have a mean value of 0.11 darwins, a rate typical for fossil vertebrate lineages measured over such short intervals. A longer lineage, *P. curtivallum-P. simpsoni,* spanning an interval of about six million years, has rates of change for two dental measurements (APL, 0.038 d; and TRW, 0.033 d) comparable to those measured in four hipparionine genera by MacFadden (1985), except that these dimensions were decreasing rather than increasing. On the other hand, the rate of crown height increase (MSCH, 0.16 d) in *Pseudhipparion* is higher than in other hipparionine genera ($\bar{X}$ = 0.10 d). The most striking example of rapid evolutionary change is the evolution of incipient hypselodonty in *P. simpsoni* which, if derived from *P.* cf. *skinneri* in the late Hemphillian, increased its crown height at the rate of 0.58 darwins over an interval of 1.5 million years.

ACKNOWLEDGMENTS

We hope that the three patronymic species introduced in this paper help acknowledge our intellectual and material debt to our paleontological predecessors. We thank the following curators for permission to study fossils under their care: Dr. Richard H. Tedford and Mr. Morris F. Skinner, American Museum of Natural History; Drs. Ernest L. Lundelius, Jr. and John A. Wilson, Texas Memorial Museum; Dr. Gerald E. Schultz and Mr. William Griggs, West Texas State University and Panhandle-Plains Historical Museum; Drs. Donald E. Savage and J. Howard Hutchison of the Museum of Paleontology, University of California at Berkeley; and Drs. Clayton E. Ray and Robert J. Emry, U.S. National Museum. We are especially grateful to Drs. Richard H. Tedford and Bruce J. MacFadden for their helpful suggestions about stratigraphic and phylogenetic problems. The illustrations were skillfully prepared by Mss. Wendy Zomlefer and Nancy Halliday of the Florida State Museum and by Mr. Raymond Gooris of the American Museum. Collection and study of Florida Miocene fossils were funded by the Florida State Museum and by National Science Foundation grants to the Univeristy of Florida. Contributions of Florida fossil specimens to the FSM by the following individuals greatly aided this study: John Waldrop, Don Crissinger, Larry Lawson, Rick Carter, Cliff Jeremiah, Eric Kendrew, Howard Converse, Frank Garcia, Danny Bryant, Larry Martin, Gale Zelnick, Joe Larned, and Roy Burgess. This report is University of Florida Contribution to Vertebrate Paleontology number 248.

REFERENCES CITED

Ameghino, F., 1904, Recherches de morphologie phylogénétique sur les molaires supérioures des ongulés: Anales del Museo Nacional de Historia Natural de Buenos Aires, 3rd. Ser., v. 3, p. 1-541.

Bernor, R. L., and Hussain, S. T., 1985, An assessment of the systematic, phylogenetic and biogeographic relationships of Siwalik hipparionine horses: Journal of Vertebrate Paleontology, v. 5, p. 32-87.

Cassiliano, M., 1980, Stratigraphy and vertebrate paleontology of the Horse Creek-Trail Creek area, Laramie County, Wyoming: Contributions to Geology, University of Wyoming, v. 19, p. 25-68.

Cope, E. D., 1889, A review of the North American species of *Hippotherium*: American Philosophical Society, Proceedings, v. 26, p. 429-458.

_____ 1892, On the permanent and temporary dentitions of certain three-toed horses: Academy of Natural Sciences of Philadelphia, Proceedings, v. 1892, p. 325-326.

_____ 1893, A preliminary report on the vertebrate paleontology of the Llano Estacado: Geological Survey of Texas, 4th Annual Report, p. 1-138.

Eisenmann, V., 1981, Les caracteres évolutifs des crânes d' *Hipparion* s.l. (Mammalia, Perissodactyla) et leur interprétation: Comptes Rendus de l'Academie des Sciences du Paris, t. 293, p. 735-738.

Forsten, A., 1975, The fossil horses of the Texas Gulf Coastal Plain: a revision: Texas Memorial Museum, Pierce-Sellards Series, v. 22, p. 1-86.

_____ 1982, The status of the genus *Cormohipparion* Skinner and MacFadden (Mammalia, Equidae): Journal of Paleontology, v. 56, p. 1332-1335.

_____ 1983, The preorbital fossa as a taxonomic character in some Old World *Hipparion*: ibid., v. 57, p. 686-704.

_____ 1984, Supraspecific groups of Old World hipparions (Mammalia, Equidae): Palaeontologische Zeitschrift, v. 58, p. 165-171.

Fortelius, M., 1985, Ungulate cheek teeth: developmental, functional, and evolutionary interrelations: Acta Zoologica Fennica, no. 180, p. 1-76.

Getty, R., 1975, Sisson and Grossman's The Anatomy of the Domestic Animals, fifth edition: Philadelphia, London, Toronto, S. B. Saunders Company, 972 p.

Gidley, J. W., 1906, New or little known mammals from the Miocene of South Dakota. Part IV. Equidae: American Museum of Natural History, Bulletin, v. 22, p. 135-154.

―――― 1907, Revision of the Miocene and Pliocene Equidae of North America: *ibid.*, v. 23, p. 865-934.

Gingerich, P. D., 1983, Rates of evolution: effects of time and temporal scaling: Science, v. 222, p. 159-161.

Gould, S. J., 1977, Ontogeny and phylogeny: Cambridge, Belknap Press, Harvard University, 501 p.

Gregory, J. T., 1942, Pliocene vertebrates from Big Spring Canyon, South Dakota: Department of Geological Sciences, University of California, Bulletin, v. 26, p. 307-446.

Haldane, J. B. S., 1949, Suggestions as to quantitative measurements of rates of evolution: Evolution, v. 3, p. 51-56.

Hesse, C. J., 1936, Lower Pliocene vertebrate fossils from the Ogallala Formation (Lavern Zone) of Beaver County, Oklahoma: Carnegie Institute (Washington), Contributions to Paleontology, v. 476, p. 47-72.

Hibbard, C. W., 1952, Vertebrate fossils from Late Cenozoic deposits of central Kansas: University of Kansas Paleontological Contributions, Vertebrata, Article 2, p. 1-14.

Hulbert, R. C., 1982, Population dynamics of the three-toed horse *Neohipparion* from the late Miocene of Florida: Paleobiology, v. 8, p. 159-167.

―――― 1984, Paleoecology and population dynamics of the early Miocene (Hemingfordian) horse *Parahippus leonensis* from the Thomas Farm Site, Florida: Journal of Vertebrate Paleontology, v. 4, p. 547-558.

―――― in press, Late Neogene *Neohipparion* (Mammalia, Equidae) from the Gulf Coastal Plain of Florida and Texas: Journal of Paleontology.

―――― in preparation, *Calippus* and *Protohippus* (Mammalia, Equidae) from the Gulf Coastal Plain and the Great Plains.

Johnston, C. S., 1938, The skull of *Nannippus gratus* (Leidy) from the Lower Pliocene of Texas: American Journal of Science, v. 19, p. 245-248.

Kovacs, I., 1971, A systematic description of dental roots, *in* Dahlberg, A. A., ed., Dental morphology and evolution: Chicago, University of Chicago Press, p. 211-256.

Leidy, J., 1869, The extinct mammalian fauna of Dakota and Nebraska: Academy of Natural Sciences, Philadelphia, Journal, 2nd. Ser., v. 7, p. 1-472.

MacFadden, B. J., 1984, Systematics and phylogeny of *Hipparion, Neohipparion, Nannippus,* and *Cormohipparion* (Mammalia, Equidae) from the Miocene and Pliocene of the New World: American Museum of Natural History, Bulletin, v. 179, p. 1-196.

―――― 1985, Patterns of phylogeny and rates of evolution in fossil horses: hipparions from the Miocene and Pliocene of North America: Paleobiology, v. 11, p. 245-257.

Macdonald, J. R., 1960, An early Pliocene fauna from Mission, South Dakota: Journal of Paleontology, v. 34, p. 961-982.

Matthew, W. D., 1924, Third contribution to the Snake Creek Fauna: American Museum of Natural History, Bulletin, v. 50, p. 59-210.

Matthew, W. D., and Stirton, R. A., 1930, Equidae from the Pliocene of Texas: Department of Geological Sciences, University of California, Bulletin, v. 19, p. 349-396.

Osborn, H. F., 1918, Equidae of the Oligocene, Miocene, and Pliocene of North America. Iconographic type revision: American Museum of Natural History, Memoirs, new series, v. 2, p. 1-331.

Patterson, B., and Pascual, R., 1972, The fossil mammal fauna of South America, *in* Keast, A., Erk, F. C., and Glass, B., eds., Evolution, mammals, and southern continents: Albany, State University of New York Press, p. 247-309.

Quinn, J. H., 1955, Miocene Equidae of the Texas Gulf Coastal Plain: Bureau of Economic Geology, University of Texas Publication No. 5516, p. 1-102.

Schultz, G. E., 1977, The Ogallala Formation and its vertebrate faunas in the Texas and Oklahoma Panhandles, *in* Schultz, G. E., ed., Field conference on late Cenozoic biostratigraphy of the Texas Panhandle and adjacent Oklahoma, Guidebook: Canyon, Texas, Kilgore Research Center, Department of Geology and Anthropology, Special Paper No. 1, West Texas State University, p. 5-104.

Simpson, G. G., 1930, Tertiary land mammals of Florida: American Museum of Natural History, Bulletin, v. 59, p. 149-211.

―――― 1953, The major features of evolution: New York, Columbia University Press, 434 p.

Skinner, M. F., and Johnson, F. W., 1984, Tertiary stratigraphy and the Frick collection of fossil vertebrates from north-central Nebraska: American Museum of Natural History, Bulletin, v. 178, p. 215-368.

Skinner, M. F., and MacFadden, B. J., 1977, *Cormohipparion* n. gen. (Mammalia, Equidae) from the North American Miocene (Barstovian, Clarendonian): Journal of Paleontology, v. 51, p. 912-926.

Skinner, M. F., and Taylor, B. E., 1967, A revision of the geology and paleontology of the Bijou Hills, South Dakota: American Museum Novitates, no. 2300, 53 p.

Skinner, M. F., Skinner, S. M., and Gooris, R. J., 1968, Cenozoic rocks and faunas of Turtle Butte, south-central South Dakota: American Museum of Natural History, Bulletin, v. 138, p. 379-436.

―――― 1977, Stratigraphy and biostratigraphy of Late Cenozoic deposits in central Sioux County, western Nebraska: *ibid.*, v. 158, p. 263-370.

Stirton, R. A., 1941, Development of characters in horse teeth and the dental nomenclature: Journal of Mammalogy, v. 22, p. 339-410.

―――― 1947, Observations on evolutionary rates in hypsodonty: Evolution, v. 1, p. 32-41.

Tabrum, A., 1981, A contribution to the mammalian paleontology of the Ogallala Group of south-central South Dakota [M.S. thesis]: Rapid City, South Dakota, South Dakota School of Mines and Technology, 408 p.

Tedford, R. H., Galusha, T., Skinner, M. F., Taylor, B. E., Fields, R. W., Macdonald, J. R., Rensberger, J. M., Webb, S. D., and Whistler, D. P., *in press,* Faunal succession and biochronology of the Arikareean through Hemphillian interval (late Oligocene through earliest Pliocene epochs), North America, *in* Woodburne, M. O., ed., Vertebrate paleontology as a discipline in geochronology: Berkeley, University of California Press.

Van Valen, L., 1960, A functional index of hypsodonty: Evolution, v. 14, p. 531-532.

Webb, S. D., 1969, The Burge and Minnechaduza Clarendonian mammalian faunas of north-central Nebraska: University of California, Publications in Geological Sciences, v. 78, p. 1-191.

Webb, S. D., MacFadden, B. J., and Baskin, J. A., 1981, Geology and paleontology of the Love Bone Bed from the Late Miocene of Florida: American Journal of Science, v. 281, p. 513-544.

Webb, S. D., and Tessman, N., 1968, A Pliocene vertebrate fauna from low elevation in Manatee County, Florida: *ibid.,* v. 266, p. 777-811.

White, T. E., 1959, The endocrine glands and evolution, no. 3: os cementum, hypsodonty, and diet: University of Michigan, Museum of Paleontology, Contributions, v. 13, p. 211-265.

MANUSCRIPT RECEIVED JANUARY 15, 1986
REVISED MANUSCRIPT RECEIVED JUNE 20, 1986
MANUSCRIPT ACCEPTED JUNE 30, 1986

Species longevity, stasis, and stairsteps in rhizomyid rodents

LAWRENCE J. FLYNN *Department of Anthropology, Peabody Museum, Harvard University, Cambridge, Massachusetts 02138*

ABSTRACT

New data from the middle and late Miocene Siwalik deposits of Pakistan provide accurate estimates of real temporal durations of extinct species of rhizomyid rodents. Most early rhizomyid species survive on the order of millions of years, with at least two spanning about five million years, and display apparent stasis in most characters. Average species duration for all Rhizomyidae of the Potwar Plateau is about 1.2 million years, a figure in line with other estimates for all Mammalia. Three closely related species show sharp differences in hypsodonty, while other traits remain static in each species or change slowly within the clade, on a scale above the species level. Evolution of this lineage shows at least one step in a staircase pattern, with descendants replacing ancestors, and entails an abrupt morphological change that provides a nonarbitrary definition for species boundaries. One ancestral morphotype appears to survive for a short time with its daughter species. Whereas early nonburrowing Rhizomyidae display longterm stasis, later species, some with burrowing adaptations, are shorter lived and at least one rhizomyine shows rapid, perhaps continuous phyletic change.

> "The time at which an evolving population became different enough from its ancestry to be called a different species cannot, even in theory, be a precise, naturally defined date unless the new, descendant species arose in a single, abrupt step."
> G. G. Simpson, 1953, p. 35

INTRODUCTION

Rhizomyid (from Greek *rhiza,* root, plus Greek *mys,* rat) rodents today are represented by three genera that are fossorial in habitus and show distinctive osteological adaptations to that mode of life (Asian *Rhizomys* and *Cannomys* and African *Tachyoryctes*). The most complete record of fossils documenting the mosaic evolution of this specialized muroid family (Flynn, 1983a) is entombed in the thousands of meters of sediments known collectively as the Siwalik Group of Pakistan and India. This Neogene sequence of molassic sediments shed from the Himalayan complex spans some eighteen million years of time and, perhaps not coincidentally, approximately the entire history of the Rhizomyidae. During the first half of this history, rhizomyids were small nonburrowers. In the late Miocene, derived forms with burrowing adaptations diversified, and some of these can be shown to be closely related to living *Rhizomys*.

An understanding of the mode and details of rhizomyid evolution has developed in direct correspondence to available fossil material. Early fossils recovered over a century ago (Lydekker, 1884) provided the first evidence for antiquity of the family. Fifty years later, further prospecting showed some diversity in the Siwalik rhizomyid record (Colbert, 1933; Hinton, 1933). All of the taxa known by that time were relatively large, surface finds, and evidently closely related to *Rhizomys*. Wood (1937) reported on additional new material, and Bohlin (1946) realized that the large Siwalik rhizomyids, although closely related to *Rhizomys,* were generically distinct. Black (1972), in analyzing the entire series of fossil rhizomyids then available, and developing an idea put forth by Hinton (1933), made an important breakthrough. He defined two clades, one including ancestors of *Rhizomys* and the other apparently encompassing the roots of *Tachyoryctes*; thus he built an argument on fossil evidence for inclusion of African *Tachyoryctes* in the Rhizomyidae.

With application of screening techniques, Jacobs (1978) demonstrated that small rhizomyids (genus *Kanisamys,* previously rarely found) were actually quite common along with murids in Siwalik localities. With large samples, Jacobs (1978) and Munthe (1980) were able to analyze the range of variation in detailed characters encountered in single rhizomyid species. Sieving efforts by de Bruijn and others (1981) and Wessels and others (1982) provided important new samples of small species, including the earliest known rhizomyid, *Prokanisamys arifi*.

A decade of collecting by field parties under the co-direction of Drs. D. Pilbeam and S. M. Ibrahim Shah inflated the number of surface finds of large rhizomyids from late Miocene horizons, and provided several new screening localities with samples of the smaller taxa. This enabled an analysis of evolution in *Kanisamys,* with recognition of stasis, lack of burrowing modifications, and slow evolution in some features (Flynn, 1982). Among the larger fossil rhizomyids, it has become evident that *Rhizomyides* is related to *Kanisamys* and *Tachyoryctes,* indicating probable early differentiation in rhizomyid evolution (Flynn, 1983b). Systematic revision of other large rhizomyids showed alliance with living Rhizomyinae and that they were diverse, evolved rapidly, and possessed many burrowing features.

The present study incorporates new microfossil samples from numerous localities in the lower Siwaliks to produce a refined knowledge of early small rhizomyid biostratigraphy. This provides the evidence for a new appraisal of evolutionary mode in early Rhizomyidae. Interpretation of the data draws heavily on evolutionary concepts introduced by Professor George Gaylord Simpson (particularly 1953). Some of the data from rodents presented here yield patterns that are especially relevent to issues developed by Eldredge and Gould (1972) and Stanley (1979, 1985).

STRATIGRAPHY

The Siwalik sequence is exceptional for its great thickness of superposed strata, abundance of land mammal horizons, and good surface exposure. Recent multidisciplinary efforts in the lower Siwaliks include measurement of long, continuous stratigraphic sections, notably by members of the Geological Survey of Pakistan, and construction of the detailed paleomagnetic reversal sequence for this composite, notably under the direction of Noye Johnson of Dartmouth College (Johnson and others, 1985, see lithostratigraphic boundaries therein). Paleomagnetic and radiometric data for a single long sequence near Chinji and Nagri villages, Pakistan, provide absolute ages for localities and an accurate estimate of time elapsed between deposition of individual strata. Fossil localities can be traced laterally into this sequence, and Barry and others (1985) offer a preliminary biostratigraphy for major faunal elements.

Figure 1 is based on the work of Johnson and others (1985), Barry and others (1985), and Tauxe and Opdyke (1982). The column on the left of Figure 1 shows a single 1500 m continuous section with localities that have yielded Rhizomyidae. The prefix "Y" signifies entry in the Yale-Geological Survey of Pakistan locality catalog. The column next to this places each fossil locality on an absolute time scale, with interpolation of several other sites from northern Pakistan. These sites, Y450 from the Bunha River area to the east, and Y491, Y259, Y310, Y182, Y367, Y388, Y24, and Y547 from the Khaur region to the north, are tied to other paleomagnetic sections and were discussed in an earlier study (Flynn, 1982). The section begins low in the Siwalik Group, with locality Y592 occurring in sediments about 16 million years old. With some gaps, temporal sampling is good up to about 10.3 Ma, and fossils seem to represent rhizomyid evolution adequately. Above the horizon of Y76, several critical intervals at times when new rhizomyids appear still are unsampled.

SYSTEMATICS

Introduction

Taxa found in the lower Siwaliks, which is the better sampled part of the composite section, are assigned to three genera, *Prokanisamys, Kanisamys,* and *Rhizomyides.* Constraints on and conclusions about their mode of evolution are the focus of this paper, and their stratigraphic ranges are indicated on Figure 1. Following here is a brief summary of the characterisitics of each genus and then a discussion on species recognition.

Prokanisamys de Bruijn and others (1981)

Prokanisamys arifi was based on a sample of small, low crowned rhizomyid teeth from the Murree Formation (de Bruijn and others, 1981), antedating locality Y592, perhaps by about two million years (Flynn and others, 1986). A diagnostic, but probably primitive feature of this taxon is the small size of M_3 relative to M_1. Length of M_3 is about 80 percent that of M_1 in *P. arifi,* but it is larger than M_1 in other genera of the family. A larger member of the genus in lower Siwalik localities has M_3 length about equal to 90 percent that of M_1. Rhizomyids have large third molars relative to other muroids; the size of M_3 is simply less derived in *Prokanisamys* than in *Kanisamys* or *Nakalimys,* for example.

Jacobs and others *(in press)* identify the lower Siwalik *Prokanisamys* as *P. benjavuni,* a species named from the Miocene of Thailand and originally placed in *Kanisamys* (Mein and Ginsburg, 1985). All of the teeth of *P. benjavuni* are considerably larger than those of *P. arifi,* but smaller than those of any other known rhizomyid (Figs. 2-4). With reference to the Siwalik sample of *P. benjavuni,* *Prokanisamys* is less lophodont and lower crowned than *Kanisamys.* This is most clearly reflected in the anterolophid of M_{2-3}, which is low, below the level of the other crests. The labial and lingual reentrants formed by the anterolophid on these teeth apparently are lost through wear earlier than in *Kanisamys.*

Kanisamys Wood (1937)

Kanisamys is a long-lived taxon of low species diversity. It is characterized by hypsodont, lophodont molars and the following trends: progressive development of mesolophs on upper molars; posterior migration of mesolophids on lower molars through time; closure of the lingual reentrant on M^3; and loss of the anterolabial reentrant on M_{2-3}. Wood (1937), Black (1972), and Flynn (1982) built a case for phyletic evolution involving the transformation of the *K. indicus* morphotype through *K. nagrii* to *K. sivalensis.* An initial attempt to quantify morphological change in this lineage included a pool of all size measurements of *K. indicus* then available (Flynn, 1982). These were compared to observed size in samples of younger, presumably descendant, species of *Kanisamys* to yield a net amount and rate of change during five million years (13 to 8 Ma). It is now apparent that *Kanisamys* did not evolve at a constant rate, but showed long-term stasis. Data from each sampled horizon are considered independently below.

Rhizomyides Bohlin (1946)

The tachyoryctines *Rhizomyides* and *Kanisamys* share the following traits: hypsodonty; well developed mesolophid; strong masseteric crest with anterior extension; and a rounded incisor bearing a longitudinal ridge.

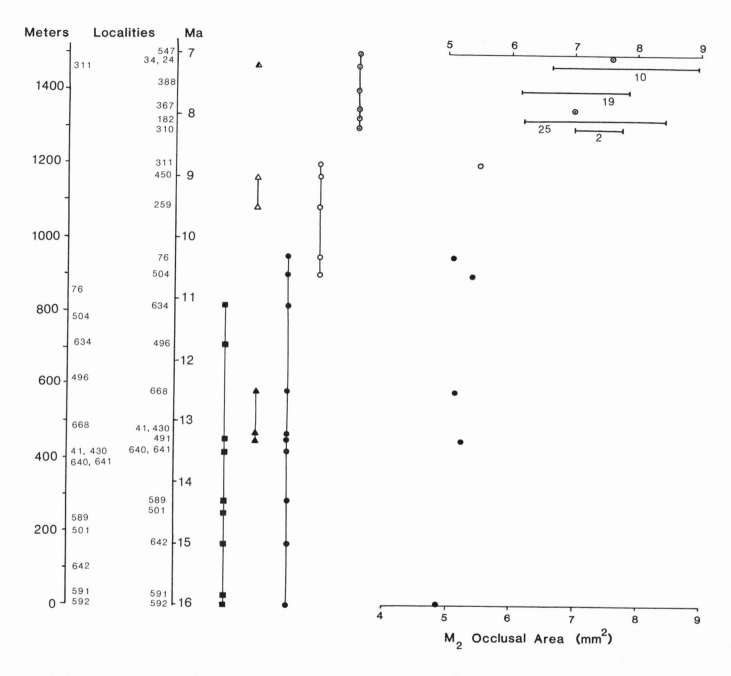

Figure 1. Small rhizomyid rodents: stratigraphic and temporal distribution of localities, biostratigraphy, and size through time in *Kanisamys*. On left, localities are arranged by stratigraphic level and by age in megaannum (Ma), which is inferred from magnetostratigraphy. Species occurrence at a locality is indicated by a symbol: *Prokanisamys benjavuni* (filled square), *Kanisamys potwarensis* (filled triangle), *Rhizomyides punjabiensis* (open triangle), *Rhizomyides* aff. *R. punjabiensis* (half filled triangle), *Kanisamys indicus* (filled circle), *Kanisamys nagrii* (open circle), *Kanisamys sivalensis* (circle with dot). Lower second molar occlusal area (length X width) is plotted for the *K. indicus-K. nagrii-K. sivalensis* clade: circles indicate single specimens; observed range for larger samples is indicated by horizontal bars with sample size in numerals below each bar (modified from Flynn, 1982, Fig. 10).

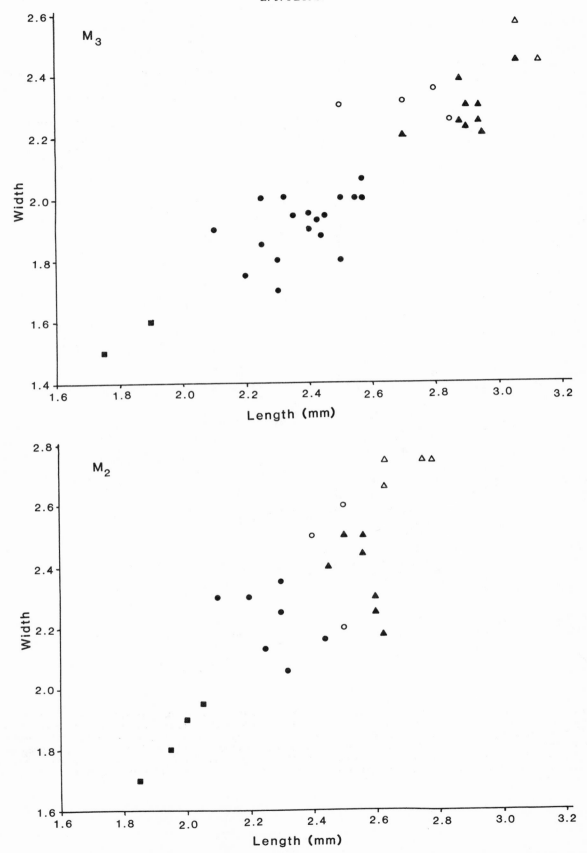

Figure 2. Length-width scatter diagram of small rhizomyid third and second lower molars. Symbols as in Figure 1.

Most species of *Rhizomyides* are quite large and, unlike *Kanisamys*, conserve the anterior position of the mesolophid on lower molars. These and other dental features, notably the protoconid-metaconid connection on M_1, are discussed in detail elsewhere (Flynn, 1983b). These traits suggest that the origin of *Rhizomyides* would have been considerably earlier than the 6.4 Ma first occurrence known previously (Flynn, 1982), and that *Rhizomyides* could have been transformed readily from an early *Kanisamys* morphotype. Here, 10 to 9 Ma *Rhizomyides punjabiensis* is classified as the oldest member of the genus, and its morphology and age are consistent with the foregoing hypothesis.

Rhizomys punjabiensis Colbert (1933) was redescribed by Black (1972); Flynn (1982) added several isolated teeth from localities Y259 and Y450. Bohlin (1946) assigned *Rhizomys punjabiensis* to his new genus *Rhizomyides;* later, Flynn (1982) split the assemblage of species placed in *Rhizomyides*, considering *R. punjabiensis* to be a primitive species of *Brachyrhizomys* Teilhard de Chardin (1942). I interpreted a worn M^1 from locality Y450 to show that *R. punjabiensis* lacks an anterolingual flexus. That absence is a rhizomyine synapomorphy that led to incorrect generic assignment of the species. A fresh specimen of the same size from Y450 has the flexus, demonstrating at least its variable presence in the species. The possibility exists that the two M^1 represent different taxa. In any case, transferral of *R. punjabiensis* to *Rhizomyides* (sensu stricto) results in more consistent distribution of other features within that taxon and of others within *Brachyrhizomys*. Like other *Rhizomyides*, *R. punjabiensis* has anteriorly placed mesolophids, a strong masseteric crest with anterior extension, a rounded incisor with a ridge, and a large posterior enamel lake on M_3.

Species Recognition

Species identification, but not definition, for individual teeth was facilitated by analysis of size in length/width scatter diagrams (Figs. 2-4). Species size increases in the order of *Prokanisamys benjavuni*, *Kanisamys indicus*, *K. nagrii*, *K. potwarensis*, *Rhizomyides punjabiensis*. The scatter diagrams include teeth studied previously (Flynn, 1982) and, by virtue of the larger samples now available, they resolve assignment of elements formerly considered indeterminate. For example, plots of M^3 from locality Y491 clearly fall into size classes that correspond to *K. indicus* and *K. potwarensis*. Similarly, a large M^1 from Y491 is now assignable to *K. potwarensis*, but could not be classified when it was the only M^1 known of that age (Flynn, 1982, p. 340).

While size helps to identify individual teeth, additional morphological characteristics are used to define the species. *Prokanisamys* has a proportionately small M_3 and has low crests. *Kanisamys potwarensis* has a distinctive sulcus on the posterolabial wall of M_3. *R. punjabiensis* is higher crowned (M^2 greater than 3 mm) than *K. potwarensis*. *K. nagrii*, although overlapping in occlusal area, is distinctly more hypsodont than either *K. indicus* or *K. potwarensis*. This contrast in hypsodonty is indicated by Table 1.

Unilateral hypsodonty is strongly developed in rhizomyid teeth, especially upper molars, where lingual sides of teeth are higher than the molars are long or wide. Maximum crown height is measurable in unworn teeth and, as for locality Y182, large samples allow estimation of maximum crown height for each tooth locus in a population. Smaller samples may include unworn teeth for few loci, but slightly worn teeth provide a reasonably close estimate of maximum crown height as well. The relative crown height for each locus in *Kanisamys sivalensis* from Y182 may be applied as a model for other similar species, such that knowing the crown height of one tooth (e. g., M^3) permits estimation of crown height in others (e. g., M^2). Therefore, samples from different localities can be compared even if unworn teeth of the same locus are not available.

Table 1 summarizes available crown height data for *Kanisamys* and *Prokanisamys* by locality. Some variance in hypsodonty among species is allometric, increasing with size along the series *Prokanisamys benjavuni*, *Kanisamys indicus*, *K. potwarensis*, *K. nagrii*, *K. sivalensis*. *Kanisamys indicus* may show increase in hypsodonty during its long history; 15 and 16 million year old teeth are shorter than 13 million year old teeth. However, samples of *K. nagrii* present a far more dramatic contrast when compared to any of *K. indicus*. Upper molars (M^2) of *K. indicus* range up to 2.5 mm high, whereas those of *K. nagrii* are uniformly 3 mm or greater. This is seen at a single site: heights of two M^3 from locality Y504 are 2.5 and 1.7 mm, which translates to estimated M^2 heights of 3.0 and 2.2 mm, using the Y182 ratios as a model. *K. sivalensis* M^2 are nearly 4 mm high. Hypsodonty, therefore, presents an unambiguous criterion for recognizing these species. Without such an abrupt change in this lineage, species limits could not be defined naturally.

Martin (1979) demonstrated how crown height plotted against the product of molar length X width successfully discriminates species of *Sigmodon*. Eventually, the same can be done for *Kanisamys*, but sample size per tooth locus is still small for most sites. Only ten of the upper second molars in Table 1 can be measured for all three dimensions, but they show no overlap.

BIOSTRATIGRAPHY

The biostratigraphic ranges on Figure 1 are based on samples collected by screen washing since 1975, and especially, through the intensive efforts of William R. Downs, since 1982. Figure 1 excludes the derived Rhizomyinae, mainly surface finds, which appear at about 8.5 Ma. The latter species have relatively short durations, the longest being about one million years (Flynn, 1982).

Two early species have relatively long durations. *Prokanisamys benjavuni* has a temporal range of about five million years (localities Y592 to Y634). *Kanisamys indicus* occurs throughout the lower Siwaliks from Y592 up through Y76, high in the Chinji Formation, a span of nearly six million years. Its range overlaps that of the

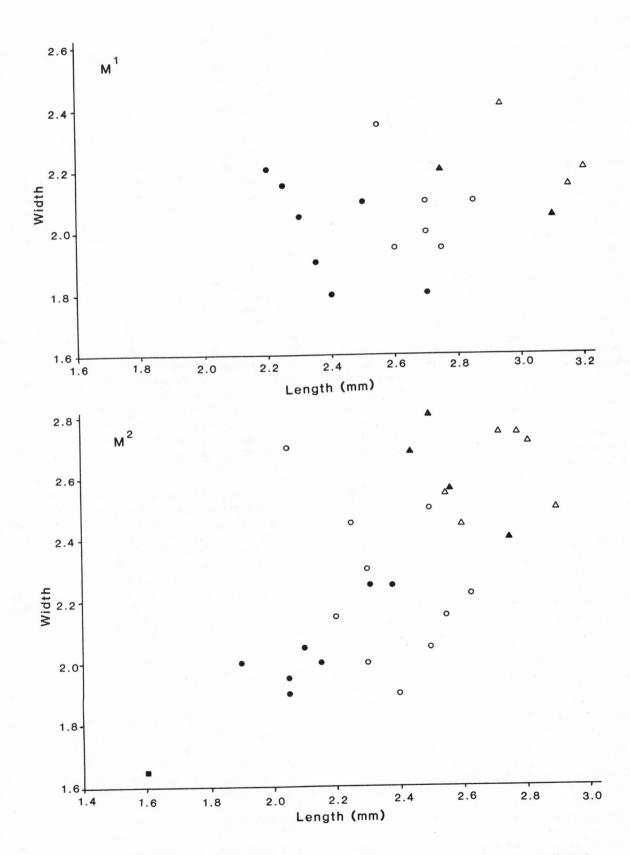

Figure 3. Length-width scatter diagram of small rhizomyid first and second upper molars. Symbols as in Figure 1.

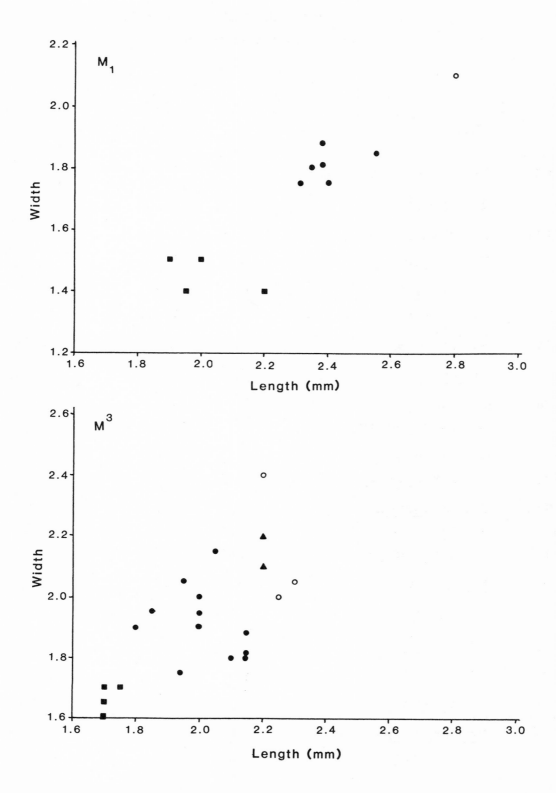

Figure 4. Length-width scatter diagram of small rhizomyid first lower and third upper molars. Symbols as in Figure 1.

TABLE 1. MAXIMUM OBSERVED CROWN HEIGHTS IN EARLY RHIZOMYIDS (MM).

Locality	Species	M^1	M^2	M^3	M_1	M_2	M_3
Y 182	K. sivalensis*	3.4	3.9	3.3	2.8	3.6	3.0
Y 311	K. nagrii		3.1	2.7			
Y 259	K. nagrii		3.1				
Y 76	K. nagrii	2.8	3.0	2.6			2.3
Y 504	K. nagrii			2.5			
504	K. indicus	1.9		1.7	1.7		2.1
Y 634	K. indicus		1.9			1.9	1.9
Y 496	Prokanisamys		1.8			1.6	
Y 668	K. indicus					1.7	1.8
668	K. potwarensis						2.0
Y 430	K. indicus		2.3			1.8	
430	K. potwarensis		2.4				
Y 491	Prokanisamys			1.8			
491	K. indicus		2.4	2.0		2.2	2.1
491	K. potwarensis	2.5	2.7	2.2		2.1	2.0
Y 640/641	K. indicus		2.5	2.3			1.9
Y 589	K. indicus			1.7			
Y 501	Prokanisamys			1.6			
Y 642	K. indicus	2.0					
Y 591	Prokanisamys			1.6			
Y 592	Prokanisamys			1.7			1.5
592	K. indicus		2.0		1.3		1.6

*Species cited are Prokanisamys benjavuni, Kanisamys indicus, Kanisamys potwarensis, Kanisamys nagrii, Kanisamys sivalensis.

more hypsodont *K. nagrii*, which spans about two million years (Y504 to Y311). The larger and more hypsodont *K. sivalensis* first appears at locality Y310, and lasts over one million years until its local (?universal) extinction at 7 Ma (sites Y457 and Y547). *Kanisamys potwarensis*, distinctive morphologically by its sulcus on M_3, survived for less than one million years in the Potwar Plateau Chinji Formation (13.2 to 12.5 Ma).

In addition to jaw fragments reviewed by Black (1972), *Rhizomyides punjabiensis* is securely identified at localities Y259 and Y450, and questionably represented at Y311 and at Kanati (see the molar identified as *Kanisamys nagrii* in Flynn, 1982), for a temporal duration of about one million years. *Rhizomyides sivalensis*, the type species of the genus, is a large rodent that appears at the end of the Miocene (about 6.4 Ma, Flynn, 1982). *R. sivalensis* also is known from 5.5 Ma rocks, and may have spanned more than one million years. A molar fragment from locality Y34, about 7.2 Ma, attests to sporadic representation of the genus throughout the Siwaliks. This specimen (M_3, 3.6 mm long) is larger than known *R. punjabiensis*, but like one of two M_3 of that species; and like *K. potwarensis*, it bears a posterolabial sulcus and an additional furrow posterior to the protoconid. The Y34 specimen is designated *Rhizomyides* aff. *R. punjabiensis*.

As postulated above, *Kanisamys indicus* embodies a suitable morphotype for the ancestry of *K. nagrii*; however, *K. indicus* persists for at least 100,000 years after *K. nagrii* appears. Although possible, it is unlikely that the Potwar Plateau was the exact site of origin of *K. nagrii*. Rather, the species may have evolved elsewhere from *K. indicus* by classical geographic speciation (Mayr, 1942), invaded the Potwar, and replaced *K. indicus*. Stanley (1985) presents a scenario of parent-daughter competition with extinction of the parent as a phenomenon associated with staircase evolution. Actual competition between these rhizomyids is possible in that they may have been forced to partition the same narrow niche. In addition to temporal overlap, there is an apparent change in species abundance: at locality Y504, 2 of 11 specimens represent *K. nagrii*, while at Y76 (100,000 years younger) 15 of 18 specimens are *K. nagrii*.

Known longevities for all Potwar Plateau rhizomyids, seventeen species, are indicated in Table 2.

TEMPO AND MODE
Species Boundaries

Kanisamys indicus, *K. nagrii*, and *K. sivalensis* are closely related, sequentially distributed in time, and known morphology is compatible with ancestral-descendant relationships. End members indicate increase in size and hypsodonty, development of mesolophs, closure of the lingual reentrant on M^3, loss of the metaconid-protoconid connection on M_1, posterior migration of metalophids, and loss of the anterolabial reentrant on M_{2-3}. A few specimens of *K. sivalensis* from Y182, Y388, and Y24 show primitive conditions (Flynn, 1982). Specimens of *K. nagrii* from locality Y76 show mostly primitive morphological features, while the poor samples of *K. nagrii* from younger sites show mostly the derived states. Distribution of these features does not unambiguously delineate species.

Similarly, tooth occlusal area increases through time, but observed ranges for species overlap. Some specimens of *K. nagrii* are as large as small *K. sivalensis* (see Flynn, 1982). The single M_2 of *K. nagrii* from locality Y311 is only slightly larger than the several M_2 plotted for *K. indicus* (Fig. 1), which show no essential change from 16 to 10 Ma. The entire range of size for *K. indicus* plus *K. nagrii* is less than that in single samples of *K. sivalensis*.

If the three species represent an ancestral-descendant lineage but the development of advanced crown morphology occurs at a different pace than increase in size, is it useful to recognize separate species? Does a non-arbitrary definition for each species exist? The degree of difference between *Kanisamys* endmember morphs makes species recognition biostratigraphically useful; furthermore, persistence of *K. indicus* with *K. nagrii* demonstrates that a splitting event took place, although one daughter remained untransformed, so far as it is known.

The fact is that there does exist an unambiguous criterion for separating *Kanisamys indicus* and *K. nagrii*. Hypsodonty is the single feature that increases abruptly and significantly in these samples (Table 1), and offers a precise and natural break for species boundaries (see Simpson, 1953, p. 35). *K. nagrii* is higher crowned than *K. indicus*, and the two species are clearly distinct at the same site (Y504, Y76). The relationship between *K. nagrii* and *K. sivalensis* could be truly anagenetic, but sites with small mammals do not occur at the time of the suspected replacement of *K. nagrii* by *K. sivalensis* to evaluate temporal overlap.

Species Longevity

Fossil rhizomyids from the Siwaliks of the Potwar Plateau indicate mean species longevity compatible with estimates for all mammals (Stanley, 1985). Simpson (1953), without availability of precise chronometry, had estimated that mean mammalian species longevity might be on the order of 10^5 to 10^6 years, while Stanley (1985) increased this estimate to 1 to 2 million years. Table 2 tabulates the temporal durations of species considered in this study and, with some modification, those cited in Flynn (1982). Species known from single localities are given a duration of 100,000 years, but in most cases these are surface finds in strata younger than 9 Ma. More fieldwork may increase some of these known ranges, but rapid speciation and replacement among later, larger rhizomyids, especially fossorial burrowers, is probably a real phenomenon (Flynn, 1985).

Average rhizomyid species duration (Table 2) is 1.2 million years, a figure close to averages estimated by Stanley (1985) for all mammals and by Gingerich (1985) for Eocene mammals. Mean species duration for small rhizomyids (*Prokanisamys* and *Kanisamys* species) is longer, about 2.9 million years, and is similar to the calculation by MacFadden (1985) for North American hipparions (3.3 million years).

TABLE 2. TEMPORAL DURATIONS OF RHIZOMYID SPECIES IN THE POTWAR PLATEAU.

SPECIES	DURATION (m.y.)
Subfamily incertae sedis	
Prokanisamys benjavuni	4.9
Subfamily Tachyoryctinae	
Kanisamys indicus	5.6
Kanisamys potwarensis	0.7
Kanisamys nagrii	1.8
Kanisamys sivalensis	1.3
Rhizomyides punjabiensis	0.5
Rhizomyides aff. R. punjabiensis	0.1
Rhizomyides sivalensis	0.9
Protachyoryctes tatroti	0.4
Eicooryctes kaulialensis	0.5
Subfamily Rhizomyinae	
Brachyrhizomys nagrii	0.5
Brachyrhizomys micrus	0.1
Brachyrhizomys blacki	0.1
Brachyrhizomys cf. B. pilgrimi	1.0
Brachyrhizomys tetracharax	1.0
Brachyrhizomys choristos	0.1
Anepsirhizomys opdykei	0.1
Mean species longevity	1.2
Mean for species Kanisamys and Prokanisamys	2.9

The density of screened localities in the lower and middle Siwaliks provides estimates of local species duration that are accurate to a few hundred thousand years. This is a good local record that will improve, but neighboring regions in Asia may never be comparably represented. Thus temporal estimates given here are minimum estimates for the longevity of species throughout their entire geographic ranges. *Prokanisamys benjavuni* and *Kanisamys indicus* endure longer than other rhizomyids, about five and six million years, respectively. These two taxa are members of the lower Siwalik chronofauna characterized by Raza (1983), who concentrated on vertebrates from the Chinji Formation. They are elements of an ecologically stable community, and show virtually no difference in size or tooth crown morphology through time. Up to about 10 Ma, stasis appears to have been the norm in rhizomyid evolution, in contrast to the late Miocene, when increasingly xeric conditions may have contributed to more frequent speciation and replacement within both tachyoryctines and rhizomyines (see also Jacobs and Flynn, 1981; Flynn and Jacobs, 1982).

Stasis and Superspecific Trends

Among rhizomyids, at least the small species typically exhibit stasis during their temporal range. Of course only a few features have been studied, and fewer can be treated statistically; the small samples can not demonstrate absence of gradual change in all traits, but size and hypsodonty seem to remain static for periods on the order of millions of years. On the other hand, populations of at least one fossorial rhizomyine attest to rapid increase in size during several hundred thousand years (Flynn, 1982). If not representing anagenetic evolution, this chronocline could possibly trace the geographical wander of populations that show clinal variation in size.

The pattern of speciation in *Kanisamys* of the Potwar Plateau does not lead to increase in diversity. Thus it differs in detail from a general model of splitting evolution by punctuation (*sensu* Eldredge and Gould, 1972), and better resembles staircase evolution with an example of short-term survival of an ancestor with its descendant (see Stanley, 1985).

Although the evolution of *Kanisamys* is punctuated by abrupt increases in hypsodonty, there are other features that show progressive change through time, and these trends transcend species boundaries. In addition to increase in size, these include the dental traits noted above (e. g., mesoloph development, migration of mesolophids, and loss of anterolabial reentrants). Whatever the mechanism for these changes, it apparently operated on all populations throughout the life of this lineage, regardless of species allocation. In the case of *Kanisamys*, selection for certain changes was sustained over a period of about ten million years.

CONCLUSIONS

The subject of this paper is the species level record of evolution in a group of rodents from the Miocene of the Potwar Plateau, Pakistan. Primary data are morphology of preserved fossils and the biostratigraphic and temporal ranges of morphologically defined units (i. e., paleontological species). The data permit conclusions about the pattern of speciation in these rodents, about species longevity, and about trends, all problems applying to Gould's (1985) "second tier." These problems can be approached instructively for some taxa discussed here, but not for others, depending on quality of the fossil samples. Given sampling density, fortuitous occurrence of wash sites, and slow turnover in early Rhizomyidae, some interesting observations on patterns of speciation can be made for lower Siwalik species, albeit with the assumption that the Potwar Plateau fossils are representative of *Kanisamys* populations in general. At present not as many conclusions can be made about late Miocene rhizomyid evolution.

An average species longevity for all Potwar rhizomyids of 1.2 million years is in line with estimates for the entire Mammalia. Observation of longevities of five or six million years for two early rhizomyids provides positive evidence bearing on the tempo of rhizomyid evolution in the middle Miocene. Long-term stasis characterizes these two early species, members of the Chinji chronofauna, while later taxa are short-lived. This provides circumstantial evidence for ecological stability in the Potwar Plateau from 16 to 10 or 11 Ma.

The mode of rhizomyid evolution changes concurrently with decrease in species longevity at 10 Ma. Earlier rhizomyids are not diverse and, in the case of *Kanisamys indicus* and *Kanisamys nagrii*, the presumed descendant appears by splitting, and soon replaces its ancestral morphotype. This scenario resembles a step in "stair case evolution," rather than the generalized splitting by punctuation model, where both morphs survive. After 10 Ma rhizomyids become diverse as well as shorter lived. This phenomenon results in part from invasion of underground niches, a lifestyle that contributes to isolation of demes and increased probability of speciation. Indeed, Rhizomyinae appear about 8.5 Ma, show numerous fossorial adaptations, and are strikingly diverse. One rhizomyine may show gradual size increase rather than stasis.

In the low diversity *Kanisamys* lineage, crown morphology and size seem to grade through time, raising the question of ambiguity in species identification. However, hypsodonty is one feature that differs sharply between species, and can be used to define species boundaries and pinpoint them in time (see Simpson, 1953). It is the existence in the same localities of teeth of markedly different crown height that shows that *Kanisamys indicus* survived for a time with *Kanisamys nagrii*. Several trends in evolution of crown morphology transcend species boundaries and were sustained over a long period of time (*circa* ten million years).

The Potwar Plateau record is not perfect, yet some powerful conclusions on tempo and mode can be drawn from it. The record is good now, and shows great potential for improvement.

ACKNOWLEDGMENTS

This study is an outgrowth of work begun for my dissertation at The University of Arizona, Tucson, under the direction of Professors E. H. Lindsay, G. G. Simpson, K. W. Flessa, E. L. Cockrum, and R. F. Wilson. Professor Simpson was inspirational to the students known as the Red Fireballs in our efforts to interpret phylogeny and understand paths of evolution in vertebrate fossils. Drs. Simpson and Roe sported a "Red Fireballs" bumper sticker on their car. This paper developed and gained direction as a result of field work and discussions with the above and with Drs. J. C. Barry, L. L. Jacobs, J. Munthe, M. J. Novacek, D. R. Pilbeam, and R. H. Tedford, and with Messrs. W. Downs, M. Norell, and A. Wyss. The biostratigraphic framework grew particularly with the painstaking care of Downs, Barry, and Lindsay.

This research was supported in part by NSF Grant BSR 8500145 and by the Carter Fellowship of The American Museum of Natural History.

REFERENCES CITED

Black, C. C., 1972, Review of fossil rodents from the Neogene Siwalik beds of India and Pakistan: Palaeontology, v. 15, p. 238-266.

Barry, J. C., Johnson, N. M., Raza, S. M., and Jacobs, L. L., 1985, Neogene mammalian faunal change in southern Asia: correlations with climatic, tectonic and eustatic events: Geology, v. 13, p. 637-640.

Bohlin, B., 1946, The fossil mammals from the Tertiary deposits of Taben-Buluk, Western Kansu. Part II: Simplicidentata, Carnivora, Artiodactyla, Perissodactyla, and Primates: The Sino-Swedish Expedition, Publication 28, v. 6, p. 1-259.

de Bruijn, H., Hussain, S. T., and Leinders, J. J. M., 1981, Fossil rodents from the Murree formation near Banda Daud Shah, Kohat, Pakistan: Koninklijke Nederlandse Akademie van Wetenschappen, Proceedings, series B, v. 84, p. 71-99.

Colbert, E. H., 1933, Two new rodents from the lower Siwalik beds of India: American Museum Novitates, no. 633, 6 p.

Eldredge, N., and Gould, S. J., 1972, Punctuated equilibria: an alternative to phyletic gradualism, in Schopf, T. J. M., ed., Models of paleobiology: San Francisco, Freeman, Cooper, and Co., p. 82-115.

Flynn, L. J., 1982, Systematic revision of Siwalik Rhizomyidae (Rodentia): Géobios, v. 15, p. 327-389.

———1983a, Mosaic evolution in a family of fossorial rodents, in Buffetaut, E., Mazin, J. M., and Salmon, E., eds., Actes du symposium paléontologique Georges Cuvier: Montbéliard, France, p. 185-195.

———1983b, Sur l'âge de la faune de vertébrés du Bassin de Bamian Afghanistan: Comptes rendus de l'Académie de Science de Paris, séries II, t. 297, p. 687-690.

———1985, Evolutionary patterns and rates in Siwalik Rhizomyidae (Rodentia): Acta Zoologica Fennica, no. 170, p. 141-144.

Flynn, L. J., and Jacobs, L. L., 1982, Effects of changing environments on Siwalik rodent faunas of northern Pakistan: Palaeogeography, Palaeoclimatology, Palaeoecology, v. 38, p. 129-138.

Flynn, L. J., Jacobs, L. L., and Cheema, I. U., 1986, Baluchimyinae, a new ctenodactyloid rodent subfamily from the Miocene of Baluchistan: American Museum Novitates, no. 2841, p. 1-58.

Gingerich, P. D., 1985, Species in the fossil record: concepts, trends, and transitions: Paleobiology, v. 11, p. 27-41.

Gould, S. J., 1985, The paradox of the first tier: an agenda for paleobiology: ibid., v. 11, p. 2-12.

Hinton, M. A. C., 1933, Diagnoses of new genera and species of rodents from Indian Tertiary deposits: Annals and Magazine of Natural History, series 10, v. 12, p. 620-622.

Jacobs, L. L., 1978, Fossil rodents (Rhizomyidae and Muridae) from Neogene Siwalik deposits, Pakistan: Museum of Northern Arizona Press, Bulletin Series 52, xi + 103 p.

Jacobs, L. L., and Flynn, L. J., 1981, Development of the modern rodent fauna of the Potwar Plateau, northern Pakistan, in Sastry, M. V. A., Kurien, T. K., Dutta, A. K., and Biswas, S., eds., Field conference Neogene-Quaternary boundary, India, 1979: Geological Survey of India, Calcutta, Proceedings, p. 79-82.

Jacobs, L. L., Flynn, L. J., and Downs, W. R., in press, Neogene rodents of southern Asia, in Black, C. C., and Dawson, M. R., eds., Albert E. Wood Festschrift: Los Angeles County Museum.

Johnson, N. M., Stix, J., Tauxe, L., Cerveny, P. F., and Tahirkheli, R. A. K., 1985, Paleomagnetic chronology, fluvial processes, and tectonic implications of the Siwalik deposits near Chinji village, Pakistan: Journal of Geology, v. 93, p. 27-40.

Lydekker, L., 1884, Rodents and new ruminants from the Siwaliks and synopsis of Mammalia: Palaeontogia Indica, series 10, v. 3, p. 105-134.

MacFadden, B. J., 1985, Patterns of phylogeny and rates of evolution in fossil horses: hipparions from the Miocene and Pliocene of North America: Paleobiology, v. 11, p. 245-257.

Martin, R. A., 1979, Fossil history of the rodent genus Sigmodon: Evolutionary Monographs, v. 2, p. 1-36.

Mayr, E., 1942, Systematics and the origin of species from the viewpoint of the zoologist: New York, Columbia University Press, 344 p.

Mein, P., and Ginsburg, L., 1985, Les rongeurs miocènes de Li (Thailande): Comptes Rendus de l'Académie de Sciences de Paris, sér. II, t. 301, p. 1369-1374.

Munthe, J., 1980, Rodents of the Miocene Daud Khel Local Fauna, Mianwali District, Pakistan. Part 1. Sciuridae, Gliridae, Ctenodactylidae, and Rhizomyidae: Milwaukee Public Museum Contributions in Biology and Geology, no. 34, 36 p.

Raza, S. M., 1983, Taphonomy and paleoecology of middle Miocene vertebrate assemblages, southern Potwar Plateau, Pakistan [Ph.D. thesis]: New Haven, Connecticut, Yale University, 414 p.

Simpson, G. G., 1953, The major features of evolution: New York and London, Columbia University Press, 434 p.

Stanley, S. M., 1979, Macroevolution: pattern and process: San Francisco, W. H. Freeman and Company, 332 p.

——1985, Rates of evolution: Paleobiology, v. 11, p. 13-26.

Tauxe, L., and Opdyke, N. D., 1982, A time framework based on magnetostratigraphy for the Siwalik sediments of the Khaur Area, northern Pakistan: Palaeogeography, Palaeoclimatology, Palaeoecology, v. 37, p. 43-61.

Teilhard de Chardin, P., 1942, New rodents of the Pliocene and Lower Pleistocene of North China: Publications de L'Institut de Géobiologie, Peking, v. 9, p. 1-101.

Wessels, W., de Bruijn, H., Hussain, S. T., and Leinders, J. J. M., 1982, Fossil rodents from the Chinji Formation, Banda Daud Shah, Kohat, Pakistan: Proceedings of the Koninklijke Nederlandse Akademie van Wetenschappen, series B, v. 85, p. 337-364.

Wood, A. E., 1937, Fossil rodents from the Siwalik beds of India: American Journal of Science, series 5, v. 34, p. 64-76.

MANUSCRIPT RECEIVED DECEMBER 15, 1985
REVISED MANUSCRIPT RECEIVED JUNE 18, 1986
MANUSCRIPT ACCEPTED JUNE 19, 1986

The late Miocene radiation of Neotropical sigmodontine rodents in North America

JON ALAN BASKIN   *Department of Geosciences, Texas A&I University, Kingsville, Texas 78363*

ABSTRACT

Abelmoschomys simpsoni, n. gen. et sp., from the latest Clarendonian of Florida, is the earliest known species of Neotropical sigmodontine rodent. Neotropical sigmodontines initially evolved and diversified in North America in the late Miocene. This group is derived from the North American genus *Copemys*, and forms the sister group of the peromyscines. The Neotropical sigmodontines entered South America in the Pliocene at about the time of formation of the Panamanian land bridge. This hypothesis is supported by evidence from physiology, karyology, molecular systematics, comparative anatomy, and paleontology, and is not contradicted by parasite data. The present day diversity of sigmodontines in South America is that expected for its continental area. Taxonomic frequency rates necessary to produce the more than 200 species of South American sigmodontines as determined by the geometric growth equation range between 0.68 and 0.82 per million years. Although these rates are higher than those previously reported for Cenozoic mammals, they are comparable to rates for other muroid radiations. These high rates may account for discrepancies between divergence times calculated by the molecular clock versus the fossil record.

"Is it of any interest to anyone but a mouse fancier? (The answer is yes.)"
George Gaylord Simpson (1980, p. 196)

INTRODUCTION

The evolution of Neotropical sigmodontine rodents is a subject that interested Dr. Simpson for many years (1950, 1969, 1980), a subject that he included in a book he published shortly before his death (1983). Not only was he interested in determining the time of arrival of these rodents in South America, but additionally he wished to determine the evolutionary rates and mechanisms needed to produce the high diversity seen in the living Neotropical sigmodontines. As will be discussed in the present paper, this subject has been a concern and indeed a subject of argument, sometimes vituperative, among a wide assortment of evolutionary biologists, whose diverse backgrounds and specializations have served to enrich and enliven the debate.

Simpson, more than anyone else, established the South American biogeographic framework into which the history of these mice must be placed. He recognized three major faunal phases in the history of mammalian interchange between North and South America (Simpson, 1940, 1950, 1969, 1980). The ancient mammalian inhabitants of South America include marsupials, edentates, and various ungulates which were present or evolved from groups that were present in South America perhaps even before it separated from North America and became an island continent, sometime before the latest Cretaceous. The second stratum consists of caviomorph rodents and platyrrhine primates, which appear in South America at the beginning of the Oligocene. These two groups arrived by overwater dispersal either from North America or Africa as waif immigrants (Ciochon and Chiarelli, 1980). The third phase, known as the Great American Interchange (Webb, 1976), marks the formation of the Panamanian Land Bridge in the late Pliocene, approximately 2.5 to 3.5 million years ago. Following this, some 11 families of North American mammals emigrated to South America and 11 families of South American mammals traveled northward. This interchange was heralded by the waif dispersal of procyonids to South America and ground sloths to North America during the late Miocene, some eight million years ago. Of all the groups that participated in the faunal interchange between North and South America, one of the most controversial is the Neotropical sigmodontines.

The following abbreviations are used in the text and tables: UF, Florida State Museum, University of Florida; mya, million years ago; mm, millimeters; km, kilometers; N, number; CV, coefficient of variation; d.f., degrees of freedom; *t*, student's t value; *r*, correlation coefficient; and *p*, probability.

MUROIDEA

As noted by Simpson (1945) and Wood (1965, 1980), among others, the classification of the Rodentia is an exceedingly complicated problem. Convergence, parallelism, and divergence both in fossil and recent rodents are some of the sources of this difficulty. The large number of rodent taxa further exacerbates the problem. Slightly more than twenty-five percent of the 4,060 species of living mammals belong to just one rodent superfamily, the Muroidea (Arata, 1967). Carleton (1980) briefly reviewed the history of classification of living muroids. Two additional classifications of interest to the problem of the relationship of the Neotropical sigmodontines are those of Reig (1980) and Jacobs and Lindsay (1984). The most recent comprehensive review of the Muroidea is that of Carleton and Musser (1984), who place all the living muroid genera in 15 subfamilies, all in a single family, the Muridae. In marked contrast, the classification of Chaline and others (1977) recognizes 19 subfamilies placed in eight families of muroids. Differences among these classifications indicate that knowledge of relationships within the Muroidea is still in formative stages. Part of the reason for this difficulty is the tremendous diversity present in this rodent superfamily, which contains over 250 living genera. The fossil record in many instances consists of only isolated teeth, or at best skulls and jaws. In addition, evidence from paleontology, comparative anatomy, biochemistry, and cytogenetics in many instances is equivocal and subject to varying interpretation.

Each of these classifications has unique advantages and difficulties. Because phylogenetic relationships of the major groups of muroids (families, subfamilies, and/or tribes according to the various classificatory arrangements listed in Carleton, 1980) remain controversial, this paper will follow the conservative classification of Carleton and Musser (1984), which makes no judgement on relationships of the muroid groups. As they so forthrightly state (*ibid.*, p. 299): "Should rats and mice be assigned to just one family, to the two families Muridae and Cricetidae, or to more than two? We don't know." Further research into muroid relationships is most definitely required.

The group of interest for this paper is the Sigmodontinae, the New World rats and mice. In the widely used classification of Simpson (1945), the Sigmodontinae of Carleton and Musser (1984) form the tribe Hesperomyini, subfamily Cricetinae, family Cricetidae. Simpson's Cricetinae also includes the Cricetini, the Old World hamsters. The Cricetinae of Carleton and Musser (1984) are restricted to this latter group. The Arvicolinae, the Holarctic voles, lemmings, and muskrats, are also included in the Cricetidae by Simpson as the Microtinae. Many recent classifications recognize this group as a distinct family. The Murinae of Carleton and Musser (1984) include approximately 117 genera and 460 species of Old World rats and mice.

Carleton and Musser (1984) recognize approximately 73 genera and 369 species of recent Sigmodontinae, which represents a greater splitting than some previous classifications. The major morphological feature that has been used to subdivide New World rats and mice is the structure of the phallus. Hooper and Musser (1964) recognize two distinct morphotypes. The complex-type penis has a tripartite baculum and terminal digits. The simple type has a unipartite, unbranched baculum and no terminal digits. Hershkovitz (1966, 1972) formally recognized two tribes of Sigmodontinae based on this dichotomy. His Sigmodontini have a complex-type penis and today are found dominantly in South America, but are well represented in Middle America and extend into North America. They include some 49 genera and 246 species (Honacki and others, 1982). His Peromyscini have a simple-type penis, and are almost exclusively North American. They include 18 genera and 117 species. Carleton (1980) cautioned against the recognition of two distinct groups, although he informally recognized the two as the South American cricetines and the neotomine-peromyscines. These two groups are referred to in the present paper as the Neotropical sigmodontines and the neotomine-peromyscines. Where cricetid is used in this paper, out of old habit, it stands for the paraphyletic group including, among others, the extinct eumyines and cricetodontines (*sensu lato*) and the living cricetines and sigmodontines.

Classically, those concerned about the evolution of New World mice have been divided into two schools. Phillip Hershkovitz, the eminent mammalogist from the Field Museum who has made many important contributions to the study of Neotropical mammals, is the main proponent of the early arrival scenario. In order to account for the more than 40 genera and 200 species of Neotropical sigmodontines, Hershkovitz (1966, 1972) hypothesized that sigmodontine rodents are members of the second phase, and arrived in South America no later than the early Miocene, at least 20 million years ago, by overwater dispersal to the then island continent from North America or perhaps even Africa. According to this hypothesis, they diversified in South America, and then, following formation of the Panamanian Land Bridge, had a few representatives, notably *Sigmodon* and *Oryzomys*, migrate to North America. The opposing school, long supported by Simpson, holds that sigmodontine mice are relatively recent invaders of South America, part of the horde that travelled south during the Great American Interchange. Simpson (1950, 1969) suggested that a rapid diversification followed the invasion by one or more genera. Patterson and Pascual (1972) elaborated on this hypothesis. They proposed that the majority of genera of Neotropical sigmodontines initially evolved in tropical North America during the late Miocene, and then further diversified in South America in the late Pliocene and Pleistocene following completion of the land bridge.

Since the early 1970s, new evidence from the North American fossil record (*e. g.*, Baskin, 1978; Jacobs and Lindsay, 1981), the South American fossil record (*e. g.*, Reig, 1978, 1980), parasites (Slaughter and Ubelaker, 1984), chromosome analysis (*e. g.*, Gardner and Patton, 1976), and comparative anatomy (*e. g.*, Carleton, 1980) has led to reanalysis, elaboration, and multiplication of hypotheses concerning the origin of Neotropical sigmodontines. A new genus and species from the late Miocene of Florida has led to this present reevaluation.

SYSTEMATIC PALEONTOLOGY

Superfamily MUROIDEA Miller and Gidley, 1918
Family MURIDAE Thomas, 1896
Subfamily SIGMODONTINAE Wagner, 1843
Abelmoschomys new genus

Diagnosis: Small, brachydont sigmodontines with weakly to moderately bifurcated anterocone of M^1 and unicusped to incipiently bifurcated anteroconid on M_1.

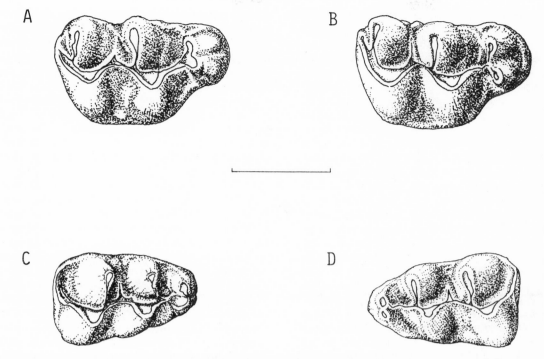

Figure 1. Occlusal views of cheek teeth of *Abelmoschomys simpsoni*, new genus and species. *A*, UF 61327, right M^1; *B*, UF 61326, right M^1, holotype; *C*, UF 61335, right M_1; and *D*, UF 61336, left M_1. Scale bar represents 1 mm.

Accessory rootlet usually present on M^1, occasionally on M_1. Non-alignment of protophule II with anterior arm of hypocone and of entolophid with posterior arm of protoconid; labial and lingual cusps are only slightly alternate.

Type species: Abelmoschomys simpsoni new species.

Etymology: Abelmoschus, the genus of okra, plus *mys* (Greek, masculine), a mouse. The type locality for this genus was discovered by Ron Love, while plowing his okra field.

Abelmoschomys simpsoni new species
Figure 1; Table 1

Diagnosis: Characters of genus.

Holotype: UF (Florida State Museum, University of Florida) 61326, right M^1.

Distribution: Love Bone Bed, Alachua Formation, Alachua County, Florida, latest Clarendonian Land Mammal Age (Webb and others, 1981).

Hypodigm: UF nos. 61326, holotype, 61327-61334, M^1; 61335-61342, M_1.

Etymology: Named for George Gaylord Simpson.

Description: M^1. The occlusal outline is suboval, longer than wide, and narrow anteriorly. The anterocone is relatively wide, but asymmetrical. The anterocone is weakly to moderately bilobed by an antero-median groove. The labial lobe is wider and higher than the lingual lobe. A labial paralophule extending from the anterocone almost to the paracone is present in five specimens. The paracone is joined to the protocone by protolophule II. There is a narrow, transverse valley between the protocone and the paracone. The mesoloph is short to medium in length. The metacone joins posteriorly to the hypocone by metalophule II. A posterior cingulum is well developed. An entostyle is present in three specimens. An accessory rootlet is present in 7 of 9 specimens.

M_1. The occlusal outline is suboval, narrowing anteriorly. The anteroconid is nearly symmetrical. In little worn specimens, an anterior groove divides the anteroconid into a larger lingual and smaller labial lobe. The lobes are incipiently to weakly developed. The metaconid is joined to the anterior arm of the protoconid by a short mesolophid. The entoconid is connected to the anterior arm by the hypoconid by a short hypolophid. The metalophid is absent on one specimen, short on one, medium on five, and long on one. The posterior cingulum is well developed, and possesses a median stylid in two specimens. An accessory rootlet is present on three specimens.

Comparisons and discussion: Abelmoschomys resembles *Copemys* in the non-alignment of protolophule II with the anterior arm of the hypocone and non-alignment of the entolophid with the posterior arm of the protoconid. In this it differs from more advanced genera such as *Peromyscus, Baiomys, Prosigmodon,* or *Calomys*. It resembles advanced species of *Copemys* in having only posterior connections between the labial and lingual cusps on the M^1 and only anterior connections on the M_1. *Calomys (Bensonomys)* and *Prosigmodon* have a much more strongly bifurcated anteroconid than does *Abelmoschomys*. In *Bensonomys* the anterocone is much more strongly bifurcated than in *Abelmoschomys*.

Jacobs and Lindsay (1984) postulate that the Eurasian genus *Megacricetodon* was ancestral to the Neotropical sigmodontines. The resemblance between *Megacrice-*

TABLE 1. MEASUREMENTS (IN MILLIMETERS) OF Abelmoschomys simpsoni.

Tooth	N	Range	Mean and Standard Error	CV
Length				
M^1	9	1.63-1.78	1.69 ± .018	3.1
M_1	8	1.40-1.45	1.48 ± .016	3.1
Width				
M^1	9	1.00-1.10	1.05 ± .011	3.2
M_1	8	0.90-1.00	0.95 ± .012	3.6

todon and *C. (Bensonomys)* is indeed striking. Additionally, the new Florida genus decreases the time gap between the last known *Megacricetodon* (greater than 12 mya) and the first known Neotropical sigmodontines (less than 9 mya). However, the morphology of *Abelmoschomys* is easier to derive from Clarendonian *Copemys* than from Eurasian *Megacricetodon*. Jacobs and Lindsay list eight dental characters primitive for the Neotropical sigmodontines and note that these are shared to varying degrees with *Megacricetodon*. They especially stress the bifurcated anterocone on the M^1. Of all the described species of *Copemys,* only *C. russelli* and *C. vasquezi* have a tendency to bifurcate the anterocone by the presence of a lingual lobe (Lindsay, 1972; Jacobs, 1977). The anteroconid of both consists of a single cusp. *Copemys esmeraldensis* has a bilobed anteroconid. They also note the presence of a short posteriorly-directed lophule on the paracone of the M^1. This feature is variably present in *Megacricetodon* and primitive *Calomys (Bensonomys)*, and is best considered a parallelism. The bilobed anterocone of *Abelmoschomys* is variably developed, and the bilobed anteroconid only incipiently developed. It seems unlikely that this form was derived from *Megacricetodon,* which has such a strongly bilobed anterocone and sometimes bilobed anteroconid (Aguilar, 1980). *Abelmoschomys* is a temporal and structural link between *Copemys* and *Bensonomys*. *Abelmoschomys* can be derived from a form such as *Copemys russelli* by continuing the tendency to bifurcate the anterocone and by developing accessory rootlets on the first molars. This trend is further developed in *Bensonomys*.

SOUTH AMERICAN FOSSIL RECORD

The purpose of this paper is to develop the simplest explanation for the present distribution and diversity of Neotropical sigmodontines, based on known evidence. New and better information from sources such as the fossil record, molecular systematics, cytogenetics, and comparative anatomy could lead to restructuring, even abandonment, of the hypothesis developed below. Discovery of sigmodontine rodents in South America significantly older than 3.5 million years, especially those of phyllotine or sigmodont generic groups, would necessitate modifications, and discovery of sigmodontines older than 16 million years there would necessitate total abandonment of this hypothesis. Presence of South American cricetids more than 3.5 million years old still remains a possibility. Intensive screen washing of sediments from well known and highly diverse fossil localities in North America and elsewhere has produced small vertebrates where none were known previously, in spite of intensive excavation. It is hoped that these techniques will be applied more extensively to South American deposits. However, small rodents can and have been discovered without wet-screening sediments, and evidence as it now stands indicates that sigmodontine rodents were not present in South America before 3.5 million years ago.

Bolomys bonapartei and *Auliscomys formosus* apparently are the two earliest sigmodontine rodents known from South America (Reig, 1978). They occur in member 3 of the Monte Hermoso Formation, northwest of Farola de Monte Hermoso, Argentina, a fossiliferous locality investigated by Darwin (Simpson, 1980). The stratigraphic section of this locality has been described by Bonaparte (1960). According to Marshall and others (1979), member 3 of Reig is equivalent to the Miembro Limolitas Claras or Bright Silt Member of Bonaparte. Bonaparte restricted the type of the Montehermosan Stage to the lower six silty beds of what he terms the Hermosense tipico. Unconformably above this unit is the Miembro de Limolitas Estratificas. The Miembro de Limolitas Claras rests uncomformably on these lower two members. Because the fossiliferous Bright Silt Member does not contain remains of large North American immigrants, such as the mustelids and tayassuids found in the Chapadmalalan, Bonaparte (1960) referred the fauna from this member to the Montehermosan, in spite of the unconformities. In any event, there are difficulties in distinguishing late Montehermosan from Chapadmalalan faunas (Simpson, 1980), and some paleontologists do not recognize a distinct Chapadmalalan Land Mammal Age (*e. g.,* Marshall and Pascual, 1978).

Marshall (1979), in order to satisfy his scenario of sigmodontine rodents as unrecorded heralds, emphasizes the possible 0.8 million year age difference between the occurrence of sigmodontine rodents from member 3 of the Monte Hermoso Formation at 3.5 mya and the occurrence of sigmodontines and other North American invaders in the Chapadmalal Formation at 2.7 mya. These two localities are approximately 300 km apart. Marshall infers a 3.5 my date for member 3 because the top of the Corral Quemodo Formation, which has produced a rich Montehermosan fauna (although without sigmodontines) has been dated at 3.54 my (Marshall and others, 1979). The 2.7 my date for the Chapadmalal fauna is based on correlating the Chapadmalalan Land Mammal Age with the better dated late Blancan, which signals the start of the Great American Interchange in North America (Webb, 1976).

There is evidence that the Panamanian isthmus began shoaling four million years ago (Keigwin, 1982), and sigmodontines could have crossed then, if not earlier. Certainly, sigmodontines are good overwater dispersers, since they are natives of the Galapagos and Antilles. However, there is no evidence to infer that sigmodontine rodents migrated considerably earlier than the rest of the horde. The 0.8 my difference may be real, but it is not significant, nor is it certain. Corral Quemado is several hundred kilometers from the type locality of the Montehermosan. The dated ash may not mark the top of the Montehermosan; unit 3 may be Chapadmalalan. Certainly there is room for refinement.

Uncertainties in the biochronology of South American terrestrial faunas are illustrated by a recent development. MacFadden and others (1985) state that the South American Oligocene Deseadan Land Mammal Age began nine million years later than Marshall and most others have interpreted. No matter who is more correct, this shows that the South American geochronologic framework requires further study.

If sigmodontines reached South America 3.5 my ago, slightly ahead of the main interval of interchange, this would make them the first members of the interchange, as the land bridge began to be established. Marshall's (1979) scenario is which sigmodontines enter South America some six million years ago and are restricted to inferred savanna-grassland habitats unrepresented in the fossil record has, as yet, no basis in real evidence (nor do hypotheses which require sigmodontines to be in South America considerably earlier).

MORPHOLOGY

Carleton's (1980) monograph is essential reading for anyone interested in phylogenetic relationships of sigmodontine rodents. He analyzes 79 qualitative characters in 75 species of muroids, most of them neotomine-peromyscines, but also includes 13 species of Neotropical sigmodontines, two Old World cricetines, six arvicolines, and five gerbillines. Data are analyzed and summarized after extensive discussion, using principal component and clustering analyses, and cladistic relationships are determined using Wagner tree analysis. Although Slaughter and Ubelaker (1984) have reinterpreted Carleton's scatter plot (his Fig. 39) to indicate an especially close relationship between Old World cricetines and Neotropical sigmodontines, they choose to ignore the association of these two with neotomine-peromyscines and the more distant relationship with microtines and gerbillines that this would imply. Carleton's own conclusion is that morphologic data, including the structure of the glans penis, do not offer convincing evidence for recognizing formal distinct assemblages of Neotropical sigmodontines and neotomine-peromyscines.

Voss and Linzey (1980) examine the male accessory gland components of 54 species of Neotropical sigmodontines. Among the 35 genera and subgenera examined, they note less variation than exists in the male reproductive anatomy of the North American genus *Peromyscus*. They hypothesize that this uniformity implies a relatively rapid phyletic proliferation following colonization of South America in the Pliocene. This conservatism is also present in the G-banding sequences (discussed below), and contrasts with the extreme variation in karyotype and moderate variation of cranio-skeletal and external morphology. This variation is hypothesized by Voss and Linzey to accompany the intense selection that attended exploitation of diverse new kinds of foods and habitats, which occurred in too short a time to have much effect on the more environmentally-buffered reproductive system.

Simpson (1950, 1980) has contrasted the great difference in higher-level taxonomic diversity between the South America caviomorphs and murids as evidence of the relatively recent entry of sigmodontines into South America. Six families of Caviomorpha are known from the Oligocene Deseadan of South America; eleven from the Recent. Wood (1983) argues that the initial high familial diversity as demonstrated by the great diversity of dental morphology indicates that caviomorphs had been present in South America at least nine million years before the start of the Deseadan. Sigmodontines in South America demonstrate a much lower morphological diversity. They have at best diversified to the tribal level from their North American relatives. A similar low level of diversification is present in the Australian murids, which entered that continent approximately 4.5 mya (Hand, 1984).

ENDOPARASITES

Slaughter and Ubelaker (1984) propose that Neotropical sigmodontines are more closely related to complex-penis, Old World muroids (cricetines of the hamster group, arvicolines, murines, *etc.*) than they are to neotomine-peromyscines. Slaughter and Ubelaker base their conclusions on penile morphology, ectoparasites and especially endoparasites. Jacobs and Lindsay (1984) agree, and add evidence from the fossil record to support this phylogeny. Jacobs and Lindsay propose that diversification of these lineages began in the middle Miocene, and that the Neotropical sigmodontines and neotomine-peromyscines represent separate middle to late Miocene dispersal events from Eurasia to North America. Although Slaughter and Ubelaker also discuss this possibility, they favor a long separate history for complex-penis versus simple-penis sigmodontines. Instead of deriving *Copemys* (the North American genus believed to be ancestral to at least the peromyscines) from a Eurasian form such as *Democricetodon* (Fahlbusch, 1964; Lindsay, 1972), they trace its ancestry to North American Oligocene *Leidymys*. They believe this long separation is needed to account for the endoparasite distribution seen in muroids.

Slaughter and Ubelaker (1984) reviewed host specificity in lungworms of the nematode family Angiostrongylidae, particularly the genus *Parastrongylus*. Although this evidence is interesting, it is by no means conclusive. Alternative *ad hoc* hypotheses can be proposed to explain the data. The host genera listed for *Parastrongylus* are mainly Old World murines, gerbillines, and arvicolines. There are some interesting anomalies that suggest common ancestry need not be the only explanation for parasite distribution. For instance, *P. malaysiensis* occurs in nine species of the murine *Rattus* and in the tree shrew *Tupaia glis*. Although the infection of *T. glis* is presumed to be accidental, it does show that host selection may not be highly specific. In Europe, *P. dujardini* occurs in *Apodemus,* a murine, and in *Clethrionomys,* an arvicoline. In South Africa, *P. sandarsae* occurs in *Gerbillus,* a gerbilline, and *Mastomys,* a murine. The sigmodontines parasitized by *Parastrongylus* are *Sigmodon, Oryzomys,* and *Zygodontomys*. All three of these are found in North America as well as South America, and close relatives of all three may make their first appearance in North America (Jacobs and Lindsay, 1984; Czaplewski, personal communication). These three genera, as well as *Liomys adspersus* and *Nasua narica,* are all parasitized by *P. costaricensis*. Both *Liomys* and *Nasua* are North American endemics. *Liomys* is a heteromyid rodent, a family known to occur only in North America, with the exception of *Heteromys* which entered northern South America some time since completion of the land bridge (Hershkovitz, 1972).

Nasua, the coati-mundi, has a Neotropical distribution. Although procyonids island-hopped to South America in the late Miocene (some eight million years ago), the South American *Cyonasua* group of procyonids is not ancestral to *Nasua* (see Baskin, 1983). *Nasua* evolved in North America and was a participant in the Great American Interchange, following completion of the land bridge. This distribution of hosts suggests that the parasite was spread in Middle America, perhaps during the Pliocene. The ultimate source of the parasite could possibly be traced to a middle Miocene Eurasian muroid that transferred *Parastrongylus* to *Copemys*. However, *Copemys* does not represent a single non-branching, aristogenic taxon that directly gave rise to all the modern sigmodontines. It contains several lineages that can be traced back at least 12 million years (Lindsay, 1972). The *Copemys* lineage ancestral to the Neotropical sigmodontines may have carried *Parastrongylus;* the lineage ancestral to the neotomine-peromyscines may have lost it. Apparently not all genera of Old World muroids are infected by *Parastrongylus*. Certainly the parasitization of only three New World genera out of more than 50 genera, or 10 out of more than 200 muroid genera worldwide, leaves much to be desired when it comes to using this distribution to formulate phylogenetic hypotheses. For that matter, the parasite may have been spread by a late Pliocene or early Pleistocene arvicoline somewhere in North America, to *Sigmodon* for instance.

ECTOPARASITES

Host relationships of certain ectoparasites have been used to support a long separate history for Neotropical and Nearctic sigmodontines (Wenzel and Tipton, 1966; Hershkovitz, 1972). Carleton (1980, p. 137-139) discusses some of the reasons host distribution is environmentally rather than phylogenetically controlled, and thus has no bearing in determining evolutionary relationships of these two groups. The ectoparasites of Neotropical sigmodontines belong to families and genera restricted or centered in South America (Wenzel and Tipton, 1966). Their closest relatives are Old World forms, especially Australian, but also Oriental and Ethiopian. This suggests these fleas, ticks, and mites have a Gondwanan origin (in contrast to Palearctic origin of the endoparasites), with a long history of adaptation to feeding on tropical and subtropical mammals. The ease of transference from one host to another is suggested by presence of the flea *Polygenis* not only on *Oryzomys* and *Sigmodon,* but also *Liomys, Peromyscus, Reithrodontomys,* and *Neotoma* (see Wenzel and Tipton, 1966; Layne, 1963). Transferrence to the last four genera, a heteromyid and three simple-penis sigmodontines, must have taken place in lowland tropical environments after completion of the Panamanian Land Bridge. Similarly, the temperate-origin ectoparasites must have been replaced by tropical-subtropical ectoparasites as Neotropical sigmodontines entered northern South America, some three or four million years ago.

PHYSIOLOGY

Mares (1975, 1980) examined adaptations for desert life in South American caviomorph and sigmodontine rodents. Deserts are harsh environments which produce extreme specializations, such as those seen in some North American heteromyids and the Old World gerbillines and dipodids. Modern deserts of the world have developed over the past 20 million years or more, and rodents from northern continents have had a long interval of time to adapt to them. Sigmodontines from the Monte Desert of South America possess few specialized traits indicative of adaptation to an arid environment. Caviomorphs have adaptations suggesting that they have been associated with the desert for a longer period than the Neotropical sigmodontines. Mares (1975, 1980) interprets this as evidence supporting a late (Pliocene) colonization of South America by the sigmodontines.

CHROMOSOME EVOLUTION

Patton and Sherwood (1983) have broadly reviewed chromosome evolution and speciation in rodents. Structural rearrangement involves chromosome breakage and reunion. Robertsonian changes involve fusion of two acrocentric chromosomes to form a metacentric, or fission of a metacentric to produce two acrocentrics, decreasing or increasing, respectively, the diploid number. Pericentric inversions change the centromere position and number of chromosome arms, but not the diploid number. There are other, less common types of rearrangements. Heterochromatin variation is determined by different staining techniques. The G-banding pattern of heterochromatin is related to arrangement of the genes (Mascarello and others, 1974). Chromosomal rearrangements can be identified by determining homologous banding patterns. Chromatin banding patterns are rarely affected by Robertsonian rearrangements of chromosome arms and, in fact, are very conservative compared to the extremely wide range of karyotypic variation present in rodents.

Gardner and Patton (1976) examined data from standard karyotypes of 22 species of oryzomyine rodents. They note that variation in diploid (32-80) and fundamental (40-136) numbers is greater than that know for any other equivalent mammalian taxonomic group. The total range they reported for all Neotropical sigmodontines was 22-92 for the diploid and 26-136 for the fundamental numbers. They conclude that the tendency has been toward evolution of lower diploid and fundamental numbers, through relatively simple chromosomal rearrangements, especially Robertsonian fusions. Gardner and Patton (1976, p. 15) state that, for the complex-penis-type sigmodontines, "chromosomal reorganization may have played a major causal role in the speciation process during what clearly appears to have been an explosive phase radiation within the group." They believe that the wide range of karyotypic characters supports a Pliocene colonization by an ancestral Neotropical sigmodontine stock, which probably did not undergo differentiation in Middle America. This is the most extreme evidence in support of a late colonization. The fossil record indicates some prior differentiation (e. g., Baskin, 1978; Jacobs and Lindsay, 1981) before dispersal.

Baker and others (1983) examined G-bands in chromosomes of 15 species of South American sigmodontines, mostly members of the genus *Oryzomys*. They conclude that most rearrangments identified in their study occurred in recent evolutionary history of the New World sigmodontines, after their separation into generic groups, such as those of *Oryzomys* or *Peromyscus*. In fact, they could identify no chromosomal synapomorphies that could distinguish neotomine-peromyscines from Neotropical sigmodontines. They could find no chromosomal characters that would align *Sigmodon* any closer to Neotropical sigmodontine genera (*Oryzomys, Holochilus,* and *Neacomys*) than to North American genera such as *Peromyscus* or *Neotoma*. Rogers and Heske (1984) state that G-banding of chromosomes indicates that *Scotinomys, Ochrotomys,* and perhaps *Baiomys* form a loose assemblage phylogenetically intermediate between Neotropical sigmodontines and neotomine-peromyscines and conclude that the dichotomy between these two groups is more imagined than real.

EVIDENCE FROM MOLECULAR SYSTEMATICS

Sarich and Cronin (1980) present preliminary data on relationships of New World cricetids based on microcomplement fixation of albumins and their analysis of the DNA-annealing data of Rice (1974). Although paleontologists have been rightly humbled in their interpretation of hominid phylogeny versus that of molecular systematics, considerable disagreement still exists concerning constancy of the molecular clock. Wu and Li (1985) demonstrate that synonymous changes in nucleotide sequences are accumulated twice as fast in rodents as in humans. Where the fossil record is much more robust than that of fossil hominids, paleontology can be used to accurately calibrate divergence dates and provide insight into possible differences in rates of molecular evolution. The divergence times for muroid rodent clades based on molecular data (Sarich and Cronin, 1980, Table 1) seem to indicate the molecular clock runs three to five times faster in muroid rodents than they estimate. As discussed elsewhere in the present paper, there is evidence that muroids may evolve at a relatively rapid rate. Thus murids diverged from cricetids some 13 to 15 million years ago (Jacobs and Lindsay, 1984) rather than 60 million years ago (which is about five million years before the first known rodent); *Mus* from *Rattus* some 8 to 14 million years ago (Jacobs and Pilbeam, 1980) rather than 35; and *Peromyscus* from *Onychomys* 4 to 8 million (Jacobs and Lindsay, 1984) rather than 20. This would indicate *Peromyscus* diverged from *Oryzomys* some 9 to 15 million years ago rather than 45.

Sarich and Cronin (1980), based on their inferred divergence dates, conclude that sigmodontines colonized South America from Africa in the late Eocene or early Oligocene, at least 40 mya, and later migrated from South to North America. Wood (1983 and *in* Patterson and Wood, 1982, p. 507) points out that this scenario does not deserve serious consideration because it does not take into account data from the fossil record.

Sarich (1985) has recalibrated the molecular clock based on his reevaluation of immunological and paleontological data. Divergence times have been adjusted downward by a factor of about one-third. Thus, for example, the *Mus-Rattus* divergence is placed at about 22 mya rather than 33 mya. Divergence times in the Muridae are now two to three times those indicated by the fossil record. This recalibration would place the divergence date of North and South American sigmodontines at about 28 mya, in the late Oligocene, at or slightly before the presumed time of arrival of sigmodontines in South America according to Hershkovitz and Reig. If the molecular clock runs two to three times faster in murids (Wu and Li, 1985), this divergence date would be 9 to 14 mya, in agreement with Simpson and his followers.

Although the actual divergence times may be debatable, the molecular evidence remains an important source of data for inferring phylogenetic relationships. Sarich and Cronin (1980) interpret that sigmodontines and microtines are more closely related to each other than they are to Old World cricetines, and that there are "several lineages among the New World cricetids that have been separated from one another for *at least* the length of time sufficient to produce a △Tm of 15° or an albumin immunological distance of 75 units. These would be *Peromyscus-Onychomys-Phyllotis-Akodon, Calomys-Oryzomys, Sigmodon, Tylomys, Ichthyomys,* and the microtines" (*ibid.*, p. 414).

A more recent study (Brownell, 1983) used DNA hybridization techniques to investigate phylogenetic relationships among 12 taxa of muroid rodents. Unfortunately, no Neotropical sigmodontines were included in this analysis. Wagner tree analysis (*ibid.*, Fig. 6) shows that *Peromyscus, Onychomys,* and *Reithrodontomys* are most closely related to *Neotoma*. These four genera are most closely related to three genera of arvicolines. *Mesocricetus,* a cricetine, is the sister group to the preceding seven genera. Murines and gerbillines form the sister group to neotomine-peromyscines, microtines, and Old World cricetines. Divergence times are three to five times greater than the fossil evidence suggests.

Sarich (1985) recognizes the neotomines, peromyscines, and Neotropical sigmodontines as a monophyletic group. All the sigmodontines, arvicolines, and cricetines, in turn, are the sister group of murines and gerbillines. Biochemical evidence supports the hypotheses discussed below that all Recent sigmodontines share a common ancestor, *Copemys,* and that *Copemys* and *Democricetodon,* the genus ancestral to cricetines, are closely related. According to Repenning (*in* Flynn and others, 1985), the arvicolines are a polyphyletic group derived from advanced cricetids. Murines and gerbillines are derived from *Megacricetodon* (Flynn and others, 1985). According to the fossil record, a separation of lineages leading to modern peromyscines and all other major murid groups 25 to 30 mya or even 20 mya (the time of origin of *Megacricetodon,* according to Aguilar, 1980), as has been proposed in some of the hypotheses concerning origin of the Neotropical sigmodontines that are discussed elsewhere in the present paper, would require murines, gerbillines, cricetines, and Neotropical sigmodontines to be more closely related to each other than to peromyscines. This is in contradiction with the molecular evidence. Phylogenetic information as provided by molecular systematics and the fossil record, as discussed below, are in accord, if the molecular clock runs two to three times faster in murids.

NORTH AMERICAN FOSSIL RECORD

In the past ten years, there have been several important contributions to the study of Hemphillian (late Miocene to early Pliocene) rodents from the southwestern United States and northern Mexico (Jacobs, 1977; Baskin, 1978, 1979; Jacobs and Lindsay, 1981; Lindsay and Jacobs, *in press;* Dalquest, 1983). These studies have elucidated the late Miocene diversification of Neotropical sigmodontine rodents in southern North America prior to their dispersal and radiation in South America, as was proposed earlier by Patterson and Pascual (1972). The ancestry of sigmodontines, like most questions about rodent systematics, remains controversial. Clark and others (1964) derive the sigmodontines from North American Oligocene and early Miocene cricetids of the subfamily Eumyinae. Fahlbusch (1964) and Lindsay (1972) derive them from the middle Miocene Eurasian genus *Democricetodon,* which gave rise to the widespread middle and late Miocene North American genus *Copemys,* regarded to be ancestral to some, if not all, of the recent peromyscines. *Copemys,* and the closely related and poorly represented *Gnomomys,* are the only known North American Barstovian and Clarendonian cricetid genera other than *Abelmoschomys*.

The first records of modern peromyscines, neotomines, and Neotropical sigmodontines occur in the Hemphillian. The phyletic transition from *Copemys* to *Peromyscus* is widely accepted (Clark and others, 1964; Lindsay 1972; Jacobs, 1977; Baskin, 1979). The earliest species of *Peromyscus, P. valensis* and *P. antiquus,* are both known from Hemphillian deposits of Oregon. By the latest Hemphillian to early Blancan, *Onychomys, Baiomys, Reithrodontomys,* as well as *Peromyscus,* are known from North America, and these presumably represent radiations from *Peromyscus* or advanced *Copemys*.

The earliest records of neotomines are *Neotoma (Paraneotoma)* from the middle Hemphillian (Dalquest, 1983) and *Repomys* from the late Hemphillian (May, 1981). The relationship of neotomines to peromyscines remains to be resolved. Jacobs and Lindsay (1984) suggest that neotomines may have originated from an Asian cricetodontine such as *Plesiodipus,* which has been considered to lie near the ancestry of myospalacines (which also have a simple penis), and possibly arvicolines and cricetines (Li and Chi, 1981). May (1981) tentatively derives *Repomys* from the early Hemphillian *"Peromyscus" pliocenicus*. This species, in turn, can be derived from *Copemys,* although this is by no means certain.

Abelmoschomys, the new taxon described in the present paper from the latest Clarendonian (late Miocene, about 9 mya) of Florida, represents the earliest record of a Neotropical sigmodontine. The fossil evidence that Neotropical sigmodontines began their radiation in North America long before they appear in South America was first presented by Baskin (1978). I described two new species of *Bensonomys* from the middle Hemphillian White Cone local fauna (approximately 6.5 mya), briefly reviewed the species of that genus, discussed similarities between *Bensonomys* and the recent South American genus *Calomys,* and concluded that *Bensonomys* should be recognized as an extinct North American subgenus of *Calomys.* Some of the similarities between *Bensonomys* and South American sigmodontines had been recognized previously, but had been attributed to parallelism (see discussion in Baskin, 1978). When there are virtually only morphological similarities, and all of the other available evidence is either neutral or supports the hypothesis, then close phylogenetic relationship is by far the most parsimonious alternative. In that paper, I also cite the early Blancan records of *Sigmodon* and the probable akodont *Symmetrodontomys* which occur in North America before establishment of the Panamanian Land Bridge at the start of the late Blancan (or Chapadmalalan in South America) in the late Pliocene as evidence favoring a North American origin and initial radiation of the Neotropical sigmodontines.

Since that paper, there has been additional evidence corroborating the hypothesis of North American sigmodontine origins. Jacobs and Lindsay (1981) describe a new genus *Prosigmodon* from the late Hemphillian of Mexico. They (Lindsay and Jacobs, *in press*) conclude that *Prosigmodon* can be derived from *Calomys (Bensonomys),* and that although there is a possibility that *Prosigmodon* may not be directly ancestral to *Sigmodon,* these latter two genera are closely related. They (*in press*) also described two new species of *Bensonomys* from Yepomera. The most common rodent from the medial Hemphillian Coffee Ranch is another species of *Bensonomys* (see Dalquest, 1983). The earliest Blancan Verde Formation promises to add even more to this pre-land bridge record, with *Bensonomys,* a new species of *Prosigmodon,* and a new genus that may be related to *Zygodontomys* (from Czaplewski, *personal communication*). May (*in* Jacobs and Lindsay, 1984) indicates the possible occurrence of *Oryzomys* from the early Blancan of New Mexico.

Shotwell (1970) assigned two unerupted teeth represented by enamel caps from the early Hemphillian of Oregon to ?*Oryzomys.* These teeth have many unusual characteristics, as noted by Shotwell, and more material is needed before affinities of this taxon can be determined. It seems more likely that this form may be a neotomine that is more closely related to *"Peromyscus" pliocenicus* than it is to *Oryzomys.* As I (1978) noted, the late Hemphillian ?*O. pliocaenicus* is generically indeterminate, but may represent *Bensonomys.* It is likely that the upper molar fragment from the early Miocene Thomas Farm that Black (1963) compared to *Sigmodon* and that I (1978) assumed was a Pleistocene contaminant, actually represents the geomyoid *Texomys.*

It should therefore be clear that the Neotropical sigmodontines have a diverse representation in the North American fossil record for at least the past 6.5, if not 9, million years. As discussed above, their record in South America includes the past 2.5 to perhaps 3.5 million years. Reig (1980) has challenged accuracy of the North American fossil record, and alluded to an undocumented pre-Pliocene "fossil record" in South America. He (1980) acknowledges similarities in enamel pattern and other aspects of molar structure between *Bensonomys* and *Calomys,* but attributes these to convergence. He takes notes of differences between Hemphillian *Bensonomys* and recent *Calomys* which argue, he states, for their generic distinction. Reig concludes (*ibid.,* p. 277) that these differences "would be also evident to Baskin if he had taken careful comparison of its fossil material with representatives of the living *Calomys.*" Clearly phylogenetic relationships, including relationships of the sigmodontines, must be based on all the data, and must be evaluated objectively to distinguish primitive from derived characters, which Reig fails to do.

The decision to recognize *Bensonomys* as an extinct subgenus of *Calomys* was not taken lightly. Contrary to Reig's (1980, p. 279) assertions, this conclusion was reached after comparison of fossil and recent specimens as noted in the materials section of my 1978 paper. The five characteristics that Reig uses to distinguish *Bensonomys* from *Calomys* are mainly primitive characters found in the earliest members of this genus and that are modified in later forms.

Bensonomys has more bunodont molars; *Calomys* more lophate. The molars of the earliest known species of *Bensonomys* from White Cone are more cuspate than later species. Blancan species such as *B. arizonae* show better connections between cusps. As noted by Jacobs and Lindsay (1981), the M^3 of living *Calomys* is especially more lophate than in *Bensonomys.* Whether this difference is sufficeint for maintaining generic distinction is debatable.

Bensonomys teeth are relatively wider than those of *Calomys.* Again, this is a primitive character. Blancan *Bensonomys* have narrower teeth that are comparable to *Calomys.*

Bensonomys has a posterior cingulum on the M^1, another primitive character. This feature is greatly reduced to absent in late Hemphillian *Bensonomys* from Yepomera (Lindsay and Jacobs, *in press*). The illustration of the late Blancan *B. meadensis* (see Hibbard, 1972, Fig. 37*A*) lacks this structure.

Bensonomys has a much better developed anteroexternal shelf on the first molars. This shelf is present in *Bensonomys* and *Calomys.* The strong development noted by Reig in *B. yahzi* may be plesiomorphic or autapomorphic for this primitive species.

The anteroconid is narrower and shorter in *Bensonomys.* This is yet another plesiomorphic feature of these Hemphillian taxa. The anteroconid is wider and longer in Blancan species, comparable to *Calomys.*

The most extreme conclusion that can be drawn from these comparisons is that perhaps *Bensonomys* should be retained as the slightly more primitive, albeit very close, relative of *Calomys.* On the other hand, based on available fossil material, these distinctions are within limits of specific variation of other genera, such as *Sigmodon* or *Oryzomys.* To emphasize the relationship, I prefer to recognize *Bensonomys* as a subgenus of *Calomys.* The name of this taxon, however, is not important. What is important is that *Bensonomys* represents a primitive phyllotine, which evolved initially in North America.

ANCESTRY OF NEOTROPICAL SIGMODONTINES

In my 1978 paper, I derived *Calomys* (and thus the phyllotines and related groups) from *Copemys,* the genus widely regarded as ancestral to *Peromyscus. Copemys,* it must be stated, does not represent a single unbranching lineage, but is a complex that includes many lineages, such as *Peromyscus s.l.* does today. The lineages in *Copemys* are poorly defined. In addition, probably only a small portion of its ancient diversity is preserved. The common ancestor of peromyscines and Neotropical sigmodontines may be Clarendonian in age, which permits at least nine million years for divergence between the two groups.

Slaughter and Ubelaker (1984) prefer to derive the neotomine-peromyscines through *Copemys* from North American Oligocene-early Miocene eumyines. They derive the Neotropical sigmodontines from late Miocene, Eurasian cricetodontines, which would align them much more closely with murines, arvicolines, and cricetines. This interpretation is in contradiction with evidence from biochemistry, karyology, comparative anatomy, and, at least according to some (e. g., Jacobs and Lindsay, 1984), the fossil record. Jacobs and Lindsay (1984) derive peromyscines from *Copemys,* which in turn they derive from a Eurasian form such as *Democricetodon. Democricetodon* usually has been considered ancestral to the cricetines as well. They derive Neotropical sigmodontines, murines, cricetines, and gerbillines from the Eurasian early and middle Miocene megacricetodonts. Although this is plausible and has much to recommend it, it still seems simpler to derive Neotropical sigmodontines from *Copemys,* as discussed above. *Abelmoschomys* can be derived from *Copemys* by initiating bifurcation of the antercone and anteroconid and by developing accessory roots. This trend is further developed in *Bensonomys.*

If phylogenies suggested through molecular systematics discussed above are correct, then neotomine-peromyscines could not have diverged from other muroids as early as Slaughter and Ubelaker (1984) and Jacobs and Lindsay (1984) suggest. Part of the rationale for the early divergence is based on penile morphology. Although Hooper and Musser (1964) suggested that a complex phallus is primitive and a simple phallus is derived, Carleton (1980) demonstrated that the reverse morphocline is equally probable.

Slaughter and Ubelaker (1984) and Jacobs and Lindsay (1984) conclude that neotomine-peromyscines are representatives of the primitive simple penis morphology, and that murines, arvicolines, Old World cricetines, Neotropical sigmodontines, and gerbillines (among others) all share a unique derived feature, the complex penis. These latter groups would therefore share a common ancestor, that presumably evolved after the neotomine-peromyscines entered North America, which the above four researchers believe is necessary to explain parasite distribution. Molecular evidence (Sarich and Cronin, 1980; Sarich, 1985; Brownell, 1983) suggests that neotomine-peromyscines and Neotropical sigmodontines are most closely related to arvicolines and cricetines, which in turn are the sister group to gerbillines and murines.

Carleton (1980) suggests the complex penis may have evolved independently in several muroid lineages. He (p. 129-130) emphasizes "the diversity of the glans penis and the numerous departures from the simple and complex plans" and that "the simple and complex glandes intergrade imperceptibly when analyzed in terms of their unit characters." When considered with other character states (*ibid.,* Table 43), it is just as parsimonious to have parallel evolution of the complex glandes as to have evolution from complex to simple in the neotomine-peromyscines and have parallel evolution of other states. The arvicoline rodents can be used to support independent evolution of the complex penis. A relatively simplified glans is present in *Dicrostonyx* and, especially, *Ellobius*. If indeed a simple penis is primitive, then complexity arose independently in the other arvicolines. As discussed by Carleton and Musser (1984), ancestry of the arvicolines remains a controversial question. Molecular systematics (Sarich and Cronin, 1980; Brownell, 1983) suggests that neotomine-peromyscines are closely related to arvicolines. Again, if a simple penis is the primitive character state, this would support independent evolution of a complex penis. In addition, it would suggest that parasite infections are more subject to ecological and biogeographical controls than to phylogentic ones, since arvicolines are infected by *Parastrongylus* and neotomine-peromyscines are not. In any event, polarity of complexity of the glans make no difference to my (1978, 1979) previous interpretation of relationships of Neotropical cricetines, as long as (if simple is primitive) complexity was independently derived.

RATES OF EVOLUTION

Evolution of Neotropical sigmodontines has been described as explosive. Rodents in general have been supposed to evolve faster then other groups of mammals. A standard model used to determine evolutionary rates is the geometric growth equation

$$\frac{dN}{dt} = rN$$

where N is the number of species and r is the intrinsic evolutionary rate of increase. The rate of increase is equal to the origination rate minus the extinction rate.

A more suitable model, except for the most recent radiations, is probably the logistic growth equation

$$\frac{dN}{dt} = rN\frac{(K-N)}{K}$$

in which K is the equilibrium value for taxonomic diversity. Logistic growth produces a sigmoid curve. All other things being equal, rates derived from the logistic equation are approximately double those of simple geometric growth. K represents the saturation taxonomic diversity, the equilibrium value of the number of taxa the habitat can support. The logistic curve assumes that initial radiation is very rapid as the taxon moves into a new adaptive zone. As the zone is exploited, competition becomes more intense, and the rate of radiation decreases and approaches zero as available niches are filled and N, the number of taxa, approaches K. The intrinsic rate of in-

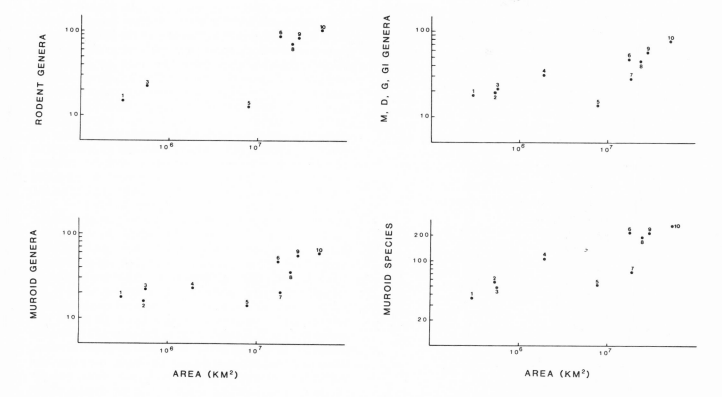

Figure 2. Continental area plotted against rodent diversity. Both axes are logarithmic. Numbers correspond to continental areas listed in Table 2. M, D, G, and Gl stand for Muroidea, Dipodoidea, Geomyoidea, and Gliroidea, respectively.

crease decreases either because of a decrease in rate of origination or an increase in the rate of extinction, or a combination of both factors. Determination of the factor(s) responsible would elucidate the speciation process. For example, if origination rate were determined to remain constant, then the ability of a taxon to speciate may be an intrinsic property of that species. In this example, extinction would increase as a result of increased competition, but origination would not be affected. Unfortunately, the fossil record, in this instance, can not yet resolve the issue.

Which model is appropriate for the South American sigmodontines? To help decide, it is necessary to determine whether South America is saturated with sigmodontine rodents. To estimate this, rodent diversity was plotted against continental area (Fig. 2), using methods discussed in Flessa (1975). This is based on MacArthur and Wilson's (1967) equilibrium theory of island biogeography, which states there is a linear relationship between the logarithm of biotic diversity and the logarithm of area. Generic and species diversities were calculated using Walker and others (1975) and Honacki and others (1982). Areas were taken from the Rand McNally atlas. North America as defined here includes the continental land area exclusive of the Caribbean and Arctic islands. Central America is defined as the political region of North America south of Mexico, which includes about half of North American Neotropica.

Regression analysis indicates a statistically significant relationship between rodent diversity and area (Tables 2, 3). The fit is improved even more by removing Australia and the United States plus Canada. Possible reasons for low diversity in Australia are discussed below. The United States plus Canada may have a lower diversity because of the absence of tropics and an abundance of land at high latitudes. Simpson (1964) noted that mammalian species density in North America increases with decreasing latitude. This has been investigated further and corroborated by McCoy and Conner (1980). Low values for the United States plus Canada versus Mexico and Central America further support Simpson. Species diversity in Mexico is increased 15-20 percent because of taxa found only on islands off Baja California. Mexico also is part of two biogeographic provinces, Nearctica and Neotropica. Muroid species diversity is especially low in the United States and Canada because of probable competition with the heteromyids, which are most diverse in arid and semi-arid environments.

In Australia there are 14 genera and 53 species of recent murines (Honacki and others, 1982). Compared to other regions, Australia has a depauperate rodent fauna (Fig. 2). Neighboring New Guinea, for instance, though much smaller, has 22 genera and 50 species of murines. The low diversity of Australian rodents may be related to their relatively recent entry there and the large area that is occupied by desert. The fossil record indicates

TABLE 2. AREA AND RODENT DIVERSITY. A= TOTAL RODENT GENERA; B = MUROID, GEOMYOID, DIPODOID, AND GLIROID GENERA; C = SPECIES OF B; D = MUROID GENERA; AND E = MUROID SPECIES. DATA FOR A FROM WALKER AND OTHERS (1975); DATA FOR B-E FROM HONACKI AND OTHERS (1982).

Region	Area $km^2 \times 10^3$	A	B	C	D	E
1. Phillipines	300	15	18	36	18	36
2. Central America	525	-	19	74	16	57
3. New Guinea	572	22	22	49	22	49
4. Mexico	1,972	-	31	163	22	108
5. Australia	7,687	13	14	53	14	53
6. South America	17,793	89	47	221	46	219
7. U.S.A. and Canada	18,028	-	29	130	20	75
8. North America	24,361	70	47	291	35	193
9. Africa	29,677	84	58	218	56	212
10. Eurasia	52,238	102	78	313	60	263

TABLE 3. CORRELATION OF LOG RODENT DIVERSITY WITH LOG CONTINENTAL AREA. LETTERS AS IN TABLE 2. LOWER CASE IS FOR REGRESSIONS PERFORMED WITHOUT DATA FOR AUSTRALIA, U.S.A., AND CANADA.

	A	a	B	b	C	c	D	d	E	e
slope	.358	.363	.224	.257	.333	.366	.200	.239	.320	.365
intcp	.243	.311	.667	.613	.834	.783	.692	.618	.812	.-724
d.f.	5	4	8	6	8	6	8	6	8	6
r	.799	.989	.758	.985	.822	.942	.708	.946	.831	.984
p	<.05	<.001	.01	<.001	<.01	<.001	.02	<.001	<.01	<.001

that murines arrived in Australia approximately 4.5 mya (Hand, 1984; Baverstock, 1984). There they found the second most arid continent in the world, second only to Antarctica. At least 44 percent of Australia is arid, and more than 70 percent is arid to semi-arid (Archer, 1984). As discussed by Mares (1980), murid rodents in Australia, as well as South America, have not yet evolved the extreme levels of xeric adaptation exhibited by counterparts living in other deserts, and therefore must be relatively recent immigrants.

Assuming that Australian murids are still in the stage of rapid diversification, as may be indicated by apparent relative undersaturation of rodents on this island continent, the geometric growth equation may be used to determine the rate of increase for these rodents. To produce 14 genera and 55 species in 4.5 million years requires r values of 0.59 and 0.89 per million years (hereafter omitted), respectively. If, on the other hand, the rate of expansion has slowed because of difficulties in adapting to the extensive desert environment, the logistic equation may be a better model. This yields r values of 1.14 and 1.77. This rate for producing new species is extremely high, over five times the highest rate Stanley (1979) calculated for different groups of Cenozoic mammals;

TABLE 4. AVERAGE MASS (IN GRAMS) OF GENERA OF SOUTH AMERICAN CAVIOMORPHA (A), ECHIMYIDAE (B), AND NEOTROPICAL SIGMODONTINES (C).

Taxon	N	Range	Mean and Standard Error
A	36	150 - 1150	335 ± 36.6
B	15	153 - 365	234 ± 14.6
C	44	82 - 256	132 ± 05.4

the estimate probably is too high. In reality, rodents reached Australia in at least three waves of dispersal (Baverstock, 1984). The first wave is represented by the Conulurini, the old endemics. The geometric rates of increase for these 8 genera and 40 species are 0.46 and 0.82, respectively. If this is a monophyletic group founded by a single colonizer (Baverstock, 1984) and if the time of arrival in Australia was 4.5 million years ago, these represent minimum rates for the tribe. Even then, however, the rate for species production is over twice Stanley's (1979) maximum, and approximately four times his average rate for Cenozoic mammals. Nevertheless, the true rates are probably higher because competition with later rodent immigrants and difficulties in expanding into desert environments has reduced the rate of production of successful new taxa. On the other hand, Stanley's estimated rates are relatively low because of the longer time intervals he uses to compute rates.

In South America, rodent diversity falls close to the expected value for its area, which could indicate that the logistic model may be more appropriate, although many factors need to be considered. The Rodentia in South America also are represented by the Caviomorpha, a diverse group consisting today of 12 families, 43 genera, and 147 species. This high familial diversity reflects the long geologic history of caviomorphs in South America. If there is competition between caviomorphs and sigmodontines, it might have decreased the rate of expansion of the latter group as it entered South America. In order to test this, body masses of caviomorphs and sigmodontines were compared. Data were taken from Walker and others (1975). Where a minimum and maximum were given, the two were averaged. Data shown in Table 4 are for the 36 genera of South American Caviomorpha, 15 genera of Echimyidae (the family of caviomorphs that contains the smallest body sizes), and 44 genera of Neotropical sigmodontines. On average, caviomorphs, as a whole, are two and one half times as large as the Neotropical sigmodontines; relatively small echimyids are nearly twice as large. There is a highly significant statistical difference between the body masses of echimyids and Neotropical sigmodontines ($t = 6.56$, $p < 0.001$).

The caviomorphs appear to occupy lagomorph and small ungulate niches in South America, niches not occupied by North American rodents (Eisenberg and Redford, 1982). Thus they should have offered little competition to the North American invaders. In fact, the sigmodontines were able to broaden their niches in many instances by becoming more insectivorous and carnivorous in the near absence of the Order Insectivora (Glanz, 1982). Sigmodontines may still be expanding their diversity, in which case the geometric model is more appropriate. To facilitate comparisons with other groups, geometric rates are used in the following discussion.

To produce 47 genera and 235 species of Neotropical sigmodontines in a time span of 8 million years would require geometric rates of increase of 0.48 and 0.68, respectively. This is certainly comparable to the Australian Conilurini, with rates of 0.46 and 0.82, or for the Holarctic arvicolines (the voles and lemmings), which would require geometric rates of 0.34 and 0.60 to produce the 17 genera and 125 species, respectively, in the past 8 million years. The arvicoline radiation was accomplished in competition with other small rodent groups, an obstacle not facing the conilurines or South American sigmodontines. Rates for the peromyscines (0.56 to produce 87 species in 8 million years) or the neotomines (0.42 to produce 29 species in 8 million years) are not overwhelmingly different from all these rates, even though their radiations are not thought explosive.

There are 16 genera and 57 species of recent muroids in Central America. Of these, some 40 percent (6 genera and 21 species) are Neotropical sigmodontines. It is unknown what proportion of the fauna was represented by these mice in the late Miocene and Pliocene, or how the present distribution was affected by Pleistocene climatic events. The fossil record (Jacobs, 1977; Baskin, 1979; Lindsay and Jacobs, in press) suggests that Neotropical sigmodontines may have dominated the late Tertiary rodent faunas of southern North America. During the main episode of the Great American Interchange (in the late Pliocene and early Pleistocene), some two dozen genera of North American large mammals dispersed southward. Because approximately 25 percent of the recent mammal fauna is composed of muroids, it is not out of line to suggest that six genera of Neotropical sigmodontines may have invaded South America, beginning 3.5 million years ago. The sigmodontines may have been represented by members of the phyllotine, sigmodont, akodont, oryzomyine, and thomasoymine generic groups. The first four have their earliest known fossil records in North America. Thomasomyines have a primitive dental morphology reminiscent of *Copemys*. *Nyctomys,* a Central American endemic, is usually considered a thomasomyine, although Carleton (1980) and Voss and Linzey (1980) note the similarities of this genus to neotomine-peromyscines.

There are 44 genera and 215 species of Recent sigmodontines in South America. To produce this diversity from an invading stock of 6 genera and 12 species from Central America beginning 3.5 million years ago would require geometric rates of 0.57 and 0.82. These rates are within the range expected for an adaptive radiation seen in other muroid groups. If this scenario is correct, development of the present-day South American diversity, starting with a multiple invasion in the Pliocene, is not a fantastic event. Van Valen (1985) estimated geometric rates of 2.9 and 3.4 for evolution of 19 genera and 29 species of condylarths in one million years separating the latest Cretaceous and early Puercan. Such estimates are more than double the geometric rates (1.1 and 1.5, respectively) required for a single species to produce the present-day South American muroid diversity in 3.5 million years.

Novacek and Norell (1982) question the accuracy and non-stochastic variability of taxonomic frequency rates within major clades. The maximum net rate of species increase calculated by Stanley (1979) was 0.39 for the early diversification of placental mammals. Novacek and Norell reevaluated Stanley's data, and recalculated the rate to be between 0.12 and 0.19. Van Valen (1985) suggested that maximum rates of evolution last only a geologic "instant," and cannot be measured over millions of years. Novacek and Norell (1982) believe that much longer intervals of geologic time are needed to measure adaptive radiations. The rodent groups studied above appear to show much higher rates than calculated by Stanley or by Novacek and Norell, but not as great as those calculated by Van Valen. All estimates do, however, support higher evolutionary rates during an adaptive radiation. Higher rates may cause the molecular clock to run at higher than average rates, and account for the more than two-fold discrepancies in estimates of divergence times for muroids calculated by the molecular clock versus the fossil record.

Neotropical sigmodontines are among the karyotypically most diverse mammals (Gardner and Patton, 1976). Although there is no known direct causal link between chromosomal change and population divergence (Patton and Sherwood, 1983), chromosomal evolution has been implicated in the high evolutionary rates of the Rodentia. High rates of chromosomal evolution and speciation have been related to inbreeding as a result of small local deme size (Wilson and others, 1975). Bush and others (1977) cite several examples of restricted social structuring and low vagility in several muroids with extensive karyotypic variation. Patton and Sherwood (1983) cite evidence indicating that inbreeding levels in such populations would be insufficient to overcome gene flow, and suggest that karyotypic differentiation would most likely result from population fragmentation caused by founder events. Fitch and Atchley (1985) have discovered that fixation rates of new mutations in inbred strains of mice appear to be several orders of magnitude higher than in natural populations. They suggest that high mutation rates may be advantageous when population sizes are greatly reduced by the bottleneck effect, and that speciation may be promoted in small isolated populations as a result of these rates. Expansion of sigmodontines into South America may have resulted in large numbers of vicariance events as these rodents rapidly expanded their range and interacted with their new environments. Vicariance events would have favored the genotypic and/or karyotypic differentiation required to produce the present diversity of South American sigmodontines since their arrival in the Pliocene.

SUMMARY AND CONCLUSIONS

1. *Abelmoschomys simpsoni* is the earliest recorded member of Neotropical sigmodontine rodents.

2. Neotropical sigmodontines are first known from the late Miocene (approximately 9 million years ago) of North America.

3. Neotropical sigmodontines began their diversification in North America.

4. Neotropical sigmodontines first appeared in South America in the late Pliocene, at about the time of formation of the Panamanian Land Bridge. At this time, several taxa dispersed to South America and underwent extensive adaptive radiation.

5. At present, no evidence except possibly data interpreted via the paradigm of the molecular clock falsifies this seemingly late (*i. e.,* Pliocene) arrival hypothesis. Morphological, chromosomal, physiological, taxonomic, and paleontological data support a late arrival. Parasite data are equivocal.

6. Neotropical sigmodontines are most closely related to peromyscines. This is supported by evidence from molecular systematics and the fossil record.

7. Neotropical sigmodontines, peromyscines, and possibly neotomines are derived from the widespread late Miocene North American genus *Copemys*.

8. *Calomys (Bensonomys)* is a member of the initial diversification of sigmodontines.

9. The diversity of muroids in South America is that expected for a continent of its size plus climatic and topographic heterogeneity.

10. Sigmodontines were able to diversify rapidly in South America because they did not have to compete with ecologically equivalent forms, and in fact were able to expand their available niches beyond those filled in North America.

11. Evolutionary rates of increase of species for

Neotropical sigmodontines were calculated to range between 0.60 and 0.82 per million years.

12. Such rates are approximately twice the maximum rate and approximately three to four times the average rate estimated by Stanley (1979) for Cenozoic mammals.

13. These high rates of speciation compare favorably with rates estimated for other muroid radiations.

14. The molecular clock gives divergence dates for muroid clades two to three times older than those seen in the fossil record.

15. High evolutionary rates such as have occurred in the muroids may cause the molecular clock to run faster during intervals of adaptive radiations.

ACKNOWLEDGMENTS

This research was supported by grants from the National Science Foundation and from Faculty Organized Research of Texas A&I University. I thank Everett Lindsay for initiating and maintaining my interest in rodent evolution, both by encouragment and stimulation from his and Louis L. Jacobs' research, and S. David Webb for his support and unflagging interest. The rodent fauna of the Love Bone Bed, the rest of which I intend to describe in the not too distant future, is mainly the result of the diligent picking of matrix by Daniel Cordier. Linda Daniels also is to be commended for her contributions to the recovery of small vertebrates. Figure 1 was drawn by Michael Vance, for whose artistic talents I am most grateful. Carol Altman assisted with reproduction of the figures. I also thank Nicholas J. Czaplewski of Northern Arizona University for permission to cite his unpublished work. N. J. Czaplewski, L. L. Jacobs, E. H. Lindsay, G. S. Morgan, and S. D. Webb offered constructive criticisms of the manuscript.

REFERENCES CITED

Aguilar, J.-P., 1980, Nouvelle interprétation de l'évolution du genre *Megacricetodon* au cours du Miocène, *in* Michaux, J., ed., Mémoire Jubilaire R. Lavocat: Montpellier, Paleovertebrata, p. 355-364.

Arata, A., 1967, Muroid, glirioid and dipodoid rodents, *in* Anderson, S., and Jones, J. K., Jr., eds., Recent mammals of the world; a synopsis of families: New York, The Ronald Press, p. 226-253.

Archer, M., 1984, Evolution of arid Australia and its consequences for vertebrates, *in* Archer, M., and Clayton, G., eds., Vertebrate zoogeography and evolution in Australasia: Carlisle, Hesperian Press, p. 97-108.

Baker, R. J., Koop, B. F., and Haiduk, M. W., 1983, Resolving systematic relationships with G-bands; a study of five genera of South American cricetine rodents: Systematic Zoology, v. 32, p. 403-416.

Baskin, J. A., 1978, *Bensonomys, Calomys,* and the origin of the phyllotine group of Neotropical cricetines (Rodentia: Cricetidae): Journal of Mammalogy, v. 59, p. 125-130.

———— 1979, Small mammals of the Hemphillian age White Cone local fauna, northeastern Arizona: Journal of Paleontology, v. 53, p. 695-708.

———— 1983, Tertiary Procyoninae (Mammalia: Carnivora) of North America: Journal of Vertebrate Paleontology, v. 2, p. 71-93.

Baverstock, P., 1984, Australia's living rodents; a restrained explosion, *in* Archer, M., and Clayton, G., eds., Vertebrate zoogeography and evolution in Australasia: Carlisle, Hesperian Press, p. 913-920.

Black, C. C., 1963, Miocene rodents from the Thomas Farm local fauna, Florida: Harvard University, Museum of Comparative Zoology, Bulletin, v. 128, p. 483-501.

Bonaparte, J. E., 1960, La sucesión estratigráfica de Monte Hermoso, (Prov. de Bs. Aires): Acta Geológica Lilloana, v. 3, p. 273-291.

Brownell, E., 1983, DNA/DNA hybridization studies of muroid rodents; symmetry and rates of molecular evolution: Evolution, v. 37, p. 1034-1051.

Bush, G. L., Case, S. M., Wilson, A. C., and Patton, J. L., 1977, Rapid speciation and chromosomal evolution in mammals: National Academy of Sciences, U.S.A., Proceedings, v. 74, p. 3942-3946.

Carleton, M. D., 1980, Phylogenetic relationships in neotomine-peromyscine rodents (Muroidea) and a reapprasial of the dichotomy within New World Cricetinae: University of Michigan, Museum of Zoology, Miscellaneous Publication, v. 157, p. 1-146.

Carleton, M. D., and Musser, G. G., 1984, Muroid rodents, *in* Anderson, S., and Jones, J. K., eds., Orders and families of recent mammals of the world: New York, John Wiley & Sons, p. 289-379.

Chaline, J., Mein, P., and Petter, F., 1977, Les grandes lignes d'une classification évolutive des Muroidea: Mammalia, v. 41, p. 245-252.

Ciochon, R. L., and Chiarelli, A. B., 1980, Evolutionary biology of the New World monkeys and continental drift: New York, Plenum Press, *xvi* + 528 p.

Clark, J. B., Dawson, M., and Wood, A. E., 1964, Fossil mammals from the lower Pliocene of Fish Lake Valley, Nevada: Harvard University, Museum of Comparative Zoology, Bulletin, v. 131, p. 29-63.

Dalquest, W. W., 1983, Mammals of the Coffee Ranch local fauna, Hemphillian of Texas: Texas Memorial Museum, Pearce-Sellards Series no. 38, 41 p.

Eisenberg, J. F., and Redford, K. H., 1982, Comparative niche structure and evolution of mammals of the Nearctic and southern South America, *in* Mares, M. A., and Genoways, H. H., eds., Mammalian biology in South America: Special Publication Series, Pymatuning Laboratory of Ecology, University of Pittsburgh, v. 6., p. 77-84.

Fahlbusch, V., 1964, Die Beziehungen zwischen einigen Criceten (Mamm. Rodentia) des nordamerikanischen und europäischen Jungtertiärs: Palaeontologische Zeitschrift, v. 41, p. 154-164.

Fitch, W. M., and Atchley, W. R., 1985, Evolution in inbred strains of mice appears rapid: Science, v. 228, p. 1169-1175.

Flessa, K. W., 1975, Area, continental drift and mammalian diversity: Paleobiology, v. 1, p. 189-194.

Flynn, L. J., Jacobs, L. L., and Lindsay, E. H., 1985, Problems in muroid phylogeny; relationship to other rodents and origin of major groups, in Luckett, W. P., and Hartenberger, J.-L., eds., Evolutionary relationships among rodents; a multidisciplinary analysis: New York, Plenum Press, p. 589-616.

Gardner, A. L., and Patton, J. L., 1976, Karyotypic variation in oryzomyine rodents (Cricetinae), with comments on chromosomal evolution in the Neotropical cricetine complex: Museum of Zoology, Louisiana State University, Occasional Papers, v. 49, p. 1-48.

Glanz, W. E., 1982, Adaptive zones of Neotropical mammals. A comparison of some temperate and tropical patterns, in Mares, M. A., and Genoways, H. H., eds., Mammalian biology in South America: Special Publication Series, Pymatuning Laboratory of Ecology, University of Pittsburgh, v. 6., p. 95-110.

Hand, S., 1984, Australia's oldest rodents; master mariners from Malaysia, in Archer, M., and Clayton, G., eds., Vertebrate zoogeography and evolution in Australasia: Carlisle, Hesperian Press, p. 905-912.

Hershkovitz, P. 1966, Mice, land bridges and Latin American faunal interchange, in Wenzel, R. L., and Tipton, V. J., eds., Ectoparasites of Panama: Chicago, Field Museum of Natural History, p. 725-751.

_____ 1972, The recent mammals of the Neotropical region. A zoogeographic and ecologic review, in Keast, A., Erk, F. C., and Glass, B., eds., Evolution of mammals and southern continents: Albany, State University of New York Press, p. 311-431.

Hibbard, C. W., 1972, Class Mammalia, in Skinner, M. F., and Hibbard, C. W., eds., Early Pleistocene pre-glacial and glacial rocks and faunas of north-central Nebraska: American Museum of Natural History, Bulletin, v. 148, p. 77-116.

Honacki, H., Kinman, K. E., and Koeppl, J. W., 1982, Mammal species of the world: Lawrence, Association of Systematic Collections, 694 p.

Hooper, E. T., and Musser, G. G., 1964, The glans penis in Neotropical cricetines (family Muridae), with comments on classification of muroid rodents: University of Michigan Museum of Zoology, Miscellaneous Publications, v. 123, p. 1-57.

Jacobs, L. L., 1977, Rodents of the Hemphillian age Redingington local fauna, San Pedro Valley, Arizona: Journal of Paleontology, v. 51, p. 505-519.

Jacobs, L. L., and Lindsay, E. H., 1981, *Prosigmodon oroscoi*, a new sigmodont rodent from the late Tertiary of Mexico: *ibid.*, v. 55, p. 425-430.

_____ 1984, Holarctic radiation of Neogene muroid rodents and the origin of South American cricetids: Journal of Vertebrate Paleontology, v. 4, p. 265-272.

Jacobs, L. L., and Pilbeam, D., 1980, Of mice and men; fossil-based divergence dates and molecular "clocks": Journal of Human Evolution, v. 9, p. 551-555.

Keigwin, L., 1982, Isotopic paleoceanography of the Caribbean and East Pacific; role of Panama uplift in late Neogene time: Science, v. 217, p. 350-353.

Layne, J. N., 1963, A study of the parasites of the Florida mouse, *Peromyscus floridanus*, in relation to host and environmental factors: Tulane Studies in Zoology, v. 11, p. 1-27.

Li C.-K., Chi H., 1981, Two new rodents from Neogene of Chilong Basin, Tibet: Vertebrata PalAsiatica, v. 19, p. 246-255.

Lindsay, E. H., 1972, Small mammal fossils from the Barstow Formation, southern California: University of California, Publications in Geological Sciences, v. 93, p. 1-104.

Lindsay, E. H., and Jacobs, L. L., *in press*, Pliocene small mammal fossils from Chihuahua, Mexico: Paleontologia Mexicana, v. 51.

MacArthur, R. H., and Wilson, E. O., 1967, The theory of island biogeography: Princeton, New Jersey, Princeton University Press, 203 p.

McCoy, E. D., and Conner, E. F., 1980, Latitudinal gradients in the species diversity of North American mammals: Evolution, v. 34, p. 193-203.

MacFadden, B. J., Campbell, K. E., Jr., Cifelli, R. L., Siles, O., Johnson, N. M., Naeser, C. W., and Zeitler, P. K., 1985, Magnetic polarity stratigraphy and mammalian fauna of the Deseadan (late Oligocene - early Miocene) Salla beds of northern Bolivia: Journal of Geology, v. 93, p. 223-250.

Mares, M. A., 1975, South American mammal zoogeography, evidence from convergent evolution in desert rodents: National Academy of Sciences, U.S.A., Proceedings, v. 72, p. 1702-1706.

_____ 1980, Convergent evolution among desert rodents, a global perspective: Carnegie Museum of Natural History, Bulletin, v. 16, p. 1-51.

Marshall, L. G., 1979, A model of paleobiogeography of South American cricetine rodents: Paleobiology, v. 5, p. 126-132.

Marshall, L. G., Butler, R. F., Drake, R. E., Curtis, G. H., and Tedford, R. H., 1979, Calibration of the Great American Interchange: Science, v. 204, p. 272-279.

Marshall, L. G., and Pascual, R., 1978, Una escala temporal radiométrica preliminar de las Edades-mamífero del Cenozoico medio y tardío sudamericano: Obra del Centenario del Museo de La Plata, v. 5, p. 11-28.

Mascarello, J. D., Stock, A. D., and Pathak, S., 1974, Conservatism in the arrangement of genetic material in rodents: Journal of Mammalogy, v. 55, p. 695-704.

May, S. R., 1981, *Repomys* (Mammalia: Rodentia gen. nov.) from the late Neogene of California and Nevada: Journal of Vertebrate Paleontology, v. 1, p. 219-230.

Novacek, M. J., and Norell, M. A., 1982, Fossils, phylogeny, and taxonomic rates of evolution: Systematic Zoology, v. 31, p. 366-375.

Patterson, B., and Pascual, R., 1972, The fossil mammal fauna of South America, in Keast, A., Erk, F. C., and Glass, B., eds., Evolution, mammals, and southern continents: Albany, State University of New York Press, p. 247-309.

Patterson, B., and Wood, A. E., 1982, Rodents from the Deseadan Oligocene of Bolivia and the relationships of the Caviomorpha: Harvard University, Museum of Comparative Zoology, Bulletin, v. 149, p. 371-543.

Patton, J. L., and Sherwood, S. W., 1983, Chromosome evolution and speciation in rodents: Annual Review of Ecology and Systematics, v. 14, p. 139-150.

Reig, O. A., 1978, Roedores cricétidos del Plioceno superior de la Provincia de Buenos Aires (Argentina): Publicaciones Museo municipal de Ciencias naturales de Mar Del Plata "Lorenzo Scaglia", v. 2, p. 164-190.

―――― 1980, A new fossil of South American cricetid rodents allied to *Wiedomys*, with an assessment of the Sigmodontinae: Zoological Society of London, Journal, v. 192, p. 257-281.

Rice, N. R., 1974, Single copy relatedness among several species of the Cricetidae (Rodentia): Carnegie Institution of Washington Yearbook, v. 73, p. 1098-1102.

Rogers, D. S., and Heske, E. J., 1984, Chromosomal evolution of the brown mice, genus *Scotinomys* (Rodentia: Cricetidae): Genetica, v. 63, p. 221-228.

Sarich, V. M., 1985, Rodent macromolecular systematics, in Luckett, W. P., and Hartenberger, J.- L., eds., Evolutionary relationships among rodents; a multidisciplinary analysis: New York, Plenum Press, p. 423-452.

Sarich, V. M., and Cronin, J. E., 1980, South American mammals, molecular systematics, evolutionary clocks, and continental drift, in Ciochon, R. L., and Chiarelli, A. B., eds., Evolutionary biology of the New World monkeys and continental drift: New York, Plenum Press, p. 399-421.

Shotwell, J. A., 1970, Pliocene mammals of southeast Oregon and adjacent Idaho: University of Oregon, Museum of Natural History, Bulletin, v. 17, p. 1-103.

Simpson, G. G., 1940, Review of the mammal-bearing Tertiary of North America. American Philosophical Society, Proceedings, v. 83, p. 639-709.

―――― 1945, The principles of classification and a classification of the mammals: American Museum of Natural History, Bulletin, v. 85, xvi + 1-350 p.

―――― 1950, History of the fauna of Latin America: American Scientist, v. 38, p. 361-389.

―――― 1964, Species density of North American recent mammals: Systematic Zoology, v. 13, p. 57-63.

―――― 1969, South American mammals, in Fittkau, E. J., Illies, J., Klinge, H., Schwabe, G. H., and Sioli, H., eds., Biogeography and ecology in South America: The Hague, Dr. W. Junk N. V., p. 879-909.

―――― 1980, Splendid isolation; the curious history of the South American mammals: New Haven and London, Yale University Press, 226 p.

―――― 1983, Fossils and the history of life: San Francisco, Scientific American Books, 239 p.

Slaughter, B. H., and Ubelaker, J. E., 1984, Relationship of South American cricetine rodents to rodents of North America and the Old World: Journal of Vertebrate Paleontology, v. 4, p. 255-264.

Stanley, S., 1979, Macroevolution; process and product: San Francisco, W. H. Freeman, 332 p.

Van Valen, L. M., 1985, Why and how do mammals evolve unusually rapidly: Evolutionary Theory, v. 7, 127-132.

Voss, R. R., and Linzey, A. V., 1980, Comparative gross morphology of male accessory glands among Neotropical Muridae (Mammalia: Rodentia), with comments on systematic implications: University of Michigan, Museum of Zoology, Miscellaneous Publications, v. 159, p. 1-41.

Walker, E. P., and others, 1975, Mammals of the world (3rd edition), Paradiso, J. L., ed., Baltimore, Johns Hopkins University Press, 1500 p.

Webb, S. D., 1976, Mammalian faunal dynamics of the Great American Interchange: Paleobiology, v. 2, p. 220-234.

Webb, S. D., MacFadden, B. J., and Baskin, J. A., 1981, Geology and paleontology of the Love Bone Bed from the late Miocene of Florida: American Journal of Science, v. 281, p. 513-544.

Wenzel, R. L., and Tipton, V. J., 1966, Some relationships between mammal hosts and their ectoparasites, in Wenzel, R. L., and Tipton, V. J., eds., Ectoparasites of Panama: Chicago, Field Museum of Natural History, p. 677-723.

Wilson, A. C., Bush, G. L., Case, S. M., and King, M. C., 1975, Social structuring of mammalian populations and rate of chromosomal evolution: National Academy of Sciences, U.S.A., Proceedings, v. 72, p. 5061-5065.

Wood, A. E., 1965, Grades and clades among rodents: Evolution, v. 19, p. 115-130.

―――― 1980, Problems of classification as applied to the Rodentia, in Michaux, J., ed., Mémoire Jubilaire R. Lavocat: Montpellier, Paleovertebrata, p. 263-272.

―――― 1983, The radiation of the Order Rodentia in the southern continents; the dates, numbers and sources of the invasions: Schriftenreihe geologische Wissenschaften Berlin, v. 19/20, p. 381-394.

Wu, C.- I., and Li, W.- H., 1985, Evidence for higher rates of nucleotide substitution in rodents than in man: National Academy of Sciences, U.S.A., Proceedings, v. 82, p. 1741-1745.

MANUSCRIPT RECEIVED DECEMBER 16, 1985
REVISED MANUSCRIPT RECEIVED MARCH 17, 1986
MANUSCRIPT ACCEPTED JUNE 8, 1986

Very hypsodont antelopes from the Beglia Formation (central Tunisia), with a discussion of the Rupicaprini

PETER ROBINSON *Museum, University of Colorado, Boulder, Colorado 80309-0315*

ABSTRACT

Two species of very hypsodont bovids are recorded from the middle Miocene (Astaracian/Vallesian) Beglia Formation of west-central Tunisia. They are presumed to be members of the Rupicaprini, a tribe that may have had Mediterranean origin. The Tunisian species show very early development of hypsodonty and loss of the basal pillars. The Rupicaprini are shown to have an important center of evolution in the Mediterranean region throughout the later Tertiary.

RESUME

Deux espèces des bovidés très hypsodontes se trouvent dans la formation Beglia (Miocène moyen, Astaracien/Vallesien) de la Tunisie ouest-centrale. Un rapport avec les Rupicaprini est envisagé. La tribu Rupicaprini est probablement d'origine mediterranéenne. Les espèces tunisiennes montrent une hypsodontie précoce et la perte des colonnes basales. La région mediterranéenne était un centre important d'évolution pour les Rupicaprini pendant le Tertiaire supérieur.

INTRODUCTION

For several seasons, beginning in 1967, the University of Colorado Museum coordinated an international effort to investigate the middle and upper Miocene rocks of Tunisia (Robinson and Black, 1969, 1974; Biely and others, 1972; Robinson, 1975). One aspect of this program was to collect and study the fossil vertebrate remains from the middle (formally considered upper) Miocene Beglia Formation of central Tunisia (Burollet, 1956). The Beglia Formation is generally sparsely fossiliferous throughout its outcrop area, but in one basin (the Bled Douarah, 45 km west of Gafsa), it has yielded an impressive collection of fossils. The commonest of these, the caprine antelope *Pachytragus solignaci*, is represented by at least 500 jaws and maxillae (Robinson, 1972). In these same collections there are some 20 isolated teeth of precociously hypsodont antelopes representing two species. Most of these specimens come from the older, or pre-*Hipparion,* fauna (Robinson and Black, 1969). In the course of studying this material, the question of antelope hypsodonty of such a developed stage at so early a time has also been considered, necessitating examination of related forms.

ABBREVIATIONS

D, decidous tooth
P, with subscript or superscript, premolar
M, with subscript or superscript, molar
R, L right and left, respectively
T-, field number of the Colorado Tunisian Expedition
SPGM, Sammlung für Paläontologie und Historische Geologie, München, Pakistan Collection

HISTORY

The fossil history of antelopes has been described in small increments, primarily because this is the manner in which the fossils were found. The Miocene (*sensu* Berggren and Van Couvering, 1974, 1978) antelopes were first found at Pikermi (Wagner, 1848; Gaudry, 1862-67), Samos (Major, 1894; Schlosser, 1904; Andree, 1926), Maragha (Rodler and Weithofer, 1890; de Mecquenem, 1924-25), and in France (Depéret, 1887). These faunas were realized to be roughly contemporaneous, and although there were differences in detail, they were considered basically similar (Pilgrim and Hopwood, 1928). Collections from India and Pakistan (several authors, but particularly Pilgrim, 1937, 1939) and from China and Mongolia (Schlosser, 1903; Bohlin, 1935; Pilgrim, 1934; Chen and Wu, 1976) added information about Asian species; the work by Crusafont and his co-workers signalled the abundant material in northeastern Spain (summarized in Crusafont-Pairo, 1973). Numerous other collections have been made, particularly in Tuscany (Major, 1872; Weithofer, 1888, 1889), but those cited are the most important, with middle or early late Miocene components found up to the discovery of the rich Turkish material (Sickenberg and others, 1975) and the Beglia collections.

Astaracian or older antelopes from the Mediterranean region are rare; the oldest forms are *Eotragus* from France and *Gazella* from Jebel Zelten (Ginsburg and Heintz, 1968; Hamilton, 1973). Most of the Miocene antelope collections are of Turolian age; good Vallesian collections are scarce, and faunas in long stratigraphic sequences are difficult to find. For these reasons the Turkish collections become very important, not only because they cover poorly known intervals and have

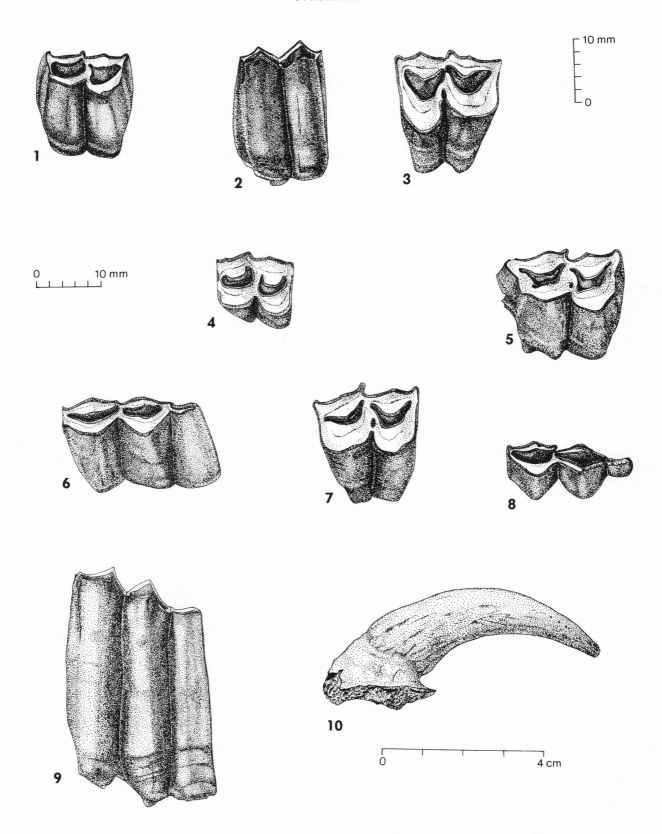

Figure 1. Species B, *1, 2, 4,* and *8—1,* T-3813, RM^3 (crown view); *2,* T-3813, RM^3 (lingual view); *4,* T-3824, RM^1 (crown view); and *8,* T-4398, LM$_3$ (crown view). Species A, *3, 5, 6, 7,* and *9—3,* T-4426, LM^2 (crown view); *5,* T-3812, RM^3 (crown view); *6,* T-3817, LM$_3$ (crown view); *7,* T-3816, RM^2 (crown view); and *9,* T-3817, LM$_3$ (buccal view). *10,* T-116, right horn core.

TABLE 1. MEASUREMENTS (IN MILLIMETERS) OF UPPER TEETH OF HYPSODONT ANTELOPES FROM THE BEGLIA FORMATION.

T-NUMBER	TOOTH	LOCALITY	LENGTH	ANT. WIDTH (Max.)	POST. WIDTH
4432	LP^4	17	7.5	----	7.6	----
3820	RDP^4	17	9.4	9.3	9.3
3821	RM^2	17	12.8	10.4	11.0
3822	RM^1	17	12.0	10.3	10.3
3823	LM^1	17	12.8	10.2	10.3
3824	RM^1	17	11.4	9.3	9.7
4425	LM^1	20	12.9	10.0	8.1
3816	RM^2	17	17.3	13.5	13.1
4426	LM^2	17	16.4	15.4	15.4
4429	RM^2	17	14.3	13.2	13.1
3812	RM^3	17	>21.50	13.8	11.0
3813	RM^3	17	15.4	9.8	7.4

demonstrated superposition, but also because their location places them in a strategic position when trying to evaluate Miocene faunal interchanges. The Tunisian collections, while having a demonstrable succession (Astaracian into Vallesian), offer most because of abundance of the material of one species, *Pachytragus solignaci* (see Robinson, 1972; Kurtén, 1983). The Beglia Formation is correlatable with marine units laterally, and is totally Serravallian in age (Wiman, 1978).

During the Miocene, the Mediterranean area was highly plastic geologically, and had a different landmass configuration than today (Bijou-Duval and others, 1977).

SYSTEMATIC PALEONTOLOGY

Family BOVIDAE Gray, 1821
Subfamily CAPRINAE Gill, 1872
Tribe RUPICAPRINI Simpson, 1945

Genus unassigned
Species A, large species
Figure 1,3; 1,5; 1,6; 1,7; and 1,9

Material: Lower Fauna, pre-*Hipparion*—M^2, T-3816, 4426, Loc. 17; M^3, T-3812, Loc. 17; M_3, T-3814, Loc. 18; T-796, 3815, 3817, 3819, Loc. 17.

Upper Fauna, with *Hipparion*—M_2, T-4399, Loc. 16; and M_3, T-3826, Loc. 5.

Description: Upper molars hypsodont, posterior part of the anterior lake expanded; metastyle of M^2 weak, metastyle of M^3 strong and directed posteriorly; base of metastyle of M^3 broken, but the preserved portion indicates that it expanded towards the base of the tooth as in some antilocaprids, *Nemorhaedus,* and *Capricornis.* Lower molars hypsodont, lacking basal pillars; hypoconulid of M_3 well developed; goat fold preserved only in upper parts of the lower molars. No cement.

Genus unassigned
Species B, small species
Figure 1,1; 1,2; 1,3; and 1,8

Material: Lower fauna—M^1, T-3822; M^2, T-3821; and M^3, T-3813 (these three teeth from Loc. 17 probably belong to the same individual). DP^4, T-3820, Loc. 17; P^4, T-4432, Loc. 17; M^1, T-3823, 3824, Loc. 17; T-4425, Loc. 20; M^2, T-4429, Loc. 17; M_3, T-3818, Loc. 20.

Upper fauna—M_{2-3}, T-184, Loc. 4; M_3, T-4398, Loc. 2; T-4435, Loc. 16.

Description: Upper molars hypsodont, lacking basal pillars, and lacking cement, even in the lakes; posterior part of anterior lake expanded; paracone rib weak; no metacone rib; mesostyle pronounced; metastyle present, but weaker than parastyle on M^{1-2}; metastyle strong and directed buccally on M^3, becoming convergent towards mesostyle near base of tooth.

Discussion: These two species (*i. e.,* species A and B) are similar except in size and in development of the metastyle on M^3, a character which might be of generic significance in a modern rupicaprine. The lack of cement may be correlated with absence of basal pillars; however, the loss of the pillars is a derived character, as basal pillars are found in primitive bovids, cervids, and giraffids. The metastyle of species B resembles that of *Maremmia haupti* (see Del Campana, 1918; Hürzeler, 1983) from Tuscany.

Genus unassigned
Species undescribed, cf. Sp. A
Figure 1,10

Material: Horn core, right: T-1116, Loc. 17.

Description: The horn core is small, curved posteriorly with a slight counterclockwise torsion, and was situated over the cranium, not the orbit. The anterior face of the pedicel is much longer than the posterior face, indicating that the horn was tilted backwards and that it may have been parallel to the facial plane of the skull,

TABLE 2. MEASUREMENTS (IN MILLIMETERS) OF LOWER TEETH OF HYPSODONT ANTELOPES FROM THE BEGLIA FORMATION.

T-NUMBER	TOOTH	LOCALITY	LENGTH	A. W.	P. W.	H.	HLD(M_3)
796	RM_3	17	27.0	8.5	8.4	----	
3814	LM_3	18	24.6	8.2	8.0	----	
3815	LM_3	17	23.9	8.0	7.6	22.8	
3817	LM_3	17	27.0	8.6	8.6	33.3	
3818	LM_3	20	21.8	7.3	7.1	----	
3819	RM_3	17	26.5	>7.9	>7.6	----	
4399	RM_2	16	16.9	7.6	8.0		
184	LM_2	4	11.0	6.0	7.7		
184	LM_3	4	21.4	6.9	6.9	----	
3826	LM_3	5	26.7	8.9	8.5	----	
4435	LM_3	16	----	---		8.5	>26.5
4398	LM_3	2	>26.7	>7.6	>7.5		26.3

NOTE: Hypoconulid heights were only measured in relatively unworn specimens. T- 3826 was collected by a local bedouin and the provenience is not absolutely certain. Localities 17, 18 and 20 are in the pre-<u>Hipparion</u> levels; localities 2, 4, 5 and 16 are in the <u>Hipparion</u> bearing beds.

as in *Nemorhaedus, Capricornis,* and *Myotragus balearicus.* If this was so, one can assume that the basicranial axis was bent in relation to the palate as in Caprini, but unlike the Antelopini. The surface of the horn core is finely textured, lacking the strong grooves (especially at the base of the horn core) characteristic of *Gazella, Oioceros,* and *Pachytragus,* and most other bovids. There is no keel.

Discussion: This is the only horn core known from the lower Beglia Formation not referable either to *Pachytragus* or *Gazella.* Lack of association prevents it from being assigned definitely to one of the two hypsodont species, but its size argues in favor of species A. Except for the curvature and torsion, it is very like the horn cores of *Nemorhaedus* and *Myotragus.* Torsion and curvature occur independently in many bovid lineages, and often vary considerably within one genus (e. g., *Gazella, Ovis,* and *Capra*). As most living rupicaprine horn cores are relatively uncurved (presumably the primitive condition), one could hypothesize that the Beglia specimen represented a more derived state. Several horn cores of *Eotragus* in the Basel collection, however, have a slight torsion, which is a less advanced stage than that found in *Oioceros (?) noverca* (see Pilgrim, 1934). The horn cores assigned to *Maremmia haupti* by Del Campana (1918) have the same fine texture and lack of grooves as in the living rupicaprines, *Myotragus,* and the Beglia specimen. The Tuscan horn cores also have curvature, torsion, and lack keels.

TUNISIAN HYPSODONT BOVIDS

The tooth and horn core morphology of the Tunisian hypsodont bovids closely resembles that of fossil and recent rupicaprines. The teeth are also similar to Antilocaprinae, particularly *Texoceros* (see Frick, 1937; Skinner and others, 1977, p. 361); antilocaprine horn cores also are smooth, and in this character resemble rupicaprines. It is worth mentioning that some antilocaprines and some rupicaprines are virtually indistinguishable on the basis of isolated teeth, especially the first and second molars.

Although similarity to rupicaprines is strong, the possibility that the Tunisian bovids are convergent with the Rupicaprini cannot be overlooked. The Tunisian forms do indicate, however, that by late Astaracian time extreme hypsodonty had already occurred in at least one bovid lineage.

The possibility that the hypsodont Beglia antelopes are antilopines has been examined and rejected. The earliest antilopines are members of the genus *Gazella* (see Gentry, 1978); Miocene *Gazella* are more brachydont than the living forms, and usually have basal pillars on the teeth. Horn cores of all gazelles are deeply furrowed, particularly at the base of the horn core, and are located over the orbits. I disagree with Gentry's suggestion (1978, p. 541) that *Rupicapra* should be placed in the Caprini because the horn core sinuses do not extend far up the horn core. On the contrary, it is possible that rupicaprines do not have a close relationship with the Caprini, particularly if *Oioceros* and *Pachytragus* are related to ancestry of the Caprini. These latter two genera were already evolving in a different direction in the middle Miocene, *Pachytragus* having already developed cement; if they were in ancestry of the Caprinae, the divergence was very old for bovids.

An alternative possibility is that the Tunisian species are related to the Antilocaprinae. Their resemblances are strong and the timing of their appearance is similar. However, the antilocaprine bifurcate horn core precludes a close relationship. Horn core morphology is basic to systematics of artiodactyls. For example, the basic difference in morphology of horn cores of antilocaprines and merycodontines is sufficient to question their taxonomic closeness. In addition, the merycodontines often developed wear facets on the bony core of their horn tines, which the keratin-protected horn cores of antilocaprines do not. O'Gara and Matson have shown (1975, p. 838-841) that at least some members of each subfamily of the Bovidae shed their horn sheaths episodically during their lifetimes. O'Gara and Matson also show that the lacrimal vacuities often considered characteristic of the antilocaprids are found in some bovids as well (p. 841).

Another possibility is that the Tunisian hypsodont bovids belong to the Alcelaphini. Thomas has recently (1984b) supported inclusion of the Tuscan genus *Maremmia* in this tribe based on its similarity to *Caprotragoides* (see Thomas, 1984a). Of the first nine dental characters listed by Thomas (1984b, p. 92) as being characteristic of the *Caprotragoides* lineage, all occur in rupicaprines as well. He cites development of the horn core sinus as occurring after separation of *Caprotragoides* and *Maremmia*, yet Vrba considers this to be a primary characteristic of the Alcelaphini (1979, p. 212, 215). Modern alcelaphines all have grooved horn cores, especially near the base, a primitive character, while those of *Maremmia* are smooth, a derived state.

The remarkable parallelism demonstrated by many bovids makes the presence of an as yet undescribed subfamily or tribe possible. However, I think that the data do not support this hypothesis because the available characters are found in the Rupicaprinae; thus I believe that the Tunisian animals are rupicaprines.

Pachytragus and *Oioceros* are more primitive than their Tunisian contemporaries in preserving the basal pillars and in being less hypsodont; they are more advanced in that cement formation had already started, at least in *Pachytragus*. If the Tunisian material is correctly assigned, the history of the Rupicaprini is extended back to the middle Miocene.

COMPARISON WITH *Oioceros*

Oioceros is the most common hypsodont antelope at the Astaracian/Vallesian boundary, and comparison of the Tunisian material with it is necessary. I will limit this discussion to species known from good material, and will exclude *O.*(?) *proaries* for reasons given by Solounias (1981, p. 196-199). The following species have been considered:

Antilope rothi Wagner, 1848. This species was made the generoholotype of *Oioceros* by Gaillard (1902). Previously known from Turolian deposits at Pikermi, Greece and Maragha, Iran; this species has been found recently in Turkey.

Antidorcas atropatenes Rodler and Weithofer, 1890. Gaillard also included this species in his concept of *Oioceros*. It has been found at Maragha and in Turkey; Turolian.

Oioceros boulei de Mecquenem, 1925. According to Heintz (1963), this is the female of *O. atropatenes*. Maragha; Turolian.

Oioceros wegneri Andree, 1926. Samos and in Turkey; Turolian.

Oioceros(?) *grangeri* Pilgrim, 1934. Tung Gur, Mongolia; Astaracian.

Oioceros(?) *noverca* Pilgrim, 1934. Tung Gur and in Turkey; Astaracian.

Oioceros mequenemi Pilgrim, 1934. Gentry (1970, p. 273) considers this a synonym of *O. rothi*. Maragha; Turolian.

Oioceros tanyceras Gentry, 1970. This species is doubtfully an *Oioceros* because of the lateral deflection of the horn cores. Fort Ternan, Kenya; Astaracian.

Oioceros(?) *jiulonkouensis* Chen and Wu, 1976. Ci Xian, China; early Astaracian.

Oioceros(?) *robustus* Chen and Wu, 1976. Ci Xian; Astaracian.

Oioceros(?) *stenocephalus* Chen and Wu, 1976. Ci Xian; Astaracian.

Of these species, four occur in the Turkish Miocene. The oldest (which resembles *O.*(?) *noverca*) is in the Çandir fauna (Berg, 1975, p. 157), and already shows the trend towards hypsodonty characteristic of the genus, a characteristic notable in the Ci Xian specimens also. Of the four pre-*Hipparion* faunal levels recognized in Anatolia (Sickenberg and others, 1975), Çandir is the second oldest. The underlying faunal group, Paşalar, is rich in isolated antelope teeth, none of which are referable to *Oioceros* (see Berg, 1975). It seems, therefore, that the first arrival of *Oioceros* in the Mediterranean region is recorded in the Çandir fauna. The specimens closely resemble *O.*(?) *noverca*; the presence of this taxon in Turkey may be good supporting evidence for the age of the Tung Gur fauna.

One of the interesting characteristics of *Oioceros* is that the relative hypsodonty of the teeth did not change significantly from the middle to late Miocene (Pilgrim, 1934; Schlosser, 1904). The teeth tend to be simple, without cement, and to lose the basal pillars from the back forward and on the upper teeth first. In *O.*(?) *noverca* and in *O.*(?) *grangeri* the basal pillars are strong in the lower molars (Pilgrim, 1934). M3/3 are significantly more hypsodont in the unworn state than M1/1. The younger samples have the lower molars more flattened mediolaterally, and have a more pronounced goat fold.

The Chinese and Mongolian species all have been questionably referred to the genus *Oioceros* by their authors. This may be because these species all are long faced, have a less developed basicranial angle than other hypsodont antelopes, and they have a relatively long diastema between the symphysial teeth and the cheek teeth. Unfortunately, these crucial characters are not preserved in the species from Samos, Pikermi, or Maragha. While the basicranial angle of the Tun Gur species is more pronounced than in the Ci Xian forms, still it is not as acute as in many later species, and the horn cores are not directed backwards as in many forms with more acute basicranial angles.

Therefore, there is some question whether the species of *Oioceros* from China and Mongolia are really related to the later species from Iran, Turkey, and Greece. The difference in time between the two samples may explain the observed morphological differences. Later species may be shorter-faced, have smaller diastemas, and more acute basicranial angles, all of which are trends that are recognized in various bovid lineages.

What should be apparent is that, although the Tunisian and Asian hypsodont species overlap in time, they are not closely related. The Tunisian specimens already had lost all traces of the basal pillars by Astaracian time, and they have significantly different horn core morphology. Moreover, the Tunisian specimens are significantly more hypsodont than *Oioceros*. Therefore, by the Astaracian, the trend already had begun towards the hypsodont condition, in at least two lineages of bovids, one of which was significantly more derived than the other in this respect.

I know of only one other pre-*Hipparion* bovid specimen with such a degree of hypsodonty: SPGM field no. 3734 from the Chinji beds of Pakistan. This specimen, an M_3, is readily comparable with the Tunisian species A, and may be conspecific. It is worth noting, however, that teeth of this type have not been found in the pre-*Hipparion* beds of Turkey; this seems a strange situation considering how well these beds have been sampled and the large size of the collections which include *Pachytragus solignaci*.

Primitive bovid horn cores (including many Miocene species) are situated over the orbit, and just in front of the frontal/parietal suture. Many bovid lineages become derived by having the horns move to the rear of the skull roof. However, the horns are still near the rear margin of the frontal bone; the frontal overrides (and often hides) the parietal from above. This is the case in *Bos, Alcelaphus,* and *Connochaetes*. In the Miocene, the horn core shape already was variable; this is demonstrated by a variety of lineages (e. g., *Gazella, Miotragocerus, Pachytragus, Oioceros,* and the Tunisian forms). The Tunisian species may have had a pronounced basicranial angle, as shown by the angle of the horn core pedicel and its location. In this regard, it was more advanced than its contemporary (Astaracian) *Oioceros,* whose horn arose vertically over the orbit.

PRE-*Hipparion* BOVID FAUNA OF BEGLIA FORMATION

Four bovids can be identified from lower beds of the Beglia Formation on the basis of teeth: (1) the extremely common *Pachytragus solignaci* (see Robinson, 1972; Kurtén, 1983; several hundred jaw and maxilla fragments are known); (2) *Gazella* sp. cf. *G. praegaudryi* (less than 1% of the former); (3) and (4) the two bovid species A and B cited above. Three species are identifiable on the basis of horn cores: *P. solignaci, Gazella* sp., and a third taxon presumed to belong to one of the hypsodont forms.

The lower Beglia strata were well-sampled at three localities: 17, 20, and 23. Localities 20 and 23 may be from the same bed, sampled on two different sides of a hill. It is unlikely that antelope taxa were missed during the excavation, due to the excellent sample of *Pachytragus solignaci* material which is of average size for an Astaracian/Vallesian antelope. Gazelle horn cores are commoner than dentitions: a situation that argues in favor of the hypothesis that gazelle teeth and dentitions were too small for inclusion in the Beglia sediments.

In a recent article, Kurtén (1983) has argued that the population of *Pachytragus solignaci* was sampled in a manner that demonstrates the presence of age groups. If he is right that natural age groups were sampled, and that the *P. solignaci* mortality resembles a sampling of a natural population in a relatively short period of time, then the *P. solignaci* members probably lived in herds, at least for part of the year. The much smaller representation of gazelles in the fauna (as determined by horn core sample) may indicate that, even in the Miocene, gazelles were widely distributed but not numerous locally, a situation similar to that of today. The extreme rarity of the hypsodont species A and B may reflect not only a rarity in the local fauna, but also that the animals did not normally inhabit a plain or flatland environment; already in the Miocene they might have been inhabitants of rocky slopes.

Minor mountain building took place in Tunisia in Astaracian time (Castany, 1951; Burollet, 1956; Robinson and Wiman, 1976). While no Astaracian tectonism has been determined in the Bled Douarah, the lower part of the Beglia Formation is missing at Dj. Sehib, some 65 kilometers to the southeast, indicating that there were locally hilly, if not mountainous, areas in the region.

VALLESIAN BOVIDS FROM BEGLIA FORMATION

Solounias will describe the diverse bovid material from the upper Beglia Formation. However, it is worth pointing out that the bovids are, paradoxically, less common relative to the whole sample, and more diverse in the upper beds. In addition to the four species found in the lower beds, there are at least four or five others. None is abundantly represented, and all have been transported some distance before burial. Evidence from marine correlations (Wiman, 1978) shows that the upper part of the Beglia Formation is still Serravallian by marine terminology and therefore, at the latest, early Vallesian in non-marine terminology. The fauna, however, resembles the Turolian faunas of Samos, Pikermi, and Maragha (among the well known bovid collections), and also that of the Vallesian of Turkey.

TUROLIAN-RUSCINIAN OCCURRENCES OF VERY HYPSODONT BOVIDS

Pre-Villafranchian bovids with extreme hypsodonty are known from a variety of sites, but only in Tuscany are they relatively common (Del Campana, 1918; Hürzeler and Engesser, 1976; Hürzeler, 1983). The two genera found there, *Maremmia* and *"Tyrrenotragus,"* have much in common with the Tunisian material, but have been assigned to the Alcelaphini (Hürzeler, 1983). These two lineages have been recovered from two faunal levels in Tuscany: levels V1 and V2 of the Baccinello Basin (Lorenz, 1968). The older collections come from Casteani (= Tatti), Ribolla (= Monte Massi), and Monte Bamboli; these localities are north and northwest of Baccinello. Hürzeler (1983, p. 248) correlates Casteani and Ribolla with level V1 and Monte Bamboli with level V2. These antelopes also occur at several other localities in Grosseto Province, but they are not common. Both levels are late Miocene (Turolian) in age.

A second important occurrence is the material from the Nagri beds of Pakistan. Pilgrim (1939) figured a maxilla, which he referred to *Dorcadoxa porrecticornis*. The holotype of *D. porrecticornis* is a horn core from the Dhok Pathan beds, and therefore is much younger than the maxilla. Taxonomic association of the maxilla with the horn core is questionable, and the generic assignment cannot be assumed to be correct. In addition to the maxilla figured by Pilgrim, there are several teeth from the Nagri in the British Museum (Natural History) which resemble the Tunisian forms.

Andrews (1902, Fig. 9) figured a M_3 from Wadi Natrun (Egypt) which is similar to the Tunisian forms, but much larger. Solounias (1981, p. 208) cites three rupicaprine horn cores from Samos, but he did not list any dental material.

Recent work on the Island of Majorca has brought to light a number of species of the peculiar rupicaprine *Myotragus*. Discovery of the Villafranchian *M. batei* (see Crusafont and Angel, 1966) was followed by discovery of *M. antiquus* (see Pons-Moya, 1977) and *M. kopperi* (see Moya-Sola and Pons-Moya, 1981) from older deposits.

Material cited above indicates that there was a number of extremely hypsodont bovids throughout most of the Mediterranean-Himalayan regions during much of the late Middle and Late Tertiary; these animals were not common, and rarely were preserved. In addition to the possibility that they lived on rocky slopes, they may not have been animals that lived in herds, and therefore may not have had large numbers of the species present in any one area.

Pilgrim (1939, p. 36, Pl. 1, Fig. 4) described a new species (?)*Gazella superba* from Hasnot, of which the type is a frontlet notable for lacking the grooves characteristic of gazelle horn cores (especially their bases). He also doubtfully assigned two maxillae to this species; one came from Chinji, the other from Hasnot. Because of the problem with the association, I prefer to consider only the type frontlet as pertaining to this species; its morphology clearly excludes it from reference to *Gazella*, but indicates its relationship to the taxa from Tunisia and Tuscany.

Pachygazella (see Teilhard and Young, 1932) is probably a junior synonym of *Capricornis*. The specimen figured by Teilhard and Young is similar to those figured by Colbert and Hooijer (1953, Pl. 38, Fig. 1). If this synonymy is correct, it extends the range of the genus backwards to the Turolian.

A Miocene North American occurrence is provided by two specimens of *Neotragocerus* from the Snake Creek Formation of Nebraska (Skinner and others, 1977). These two horn cores are similar to *Oreamnos*.

POST-RUSCINIAN OCCURRENCES

During the Villafranchian (and Blancan), the Rupicaprini became widespread in Eurasia and occurred rarely in North America. Both *Capricornis* and *Nemorhaedus* are known from China (Guérin, 1965), *Hesperoceros* from Spain (de Villalta-Comella and Crusafont-Pairo, 1956), *Gallogoral* from Italy, France, and Spain (Guérin, 1965; Schaub, 1922), *Myotragus* from Majorca (Crusafont and Angel, 1966) and possibly Sardinia (Dehaut, 1911; Kotsakis, 1980; Kotsakis and Palombo, 1979), and *Procamptoceras* from Italy, France, and Hungary (Schaub, 1944). A possible Villafranchian or younger occurrence of *Capricornis* from Tibet is cited by Pilgrim (1939, p. 57). The American occurrence is *Neotragocerus lindgreni* from Idaho (Harington, 1971).

The two reported occurrences of Villafranchian rupicaprines from North Africa are questionable. *Rabaticerus* (see Ennouchi, 1953) has been reallocated to the Alcelaphini (Gentry, 1978, p. 556-557), and *Numidocapra* (see Arambourg, 1949) is enigmatic and might best be placed in Caprinae *incertae sedis*.

SYSTEMATIC POSITION OF TUNISIAN HYPSODONT BOVIDS

Tunisian bovids have three apomorphies which single them out among Miocene bovids: (1) they have lost basal pillars; (2) they have extreme hypsodonty; and (3) they have a smooth horn core surface unlike their contemporaries. They retain a primitive condition in not developing cement, and they have the goat fold, a derived character that they share with *Pachytragus* and *Oioceros* but not with *Eotragus* and *Gazella*. Many of these characters are repeated throughout history of the bovids, but Tunisian forms specialized early. The paucity of material, especially post-cranial, makes comparison with modern groups difficult; furthermore, since our concepts of bovid classification are based on projecting the modern arrangement back in time, lack of the rest of the skeleton necessarily makes any arrangement temporary. Of modern groups, only the rupicaprines have similarities which make them the best fit. They have extremely hypsodont teeth without cement or basal pillars, and their horn cores are smooth. In addition, the rupicaprines have their horns located at the anterior part of the braincase, behind the orbits or medial to the orbits *(Rupicapra)*. The one Tunisian horn core is located over the anterior braincase.

Since Andrews' work (1915), most students agree that *Myotragus* is a rupicaprine. The work of various Spanish paleontologists has shown that this animal developed a progressively shorter face with fewer premolars and incisors, eventually ending up with a single, open-rooted incisor in each jaw ramus.

Hürzeler (1983) has shown that *Maremmia* progressively lost premolars and that it developed a series of three incisors and the incisiform canine into a battery that must have functioned as a single tooth developed for close cutting or cropping. Hürzeler assigns the Tuscan antelopes to the Alcelaphini. Modern alcelaphines are either relatively long-faced *(Alcelaphus)* or very long-faced *(Connochaetes)*. There may be some reduction in relative size of the premolars, but not in the number. In addition, their teeth have cement, and the horn cores are situated near the rear of the expanded frontal bone at the back of the skull. The teeth of both *Alcelaphus* and *Connochaetes* lack basal pillars, but this may be a more recent convergence. Modern alcelaphines do not have mediolaterally compressed teeth with small lake area. The evidence, scanty as it is, suggests that the Tunisian and Tuscan Miocene antelopes were related, and that they have more in common with Rupicaprini than any other group, contrary to the opinion of Hürzeler (1983) and Thomas (1984b). The extreme similarity between these forms and Antilocaprinae (such as *Texoceros*), should be examined, as should the traditional association of Rupicaprini with Caprini. Gentry and Gentry have suggested (1978, p. 296) that *Rupicapra* is misallocated, and should be in the Caprini. I suggest, instead, that the entire tribe may be more distantly related to the Caprini than is usually thought. The Caprini/Rupicaprini divergence appears to go back to the middle Miocene (Astaracian); if one considers the extreme hypsodonty of the Beglia teeth, the divergence could be even older.

The evidence could also support an argument that these forms represent a Miocene radiation that had no survivors; if that is true, then the Mediterranean region prompted repeated, parallel developments in antelope evolution.

The abundance of rupicaprines in the Villafranchian of the Mediterranean region argues for the Mediterranean being a center of rupicaprine evolution with occasional outward migrations, even as far afield as North America. The Tuscan antelopes, if correctly allocated to the Rupicaprinae, could have migrated from Africa by island hopping, a suggestion made by Azzaroli (Hürzeler, 1975, p. 876; Thomas, 1984b). No post-Miocene specimens assignable to the Rupicaprini without doubt are known from Africa; the Miocene forms are all peri-Mediterranean, and could be uninvolved in sub-Saharan antelope evolution.

CONCLUSIONS

Two species of very hypsodont antelopes occur in the middle Miocene Beglia Formation of Tunisia. These animals already show certain evolutionary characters which were apomorphies for the time: loss of basal pillars and mediolateral compression of the molars, extreme hypsodonty, and loss of grooves on the horn core. Though scanty, preserved material indicates that the Tunisian species were more closely related to the Rupicaprini than to any other extant group of bovids.

The geologic history of Rupicaprini is reviewed, and it is shown that they may have had a more important center of evolution in the Mediterranean region than elsewhere. It is suggested that separation of the Rupicaprini from caprine stock is at least as old as middle Miocene, and that their similarities to the Caprini may be due to convergence. If assignments are correct, the oldest known rupicaprines are north African and the oldest known caprines are Eurasian (Ci Xian, China), though slightly younger specimens are known from Africa *(Pachytragus)*. The Caprini retained basal pillars on the molars until well into the later Tertiary.

ACKNOWLEDGMENTS

In any research concerned with taxonomy of mammals, the classification proposed by G. G. Simpson (1945) is used either directly or in a secondary manner by referring to papers which cite it as a reference. In a volume dedicated to the memory of George Simpson, this classic work will be cited many times. However much ideas of classification may have changed since 1945, everyone working on mammalian relationships owes Simpson a vote of thanks for that impressive study.

I thank the following people and institutions for opportunity to examine material: in London, Dr. A. Gentry and the late Dr. R. Hamilton; In Mainz, Prof. Dr. H. Tobien and Dr. D. Berg; in Munich, Prof. Dr. R. Dehm and Drs. V. Fahlbusch and K. Heissig; in Basel, Prof. Dr. J. Hürzeler and Dr. B. Engesser; in Paris, Dr. E. Heintz; in Florence, Prof. A. Azzaroli and Dr. M. M. Mazzarini; and in New York, Drs. M. C. McKenna and R. Tedford.

Many colleagues, particularly N. Solounias, have suffered through my thoughts about Miocene antelopes, and their patience is acknowledged. Drs. Heintz and Gentry have been kind enough to discuss their opinions on bovids, and Dr. McKenna and Dr. L. Krishtalka discussed cladistics with me. The drawings were made by Emmett Evanoff.

This study was supported by grants for field work from the Smithsonian Institution and by a grant for study in Germany from the Deutsche Akademische Austauschdienst (DAAD). Additional support has been provided by the University of Colorado and the Carnegie Museum of Natural History.

Figured specimens are part of the collections of the Geological Survey of Tunisia (Service Géologique de Tunisie).

REFERENCES CITED

Andree, J., 1926, Neue Cavicornier aus dem Pliocän von Samos: Palaeontographica, v. 67, p. 135-175.

Andrews, C. W., 1902, A Pliocene vertebrate fauna from the Wadi Natrun, Egypt: Geological Magazine, New Series, Decade 4, v. 9, p. 433-439.

―――― 1915, A description of the skull and skeleton of a peculiarly modified rupicaprine antelope (*Myotragus balearicus*, Bate), with a notice of a new variety, *M. balearicus* var. *major*: Royal Society, London, Philosophical Transactions, Ser. B., v. 206, p. 281-305.

Arambourg, C., 1949, *Numidocapra crassicornis* nov. gen., nov. sp., un Ovicapriné nouveau du Villafranchien constantinois: Comptes Rendus sommaires, Societé Géologique de France, v. 13, p. 290-291.

Berg, D., 1975, Miocäne Boviden (excl. Ovibovinen) aus der Türkei, *in* Sickenberg, O., (compiler) and others, 1975, Die Gliederung des höheren Jungtertiärs und Altquatärs in der Türkei nach Vertebraten und ihre Bedeutung für die internationale Neogen-Stratigraphie: Geologische Jahrbuch, Reihe B, Heft 15, p. 157-158.

Berggren, W. A., and Van Couvering, J. A., 1974, The Late Neogene-biostratigraphy, geochronology and paleoclimatology of the last 15 million years in marine and continental sequences: Palaeogeography, Palaeoclimatology, Palaeoecology, v. 16, p. 1-216.

―――― 1978, Biochronology, *in* Cohee, G. V., Glaessner, M. F., and Hedberg, H. D., eds., Contributions to the geologic time scale: American Association of Petroleum Geologists, Studies in Geology, no. 6, p. 39-55.

Biely, A., Rakus, M., Robinson, P., and Salaj, J., 1972, Essai de correlation des formations miocenes au Sud de la Dorsale tunisienne: Service Géologique de Tunisie, Notes, no. 38, p. 73-92.

Biju-Duval, B., Dercourt, J., and Le Pichon, X., 1977, From the Tethys Ocean to the Mediterranean Seas: a plate tectonic model of the evolution of the western Alpine System: Paris, International Symposium on the Structural History of the Mediterranean Basins, Editions Technip, p. 143-164.

Bohlin, B., 1935, Cavicornier der Hipparion-Fauna Nord-Chinas: Palaeontologica Sinica, Series C, v. 9, fasc. 4, p. 1-166.

Burollet, P. F., 1956, Contribution à l'étude stratigraphique de la Tunisie centrale: Annales des Mines et de la Géologie, Tunis, no. 18, p. 1-345.

Castany, G., 1951, Etude Géologique de L'Atlas tunisien orientale: *ibid.*, no. 8, p. 1-632.

Chen, G., and Wu, W., 1976, Miocene mammalian fossils of Jiulongkou, Ci Xian District, Hebei: Vertebrata PalAsiatica, v. 14, p. 6-15 (in Chinese, English translation prepared by S.-K. Wu).

Colbert, E. H., and Hooijer, D. A., 1953, Pleistocene mammals from the limestone fissures of Szechwan, China: American Museum of Natural History, Bulletin, v. 102, p. 1-134.

Crusafont-Pairo, M., 1973, Mammalia Tertiariae Hispaniae: Fossilium Catalogus I, Animalia, no. 121, p. 1-198.

Crusafont-Pairo, M., and Angel, B., 1966, Un *Myotragus* (mammifère ruminant), dans le Villafranchien de l'ile de Majorque: *Myotragus batei* nov. sp.: Academie des Sciences, Paris, Comptes Rendus, Series D., v. 262, p. 2012-2014.

Dehaut, E.-G., 1911, Animaux fossiles du Cap Figari, matériaux pour servir à l'histoire zoologique et paléontologique des iles de Corse et de Sardaigne: Paris, G. Steinheil, fasc. 3, p. 53-59.

de Mecquenem, R., 1924-25, Contribution à l'étude des fossiles de Maragha: Annales de Paléontologie, v. 13, p. 135-160, v. 14, p. 1-36.

de Villalta-Comella, J. F., and Crusafont-Pairo, M., 1956, Un nuevo ovicaprino en la fauna villafranquiense de Villaroyo (Logrono): Actes 4 Congres International du Quaternaire (INQUA), v. 1, p. 426-432.

Del Campana, D., 1918, Considerazioni sulle antilopi terziarie della Toscana: Palaeontographia Italica, v. 24, p. 147-233.

Depéret, C., 1887, Etudes paléontologiques dans le bassin du Rhône: Museum Sciences naturelles, Lyon, Archives, v. 4, p. 45-313.

Ennouchi, E., 1953, Un nouveau genre d'Ovicapriné dans un gisement Pléistocene de Rabat: Societé Géologique de France, Comptes Rendus Sommaires, v. 8, p. 126-128.

Frick, C., 1937, Horned ruminants of North America: American Museum of Natural History, Bulletin, v. 69, p. i-xxvii + 1-669.

Gaillard, C., 1902, Le Bélier de Mendés ou le mouton domestique de l'ancienne Egypte: Societé Anthropologique et Biologique de Lyon, Bulletin, v. 20, p. 69-102.

Gaudry, A., 1862-67, Animaux fossiles et Géologie de l' Attique: Paris, p. 1-476.

Gentry, A. W., 1970, The Bovidae (Mammalia) of the Fort Ternan fossil fauna, *in* Leakey, L. S. B., and Savage, R. J. G., eds., Fossil vertebrates of Africa, v. 2: New York, Academic Press, p. 243-323.

―――― 1978, Bovidae, *in* Maglio, V. J., and Cooke, H. B. S., eds., Evolution of African mammals: Cambridge, Harvard University Press, p. 540-572.

Gentry, A. W., and Gentry, A., 1978, Fossil Bovidae (Mammalia) of Olduvai Gorge, Tanzania: British Museum (Natural History), Bulletin, Geology, part 1, v. 29, p. 289-446; part 2, v. 30, p. 1-83.

Ginsburg, L., and Heintz, E., 1968, La plus ancienne antilope d'Europe, *Eotragus artenensis* du Burdigalien d'Artenay: Museum National d' Histoire Naturelle, Bulletin, Series 2, v. 40, p. 837-842.

Guérin, C., 1965, *Gallogoral* (nov. gen.) *meneghinii* (Rütimeyer, 1878) un Rupicapriné du Villafranchien d'Europe occidental: Documents, Laboratoire de Géologie, Faculté de Science, Lyon, no. 11, p. 1-353.

Hamilton, R., 1973, The lower Miocene ruminants of Gebel Zelten, Libya: British Museum (Natural History), Bulletin, Geology, v. 21, p. 75-150.

Harington, C. R., 1971, A Pleistocene mountain goat from British Columbia and comments on the dispersal history of *Oreamnos:* Canadian Journal of Earth Science, v. 8, p. 1081-1093.

Heintz, E., 1963, Complément d'étude sur *Oioceros atropatenes* (Rod. et Weith.), Antilope du Pontien de Maragha (Iran): Societé Géologique de France, Bulletin, Series 7, v. 5, p. 109-116.

Hürzeler, J., 1975, L'age géologique et les rapports géographiques de la faune de mammiferes du lignite de Grosseto (note preliminaire): Problemes actuels de paléontologie; evolution des vertébrés, Colloque International, Centre National de la Recherche Scientifique, no. 218, p. 873-876.

―――― 1983, Un alcélaphiné aberrant des "Lignites de Grosseto" en Toscane: Academie des Sciences, Paris, Comptes Rendus, Series III, v. 296, p. 243-249.

Hürzeler, J., and Engesser, B., 1976, Les faunes de mammifères néogenes du bassin de Baccinello (Grosseto, Italie): *ibid.,* Series D, p. 333-336.

Kotsakis, T., 1980, Osservazioni sui vertebrati quaternari della Sardegna: Societa Geologica Italiana, Bolletino, v. 99, p. 151-165.

Kotsakis, T., and Palombo, M. R., 1979, Vertebrati continentali e paleogeografia della Sardegna durante il Neogene: VIIth International Congress on Mediterranean Neogene, Annales Géologiques des Pays Helleniques, Tome Hors Serie, fasc. II, p. 621-630.

Kurtén, B., 1983, Variation and dynamics of a fossil antelope population: Paleobiology, v. 9, p. 62-69.

Lorenz, H. G., 1968, Stratigraphische und micropaläontologische Untersuchungen des Braunkohlengebietes von Baccinello (Provinz Grosseto-Italien): Rivista Italiana di Paleontologia e Stratigrafia, v. 74, p. 147-270.

Major, C. I. F., 1872, La Faune des Vertébrés de Monte Bamboli: Societa Italiana Scienze naturale, Atti, v. 15, p. 290-303.

―――― 1894, Le gisement ossifère de Mitylini et Catalogue d'ossements fossiles: Lausanne, p. 1-51.

Moya-Sola, S., and Pons-Moya, J., 1981, *Myotragus kopperi,* une nouvelle espèce de *Myotragus* Bate, 1909, (Mammalia, Artiodactyla, Rupicaprini): Koninklijke Nederlandse Akademie van Wetenschappen, Proceedings, Series B, v. 84, p. 57-69.

O'Gara, B. W., and Matson, G., 1975, Growth and casting of horns by prongbucks and exfoliation of horns by bovids: Journal of Mammalogy, v. 56, p. 829-846.

Pilgrim, G. E., 1934, Two new species of sheep-like antelopes from the Miocene of Mongolia: American Museum Novitates, no. 716, 29 p.

―――― 1937, Siwalik antelopes and oxen in the American Museum of Natural History: American Museum of Natural History, Bulletin, v. 72, p. 729-874.

―――― 1939, The fossil Bovidae of India: Palaeontologia Indica, new series, v. 26, p. 1-356.

Pilgrim, G. E., and Hopwood, A. T., 1928, Catalogue of the Pontian Bovidae of Europe in the Department of Geology, B. M. (N. H.): London British Museum (Natural History), *viii* + p. 1-106.

Pons-Moya, J., 1977, La Nouvelle espèce *Myotragus antiquus* de l'ile de Majorque (Baleares): Koninklijke Nederlandse Akademie van Wetenschappen, Proceedings, Series B, v. 80, p. 215-221.

Robinson, P., 1972, *Pachytragus solignaci,* a new species of caprine bovid from the Late Miocene Beglia Formation of Tunisia: Service Géologique de Tunisie, Notes, no. 37, p. 73-92.

―――― 1975, Neogene continental rock units of Tunisia: Bratislava, VI Congress Committee Mediterranean Neogene Stratigraphy, Proceedings, p. 415-419.

Robinson, P., and Black, C. C., 1969, Note préliminaire sur les Vertébrés fossiles du Vindobonien (Formation Béglia) du Bled Douarah, Gouvernorat de Gafsa, Tunisie: Service Géologique de Tunisie, Notes, no. 31, p. 67-70.

―――― 1974, Vertebrate faunas from the Neogene of Tunisia: Geological Survey of Egypt, Annals, v. 4, p. 319-332.

Robinson, P., and Wiman, S. K., 1976, A revision of the stratigraphic subdivision of the Miocene rocks of sub-dorsale Tunisia: Service Géologique de Tunisie, Notes, no. 46, p. 71-86.

Rodler, A., and Weithofer, K. A., 1890, Die Wiederkäuer der Fauna von Maragha: Akademie der Wissenschaften, Wien, Denkschrift, v. 57, p. 753-773.

Schaub, S., 1922, *Nemorhaedus philisi* nov. spec., ein fossiler Goral aus dem Oberpliocän der Auvergne: Eclogae geologicae Helveticae, v. 16, p. 558-563.

―――― 1944, Die oberpliocäne Säugetierfauna von Seneze (Haut- Loire) und ihre verbreitungsgeschichte Stellung: *ibid.,* v. 36, p. 270-289.

Schlosser, M., 1903, Fossilen Säugethiere Chinas nebst einer Odontographie der recenten Antilopen: Bayerisches Akademie der Wissenschaften, II Classe, Abhandlungen, v. 22, p. 1-221.

―――― 1904, Die fossilien Cavicornia von Samos: Beitráge Paläontologie und Geologie der Osterreich-Ungarns, Wein, v. 17, p. 21-118.

Sickenberg, O., Becker-Platen, J. D., Benda, L., Berg, D., Engesser, B., Gaziry, W., Heissig, K., Hünermann, K. A., Sondaar, P. Y., Schmidt-Kittler, N., Staesche, K., Staesche, U., Steffens, P., and Tobien, H., 1975, Die Gliederung des höheren Jungtertiärs und Altquartärs in der Türkei nach vertebraten und ihre Bedeutung für die internationale Neogen- Stratigraphie: Geologische Jahrbuch, Reihe B, Heft 15, p. 1-167.

Simpson, G. G., 1945, The principles of classification and a classification of the mammals: American Museum of Natural History, Bulletin, v. 85, p. *i-ix* + 1-350.

Skinner, M. F., Skinner, S. M., and Gooris, R. J., 1977, Stratigraphy and biostratigraphy of late Cenozoic deposits in central Sioux County, western Nebraska: American Museum of Natural History, Bulletin, v. 158, p. 263-370.

Solounias, N., 1981, The Turolian fauna from the island of Samos, Greece, with special emphasis on the hyaenids and the bovids: Basel, S. Karger, Contributions to Vertebrate Evolution, no. 6, p. 1-232.

Teilhard de Chardin, P., and Young, C. C., 1932, Fossil mammals from the late Cenozoic of northern China: Paleontologica Sinica, Series C, v. 9, p. 1-66.

Thomas, H., 1984a, Les Giraffoidea et les Bovidae miocenes de la formation Nyakach (Rift Nyanza, Kenya): Palaeontographica, Abteilung A, v. 183, p. 64-89.

―――― 1984b, Les origines africaines des Bovidae (Artiodactyla, Mammalia) miocenes des lignites de Grosseto (Toscane, Italie): Museum national d'Histoire naturelle, Bulletin, Series 4, v. 6, p. 81-101.

Vrba, E. S., 1979, Phylogenetic analysis and classification of fossil and recent Alcelaphini, Mammalia: Bovidae: Linnean Society, Biological Journal, v. 11, p. 207-228.

Wagner, A., 1848, Urweltliche Säugethier-Ueberreste aus Griechenland: Bayerische Akademie der Wissenschaften, Abhandlungen, v. 5, p. 333-378.

Weithofer, K. A., 1888, Alcune osservazioni sulla fauna delle ligniti di Casteani e di Montebamboli (Toscana): Bolletino del Reale Comitato Geologico d'Italia, v. 19, p. 363-368.

―――― 1889, Ueber die Tertiären Landsäugethiere Italiens: Jahrbuch der Kaiserliche und Königliche Reichsanstalt, v. 39, p. 55-82.

Wiman, S. K., 1978, Mio-Pliocene foraminiferal biostratigraphy and stratochronology of central and northeastern Tunisia: Revista Espanola de Micropaleontologia, v. 10, p. 87-143.

MANUSCRIPT RECEIVED SEPTEMBER 16, 1985
REVISED MANUSCRIPT RECEIVED MAY 23, 1986
MANUSCRIPT ACCEPTED JUNE 24, 1986

Faunal provinces and the Simpson Coefficient

JOHN J. FLYNN *Department of Geological Sciences, Rutgers University, New Brunswick, New Jersey 08903*

ABSTRACT

Biogeographers have used biotic resemblances and differences to distinguish discrete biogeographic areas. One of the most useful and commonly applied measures of biotic resemblance is a simple binary similarity coefficient developed by G. G. Simpson (1936). In the present study I apply the Simpson Coefficient of faunal similarity to three distributional data sets (at several taxonomic levels): (1) Recent North American mammals; (2) Recent global mammals; and (3) Early Eocene mammals of North America, Europe, and Asia. These analyses yield the following results.

Faunal realms of large geographic scale can be distinguished both by within-realm faunal similarities/between realm faunal dissimilarity, and by relatively sharp gradients of change of similarity across faunal realm boundaries.

Geographic distribution patterns of Recent North American mammals show several important trends: (1) an inverse relationship between faunal similarity and both longitudinal and latitudinal separation between sites; (2) a latitudinal asymmetry in similarity comparisons; (3) endemism of western North American faunas; and (4) differentiation of lower level faunal provinces is the result of complex interplay between major latitudinal climatic gradients and less influential (frequently longitudinal) regional climatic, tectonic, and geographic factors.

Significant differences in taxonomic diversity between localities makes the Simpson Coefficient a more useful measure of faunal resemblance than other binary similarity coefficients, both for Recent and fossil assemblages.

High similarities for Early Eocene faunal comparisons indicate presence of a single North American-European faunal realm; low resemblances of Asian Early Eocene faunas both to European and North American faunas indicate a distinct Asian faunal realm.

Available, but incomplete, evidence from faunal resemblances indicates that Ellesmere Island was part of a single, continuous European/North American faunal realm.

INTRODUCTION

Throughout his career, George Gaylord Simpson was concerned with measurement of faunal resemblance and recognition of effects of geography, climate, tectonics, ecology, and time on mammalian faunal composition. One of Simpson's most significant contributions to paleobiogeographic analysis was his (Simpson, 1943, 1947a, 1960) development and promotion of a simple statistic for measurement of faunal resemblance, now commonly called the Simpson Coefficient (e. g., Cheetham and Hazel, 1969). The Simpson Coefficient is $C/N_1 \times 100$, where C is the number of taxa (at a specified taxonomic level) in common between the two faunas or samples being compared, and N_1 is the total number of taxa (at the same taxonomic level) present in the *smaller* of the samples; the coefficient can be applied at any taxonomic level. In several papers, Simpson (1936, 1943, 1947a,b, 1953, 1960, 1965) applied this coefficient to analysis of faunal resemblance between geographic areas through time.

Subsequent workers (Hagmeier and Stults, 1964; Hagmeier, 1966; Cheetham and Hazel, 1969; Fallaw, 1979; Flessa and others, 1979) have used and tested the Simpson Coefficient both in theoretical and empirical studies. Simpson (1947a, 1960) and others (cited above) outlined several important properties of the coefficient. These include: (1) emphasis of similarities, rather than differences, between faunas; (2) less sensitivity to ecological differences between faunas; and (3) reduction of the significance of discrepancies in sample size (due to sampling biases or true diversity differences). Simpson (1947a) stressed that the smaller sample ideally should represent the true regional faunal diversity or, in the case of fossil faunas, must be from a similar facies.

Sclater (1858), Huxley (1868), Allen (1871), Wallace (1876), and other biogeographers have recognized a number of large geographic regions that are readily distinguishable by their distinctive mammalian (and other faunal/floral) assemblages. Boundaries between these major biogeographic regions may be sharp or gradational (transition zones; see Pielou, 1979; Simpson, 1961). Pielou (1979, p. 5) suggested that provinces are distinguished by their differences (most obvious across the boundaries). Smith (1960, p. 43) hoped it might be possible to establish ". . . a scale for measurement of faunistic relationship or distinction of two or more localities or areas on the basis of proportion of shared and excluded forms." Biogeographic relationships between faunal provinces, and historical/ecological effects on biogeographic patterns also have been considered by many other authors (e. g., Simpson, 1936, 1943, 1947a, 1953; Smith, 1960; Hagmeier and Stults, 1964; Hagmeier, 1966; Anderson, 1974, 1977; Fallaw, 1979; Flessa and others, 1979; McKenna, 1983a,b; Smith, 1983a,b).

Recently, vicariance biogeographers (*e. g.*, Croizat and others, 1974; Patterson, 1981; Cracraft, 1983) have argued that biogeographic history and relationships must be based on area cladograms derived from cladistic phylogenies. Yet, because of the small number of available comprehensive cladistic phylogenies, many studies (*e. g.*, Rosen, 1975, 1985) must still employ the more phenetic vicariance methodology, advocated by Croizat (1958, 1982), of compositing many individual distribution tracks. However, most of these studies do not consider the within province geographic distribution of the taxa under study (but see Cracraft, 1983); they assume the existence and identification of major biogeographic regions.

In this paper I apply the Simpson Coefficient of faunal similarity to three distributional data sets: (1) Recent North American mammalian species, genera, and families; (2) Recent global mammalian genera, families, and orders; and (3) Early Eocene mammalian species (North America), genera (North America, Europe, Asia), and families (North America, Europe, Asia). Results of these initial analyses are then used to address the following four questions.

1) How do we recognize large scale faunal provinces (past and present) on the basis of faunal similarity data, and what are the levels of faunal similarities both within and between these faunal provinces?

2) Simpson (1936, 1947a,b, 1953) emphasized comparisons of faunal similarity between areas at similar latitudes. Latitudinal diversity gradients yield large diversity differences between areas at different latitudes. What effect do latitudinal diversity gradients have on similarity comparisons between areas at similar longitudes but different latitudes? The Simpson Coefficient should be particularly useful for comparing these areas with large sample size (diversity) differences.

3) Are there advantages of using the Simpson Coefficient, relative to other similarity coefficients, for comparisons of both fossil and Recent faunas?

4) The Early Eocene was a time of high faunal resemblance between North America and Europe. How do Early Eocene intra- and inter-continental faunal similarities compare with corresponding Recent faunal similarities?

METHODS

Introduction

The first step in this analysis was to compile taxonomic distribution information for Recent and Early Eocene mammals, and to construct a series of faunal similarity matrices (at several taxonomic levels) using the Simpson Coefficient. A series of computer programs was created to construct these similarity matrices. Distribution maps, ranges, and taxonomy were derived from Hall (1981) and Nowak and Paradiso (1983) for Recent mammals, and from the sources listed in Table 1 and M. C. McKenna (*personal communication*, mammalian classification in preparation) for fossil mammals.

North American Distribution, Recent Mammals

The analysis for Recent North American mammals consisted of: (1) selection of a locality grid (Fig. 1) of 82 localities providing detailed geographic coverage of the North American continent (Nearctic, and northernmost Neotropical Regions of Pielou, 1979, Fig. 1.1; these localities provide a North American sampling grid of 5° latitude by 10° longitude—additional localities were added to sample areas of particular importance for Recent or fossil comparisons or to provide sampling of areas poorly covered by the main grid); (2) determination of individual species occurrences at each locality using the range maps in Hall (1981; a species was considered present at any locality that occurred within, or at the border of, the shaded geographic range for the entire species [composite of all subspecies] on those maps); (3) determination of species' status—species considered extinct, introduced, or of dubious status by Hall (1981) were not included in the analysis, nor were the aquatic Pinnipedia, Cetacea, and Sirenia; (4) tabulation of data—data on species occurrences were tabulated for each locality using a newly created computer program—species occurrence tabulations were done twice, once excluding occurrences of chiropteran species (as done by Hagmeier and Stults, 1964 and Hagmeier, 1966), and once including chiropterans; (5) determination of species similarities—computer programs were used to create a species similarity matrix, for a locality by locality comparison using the Simpson Coefficient; (6) generic and familial occurrences were tabulated for each locality by computer programs compositing occurrences of all individual species included within each genus and family by Hall (1981); and (7) step 5 (construction of a similarity matrix) was repeated for the generic and familial levels of analysis.

Step 1 yielded 374 species (536 species including Chiroptera) present in at least one of the 82 localities (210 species [245 species including Chiroptera] occur in North America but not at any grid locality; and 35 terrestrial species were not considered [7 introduced, 26 extinct, 1 of doubtful status, and *Homo sapiens*]). Step 6 yielded 127 genera (196 with Chiroptera) and 33 families (42 including Chiroptera) present in at least one of the 82 localities.

Complete matrices were constructed for the 82 localities; however, because of space limitations, abbreviated matrices of similarity comparisons for the North American Recent fauna are presented in Tables 3 and 4. These matrices include localities: (1) occurring on a grid of 70°, 80°, 90°, 100°, 110°, and 120° W longitude by 30°, 35°, 40°, 45°, 55°, and 65° N latitude; (2) in the Neotropics (localities 1-5, 7); (3) at high latitude (localities 71, 73, 80-82); (4) in western North America for Eocene comparisons (localities 16, 21, 31, 34-36, 43); and (5) in Asia and South America (localities 83, 84).

World Distribution, Recent Mammals

Analysis of distribution of Recent mammals on a global scale consisted of: (1) tabulation of generic, familial, and ordinal occurrences for each of seven ma-

TABLE 1. FOSSIL LOCALITIES.

	Locality	References
1)	Composite Clarkforkian L.M.A.	Savage and Russell, 1983
2)	Composite Graybull (Wasatchian L.M.A.)	"
3)	Composite Lysite (Wasatchian L.M.A.)	"
4)	Composite Lostcabin (Wasatchian L.M.A.)	"
5)	Baja California, Punta Prieta fauna	Flynn and Novacek, 1984; Novacek and others, in press
6)	San Juan Basin, New Mexico - Almagre	Lucas and others, 1981
7)	San Juan Basin, New Mexico - Largo	"
8)	Piceance Basin, Colorado - Early Wasatchian	Kihm, 1984
9)	Piceance Basin, Colorado - Late Wasatchian	"
10)	Piceance Basin, Colorado - Middle Wasatchian	"
11)	Four Mile Local Fauna, NW Colorado	McKenna, 1960; Delson, 1971
12)	Willwood Fm., Wyoming - No Water Fauna, Sand Creek Facies	Bown, 1979, 1980
13)	Willwood Fm., Wyoming - No Water Fauna, Elk Creek Facies	"
14)	Wind River Fm., Wyoming - Lysite	Guthrie, 1967
15)	Wind River Fm., Wyoming - Lost Cabin	Guthrie, 1971; Stucky, 1984
16)	Wind River Fm., Wyoming - *Lambdotherium* Zone	Stucky, 1984
17)	Wind River Fm., Wyoming - *Paleosyops borealis* Zone	"
18)	Willwood Fm., Wyoming - *Haplomylus* - *Ectocion* Zone	Schankler, 1980
19)	Willwood Fm., Wyoming - *Bunophorus* Zone	"
20)	Willwood Fm., Wyoming - *Heptodon* Zone	"
21)	Powder River Basin, Wyoming	Delson, 1971
22)	Eureka Sound Fm., Ellesmere Island, Canada	Hickey and others, 1983; M.C. McKenna, pers. comm.
23)	Composite Sparnacian	Savage and Russell, 1983
24)	Composite Cuisian	"
25)	Mutigny Local Fauna (Sparnacian), France	"
26)	Abbey Wood and Suffolk Pebble Beds (Sparnacian), England	Hooker, 1980; Hooker and others, 1980
27)	Blackheath Beds, England	Russell and others, 1982
28)	Suffolk Pebble Beds, England	"
29)	Dormaal, France	Godinot and others, 1978
30)	Dormaal = Upper Landenian, NW Europe	Russell and others, 1982
31)	Sables a Unios et Teredines, France	"
32)	Argile a lignites and Argiles Plastique, France	"
33)	Rians, France	Godinot 1978, 1981
34)	Early Eocene of Asia (Composite)	Savage and Russell, 1983; Li and Ting, 1983

jor zoogeographic regions inhabited by mammals, using the generalized distribution compiled in Nowak and Paradiso (1983, p. *xix-xliv*; any taxon noted as occurring within some part of the region was considered present throughout the region); (2) occurrence tabulations for each locality and each taxonomic level were done twice, both excluding and including the volant (chiropteran/dermopteran) taxa (non-terrestrial and extinct taxa were excluded from all analyses); and (3) the computer programs mentioned above were used to create Simpson Coefficient similarity matrices for the regions at the generic, familial, and ordinal levels.

Holarctic and North America Distributions, Early Eocene Mammals

Determination of taxonomic distributions for Early Eocene fossil mammals was slightly more complicated than the above analyses. Information from a wide variety of primary and secondary sources had to be compiled (see Table 1). Although geographic coverage of available fossiliferous localities was less complete than for other analyses, a broad latitudinal distribution of localities is available within North America (see Fig. 2). Comparisons between major Holarctic faunal regions (North America, Europe, Asia) provide broad longitudinal sampling of Early Eocene faunas. The procedure for analysis of Early Eocene mammalian distributions consisted of: (1) selection of 34 "localities" or samples (Table 1), consisting of 7 composite faunas (representing broad geographic and temporal sampling) and 27 individual faunas occurring at different places, or at different stratigraphic/temporal horizons within a sequence at one place; (2) tabulation of species within each of the 22 North American Early Eocene localities/samples based on data given in sources listed in Table 1; (3) tabulation of generic and familial occurrences for each of the 34 Holarctic Early Eocene localities/samples listed in Table 1 (generic, familial, and ordinal taxonomy follows the classification in prepara-

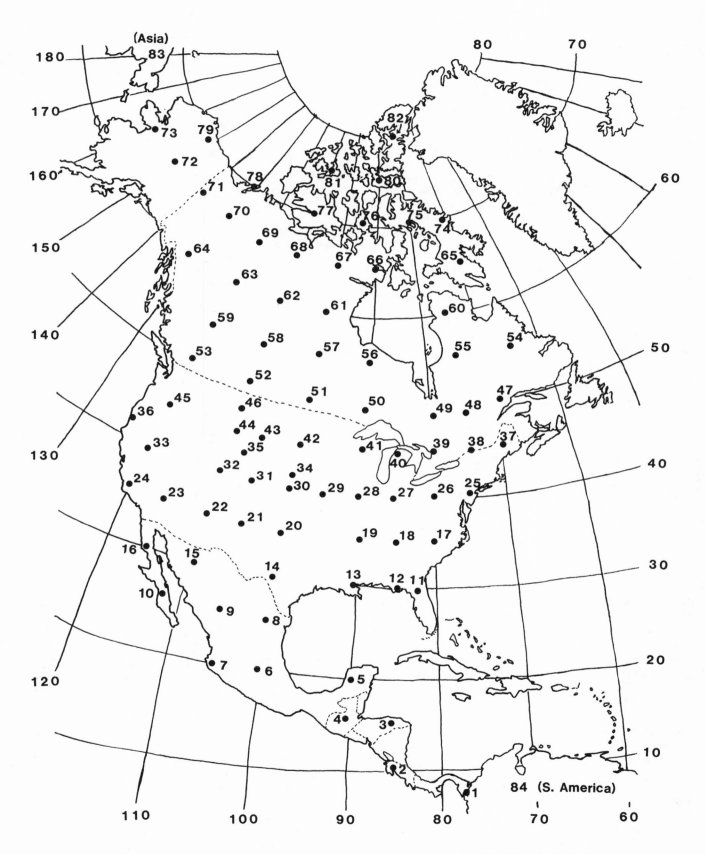

Figure 1. Geographic locations of Recent North American localities (map modified from Hall, 1981).

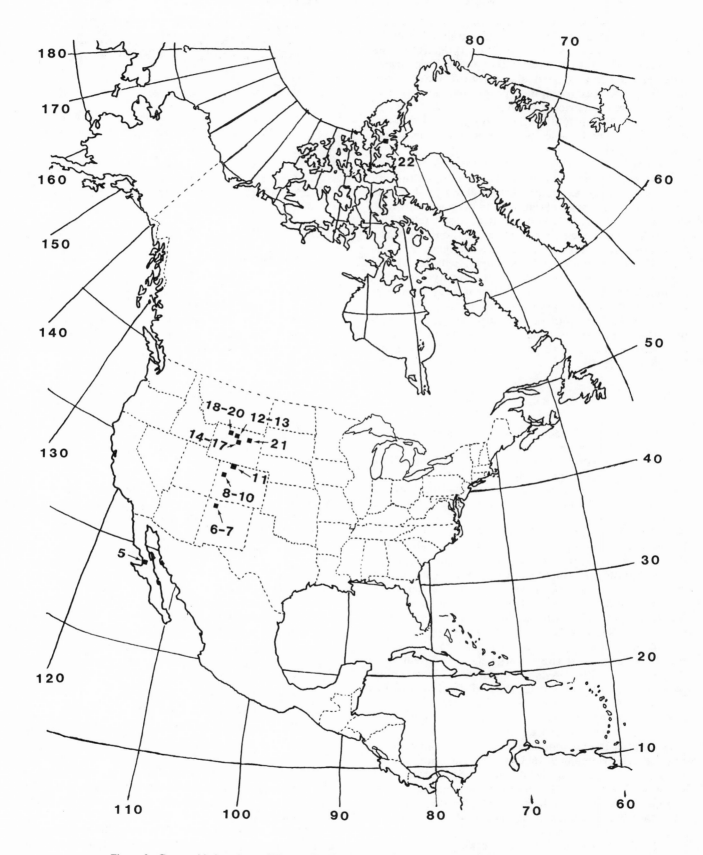

Figure 2. Geographic locations of Eocene North American localities (map modified from Hall, 1981).

tion by M. C. McKenna [American Museum of Natural History] to standardize the widely disparate taxonomies used in the studies cited in Table 1); (4) determination of taxonomic status (taxa referred to as "cf." or "near" were assigned to that particular taxon); (5) creation of similarity matrices (computer programs were used to create similarity matrices of Simpson Coefficients for 20 of 22 North American localities at the species level, and for all 34 Early Eocene localities at the generic and familial levels).

Fossil faunal samples probably are less complete than Recent samples. This is true for several reasons, including incomplete sampling of skeletal morphology and individual morphologic variation, and because many non-skeletal characters used to differentiate closely related living species cannot be detected in fossils. Sampling biases (*e. g.*, size sorting, limited collecting techniques, restricted facies/habitats sampled in preserved assemblages) also may yield underestimation of species diversity. However, balanced sampling for taxa of different body sizes, time averaging of faunal composition in some sequences, and possible preservation of taphonomic samples representing broad composites from several different habitats/facies will tend to have the opposite effect (Novacek and Lillegraven, 1979).

RESULTS

World, Recent

Simpson Coefficient similarity matrices for World distributions of Recent mammals are given in Table 2.

In generic comparisons (without Chiroptera/Dermoptera), both Australia and South America (except South America/North America) have low similarities to all other areas. In comparison, Europe/Asia and Asia/SE Asia Islands similarities are high. Similarity of the North American fauna to faunas from other areas varies widely. Inclusion of chiropterans/dermopterans yields a slight similarity decrease in only three of 21 comparisons; remaining similarity values increase. The most marked increase is in comparison of Australia to *all* other areas.

In familial comparisons (without Chiroptera/Dermoptera), Australia still has low similarities to other regions, but South American faunas show large increases in similarity (relative to generic levels) to all regions. North America also shows large increases relative to Africa and Southeast Asian Islands. As in generic comparisons, familial similarities are high for Europe/Asia and Asia/SE Asian Islands, but the North American/South American faunas also are high in similarity. All other familial similarities are much lower. North American familial faunal similarities are more homogeneous than the generic values. Inclusion of chiropterans/dermopterans in familial level comparisons yields nine decreases and 11 increases in similarities, but most of the changes are small. As in generic comparisons, inclusion of volant mammals results in increased similarities between Australia and six of the other regions. Three of the remaining five increases are between the southeast Asian Islands and nearby regions (Africa, Asia, Europe).

TABLE 2. SIMPSON COEFFICIENTS OF SIMILARITY; RECENT, WORLD (GENUS, FAMILY, ORDER)[1].

A) GENERA

	1	2	3	4	5	6	7
1	--	44	8	40	23	8	0
2	57	--	2	7	4	5	0
3	8	3	--	40	24	16	0
4	41	10	45	--	94	20	0
5	18	4	29	95	--	73	2
6	8	5	20	24	75	--	5
7	6	4	12	8	16	20	--

B) FAMILIES

	1	2	3	4	5	6	7
1	--	81	55	65	48	40	6
2	80	--	46	45	35	40	6
3	58	48	--	60	53	32	19
4	64	41	59	--	100	65	6
5	43	30	52	100	--	84	6
6	35	32	35	68	88	--	19
7	10	10	19	10	29	38	--

C) ORDERS

	1	2	3	4	5	6	7
1	--	100	75	100	71	86	67
2	100	--	86	100	71	86	67
3	78	88	--	100	100	86	67
4	100	100	100	--	100	100	33
5	67	75	89	100	--	86	33
6	88	88	88	100	88	--	67
7	67	67	67	33	.33	67	--

[1] Including (lower half) and excluding (upper half) chiropterans/dermopterans. Locality numbers on margins of matrices refer to: 1) N. America, 2) S. America, 3) Africa, 4) Europe, 5) Asia, 6) SE Asia Islands, 7) Australia

Again, the only large changes resulting from inclusion of chiropterans/dermopterans are Australian, between Australia and Asia and between Australia and the Southeast Asian Islands.

In the ordinal matrix (without chiropterans/dermopterans), similarities between most regions are high, except for Australia (with only three orders present). In-

TABLE 3. SIMPSON COEFFICIENTS OF SIMILARITY; RECENT, NORTH AMERICA, SPECIES (UPPER HALF); EOCENE, NORTH AMERICA, SPECIES (LOWER HALF).
LOCALITY NUMBERS ON MARGINS OF MATRICES, DOUBLE ASTERISK (**) DENOTES SIMILARITY OF 100.

LOC	1	2	3	4	5	6	7	11	13	14	15	16	17	19	20	21	22	24	26	28	30	31	32	33	34	35	36	37	39	41	42	43	44	45	55	56	57	58	59	65	66	67	68	69	71	73	80	81	82	84	83
1	--	64	74	45	50	10	26	14	19	18	20	5	12	13	10	10	9	7	8	10	9	6	5	3	9	6	6	9	6	7	9	6	5	7	0	0	0	7	6	0	0	0	0	0	0	0	0	0	0	98	0
2		--	88	75	74	23	41	31	39	31	28	8	30	31	23	20	13	13	16	21	20	14	9	7	18	8	12	13	13	15	15	9	9	13	0	3	3	12	8	0	0	0	0	0	3	3	0	0	0	64	0
3			--	79	62	21	41	23	35	29	26	3	21	21	19	21	12	10	12	12	47	12	8	8	12	7	10	10	10	12	10	10	10	12	0	3	3	10	7	0	0	0	0	0	0	0	0	0	0	76	0
4				--	79	39	49	34	39	35	39	11	30	31	25	23	18	18	18	21	20	17	11	9	20	10	14	13	13	13	15	11	10	14	0	3	3	12	8	0	0	0	0	0	3	3	0	0	0	46	0
5					--	17	38	20	23	24	21	5	18	18	19	17	14	12	14	12	17	12	10	10	14	7	12	10	10	12	12	12	10	12	0	3	3	7	5	0	0	0	0	0	0	0	0	0	0	50	0
6						--	47	37	42	47	46	30	33	36	42	57	39	33	25	33	41	45	33	29	36	18	27	20	21	25	30	27	23	31	4	9	8	20	12	13	6	6	4	7	8	6	10	11	10	10	0
7							--	39	39	43	46	19	30	33	31	33	28	24	23	27	26	26	22	16	24	14	20	17	17	20	21	46	46	20	0	6	5	15	10	0	0	0	0	3	5	3	0	0	0	26	0
11								--	85	54	29	11	79	83	49	43	23	23	63	57	60	43	31	29	46	29	31	34	37	49	43	34	31	37	12	15	14	26	23	0	0	0	9	10	9	9	0	0	0	14	0
13									--	65	35	8	81	96	58	58	31	23	69	62	65	54	39	35	42	39	35	42	50	54	50	46	42	46	19	23	23	39	35	0	0	0	9	19	19	15	0	0	0	19	0
14										--	61	32	55	56	67	67	47	36	45	48	63	57	41	33	53	35	33	33	32	37	49	45	37	41	23	24	22	29	25	25	12	13	17	19	18	15	20	11	20	18	8
15											--	49	30	28	50	63	61	49	28	33	44	54	46	39	36	35	37	24	24	30	37	37	39	44	12	15	19	29	24	13	12	13	13	13	15	12	10	11	10	20	8
16												--	9	14	35	49	54	70	14	49	38	43	43	54	32	32	41	14	11	16	32	35	32	35	0	3	5	14	14	0	6	6	4	7	8	6	0	0	0	5	8
17													--	94	55	49	33	27	79	79	73	58	39	30	55	33	36	55	61	73	61	52	42	46	27	30	33	39	36	13	12	13	47	23	24	48	10	0	10	12	8
19														--	56	51	33	28	74	74	69	51	41	33	54	36	36	51	56	67	54	46	41	46	27	29	30	39	39	13	6	6	13	23	23	18	10	0	10	13	8
20															--	88	58	42	44	60	88	79	54	44	78	48	38	33	36	48	64	65	48	52	23	27	30	42	31	13	12	13	22	26	23	18	10	11	10	10	8
21																--	76	56	43	57	76	79	59	44	80	57	46	37	38	43	68	64	46	49	27	29	32	44	35	13	12	13	26	29	28	21	10	11	10	10	8
22																	--	53	31	41	53	73	73	58	58	55	46	30	30	30	53	56	51	56	23	27	32	37	35	13	12	13	22	26	28	21	10	11	10	9	8
24																		--	24	31	44	49	47	53	38	36	53	20	20	27	38	40	38	44	4	12	46	24	20	0	6	6	4	10	10	6	0	0	0	7	8
26																			--	81	61	61	49	39	56	41	37	80	83	82	62	55	55	49	69	68	65	66	53	38	35	37	48	55	48	39	30	22	30	8	25
28																				--	81	71	50	43	67	43	41	62	69	91	76	64	55	52	46	47	46	54	48	38	29	31	35	39	38	30	30	22	30	10	17
30																					--	86	55	47	96	53	40	48	51	59	89	73	55	55	42	44	46	54	41	25	35	38	39	42	38	33	20	11	20	9	33
31																						--	82	62	87	92	56	61	60	69	89	95	83	76	69	71	73	78	74	50	41	44	57	58	58	52	10	33	40	6	33
32																							--	74	62	82	52	48	47	46	66	75	77	76	54	56	60	66	61	38	24	25	39	48	48	39	30	22	30	5	17
33																								--	56	73	60	46	43	43	55	64	71	83	46	56	57	63	61	50	29	31	44	45	45	39	40	33	40	3	25
34																									--	58	42	44	47	58	78	73	56	64	35	35	38	44	40	25	24	25	30	36	30	30	20	11	20	9	25
35																										--	40	46	45	47	62	80	90	77	50	53	57	66	61	25	18	19	39	45	45	39	20	22	20	6	25
36																											--	39	38	40	40	42	52	60	31	38	43	46	43	25	18	19	30	36	35	27	20	22	20	6	17
37																												--	89	87	52	54	63	50	85	85	78	76	67	63	41	44	61	71	63	55	50	44	50	9	33
39																													--	89	55	55	62	50	85	85	81	78	68	63	47	50	65	74	65	58	50	44	50	6	50
41																														--	68	57	61	50	81	88	84	85	67	50	41	44	61	71	65	55	40	33	40	7	50
42																															--	87	68	62	50	50	51	61	47	38	41	44	44	48	43	39	30	22	30	9	33
43																																--	80	69	58	59	60	68	57	50	44	44	48	52	50	42	40	33	40	6	33
44																																	--	78	69	77	78	85	80	50	41	44	61	65	60	55	40	33	40	5	42
45																																		--	46	56	60	68	65	38	24	25	30	42	43	33	30	22	30	7	25
55																																			--	92	92	81	89	75	59	63	70	85	81	69	70	67	70	0	42
56																																				--	**	91	91	75	53	56	70	81	79	61	60	56	60	0	58
57																																					--	92	92	75	59	63	74	84	78	67	60	56	60	0	67
58																																						--	90	38	41	44	61	74	70	61	30	33	30	7	50
59																																							--	63	53	56	70	84	80	70	50	44	50	6	58
65																																								--	**	**	**	88	75	**	88	75	88	0	38
66																																									--	**	94	82	71	88	**	**	**	0	67
67																																										--	**	88	75	88	90	89	90	0	67
68																																											--	91	83	87	90	89	90	0	75
69																																												--	94	77	70	67	70	0	83
71																																													--	88	50	44	50	0	92
73																																														--	80	78	80	0	**
80																																															--	**	**	0	30
81																																																--	**	0	22
82																																																	--	0	30
84																																																		--	0
83																																																			--

LOC	1	2	3	4	6	7	8	9	10	11	12	13	14	15	16	17	18	19	20	21
1	--																			
2	40	--																		
3	7	43	--																	
4	10	27	49	--																
6	11	41	37	44	--															
7	6	34	40	40	83	--														
8	25	67	44	36	39	20	--													
9	12	44	56	52	42	40	47	--												
10	6	32	35	68	26	16	13	48	--											
11	31	67	26	26	38	17	44	26	10	--										
12	37	68	21	24	22	20	50	31	6	50	--									
13	35	50	27	21	33	34	58	38	23	57	84	--								
14	9	42	75	49	33	31	36	42	39	24	19	28	--							
15	8	24	41	61	35	29	22	37	65	24	12	18	53	--						
16	12	20	25	51	22	14	11	17	35	10	12	18	27	80	--					
17	7	15	20	41	15	11	8	11	23	7	7	11	22	76	61	--				
18	26	68	26	25	32	23	53	31	23	55	57	60	33	21	18	17	--			
19	14	55	39	45	29	23	39	27	23	29	35	51	37	37	12	13	71	--		
20	12	43	41	45	33	31	36	35	58	36	28	40	53	40	29	20	51	84	--	
21	35	77	32	32	45	19	48	35	10	90	61	77	26	29	13	6	68	42	42	--

TABLE 4.
RECENT NORTH AMERICA– EAST-WEST COMPARISONS ALONG LATITUDES, AND WESTERN ENDEMISM.

Species–

	Locality	Comparative Localities/Similarities								
		15	14	13	12	11				
1) 30°N: W to E	16	49	32	8	13	11				
		12	13	14	15	16				
E to W	11	100	85	54	29	11				
		33	32	31	30	29	28	27	26	25
2) 40°N: W to E	36	39	60	52	56	40	41	38	37	33
		25	27	28	29	30	31	32	33	36
E to W	25	87	82	74	69	58	60	44	36	33

Genus–

	Locality	15	14	13	12	11				
1) 30°N: W to E	16	96	79	41	46	50				
		12	13	14	15	16				
E to W	11	100	91	67	57	50				
		33	32	31	30	29	28	27	26	25
2) 40°N: W to E	36	78	84	86	68	65	68	69	76	68
		25	27	28	29	30	31	32	33	36
E to W	25	92	92	85	81	73	76	70	65	68

clusion of chiropterans/dermopterans results in one large and one small decrease, and seven small increases in area similarities. Similarities of Australia and Europe to all other areas remain unchanged by inclusion of volant mammals. Most regions show large increases in similarities at the ordinal level, relative to the familial level.

North America, Recent

Species

At the species level (Table 3), similarities between North American localities range from 0 to 100, and several trends can be observed in the data. Low latitude localities lying within the Neotropical Realm (Pielou, 1979; Hall, 1981; localities 1-5 and 7—although localities 6 and 8 have been assigned to the Neotropical Realm by other workers using different provincial boundaries) have similarities less than, or equal to, 20 when compared to all localities at latitudes above 35°. These Neotropical localities have high similarities (40-88) when compared to the nearest localities, and similarity between localities decreases rapidly over short distances (less than 15° latitude), particularly when compared to nearby localities in the Nearctic Province. Localities 1-5 form a distinct grouping on the basis of their high similarities to one another, and distinctly lower similarities to other localities. Further, these are the only North American localities that have high similarities to South American Neotropical faunas (localities 46-98). The affinities of locality 7 are equivocal; its similarities to all other

TABLE 5.
RECENT NORTH AMERICA— NORTH-SOUTH COMPARISONS ALONG LONGITUDES.

Species—

Locality	Comparative Localities/Similarities										
	4	5	13	19	28	41	50	56	66	80	82
1) 90°W 2	75	74	39	31	21	15	10	4	0	0	0
	22	32	44	52	58	62	68	77	81	82	
2) 110°W 15	61	46	39	33	29	18	13	10	11	10	

Genus—

Locality											
	4	5	13	19	28	41	50	56	66	80	82
1) 90°W 2	87	88	59	48	38	27	21	17	13	20	20
	22	32	44	52	58	62	68	77	81	82	
2) 110°W 15	91	82	65	68	47	37	35		44	40	

localities are relatively low (values of Simpson Coefficient maximum of 55). However, the relatively high similarities between the definite Neotropical localities and locality 7, and the much lower similarities between the Neotropical localities and all other localities indicate a Neotropical affinity for locality 7. Low latitude localities 6 and 8-10 all exhibit much higher similarities to Nearctic than to Neotropical localities.

There are also geographic trends in data for Recent species. Along any latitudinal transect, for example, similarity is directly proportional to longitudinal separation (Table 4). However, similarities decrease much more strongly with increasing longitudinal separation along low latitude transects than along high latitude transects. At all latitudes there is an asymmetry in similarity comparisons along lines of latitude; western North American faunas are more endemic. This is indicated by the relatively low similarities of far western faunas to both nearby and distant faunas, as supported by the asymmetry of latitudinal similarity values when comparisons are from W to E rather than E to W (see Table 4).

Comparisons along lines of longitude show a regular decrease in similarity with increasing latitudinal separation (Table 5). Further, for any given areal distance separation, similarities between localities at the same latitude are much greater than between localities at the same longitude. In comparisons of Nearctic localities at different latitudes: (1) high latitude faunas have high similarities to other high latitude faunas, moderate similarities to middle latitude faunas, and relatively low similarities (values of Simpson Coefficient <20) to low latitude Nearctic faunas; (2) middle latitude faunas have high similarities to other middle latitude faunas, moderate similarities to high latitude faunas, and relatively low similarities to low latitude faunas; and (3) low latitude faunas have moderate similarities (Simpson Coefficient 40-70) to other low latitude faunas, relatively low similarities (Simpson Coefficient 30-40) to middle latitude faunas, and low similarities (Simpson Coefficient <30) to high latitude faunas. Low latitude Neotropical localities have relatively low similarities (Simpson Coefficient 20-40) to low latitude Nearctic faunas, low similarities (Simpson Coefficient 5-20) to middle latitude Nearctic faunas, and very low similarities to high latitude Nearctic faunas. There is a large asymmetry of similarity at low latitudes; for any one low latitude locality, similarity comparisons across equal latitudinal distances yield much lower similarity values towards lower latitude localities than towards higher latitude localities.

Addition of chiropterans to this comparison reveals similar trends.

Genus

Similarity values at the generic level between North American localities range from 7 to 100 (Table 6). If one excludes the Neotropical localities (1-5 and 84), all similarities among Nearctic localities are greater than 20, and most are greater than 30. Results for the generic matrix are similar to results for the species matrix, although similarities for the generic matrix are higher.

Faunas of localities 1, 2, 4, and 5 are similar to one another (values of Simpson Coefficients 87-94) and to the Neotropical South America fauna (Simpson Coefficients 66-82). Each of these localities also is more similar to

TABLE 6. SIMPSON COEFFICIENTS OF SIMILARITY; RECENT, NORTH AMERICA, GENUS (UPPER HALF) AND FAMILY (LOWER HALF). LOCALITY NUMBERS ON MARGINS OF MATRICES, DOUBLE ASTERISK (**) DENOTES SIMILARITY OF 100.

locality 7 than to any other locality outside the Neotropical group (by at least 5-13 Simpson Coefficient units). However, the greater similarity between locality 7 and the Neotropical localities (relative to Nearctic localities) is not as pronounced in the generic as in the specific matrix.

In comparisons along lines of latitude (Table 4), similarities generally decrease with increasing longitudinal separation, although this trend is not as regular or as pronounced as in the species similarity matrix. As in the species similarity matrix, this pattern is much more pronounced at lower latitudes; generic similarities between localities along lines of latitude increase with increasing latitude (Table 5). Endemism of western localities, noted in the species similarity matrix, is present in the generic matrix, but it is not as strong.

For a given areal separation, localities along the same latitudinal transect have much greater similarities than localities along the same longitudinal transect. Comparisons of Nearctic localities yield: (1) high similarities (values of Simpson Coefficient >80) among faunas of high latitudes, those of middle latitudes, and those of low latitudes; (2) high similarities (Simpson Coefficient 35-80) in comparisons between high and middle latitudes, and those of middle and low latitudes; and (3) moderate similarities (Simpson Coefficient 20-40) between those of high and low latitudes. Comparisons of Neotropical North American localities (1-5) to Nearctic localities yield similarities that are moderate to high (Simpson Coefficient 20-70) for Nearctic localities of low latitudes, moderate (Simpson Coefficient 15-50) for those of middle latitudes, and low (Simpson Coefficient 7-31) for those of high latitudes.

The longitudinal asymmetry in similarity values (higher similarities, across equal latitudinal distances, toward higher latitude localities than toward lower latitude localities) noted for low latitude localities in the specific comparisons is also present in the generic matrix. This asymmetry of similarity also may be seen for some middle latitude localities.

Family

Comparisons of North American localities at the familial level yield similarities in terms of Simpson Coefficients of 40-100; exclusion of Neotropical localities yields similarities among Nearctic localities of 57-100 (only 2 comparisons have similarities less than 60). More than one in three of the familial similarity comparisons in Table 6 are 100.

There is no consistent relationship between similarities and longitudinal separation along latitudinal transects, in contrast to the patterns exhibited in the specific and generic matrices. There are slight increases in similarities (along latitudinal transects) with increasing latitude. The endemism of western faunas, noted in the specific and generic matrices, is weak or absent at the familial level.

Although increasing latitudinal separation of localities along longitudinal transects tends to yield decreases in similarities, this does not occur in a regular pattern. Comparisons of low latitude faunas (particularly Neotropical localities) to other faunas appear to show the most regular decreases in similarities with increasing latitudinal separation.

The asymmetry in similarity values across equal latitudinal distances noted for the specific and generic matrices also is pronounced in the low and middle latitude familial comparisons. Similarities of low and middle latitude localities to the low diversity high latitude localities are high, whereas comparisons of these low and middle latitude localities to lower latitude localities yield much lower values of similarity.

Holarctic, Early Eocene

Species

Simpson Coefficients of similarity for Early Eocene faunas (of North America only, as few, if any, species occurred in both Europe and North America during the Eocene) range from 6 to 90 (Table 3). Most of the highest similarities are between stratigraphically successive samples from the same geographic location, or between samples from different areas that are similar in age. Approximately one in four of the comparisons between the Wasatchian localities (Table 3, excluding the composite Clarkforkian) are greater than 50 in terms of the Simpson Coefficient. Similarity comparisons of the composite Clarkforkian versus other localities range from 6-40, while 11 of 19 comparisons are less than 15. The Clarkforkian fauna has the highest similarities to faunas that generally are considered to be very early in the Wasatchian (*e. g.*, localities 11, Four Mile; 21, Powder River; 8, basal Piceance; 12 and 18, lower Willwood; 2, composite "Greybullian"). Locality 17 (upper Wind River Formation, Stucky (1984) has very low similarities (most less than 25) to all other localities except those of the immediately underlying Wind River Formation (localities 15 and 16). Comparisons of localities arranged in a general north-south transect (Table 7; New Mexico, localities 6-7; Piceance, Colorado, 8-10; Four Mile, Colorado, 11, Wind River Formation, Wyoming, 14-17; Willwood Formation, Wyoming, 12-13 and 18-20; and Powder River Basin, Wyoming, 21) yield geographic trends similar to those found in the Recent species data for a similar geographic transect. However, the latitudinal span for this Early Eocene north-south transect is relatively limited, and the species similarities are somewhat variable both in the Recent and fossil transects.

Genus

Similarity comparisons (via the Simpson Coefficient) of North American, European, and Asian Early Eocene faunas at the generic level range from 0 to 100, while exclusion of the Asian Early Eocene yield similarities of 13-100 (Table 8).

Similarity comparisons involving the Asian composite Early Eocene fauna (locality 34) range from 0-27, and only four of 33 comparisons have Simpson Coefficient values greater than 20. Comparisons of the Asian composite to the North American composite faunas yield

TABLE 7.
NORTH-SOUTH TRANSECT COMPARISONS- RECENT/EARLY EOCENE (1).

Species-

Comparative Localities/Similarities

```
                Loc.  (12 13 18 19 20)  (14 15 16 17)  11  ( 8  9 10)  ( 6  7)
1)N-S:Eocene    21    61 77 68 42 42    26 29 13  6    90   48 35 10    45 19
                                           35            31      32       19
   Recent       43          80               95          75              64

                      21  ( 8  9 10) 11  (14 15 16 17)  (12 13 18 19 20) 21
2)S-N:Eocene    6     83   39 42 26  38   33 35 22 15    22 33 32 29 33  45
                           32        31                  35               36
   Recent       21    59             79                  57              64
```

Genus-

```
                Loc. 21 (12 13 18 19 20)  (14 15 16 17)  11  ( 8  9 10)  ( 6  7)  5
1)N-S:Eocene    22   38  50 50 50 50 63    50 75 63 63   50   50 50 50    38 25  13
                         43                   35             31    32       19    14
   Recent       82   60                       50             60    60       50    40

                     ( 6  7) ( 8  9 10) 11  (14 15 16 17)  (12 13 18 19 20) 21 22
2)S-N:Eocene    5     73 73   64 73 64  64   64 64 73 55 45  64 73 64 73 64  55 13
   (a)                   21        32   31                   35              36 76
   Recent       16    92          96     88                  75              83 40
   (a)

                     5   7 ( 8  9 10) 11  (14 15 16 17)  (12 13 18 19 20) 21 22
   Eocene       6    73  93  76 70 64 64   64 64 62 51 42  67 82 78 76 73  74 38
   (b)                  16        32   31                  35              36 76
   Recent       21   92         84     86                  84              85 50
   (b)
```

[1] Approximate latitudes for Eocene localities: 1)22, 78°N; 2)21, 45°N;
3)12-20, 43°N; 4)8-11, 39-40°N; 5)6-7, 36°N; 6)5, 29°N; localities
from the same basin sequence are placed together, in approximate superposition,
in parentheses

TABLE 8. SIMPSON COEFFICIENTS OF SIMILARITY; EOCENE, NORTH AMERICA, EUROPE, AND ASIA; GENUS (UPPER HALF) AND FAMILY (LOWER HALF); LOCALITY NUMBERS ON MARGINS OF MATRICES, DOUBLE ASTERISK (**) DENOTES SIMILARITY OF 100.

LOC	1	2	3	4	5	6	7	8	9	10	11	12	13	14	15	16	17	18	19	20	21	22	23	24	25	26	27	28	29	30	31	32	33	34	LOC
1	--	65	44	31	64	56	52	55	48	43	58	60	65	45	29	32	18	59	59	47	65	50	32	25	33	38	41	33	32	28	24	36	35	8	1
2	85	--	75	59	82	82	78	90	83	71	80	85	81	82	52	55	40	89	91	77	87	50	50	43	53	66	82	50	45	48	42	55	65	13	2
3	74	83	--	75	73	80	85	76	80	86	69	62	71	98	58	68	48	66	85	83	68	50	42	43	56	59	76	50	41	38	44	51	60	13	3
4	74	84	94	--	83	78	89	76	83	93	58	55	60	86	78	72	66	57	88	81	61	75	37	40	47	55	71	50	36	34	38	43	50	18	4
5	75	83	92	92	--	73	73	64	73	64	64	64	73	64	73	55	45	64	73	64	55	13	64	45	36	55	45	18	27	36	36	45	27	27	5
6	79	96	**	**	92	--	73	76	64	64	64	67	82	64	62	51	42	78	76	73	74	38	49	42	44	48	65	50	36	34	38	44	55	11	6
7	78	94	**	**	75	--	--	52	--	83	64	70	56	74	70	52	52	70	78	70	44	25	44	37	30	37	47	33	18	22	30	37	30	11	7
8	68	91	**	**	83	91	78	--	83	64	83	79	90	69	72	59	48	83	66	72	62	50	55	41	34	38	47	33	27	24	41	55	35	14	8
9	73	92	**	**	92	92	89	95	--	82	60	65	67	85	59	50	43	73	70	78	44	50	48	40	36	48	65	50	18	31	38	45	40	13	9
10	68	86	**	**	75	91	83	86	91	--	57	--	76	80	53	47	36	68	65	62	46	50	50	39	29	43	53	42	18	36	36	46	35	18	10
11	82	96	93	93	83	86	78	86	85	91	--	--	76	50	47	47	32	82	71	60	84	50	51	42	50	48	65	42	36	38	40	49	60	11	11
12	83	**	93	93	83	89	83	86	85	96	--	--	91	57	47	36	32	82	71	60	84	50	45	38	47	59	76	50	41	38	36	47	60	8	12
13	82	97	89	91	83	96	94	91	96	91	91	--	--	84	55	46	46	92	94	83	71	37	47	42	50	59	71	50	41	41	36	53	70	16	13
14	75	89	**	**	75	82	89	91	85	95	**	97	94	--	--	74	70	51	74	64	61	75	30	32	36	48	65	50	36	48	42	41	50	13	14
15	71	81	92	94	92	82	89	91	**	75	89	90	93	**	--	74	--	70	55	59	60	42	63	45	50	38	47	42	36	34	34	38	45	21	15
16	67	83	90	93	86	79	83	86	91	79	87	84	82	93	85	--	87	--	42	53	44	35	45	32	30	39	41	42	36	34	38	43	55	13	16
17	65	81	90	94	83	82	83	82	88	82	77	76	78	82	83	81	--	--	88	72	90	50	59	43	50	34	59	50	32	28	32	50	16	8	17
18	79	94	88	91	83	89	94	91	88	93	97	97	**	94	91	88	91	--	96	94	94	72	45	38	47	55	71	50	41	41	42	53	65	9	18
19	81	**	92	96	**	**	**	93	97	**	92	97	97	94	88	92	**	92	--	88	**	94	59	53	50	59	**	50	42	34	42	53	45	16	19
20	75	88	97	92	93	88	94	95	92	**	92	93	94	**	92	77	81	**	**	--	94	--	40	36	42	48	65	50	32	34	31	40	50	13	20
21	78	**	91	91	67	91	72	68	78	68	96	**	91	74	91	78	87	83	**	87	--	--	38	52	42	41	53	42	13	25	38	48	50	25	21
22	67	67	83	**	50	83	67	83	82	83	67	67	79	83	77	83	77	71	79	83	67	--	--	38	25	25	25	13	13	25	38	13	**	11	22
23	70	82	79	76	83	82	83	82	81	77	86	79	75	73	62	67	71	79	81	72	83	67	--	74	--	97	97	58	59	52	94	74	85	5	23
24	56	62	62	62	75	64	67	59	62	59	68	69	62	57	62	70	74	62	67	56	65	67	82	--	86	62	71	67	55	48	81	83	80	6	24
25	70	81	78	74	67	67	67	59	59	70	68	74	78	59	70	74	70	78	62	67	70	50	**	93	--	59	71	**	50	48	62	76	65	10	25
26	80	95	95	95	90	90	80	75	80	80	90	95	95	90	90	90	80	95	85	90	75	67	95	85	80	--	--	**	50	48	62	76	65	10	26
27	76	**	**	**	**	**	**	83	**	94	**	**	**	**	94	94	82	**	88	94	92	67	**	88	82	**	--	50	35	47	59	65	47	18	27
28	67	92	92	92	92	92	92	83	92	83	92	92	92	92	92	92	83	92	92	92	83	33	92	75	75	**	--	--	42	67	50	67	50	25	28
29	87	93	93	93	93	93	93	77	93	73	93	93	93	93	93	87	80	93	93	87	93	17	**	93	87	80	67	**	--	95	59	68	30	0	29
30	82	**	**	**	75	**	82	**	**	82	**	**	**	82	**	88	88	**	82	94	82	33	**	82	**	81	75	67	93	--	95	68	30	0	30
31	57	67	63	63	67	57	61	55	58	50	61	62	63	50	63	60	60	63	54	53	56	67	80	71	**	75	76	67	80	71	--	55	62	45	10	31
32	79	90	83	79	83	75	83	77	77	68	82	83	86	68	76	72	72	86	77	76	78	67	97	83	89	85	88	75	87	88	79	--	79	85	5	32
33	75	94	93	94	94	88	75	63	88	63	88	94	94	75	88	94	88	94	69	81	75	33	94	94	94	81	69	67	60	63	94	94	--	10	33
34	25	33	42	46	33	33	38	36	29	36	25	25	38	33	46	38	42	25	25	38	30	67	33	25	21	30	29	50	27	35	25	21	25	--	34
LOC	1	2	3	4	5	6	7	8	9	10	11	12	13	14	15	16	17	18	19	20	21	22	23	24	25	26	27	28	29	30	31	32	33	34	LOC

similarities of 8-18 (highest value for the "Lostcabinian"), and to the European composite faunas yield similarities of 5 (Cuisian) and 11 (Sparnacian).

Comparisons of North American with European composite faunas yield similarities in terms of the Simpson Coefficient of 37-50 (25-50, if the composite Clarkforkian is included). The European Sparnacian fauna is most similar to the "Greybullian," and the European Cuisian fauna is about equally similar to all of the composite Wasatchian North American faunas. As would be expected, the North American Clarkforkian fauna is more similar to the older Sparnacian than to the younger Cuisian fauna. Comparisons of individual North American localities to the European composites yield similarities of 30-64, and of individual European localities to North American composites (including Clarkforkian) yield similarities of 24-82.

Comparisons of individual European to individual North American localities yield similarities in terms of the Simpson Coefficient of 13 to 76. The highest similarity values are for comparisons of the European Blackheath Beds (locality 27) and Rians (locality 33) localities to North American localities, and for North American localities 6, 11-13, and 18 to European localities. These similarities range from 25-76 (41-76, excluding Ellesmere Island) for the Blackheath Beds, from 13-70 (27-70, excluding Ellesmere Island) for the Rians fauna, and 34-76 for the North American localities listed above.

Values of the Simpson Coefficient for intra-European comparisons of individual localities range from 30-100; comparisons of individual to composite European localities range in value from 52-100.

Intracontinental similarity comparisons in terms of the Simpson Coefficient of North American localities (excluding the Baja California, locality 5, and Ellesmere Island, locality 22, faunas) range from 32-94 (for individual localities only) or 32-98 (including composite North American faunas). Comparison of the Baja California fauna to other North American localities yields similarities of 45-83 (excluding Ellesmere Island; similarity of 13), while comparison of the Baja California fauna to European localities yield similarities of 18-64. Ellesmere Island/North American similarity comparisons range from 25-75 (excluding Baja California; similarity of 13), and 17 of 21 comparisons have similarities greater than 50. Ellesmere Island/European comparisons range from 13-38.

Similarity comparisons in terms of the Simpson Coefficient of the composite Clarkforkian fauna to other areas range from 8 (Asian composite), 24-41 (European localities), to 18-65 (North American localities). These relatively low similarities indicate that age differences between faunas will significantly decrease faunal similarity, both for localities within the same continent and for those in different continents.

Arrangement of the North American localities in a north-south transect, adding the southernmost locality 5 (Baja California) and the northernmost locality 22 (Ellesmere Island) to the array given in the specific comparisons (Table 7; and see above), yields no simple geographic pattern of general similarities. Again, the similarity comparisons yield similar geographic trends in the Recent and Eocene data sets, although the Eocene similarity values generally are lower than comparable Recent values. There clearly is a complex interplay between temporal and geographic similarities/differences between localities, although the effect of temporal differences/similarities between localities appears to be less in the generic than in the specific comparisons. Similarities are greater and faunas are more homogeneous between localities at the generic than at the specific level.

Family

Similarity values of the Simpson Coefficient for familial comparisons of North American, European, and Asian localities range from 17-100 (42-100, excluding Asia, locality 34 and Ellesmere Island, locality 22; Table 8).

Similarity comparisons via the Simpson Coefficient of the Asian composite Early Eocene locality to all other localities range from 21-67, with only 7 of 33 comparisons greater than 40. Comparisons to North American composites (Clarkforkian to Wasatchian) range from 25-46; comparisons to the European Sparnacian (33) and Cuisian (25) yield similar values.

Comparisons by way of Simpson Coefficients of the European composite Sparnacian and Cuisian faunas to the North American composite (Clarkforkian to Wasatchian) faunas range from 70 to 82 (for the Sparnacian) and from 56 to 62 (Cuisian). Comparisons of individual North American samples to composite European faunas range from 56-86, and comparisons of individual European samples to composite North American faunas range from 57-100 (localities 27 [Blackheath Beds] and 30 [Dormaal, France] both have values of 100 to all three North American composite Wasatchian faunas). Individual European versus individual North American locality comparisons range from 17-100 (42-100, excluding Ellesmere Island).

Intracontinental comparisons via the Simpson Coefficient of individual North American localities (excluding Baja California [locality 5] and Ellesmere Island [locality 22]) have similarities of 68-100; comparison of these same localities to the composite North American faunas ranges from 81-100 (for the three Wasatchian composites) and 65-83 (for the Clarkforkian).

Comparison of the Baja California fauna to the faunas of other individual North American samples yield similarities via the Simpson Coefficient of 67-92 (excluding Ellesmere Island, similarity of 50); comparisons to European localities range from 42-83. The Ellesmere Island comparisons to other individual North American localities range from 67-100 (excluding Baja California; similarity of 50); comparisons to European localities range from 17-67.

Comparisons of the composite Clarkforkian familial fauna to other samples via the Simpson Coefficient range from 25 (Asia), to 56-87 (Europe), to 65-100 (North America).

Arrangement of the North American localities in a north-south transect (as for the specific and generic comparisons) does not yield any geographic pattern.

DISCUSSION

Analyses of distribution patterns of Recent North American mammals reveal several interesting results. Increasing either latitudinal or longitudinal separation yields decreases in similarities between localities; but the decrease for a given distance is much greater with latitudinal than with longitudinal separation. Thus, mammalian provinces tend to be oriented east-west (see Hagmeier, 1966; Hagmeier and Stults, 1964). Further, the amount of faunal change per unit distance is greater in all directions for lower latitude faunas. Various climatic and ecologic factors influence latitudinal gradients in mammalian diversity and the taxonomic composition of provincial regions (Simpson, 1964; Wilson, 1974; Flessa, 1975; McCoy and Connor, 1980); they strongly influence formation of latitudinal faunal barriers as well.

Superimposed upon this pattern of latitudinal gradients is a pattern of longitudinal faunal differentiation, for example the endemism of western North American faunas, and fragmentation of the southwestern United States into many small faunal provinces (Hagmeier, 1966). These longitudinal gradients most probably are due to geographic and climatic isolation caused by the north-south orientation of the Pacific Coast, Rocky Mountains and other mountains, and western deserts. Another example is the relatively high faunal similarities from north to south in the eastern coastal regions, which results in northward deflection along the eastern seaboard of the generally east-west North American faunal provinces (Hagmeier, 1966; Hagmeier and Stults, 1964). In coastal Mexico the Neotropical/Nearctic Realm boundary also is oriented north to south.

Simpson's (1936, 1943, 1947a, 1953) studies of patterns of geographic faunal differentiation and similarities emphasized comparisons along the same latitudes and across longitudes. In particular, his 1936, 1943, and 1953 discussions presented a series of comparisons (at different taxonomic levels) between temporally equivalent, geographically separated faunas; only one of the seven comparisons was between areas separated by significant latitudinal distances (North America-South America from the Miocene to Recent). For most of these comparisons there was a clear inverse relationship between longtudinal separation and faunal similarity; comparison of localities in separate faunal realms further accentuated the inverse relationship. Simpson also noted that, within a single faunal realm, the faunal composition of one locality could be used as a relatively accurate predictor of the faunal composition of other localities (accuracy increases in comparisons at higher taxonomic levels). My analyses support Simpson's observations, and indicate that such predictors are accurate because of the large faunal homogeneity along lines of latitude within North America. However, comparisons within the same faunal realm *across* latitudes yield more complex patterns of similarity. Because of the presence of latitudinal gradients in faunal diversity and composition, such comparisons usually will yield lower similarities (and therefore less accurate predictions of faunal composition) between localities, than between localities at the same latitude.

Although this latitudinal heterogeneity exists, large scale faunal realms still can be recognized on the basis of faunal similarity (see discussion below). Further, use of the Simpson Coefficient is important in comparisons between localities at different latitudes, because of latitudinal gradients in the number of taxa present. Many of the conclusions drawn by Simpson, only on the basis of longitudinal comparisons at similar latitudes, regarding faunal similarity between different geographic regions are still valid.

One goal of my study was to compare analyses for North American Recent mammal species using the Simpson Coefficient (this study) and the Jaccard Coefficient (Hagmeier and Stults, 1964; Hagmeier, 1966). Hagmeier and Stults (1964) and Hagmeier (1966) used the Simpson Coefficient only for comparison of island to mainland faunas, and used the Jaccard Coefficient of Community for comparison of intracontinental North American mammal provinces. The localities in the present study sampled 24 of the 35 provinces of Hagmeier (1966); Hagmeier (1966) and Hagmeier and Stults (1964) did not analyze North American mammal distributions south of the United States/Mexico border. Comparison of the similarity matrix between localities representing these 24 provinces, and the 35 province similarity matrix of Hagmeier (1966), indicates that use of the Simpson Coefficient yields much greater similarities than the Jaccard Coefficient for comparisons between both nearby and distant localities (Table 9). For instance, many of the similarity indices in the Hagmeier (1966) matrix are less than 10, but less than 0.25 percent of the values in my representative matrix are less than 10.

Although differences between results of the two techniques are present in all comparisons, they are especially significant for comparisons of high latitude faunas to faunas at all other latitudes. This would be expected, as comparisons between low diversity, high latitude localities and high diversity, lower latitude localities always will yield lower similarities using the Jaccard rather than the Simpson Coefficient, simply as a consequence of the statistic. The Jaccard Coefficient does not compensate for differences in taxonomic diversity between localities. Although the Jaccard Coefficient may be useful for clustering closely related faunas of comparable diversity into local faunal provinces, its emphasis on differences between faunas makes it a poor tool for clustering widely separated localities and recognizing faunal realms on a continental scale. Low values of the Jaccard Coefficient may arise because faunal similarities are low or because of large diversity differences between localities; diversity differences in contrast have little effect on the value of the Simpson Coefficient. However, the emphasis on similarity with the Simpson Coefficient may mark biases of incomplete sampling of true diver-

TABLE 9. SIMPSON COEFFICIENT (LOWER HALF; THIS STUDY) VERSUS JACCARD COEFFICIENT (UPPER HALF; PARTIAL MATRIX OF HAGMEIER, 1966) OF SIMILARITY, RECENT, NORTH AMERICA, SPECIES. COMPARISON OF REPRESENTATIVE MATRIX OF LOCALITIES (THIS STUDY) OCCURRING WITHIN ONE OF THE 35 PROVINCES OF HAGMEIER (1966); CONSIDERED ONLY NORTH AMERICAN FAUNAS NORTH OF THE U.S./MEXICO BORDER); LOCALITY AND PROVINCE (IN PARENTHESES) NUMBERS ON MARGINS OF MATRICES.

LOC	(PROV)	(1)	(2E)	(3)	(5)	(6W)	(7W)	(8)	(9)	(11)	(13)	(14)	(15)	(16)	(17)	(18)	(19)	(20)	(22)	(23)	(26)	(29)	(30)	(32)	(34)	LOC	(PROV)
60	(1)	--	50	41	21	22	16	9	10	7	14	11	6	3	9	2	2	4	7	2	3	2	3	7	2	0	(1)
67	(2E)	60	--	46	26	42	23	14	10	9	10	8	4	3	9	3	1	2	6	3	3	3	6	4	2	(2E)
73	(3)	80	88	--	55	62	39	28	19	15	13	22	16	19	6	1	5	7	10	3	5	7	9	7	(3)
70	(5)	73	71	88	--	58	57	49	29	25	18	34	25	15	14	6	5	7	16	7	12	16	7	4	(5)
69	(6W)	87	82	77	58	--	56	40	27	21	16	31	23	16	13	5	4	8	15	7	10	14	7	2	(6W)
58	(7W)	67	44	61	56	74	--	55	40	28	23	45	33	23	19	7	6	10	24	9	13	21	8	5	(7W)
53	(8)	73	50	64	55	74	88	--	58	45	37	47	36	31	24	14	12	13	40	16	24	29	15	14	(8)
44	(9)	60	44	55	65	65	85	82	--	41	43	35	31	25	29	17	15	14	56	17	30	41	15	14	(9)
36	(11)	20	18	27	35	35	50	50	52	--	44	31	25	21	26	15	13	16	34	17	20	29	11	22	(11)
33	(13)	40	31	39	47	45	63	67	71	60	--	24	19	22	22	20	16	12	41	20	29	39	21	46	(13)
38	(14)	73	50	58	61	74	83	64	58	40	42	--	60	45	30	27	20	16	25	23	15	18	14	22	10	11	(14)
28	(15)	47	31	30	36	39	54	52	55	40	43	76	--	74	50	45	64	30	33	43	23	23	35	58	18	23	13	(15)
25	(16)	53	31	39	44	48	61	56	56	33	36	87	74	--	64	45	50	43	36	32	23	28	58	26	16	15	13	(16)
34	(17)	40	24	30	33	35	44	44	56	42	56	51	67	74	--	53	45	51	45	32	36	27	24	19	20	(17)
43	(18)	60	44	42	50	52	68	60	80	46	64	49	64	67	73	--	80	64	42	38	47	22	16	37	33	43	20	(18)
21	(19)	20	13	21	28	29	44	40	46	33	44	38	57	51	80	--	54	21	33	23	24	16	25	17	(19)
14	(20)	13	13	15	22	19	29	27	37	33	33	35	48	42	53	64	--	28	21	10	15	13	19	11	(20)
17	(22)	27	13	18	21	23	39	45	42	36	30	70	79	82	42	55	45	67	--	55	15	24	34	52	22	11	(22)
45	(23)	47	25	33	44	42	68	69	78	33	39	49	52	49	55	64	49	48	41	--	45	30	34	22	21	(23)
15	(26)	7	13	12	17	13	29	33	39	37	58	28	33	27	36	37	52	49	55	61	--	24	--	30	41	25	(26)
22	(29)	20	13	21	28	26	37	40	51	46	58	28	40	29	56	47	69	41	30	33	43	--	58	32	25	(29)
32	(30)	47	25	39	50	48	66	65	77	52	74	45	50	44	62	75	59	65	33	56	76	61	--	73	30	(30)
23	(32)	20	12	12	17	23	28	35	43	40	68	25	30	23	40	48	45	49	39	48	46	73	--	60	(32)
24	(34)	0	6	8	10	8	24	31	38	53	53	24	31	20	38	44	40	56	36	27	44	49	53	47	60	--	(34)
LOC	(PROV)	60	67	73	70	69	58	53	44	36	33	38	28	25	34	43	21	14	17	45	15	22	32	23	24	LOC	
		(1)	(2E)	(3)	(5)	(6W)	(7W)	(8)	(9)	(11)	(13)	(14)	(15)	(16)	(17)	(18)	(19)	(20)	(22)	(23)	(26)	(29)	(30)	(32)	(34)	(PROV)	

J. J. FLYNN

sity for individual faunas (often reflected in small sample sizes), and real diversity differences between faunas.

Fallaw (1979) and Cheetham and Hazel (1969) also emphasized the greater utility of the Simpson Coefficient, relative to the Jaccard Coefficient (and other binary coefficients that emphasize dissimilarity), in geographic comparisons of similarity between faunas of widely disparate diversity. Fallaw's (1979) study showed that the Simpson Coefficient was the only binary similarity coefficient that documented the change from cosmopolitan faunal distributions to increasing longitudinal faunal provinciality in Mesozoic to Cenozoic marine invertebrate assemblages (of widely differing diversities) of the North Atlantic. This increasing differentiation and longitudinal provinciality developed in response to plate tectonic widening of the North Atlantic basin, following an early Mesozoic rifting event. The Resemblance Equation and the Second Kulczynski Coefficient indicated slight agreement with the documented decreases in faunal similarity through time, while nine other binary coefficients (including the Jaccard Coefficient) yielded erratic or contradictory results (Fallaw, 1979).

I also wanted to examine criteria for recognizing large scale faunal realms, and localities belonging to the same faunal realm, on the basis of faunal similarity. Global biogeographic regions (Realms) have been recognized at least since Sclater (1858); these realms have been defined primarily by the distribution of plants. The global distribution of Recent mammals reinforces recognition of these biogeographic realms; each realm can be clearly distinguished from all others on the basis of its mammalian faunal composition. Inter-realm similarities are relatively low, and intra-realm similarities are relatively high. In generic level comparisons between the major continents (excluding Chiroptera/Dermoptera, Table 2), only two intercontinental comparisons have a Simpson Coefficient of faunal similarity greater than 45, and most comparisons are less than 25. The familial comparisons show only three intercontinental comparisons with similarities greater than 65. Comparisons between areas 4 (Europe) and 5 (Asia) have a generic similarity of 94.3 (100 for familial), which is indicative of the fact that both "continents" belong to the Palearctic Realm. The only anomalies are the generic similarity of 73 (84 for familial) between areas 5 (Asia) and 6 (SE Asian Islands), and the familial similarity of 81 for areas 1 (North America) and 2 (South America). The relatively high similarity between areas 5 and 6 probably is due to the fact that area 6 is part of the Oriental Realm, while area 5 is a composite of faunas from *both* the Oriental and Palearctic Realms. The relatively high familial (but not generic) similarity between areas 1 and 2 may be due to the fact that North America is a composite of Nearctic and Neotropical Realm faunas, and possibly the complex influence of the Plio-Pleistocene faunal interchange between North and South America.

Distributional information for North American Recent mammals also helps refine criteria for recognition of major faunal realms. Species distributions are particularly useful. As discussed above, several criteria may be used to distinguish Nearctic from Neotropical localities within North America. Neotropical localities within North America form a distinct grouping based on their high specific similarities to each other (values of Simpson Coefficient 40-88) and much lower similarities to other (Nearctic) localities, and because they are the only North American localities with high similarities (Simpson Coefficient 46-98) to South American Neotropical faunas. Further, there are sharp similarity gradients between these localities and Nearctic localities. Such gradients probably are the best criteria for recognizing boundaries between major faunal realms, both in Recent and fossil assemblages. Figure 3 illustrates the similarity gradient between Neotropical and Nearctic localities (recognized solely on sharpness of the gradient of change in similarity between locality 84 and other localities). A sharp gradient also marks the boundary between Nearctic and Palearctic Realms, at the North America/Europe border, and the North America/Asia border, both for the comparisons of specific and generic similarities.

Similarity comparisons in terms of the Simpson Coefficient among Nearctic localities, at the generic level are all greater than 25, and most are greater than 30. Inclusion of Neotropical localities in the comparison yields many similarities that are much lower; some similarity comparisons have values as low as 7. The similarity gradient across the Nearctic/Neotropical Realm boundary region also is present for the generic similarity matrix.

Familial comparisons of Nearctic localities all yield similarities by way of Simpson Coefficients greater than 60; most are greater than 70-80, and most high latitude sites (above 60°) have similarities close to 100 in comparison to all other Nearctic sites. Inclusion of Neotropical localities brings the range of similarities as low as 40, but many of the similarity comparisons between individual Nearctic and Neotropical localities are as high as comparisons between localities only within the Nearctic. Familial similarity comparisons between localities on either side of the Nearctic/Neotropical Realm border are quite high, and there is no sharp familial similarity gradient to mark location of the boundary between realms.

Comparisons of distributions of Recent and Early Eocene mammals indicate that both time and space influence distributions and similarity comparisons of Early Eocene mammals. No, or few, species are shared by North American and European mammal assemblages in either the Early Eocene or Recent. This would support differentiation of the two continental regions into two distinct biogeographic regions. However, it should be noted that many widely separated, individual localities within the Recent Nearctic Realm also lack species in common.

Comparisons between localities along north-south transects show subtle or no geographic trends in similarities for both the Recent and Early Eocene data (Table 7; comparisons are restricted to localities from New Mexico to northern Wyoming because of the limited geographic coverage of Early Eocene localities with taxa identified to species). However, similarity values are much higher for the Recent than for the Early Eocene between

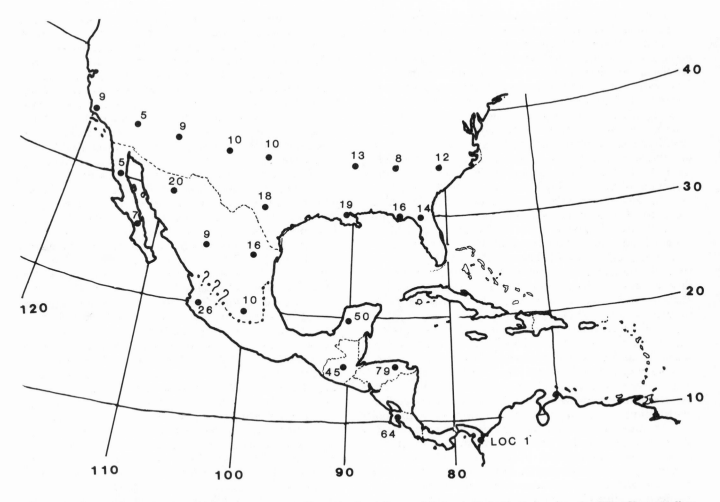

Figure 3. Simpson Coefficients for similarity comparisons of locality 1 (Panama) to low latitude North American localities. Sharp gradient of change in similarities defines Neotropical/Nearctic faunal realm boundary (dashed-dotted line; ? denote uncertainty in precise position of boundary).

equivalent geographic localities, presumably because of the influences of time, facies, and sampling plus taxonomic biases in the Early Eocene data base.

As indicated in the "Results" section, Eocene generic faunal similarity values overlap significantly for intra-European, intra-North American, and European/North American comparisons. However, intracontinental comparisons tend to have higher average and observed ranges of similarity values than do intercontinental comparisons. These results indicate that North America and Europe were distinct faunal regions in the Eocene, but there also was significant faunal continuity between the two continents. The high levels of intercontinental faunal similarity indicate that the faunas of Europe and North America probably were not sufficiently distinct to be considered equivalent to Recent "Realms."

Both Baja California and Ellesmere Island have ranges of similarities comparable to other North American localities (except for the anomalously low similarity value between the two, which may be an artifact of non-representative, incomplete sampling of the total diversity for both localities; non-representative sampling would make any similarity comparison statistic of dubious validity, as noted by Simpson, 1947a). Further, both localities show closer resemblance and biogeographic affinity to North American than to European Early Eocene localities.

Comparison of generic similarity values between localities along north-south transects (Table 7; from Baja California to Ellesmere Island) for Recent and Early Eocene faunas show similar trends in geographic variation of faunal similarity for both data sets. However, some of the individual similarity comparisons yield relatively lower values for the Early Eocene, and there are significant variations in similarity comparisons for localities at approximately the same latitude. Many of these variations may be due to significant temporal differences between some localities, both between different geographic areas and within stratigraphic sequences from the same basin.

Familial comparisons of Early Eocene localities yield the same patterns of similarity as generic comparisons. There is a significant overlap of similarity values for intra-European, intra-North American, and European/

North American comparisons. Again, intracontinental similarities generally are somewhat greater than intercontinental similarities. These faunal resemblances support recognition of the close biogeographic affinity of North America and Europe during the Early Eocene.

Both the Baja California and Ellesmere Island faunas have similarities to other North American localities that are comparable to values for other intra-North American comparisons, although the Baja California/Ellesmere Island comparisons are slightly lower. Both the Baja California and Ellesmere Island faunas have greater average (and observed ranges) similarities to North American than to European localities.

Intra-North American familial comparisons of localities along north-south transects yield no geographic trends for either the Recent or Early Eocene data. The Recent similarity values for Simpson Coefficients are particularly homogenous (all greater than 92) for this transect; the Early Eocene comparisons also are generally high, but more variable. Variability in Early Eocene similarities presumably is due to the temporal and sampling biases discussed above. The Baja California/Ellesmere Island comparison is particularly low, and may reflect real geographic differences between these widely separated localities, or differences due to the same biases.

The Asian Early Eocene fauna is clearly distinct, at all taxonomic levels, both from the European and North American faunas. There is no doubt that Asia represented a separate biogeographic realm during this time period.

Simpson (1943, 1947a, 1953) attempted to use available faunal similarity information for Early Eocene faunas from Asia, Europe, and North America to determine the nature of faunal connections among these areas, and whether available biogeographic evidence supported mobilist or stabilist models of the earth's crust. In the earlier works (1943, 1947a) Simpson stated that available evidence favored stabilist earth models, and close similarity between European and North American Early Eocene faunas was due to a land bridge connection between the two continents, through Asia and the Bering region. In these papers, however, Simpson stressed that the evidence could not rule out possibility of a connection across the North Atlantic (which would support mobilist models of continental drift). Critically important to solution of this question was much more detailed faunal information from Asia: "The needed direct evidence is knowledge of the early Eocene mammals of Asia, and especially of northeastern Asia" (Simpson, 1947a, p. 658). Simpson later (1953) more definitively rejected Wegener's continental drift theory and Cenozoic paleogeographic reconstructions, including a statement that Eocene faunal resemblances/differences precluded a North Atlantic and supported a Bering/Asia connection for Europe and North America.

The information presented in the present study supports an alternative interpretation of Early Eocene biogeographic relationships. It is clear from the faunal similarity matrices (Tables 3 and 8), at all taxonomic levels, that the now more completely known Early Eocene fauna of Asia was distinct and isolated in a separate faunal realm from the faunas of Europe and North America. The high faunal similarities between European and North American faunas argue for a close connection and biogeographic affinity between the two areas. They probably represented major provincial regions within the same, geographically broad faunal realm, based on comparisons to Recent patterns of faunal similarity and distribution patterns within and between distinctive Realms. Although the high similarity values alone say nothing about location of geographic connection between the two areas (as was clearly recognized by Simpson, 1943, 1947a,b, 1953), the distinctiveness and strong faunal dissimilarity of Asia both to European and North American faunas, and the presence of a significant Turgai Strait, indicate the geographic isolation of Asia and a more probable North Atlantic connection (see McKenna, 1983a,b). This conclusion is further supported by lack of indication of a strong faunal similarity gradient and biogeographic barrier between any North American and European localities, as well as by much recent geophysical evidence supporting a continuous North Atlantic terrestrial connection between the two areas in the Early Eocene (McKenna, 1983a,b). With benefit of hindsight (conclusive support for a mobilist, plate tectonic model of the earth's surface) and new, more detailed understanding of composition of the Asian Early Cenozoic mammalian fauna, we are now able to choose between the two competing explanations presented by Simpson (1943, 1947a) for the extremely high faunal similarities between Europe and North America during the Early Eocene.

The Early Eocene is particularly interesting because of recent speculation that the North American Arctic was a separate faunal region in which faunas evolved earlier, and then spread to more southerly latitudes (Hickey and others, 1983; but see Flynn and others, 1984). Although the complete taxonomic composition of the Arctic Ellesmere Island Eocene mammalian fauna remains poorly known, available faunal resemblance evidence indicates that Ellesmere Island was part of a single, continuous European/North American faunal realm. More tentatively, the Eocene Ellesmere fauna more closely resembles other North American than European localities. More conclusive statements about Eocene biogeographic relationships of the Ellesmere Island fauna await more detailed taxonomic descriptions and more complete faunal sampling.

CONCLUSIONS

1. Faunal realms, of large geographic scale, can be distinguished using two criteria—within realm faunal similarities versus among realm faunal similarity, and relatively sharp gradients of change of faunal similarity across faunal realm boundaries.

2. Geographic distribution patterns of Recent mammals within North America show several important trends. First, there is an inverse relationship between faunal similarity and both longitudinal and latitudinal separation between sites. However, latitudinal separation

yields greater dissimilarities than longitudinal separation for sites separated by equal areal distances. Presumably, this is caused by presence of strong latitudinal diversity gradients. Second, there is an asymmetry in similarity comparisons, in which similar areal distance from a locality yields lower similarities in comparisons towards lower latitudes than towards higher latitudes. This probably also is the result of latitudinal diversity gradients. Third, western North American faunas exhibit high endemism, and the southwestern United States is fragmented into many small faunal provinces. Differentiation of lower level (intra-realm) faunal provinces is the result of complex interplay between major latitudinal climatic gradients and less influential (frequently longitudinal) regional climatic, tectonic, and geographic factors.

3. The existence of strong differences in taxonomic diversity, and incomplete sampling of diversity for fossil assemblages, between localities makes the Simpson Coefficient a more useful measure of faunal resemblance than other binary similarity coefficients, both for Recent and fossil assemblages.

4. High similarities for intra-European, intra-North American, and European/North American Early Eocene faunal comparisons indicate presence of a single North American-faunal realm. More subtle patterns of faunal similarity support differentiation of Europe and North America into separate sub-realm provinces. Low resemblances of Asian Early Eocene faunas both to European and North American faunas indicate a separate Asian faunal realm. The Asian Realm most probably was separated from Europe by the Turgai Sea, and from North America by Bering Strait.

5. Although complete taxonomic composition of the Eocene mammalian fauna of Ellesmere Island remains poorly known (only eight mammalian taxa identified), available evidence from faunal resemblances indicates that Ellesmere Island was part of a single, continuous European/North American faunal Realm. It is possible to conclude tentatively that Ellesmere Island also was part of a North American sub-realm province.

ACKNOWLEDGMENTS

I thank George Gaylord Simpson for his extensive scientific contributions, and his active integration of vertebrate paleontology into studies of all areas of evolution and earth history. I also thank Kathy Flanagan and Dr. Anne Roe Simpson for the invitation to contribute to this volume. Chris Jesinkey and Ross Dimmick provided invaluable assistance in creating the computer programs to do much of the data manipulation for the similarity matrices. Carol Vadnos typed parts of the manuscript. Malcolm C. McKenna graciously allowed me to use his mammalian classification that is in preparation. Alison Gold gave much needed help in preparing figures and tables. Sydney Anderson, Michael Novacek, Andre Wyss, and Mark Norell provided critical commentary that greatly improved the manuscript.

REFERENCES CITED

Allen, J. A., 1871, On the mammals and winter birds of east Florida: Museum of Comparative Zoology, Bulletin, v. 2, p. 161-450.

Anderson, S., 1974, Patterns of faunal evolution: The Quarterly Review of Biology, v. 49, p. 311-332.

────── 1977, Geographic ranges of North American terrestrial mammals: American Museum Novitates, no. 2629, 15 p.

Bown, T. M., 1979, Geology and mammalian paleontology of the Sand Creek Facies, lower Willwood Formation (Lower Eocene), Washakie County, Wyoming: The Geological Survey of Wyoming, Memoir No. 2, 151 p.

────── 1980, The Willwood Formation (Lower Eocene) of the southern Bighorn Basin, Wyoming and its mammalian fauna, in Gingerich, P. D., ed., Early Cenozoic paleontology of the Bighorn Basin, Wyoming: University of Michigan, Papers on Paleontology, no. 24, p. 127-138.

Cheetham, A. H., and Hazel, J. E., 1969, Binary (presence-absence) similarity coefficients: Journal of Paleontology, v. 43, p. 1130-1136.

Cracraft, J., 1983, Cladistic analysis and vicariance biogeography: American Scientist, v. 71, p. 273-281.

Croizat, L., 1958, Panbiogeography. Published by the author, Caracas, 3 vols.

────── 1982, Vicariance/vicariism, panbiogeography, "vicariance biogeography," etc.: a clarification: Systematic Zoology, v. 31, p. 291-304.

Croizat, L., Nelson, G., and Rosen, D. E., 1974, Centers of origin and related concepts: ibid., v. 23, p. 265-287.

Delson, E., 1971, Fossil mammals of the early Wasatchian Powder River Local Fauna, Eocene of northeast Wyoming: American Museum of Natural History, Bulletin, v. 146, p. 305-364.

Fallaw, W. C., 1979, A test of the Simpson Coefficient and other binary coefficients of faunal similarity: Journal of Paleontology, v. 53, p. 1029-1034.

Flessa, K. W., 1975, Area, continental drift and mammalian diversity: Paleobiology, v. 1, p. 189-194.

Flessa, K. W., Barnett, S. G., Cornue, D. B., Lomaga, M. A., Lombardi, N., Miyazaki, J. M., and Murer, A. S., 1979, Geologic implications of the relationship between mammalian faunal similarity and geographic distance: Geology, v. 7, p. 15-18.

Flynn, J. J., MacFadden, B. J., and McKenna, M. C., 1984, Land-mammal ages, faunal heterochrony, and temporal resolution in Cenozoic terrestrial sequences: Journal of Geology, v. 92, p. 687-705.

Flynn, J. J., and Novacek, M. J., 1984, Early Eocene vertebrates from Baja California: evidence for intracontinental age correlations: Science, v. 224, p. 173-174.

Godinot, M., 1978, Diagnoses de trois nouvelles especes de mammiferes du Sparnacien de Provence: Compte Rendu Sommaire des Seances, Societe Geologique de France, fasc. 6, p. 286-288.

———— 1981, Les mammiferes de Rians (Eocene inferieur, Provence): Palaeovertebrata, v. 10, p. 43-126.

Godinot, M., de Broin, F., Buffetaut, E., Rage, J.-C., and Russell, D., 1978, Dormaal: une des plus anciennes faunes eocenes d'Europe: Academie des Sciences, Comptes Rendus, t. 287, serie D, p. 1273-1276.

Guthrie, D. A., 1967, The mammalian fauna of the Lysite Member, Wind River Formation, (Early Eocene) of Wyoming: Southern California Academy of Sciences, Memoirs, v. 5, 53 p.

———— 1971, The mammalian fauna of the Lost Cabin Member, Wind River Formation (Lower Eocene) of Wyoming: Annals of Carnegie Museum, v. 43, p. 47-113.

Hagmeier, E. M., 1966, A numerical analysis of the distributional patterns of North American mammals II. Re-evaluation of the provinces: Systematic Zoology, v. 15, p. 279-299.

Hagmeier, E. M., and Stults, C. D., 1964, A numerical analysis of the distributional patterns of North American mammals: *ibid.,* v. 13, p. 125-155.

Hall, E. R., 1981, The mammals of North America: New York, John Wiley and Sons, 2 vols., 1179 p. + 90 p. index.

Hickey, L. J., West, R. M., Dawson, M. R., and Choi, D. K., 1983, Arctic terrestrial biota: paleomagnetic evidence of age disparity with mid-northern latitudes during the Late Cretaceous and early Tertiary: Science, v. 221, p. 1153-1156.

Hooker, J. J., 1980, The succession of *Hyracotherium* (Perissodactyla, Mammalia) in the English early Eocene: British Museum Natural History, Bulletin (Geology), v. 33, p. 101-114.

Hooker, J. J., Insole, A. N., Moody, R. T. J., Walker, C. A., and Ward, D. J., 1980, The distribution of cartilaginous fish, turtles, birds and mammals in the British Palaeogene: Tertiary Research, v. 3, p. 1-45.

Huxley, T. H., 1868, On the classification and distribution of the Alectoromorphae and Heteromorphae: Zoological Society of London, Proceedings, p. 294-319.

Kihm, A. J., 1984, Early Eocene mammalian faunas of the Piceance Creek Basin northwestern Colorado [Ph.D. thesis]: Boulder, Colorado, University of Colorado, 381 p.

Li C.-K., and Ting S.-Y., 1983, The Paleogene mammals of China: Carnegie Museum of Natural History, Bulletin, no. 21, 98 p.

Lucas, S. G., Schoch, R. M., Manning, E., and Tsentas, C., 1981, The Eocene biostratigraphy of New Mexico: Geological Society of America, Bulletin, v. 92, p. 951-967.

McCoy, E. D., and Connor, E. F., 1980, Latitudinal gradients in the species diversity of North American mammals: Evolution, v. 34, p. 193-203.

McKenna, M. C., 1960, Fossil Mammalia from the Early Wasatchian Four Mile Fauna, Eocene of northwest Colorado: University of California, Publications in Geological Sciences, v. 37, p. 1-130.

———— 1983a, Holarctic landmass rearrangement, cosmic events, and Cenozoic terrestrial organisms: Missouri Botanical Gardens, Annals, v. 70, p. 459-489.

———— 1983b, Cenozoic paleogeography of North Atlantic land bridges, *in* Bott, M. H. P., Saxov, S., Talwani, M., and Thiede, J., eds., Structure and development of the Greenland-Scotland Ridge: New York, Plenum, p. 351-399.

Novacek, M. J., and Lillegraven, J. A., 1979, Terrestrial vertebrates from the later Eocene of San Diego County, California: a conspectus, *in* Abbott, P. L., ed., Eocene depositional systems San Diego, California: Society of Economic Paleontologists and Mineralogists, Pacific Section, p. 69-79.

Novacek, M. J., Flynn, J. J., Ferrusquia-Villafrancha, I., and Wyss, A., *in press,* The Wasatchian (Early Eocene) Punta Prieta Local Fauna, Baja California, Mexico: Mammalia: American Museum Novitates.

Nowak, R. M., and Paradiso, J. L., 1983, Walker's mammals of the world (4th edition): Baltimore, The Johns Hopkins University, 1362 p.

Patterson, C., 1981, Methods of paleobiogeography, *in* Nelson, G., and Rosen, D. E., eds., Vicariance biogeography: a critique: New York, Columbia University Press, p. 446-500.

Pielou, E. C., 1979, Biogeography: New York, John Wiley and Sons, 351 p.

Rosen, D. E., 1975, A vicariance model of Caribbean biogeography: Systematic Zoology, v. 24, p. 431-464.

———— 1985, Geological hierarchies and biogeographic congruence in the Caribbean: Missouri Botanical Gardens, Annals, v. 72, p. 636-659.

Russell, D. E., and others, 1982, Tetrapods of the Northwest European Tertiary Basin: Geologisches Jahrbuch, Reihe A, Heft 60, p. 5-74.

Savage, D. E., and Russell, D. E., 1983, Mammalian paleofaunas of the world: Reading, Massachusetts, Addison-Wesley Publishing Co., 432 p.

Schankler, D. M., 1980, Faunal zonation of the Willwood Formation in the Central Bighorn Basin, Wyoming, *in* Gingerich, P. D., ed., Early Cenozoic paleontology of the Bighorn Basin, Wyoming: University of Michigan, Papers on Paleontology, no. 24, p. 99-114.

Sclater, P. L., 1858, On the general geographical distribution of the members of the class Aves: Journal of the Linnean Society (Zoology), v. 2, p. 130-145.

Simpson, G. G., 1936, Data on the relationships of local and continental mammalian faunas: Journal of Paleontology, v. 10, p. 410-414.

———— 1943, Mammals and the nature of continents: American Journal of Science, v. 241, p. 1-31.

———— 1947a, Holarctic mammalian faunas and continental relationships during the Cenozoic: Geological Society of America, Bulletin, v. 58, p. 613-688.

———— 1947b, Evolution, interchange, and resemblance of North American and Eurasian Cenozoic mammalian faunas: Evolution, v. 1, p. 218-220.

_____ 1953, Evolution and geography: Eugene, Oregon, Condon Lectures, Oregon State System of Higher Education, 64 p.

_____ 1960, Notes on the measurement of faunal resemblance: American Journal of Science, v. 258-A, p. 300-311.

_____ 1961, Historic zoogeography of Australian mammals: Evolution, v. 15, p. 431-446.

_____ 1964, Species density of North American Recent mammals: Systematic Zoology, v. 13, p. 57-73.

_____ 1965, The geography of evolution: Philadelphia, Chilton Books, 249 p.

Smith, H. M., 1960, An evaluation of the biotic province concept: Systematic Zoology, v. 9, p. 41-44.

_____ 1983a, A system of world mammal faunal regions. I. Logical and statistical derivation of the regions: Journal of Biogeography, v. 10, p. 455-466.

_____ 1983b, A system of world mammal faunal regions. II. The distance decay effect upon interregional affinities: *ibid.*, v. 10, p. 467-482.

Stucky, R. K., 1984, Revision of the Wind River faunas, Early Eocene of Central Wyoming. Part 5. Geology and biostratigraphy of the upper part of the Wind River Formation, Northeastern Wind River Basin: Annals of Carnegie Museum, v. 53, p. 231-294.

Wallace, A. R., 1876, The geographical distribution of animals: London, MacMillan, 2 vols.

Wilson, J. W., III, 1974, Analytical zoogeography of North American mammals: Evolution, v. 28, p. 124-140.

MANUSCRIPT RECEIVED JANUARY 8, 1986
REVISED MANUSCRIPT RECEIVED JUNE 23, 1986
MANUSCRIPT ACCEPTED JULY 3, 1986

Evolutionary epicycles

JON MARKS *Department of Genetics, University of California, Davis, California 95616*

ABSTRACT

Investigation of organismal evolution has been augmented in the last few decades by the development of techniques permitting the genetic material to be studied directly across extant taxa. The relationship among evolutionary processes governing genes, chromosomes, and organisms is explored here. While their rates and modes of evolution are usually greatly dissimilar, genotypes, karyotypes, and phenotypes trace the same phylogenetic path, and can therefore be used to illuminate different aspects of that path. A useful metaphor to express the relationship among these diverse evolutionary systems is the Ptolemaic epicycle, as each epicycle (*i. e.*, evolutionary system) represents a unique detour off the main orbit (*i. e.*, phylogeny). The hominoid primates are taken as an illustration of how these different systems can be used to track aspects of the common phylogeny.

PROLOGUE

There are two paradoxes of primate biology which have surfaced in the last decade, with important consequences for a comprehensive understanding of the evolutionary processes. First, King and Wilson (1975) showed that genetic and anatomical evolution in humans and chimpanzees has occurred at radically different rates. Second, we have in the gibbons on the one hand and in the macaques and baboons on the other, two groups of animals similar in their genetic diversity, and similar in their taxonomic diversity as well.[1] Interspecific hybridization within both taxa occurs (*e. g.*, Myers and Shafer, 1978; Nagel, 1973). Nevertheless, among the hylobatids, chromosomal analysis reveals a vast amount of difference among species — including those that can hybridize (Stanyon and Chiarelli, 1983; van Tuinen and Ledbetter, 1983); while among the papionins, all taxa studied to date have an identical karyotype, or a minor variant of it (Marks, 1983, and references therein). Thus, chromosomal change can be shown to be divorced from both genotypic change and phenotypic change — as both of the latter are effectively held constant in the comparison between the hylobatids and papionins.

I have adopted an anachronistic metaphor as the title of this paper because its imagery gives an approximation of what I believe to be the main relationships among genetic, chromosomal, and organismal evolution. Namely, they follow a central trajectory, the phylogeny of the species in question, but each involves a unique detour off that central trajectory.

With respect to recent debates about new evolutionary synthesis (Gould, 1980; Stebbins and Ayala, 1982), this analysis will be explicitly anti-synthetic. Not anti- "the synthetic theory of evolution," but anti- the prospects for unifying evolutionary studies of genes, chromosomes, and organisms within one comprehensive theory. The only harmony which exists among these three systems is that the genes, the chromosomes, and the organisms are all relics of the same historical reality: the unique phylogeny of the species.

AN HISTORICAL INTRODUCTION

Evolutionary studies traditionally, indeed from the outset, have addressed primarily the question of how organisms adapt. Involved in this is the slightly different question on how organismal forms may be altered over time. The literature on the question of how form and adaptation relate is becoming voluminous (*e. g.*, Lewontin, 1978; Gould and Lewontin, 1979; Mayr, 1983), but for the present purposes it need only be recognized that changes in form and adaptation from an evolutionary standpoint are somehow the results of changes in the genetic compositions of populations (Chetverikoff, 1926; Fisher, 1930; Wright, 1931; Haldane, 1932). The specific genetic changes themselves are rarely made explicit, as so little is known about the composition of the genome.

Through the 1960s, the dominant paradigm of genomic structure was the "beads on a string" model, derived early in this century through studies of the fruitfly, *Drosophila* (see Sturtevant, 1913). This view received molecular support from the "one gene-one cistron" hypothesis (Beadle and Tatum, 1941), whereby each gene (at least in bacteria) was indeed shown to occupy a unique position and to be responsible for the production of a single product, a protein.

There was an idiosyncratic holdout against the linear, indeed against the physical model of the genome adopted so nearly unanimously, in the geneticist Richard Goldschmidt. In a series of publications spanning several

---

[1]. While we generally recognize the gibbons at the family level (Hylobatidae) and the macaques and baboons at the tribe level (Papionini), the classification of Simpson (1945) allotted the gibbons a subfamily rank, and that of Chiarelli (1966) gave the papionins a subfamily rank. Both of these classifications are now non-standard, the latter considerably more so.

decades, Goldschmidt (*e. g.,* 1937, 1940, 1952) developed a theory basically extrapolated from the "position effect" phenomenon known from the Bar locus in *Drosophila*. To Goldschmidt, there were no point mutations, and no beads on a string; there were merely shufflings of the spatial arrangements of genetic material. Goldschmidt recognized limitations of the existing paradigm of genome structure, but in his attempt to describe that structure and the evolutionary processes which would emerge from it, he was considerably more wrong than he was right (Wright, 1941; Dobzhansky, 1941; Mayr, 1942; White, 1945; Templeton, 1982). However, Goldschmidt did call attention to an aspect of the genetic and evolutionary machinery which had not attracted much attention in that context previously, the chromosome.

By the turn of the twentieth century, it was acknowledged that each species had a characteristic number of chromosomes (Wilson, 1902). By 1902, the chromosomes had independently been recognized as bearers of the hereditary material by Sutton and Boveri. The morphologies of chromosomes were first used as phylogenetic markers by Sturtevant and Dobzhansky (1936) in *Drosophila*.

Nevertheless, the role of chromosomes in the evolutionary process remained virtually a total mystery. In the seminal work of the evolutionary synthesis, Dobzhansky discussed chromosomes in some detail. Yet there was candor, even if understated, in his acknowledgment that precious little was actually known about their evolutionary role: "The nature and causation of the chromosomal differences are subjects which have proved rather more elusive than the straight descriptive picture of the situation" (Dobzhansky, 1937, p. 100).

Thus, while evolution was becoming a study of processes, comparative cytogenetics, and particularly mammalian cytogenetics, remained a chaste descriptive endeavor. Indeed, there remained no major attempts to synthesize cytogenetics and evolutionary theory outside of the work of Stebbins (1950) for plants, until the third edition (1973) of M. J. D. White's *Animal Cytology and Evolution*.

In the meanwhile, the molecular revolution overtook evolution. Beginning in the early 1960s with protein studies, and presently continuing with recombinant DNA technologies, it has become clear that rates of molecular evolution are not easily correlated with rates of anatomical evolution (Wilson and others, 1974; King and Wilson, 1975). The modern inference that genetic and anatomical evolution are "decoupled" has led to a renewed interest in the problem of specifically what the genetic causes of anatomical variations are (Bonner, 1980). This, I believe, obscures an important consideration. The evolutionary variations in form and adaptation do, indeed, have a genetic etiology; however, any description of the genetic basis of morphological change is limited by the state of knowledge of the genome and its processes. In this case, attempts to explain the genetic basis of anatomical evolution involve not so much a reduction of epistemological explanatory levels, but

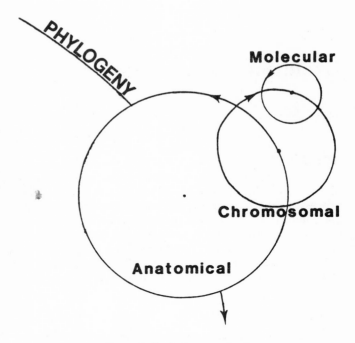

Figure 1. Evolutionary epicycles.

rather, descriptions of the poorly known in terms of the less known.

What I wish to do in the following sections is to set out a contrast among the evolutionary systems of the genes, the chromosomes, and the body — and to demonstrate that they share few, if any, qualities. They share one profound property, however, in tracking a common phylogeny. Before these three systems can be integrated into a "new and general theory of evolution" (Gould, 1980), it will be necessary to understand the properties of each evolutionary system.

I present, as a heuristic device, the Ptolemaic epicycle (Fig. 1), based on a clever, if incorrect, description of planetary motion. This was originally an attempt to account for the retrograde motion of planets by designing a (geocentric) orbital system based purely on the circle. The greater circular orbit had a smaller one imposed upon it, whose direction and speed were different. The smaller orbit could have a still smaller one imposed upon it, and so forth. As a metaphor for visualizing relationships among the evolutionary systems, the epicycle model emphasizes independence of the systems, yet binds them to a common main track. There are three main evolutionary systems or epicycles; the genetic, chromosomal, and anatomical (Fig. 2). All differ in their modes of variation, the effects of those variations, the rates at which those variations occur, and the expression of those variations in the dynamic relations of organism and environment. With examples from the primate superfamily Hominoidea, the apes and humans, I will discuss these epicyclic systems, and finally show that they track a single historical reality, the common phylogenetic relationships within the taxon.

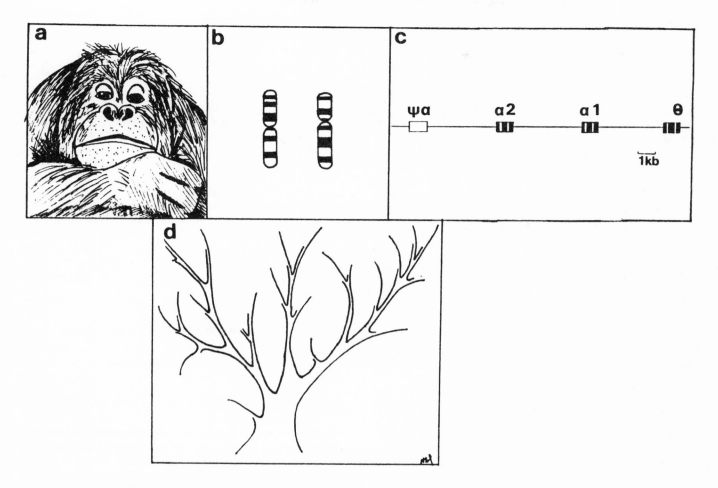

Figure 2. Examples of data obtained from each evolutionary system or epicycle. *a*, anatomy; *b*, G-banded chromosomes; *c*, adult alpha-globin gene cluster; and *d*, phylogenetic tree, the common path traversed by all epicycles.

EPICYCLE 1: GENETIC EVOLUTION

Genetic evolution became a major focus of debate in anthropology when, in 1967, Sarich and Wilson applied the "molecular clock" and challenged the mid-Miocene divergence of human and chimpanzee. Yet when Selander (1982, p. 49-50) writes that "the credibility of primate paleontologists has been weakened by the advent of molecular techniques," he tells only part of the story. The molecular clock challenged not merely paleoanthropology, but paleontology generally, yet few paleontologists besides Simpson (*e. g.*, 1964) recognized this challenge, coming as it did specifically against the study of primate divergences. And while Sarich and Wilson properly called attention to the poor fossil support for the mid-Miocene human-chimpanzee divergence, the apparent slowdown in the rate of molecular evolution in hominoids (Goodman and others, 1982, 1983) undermines the assumptions of the molecular calculations, making it likely that the molecular data yield an underestimate of the actual divergence (cf. Britten, 1986).

It is clear that molecular evolution is governed by horotelic (Simpson, 1944) evolutionary rates, strongly modal and not fluctuating widely across taxa. It therefore provides a gross picture of the relative branching sequences among taxa. However, the processes which govern molecular evolution are far from clear; therefore the fine branching sequences may be more obscured than clarified.

The known processes of genomic evolution can be found to operate in the alpha-globin cluster of the primates. Located on the short arm of chromosome 16, the human α-cluster contains two embryonic (zeta) and four adult (alpha) genes. If we focus on the four adult genes, we find they occur in the order $\psi\alpha$-$\alpha 2$-$\alpha 1$-θ (Fig. 1C). All four genes possess two introns (non-coding sequences) separating three exons (coding sequences). If we compare the $\alpha 1$ gene from the promoter sequence to the polyadenylation signal between human and orangutan, we find predominantly base substitutions differentiating them. Indeed, we can tabulate the differences and note that between human and orangutan, in the $\alpha 1$ globin gene, about 95 out of every 100 nucleotides are identical (Marks and others, 1986a,b).

Upstream from $\alpha 1$ in the human genome is $\alpha 2$, also functional. As all anthropoid primates studied have been found to possess $\alpha 2$ and $\alpha 1$, we recognize that these genes share an ancient phylogenetic history, having been derived from a single ancestral gene via duplication (Zimmer, 1980). The genes are, in Owen's original terminology, serially homologous; in modern genetic parlance, paralogous. But these genes do not stand alone; for pseudo-alpha ($\psi\alpha$) and θ flank the $\alpha2$-$\alpha1$ dyad, and bear striking similarities. Pseudo-alpha is non-functional; θ has been located but not yet sequenced in the human, although it has been sequenced in the orangutan (Marks and others, 1986b). It is certain, however, that the original α-globin gene has been duplicated repeatedly through evolutionary history of the vertebrates. On a different chromosome (#11), one finds the human beta-globin cluster (even more complex than the alpha-cluster), yet each gene bears a fundamental likeness to the alpha-globin genes, indicative of a remote common ancestry (Goodman and others, 1975).

Using the α-globins as a paradigm case, we can detect the following five processes as sources of genotypic variation.

1. *Base substitutions.* These constitute the predominant mode of genotype variation, and are particularly useful in that they can be tabulated and quantified. However, insofar as there are only four nucleotides, homoplasy presents a major difficulty in deriving phylogenetic information from these mutations.

2. *Minor insertions and deletions.* These are common in non-coding DNA. Comparing the $\alpha1$ globin gene between human and orangutan, in addition to the base substitutions, we find 2 single base deletions (or insertions) and one of 7 base-pairs (Marks and others, 1986a,b). All occur in non-coding regions of the gene.

3. *Tandem duplication.* Structural similarity among genes of the alpha-cluster suggest that they are descended from a single ancestral gene; indeed, that all globin genes are so descended. Protein studies by Doolittle (1981) may indicate that most genes are the result of serial duplications of a small number of primordial genes.

4. *Gene correction.* When we compare the tandemly duplicated $\alpha1/\alpha2$ pair with homologous genes in the orangutan, an interesting pattern emerges. The duplication is an ancient one in the higher primates, yet human $\alpha1$ is more similar to human $\alpha2$ than to orangutan $\alpha1$. For example, in the coding sequence of orangutan $\alpha1$ and $\alpha2$, a sequence CCATGG appears, which is a recognition site for the restriction enzyme Nco I. At the same position in both human genes, the sequence reads CCACGG. The chance of an identical mutation occurring in the same position is small, and yet this situation is encountered no less than 20 times in the gene (Marks and others, 1986a). This indicates that one gene has been used as a template to "correct" the other. The pattern of similarity breaks down in the 3'-untranslated region of the gene, where $\alpha1$ of human and orangutan are most similar to each other. In this case, we can not only identify gene correction, but show as well its boundaries (Marks and others, 1986a).

5. *Insertion/deletion of repetitive elements.* There are at least eight members of the Alu repeat family scattered through the human alpha-globin cluster, and probably more (Lauer and others, 1980). The Alu family itself consists of a consensus sequence about 300 bases long, existing millions of times throughout the human genome (Schmid and Jelinek, 1982). Other families of repeated sequences are known (Shafit-Zagardo and others, 1982). Similarly, viral elements may be able to enter the genome. In 1976, Benveniste and Todaro reported that DNA hybridization of Type C retroviral DNA to the genomes of higher primates yielded positive results for African species and negative results for Asian species. While they used these data to infer an "Asian origin of man," and not to infer genomic process, the data seem to indicate a propensity for this retroviral DNA to interpose itself into the germ-line genomes of African primates.

In sum, genomic evolution is not well correlated with anatomical evolution (Wilson and others, 1977), and its complexity is just beginning to be perceived (Nei and Koehn, 1982). What relation, if any, these processes have to changes in anatomical form and adaptation is unknown.

EPICYLE 2: CHROMOSOMAL EVOLUTION

The chromosomes, cellular structures housing the genetic information, are morphologies insofar as tracing their evolution across lineages involves tracing changes in form and structure. Nevertheless, there is little similarity between anatomical changes across taxa and chromosomal changes across the same taxa, outside of the obvious fact that they frequently accompany each other. For example, as I mentioned at the outset, the obvious anatomical differences among the great apes are paralleled by some chromosomal differences, but those among the macaques and baboons are not; simultaneously, the subtler anatomical differences among the gibbons are paralleled by extreme differences in chromosome morphology.

Much more is known about chromosomal evolution in other organisms such as plants and fruitflies than in mammals. Unfortunately, it does not seem as though the principles which guide chromosomal evolution are strictly comparable across these higher taxa. In plants, for example, polyploidy may play a major role, along with other gross variations in genome size, while these do not play major roles in mammalian chromosomal evolution. Similarly, chromosomal inversions represent one of several adaptations in *Drosophila* for maintaining allelic combinations, and are visible with very high resolution as reduplicated polytene chromosomes of the larval salivary glands; comparable situations seem not to exist in mammals.

When we compare genic and chromosomal evolution, we find that their modes of change are significantly different. While the dominant mode of genic mutation is the base substitution (which presents only three possibilities at any time and occurs with a predictable frequency), the major avenues of chromosomal change involve breakage and reunion of chromosomes in two places (an effectively unique circumstance). Chromosomal mutations alter the meiotic pairing of the chromosomes, and consequently have direct effect upon the reproductive machinery. As with genic changes, it appears as though the great majority of chromosomal rearrangements have no effect upon the phenotype. From this, coupled with the direct effect of chromosomal changes upon the reproductive system, we may infer that chromosomal changes, in contrast to the great bulk of anatomical and genic changes, are predominantly involved in the generation of reproductive differences between species (White, 1978). While the rate of genomic evolution is largely time-dependent, and that of phenotypic evolution is dependent upon adaptive opportunities and environmental factors, chromosomal evolution (at least in primates, and perhaps in mammals generally) may be dependent primarily upon social and ecological factors, such as territoriality and mating structure (Arnason, 1972; Marks, in press).

Chromosomal variations, which are the raw materials for karyotype evolution, are of five main types.

1. *Inversions.* An inversion involves breakage of a chromosome in two places, followed by a reversal of polarity of the internal segment. Inversions constitute the predominant mode of karyotype change among great apes (Miller, 1977).

2. *Translocations.* A translocation involves breakage of two chromosomes and their mis-reunion to form two "hybrid" chromosomes. The segregational consequences to meiosis are more drastic than those produced by inversions.

3. *Fusions and fissions.* These change chromosome numbers dramatically, and are the predominant mode of change in karyotypes of lemurs (Hamilton and Buettner-Janusch, 1977).

4. *Amplifications and deletions.* Through processes poorly understood, constitutively heterochromatic (*i. e.,* darkly-staining by the process of C-banding) regions on chromosomes often can be found to vary inter-specifically in location and to vary intra-specifically in size. In the human karyotype, these regions are the long arm of the Y-chromosome and sub-centromeric regions of chromosomes 1, 9, and 16. These are uniquely derived features of the human karyotype, but vary widely in size within the human species. The chimpanzee lacks these heterochromatic regions, but has others, on the tips of most chromosomes and in the middle of the long arms of chromosomes 6 and 14.

5. *Ploidy.* Although this has not been found to occur in mammalian taxa, direct multiplication of the genome, doubling the chromosome number and DNA content per cell, is common among plants (Stebbins, 1950). I mention it for the sake of completing this list, although it is not significant for the study of primate karyotypes.

The greatest advantage afforded by study of chromosome change in evolution is that the problem of homoplasy is often avoided. While genomic sequences reveal evolutionary conservation of nucleotides when compared across taxa, it must be borne in mind that two nucleotides have a 25 percent chance of being identical purely at random. And in organismal evolution, parallel evolution, or related but separate lineages responding similarly to environmental pressures, is common. Chromosomal changes, however, are effectively unique events by virtue of requiring two independent breaks, or the generation of a unique heterochromatic novelty. Indeed, among hominoid primates, only two examples of chromosomal homoplasy are identifiable: (1) the terminal heterochromatic bands which have evolved independently in gibbons and African apes; and (2) the conversion of an ancestral acrocentric chromosome in the great apes (#IX) to the submetacentric chromosome found in human and chimpanzee (Dutrillaux, 1977).

EPICYCLE 3: ANATOMICAL EVOLUTION

Anatomical variation among groups of organisms is evolution in the classical sense, the change of interest to Lamarck, to Darwin, and to their successors. As has been widely recognized since *The Origin of Species,* adaptive anatomical change proceeds by accumulation of successive favorable variations, and elimination of unfavorable variation — the process its author called Natural Selection. And, of course, adaptive modifications can be followed across time and space to reveal a pattern of common ancestry for the organisms under study.

Preservation of favorable variations reduces to the preservation of favorable alleles only insofar as anatomical variations may reduce to genetic variations, or phenotypes to genotypes. Phenotypes, except in the simplest cases in the simplest organisms or in pathologies of complex organisms, are virtually never reducible to genotypes. Therefore, it is unreasonable to expect that genetic changes should translate easily into anatomical variations on which natural selection can operate. It is easy to make this observation with aid of hindsight, but this was not apparent in the 1960s.

Because so little was known about the genetics of morphological change, and the newly-developed biochemical methods could reveal taxonomic variation at the genetic level, it was thought in the 1960s that measurable genetic change might, or indeed would, correlate well with anatomical change. This idea, however, was falsified in the mid-1970s (Wilson and others, 1974; King and Wilson, 1975). The falsification, however, produced a paradox: if measurable genetic changes do not correlate well with measureable anatomical changes, and we recognize that anatomical evolution is fundamentally a genetic process, then what genetic phenomena do correlate with and cause the perceptible differences among species?

Answers have been plentiful: chromosomal rearrangements (Wilson and others, 1974); the spread of repetitive DNA fractions (Britten and Davidson, 1971; Dover, 1982) mutator genes (Thompson and Woodruff, 1978); transposable elements (Nevers and Saedler, 1977); "regulator" genes (Valentine and Campbell, 1975); and extra-nuclear inheritance (Ho and Saunders, 1979). What these suggestions have in common is that they are all recently-discovered genomic phenomena which are very poorly understood, and they are all mutational theories, attempting to explain morphological change via a direct linkage to genic change.

It is a truism that to be significant in an evolutionary context, any anatomical variation must have an hereditary basis. Yet phenotypic Mendelian mutants, or "sports" as they were known to Darwin, do not play a significant role in the evolutionary process. And those biologists who attempted to explain evolution by explicitly mutational theories (R. B. Goldschmidt and C. B. Davenport, for example) were roundly rejected. Although the ultimate source of phenotypic variation in evolution undoubtedly lies in genomic variation (the ultimate source of which is mutation), precious little is known about translation of genotypic into phenotypic variation. If the predominant phenotypic variants operated upon by natural selection are quantitative characters with complex polygenic foundations (Wright, 1968), it would seem unlikely that recourse to directly expressed mutations would yield an accurate account of the genetic basis of morphological change. Further, if a major mode of action of natural selection is not so much on the variability of characters themselves, but on variability of expression of characters, that is, on an organism's adaptability (Waddington, 1957), it seems even less fruitful to seek answers in simple gene-character correlations.

Thus, the problem of reducing anatomical change to its genetic basis is a false problem for evolutionary biologists. It is merely the redressing of a central problem in genetics, the relationship between genotype and phenotype, a problem of considerable complexity. The genotype is equally (if not more) as complex as the phenotype in its own right, and is subject to unique variations and evolutionary patterns. Consequently, it seems less important to pass off anatomical variation as the direct effect of an arbitrary and arcane portion of the genome, as it is to obtain some sort of adequate description of the different properties and processes of anatomical evolution *sui generis*.

The first major contrast between anatomical and genomic evolution is that the genome does not interact directly with the environment except through its expression. Therefore, insofar as evolution may be considered a process of adapting, a considerably greater part of anatomical change than genomic change is likely to be adaptive.

Second, while we have a fairly clear idea of what the evolutionary sources of genotypic variation are, we really do not have such a clear idea of what the evolutionary sources of phenotypic variation are. Four non-exclusive categories of these sources are likely.

1. *Single-locus mutations, directly translated into anatomical differences.* These, as mentioned above, are unlikely to play a major role in evolution, as they produce the "hopeful monster" scenario. Nevertheless, they may well play isolated minor roles (*e. g.*, Frazzetta, 1970), and are tempting to implicate in developmental "on/off" switching. Such single gene variants cause discrete effects; consequently, we must rely on polygenic etiologies for the other categories, more likely to be non-discrete variations of the kind upon which natural selection has been conceived to operate.

2. *Genetic assimilation and subsequent canalization of an originally facultative response to environmental stimulus.* This is a major contribution of the developmental biologist Waddington (1942, 1953, 1957); it is an explanation that bridges the individual's behavioral response to environmental stresses and the responses of future generations, yet is consistent with the Weismann-Mendel theories of heredity. In Waddington's view, individuals vary genetically in their capacities to respond facultatively to stresses, both in intensity of response and in the ability to have such response triggered. This cryptic genetic variation may, therefore, be subject to selection through exigencies of the environment, and the resulting pattern may mimic "Lamarckian" inheritance. Perhaps major behavioral shifts, such as bipedalism, may have originated in this way.

3. *Heterochrony.* Altering the relative growth rates of body parts may result in phenotypic change. Heterochrony is thus the dissociation of developmental patterns between two characters in an organism. Gould (1977) plausibly interprets the human chin as a result of the retardation of growth in the two growth fields of the anterior human mandible, one more so than the other.

4. *Allometry.* Statistical methods have been refined to a degree which makes possible interpretation of some anatomical differences among taxa as consequences of size differences. This is often presented as one aspect of heterochrony, but I separate it to stress the effects of simple size changes across taxa, as opposed to the relation of two growth rates to each other within the same taxon. Giles (1956), Shea (1981, 1983, 1985), and Jungers and Susman (1984) have shown that many anatomical relations among the pygmy chimpanzee, common chimpanzee, and gorilla are of this nature.

HOMINOID PHYLOGENY

Relationships between gene and organism, gene and chromosome, and chromosome and organism are, it should be apparent, poorly defined. Evolution proceeds via a different set of processes for genotype, karyotype, and phenotype; and changes in any of the three have different effects accruing to the organism and population. While I have attempted to outline some evolutionary principles to contrast these three epicycles of evolution, it nevertheless remains that there is a fundamental connection among them. That connection is phylogeny, the historical linkage among the taxa under scrutiny. Anatomical, cytological, and genetical aspects of organisms all change through time, and change in different ways, constrained by different rules, and often at radically different rates. All three aspects, however, are potentially phylogenetically informative — depending, of course, upon the taxa selected.

The taxa I shall discuss in the remainder of this paper constitute the primate superfamily Hominoidea, consisting of five genera, *Homo* (humans), *Pan* (chimpanzees), *Gorilla*, *Pongo* (orangutans), and *Hylobates* (gibbons).

In the early twentieth century, the study of primate relationships was dominated by an academic tradition emanating not so much from systematic evolutionary biology as from comparative anatomy. This historical fact, and its legacy to this day, are largely responsible for the generally poor systematic and phylogenetic conclusions which have plagued physical anthropology (see, for example, Simpson, [1945] bemoaning the state of primate systematics).

By the 1930s, the picture of hominoid phylogeny was almost completely obscured. The fundamental question, "what is the closest living relative of our species," was to be answered in any number of ways. Sonntag (1924) argued for a chimpanzee-gorilla branch as the closest relative of humans; Elliot Smith (1924) saw the gorilla alone in that position; Weinert (1932) saw the chimpanzee alone in that position; and most others (Keith, 1915; Schultz, 1930; Le Gros Clark, 1934) united the chimpanzee, gorilla, and orangutan as a branch whose stem represented that closest relative to man. In 1949, no closer to resolution, the problem was reviewed by Straus as "The Riddle of Man's Ancestry."

Granted, the problem is not an easy one. It requires recognition that one genus *(Homo)* has entered a new adaptive zone and has diverged radically in form and behavior (Simpson, 1963). As a consequence, most anatomical similarities shared between humans and their closest relative are likely to be ancestral retentions. The result is a cladist's nightmare. Chimpanzee, gorilla, and orangutan are, indeed, most similar to one another overall, presumably sharing ancestrally retained characters among themselves, from which humans diverge.

Pan and *Gorilla* share a number of anatomical characters widely regarded as derived, and indicative of an especially intimate phylogenetic connection between them. These characters are primarily molar enamel morphology (Pilbeam, 1982; Martin, 1985), and most aspects of locomotor anatomy (Tuttle, 1969; Stern, 1975), including aspects of the hands, arms, and shoulders. As these are fairly broad character complexes, presumably of functional and adaptive significance, they seem to be robust as indicators of phylogeny.

Beyond this branch, however, anatomical relationships are still hazy. For the presently dominant view that the closest living relative to *Homo* is the *Pan-Gorilla* clade (cf. Sonntag, 1924; Morton, 1927), Ciochon (1983) provides the following derived characters: true incisive canal, large sphenopalatine fossae, fronto-ethmoidal sinus, reduction of trigonid in lower molars, early and complete fusion of the os centrale and scaphoid in the wrist, shortening, widening, and lateral projection of the iliac blades; aspects of the hip joint; and reduction of mid-digital hair. Additional characters of the skin include: high elastic fiber content; predominance of eccrine over apocrine sweat glands; and the axillary organ (Montagna, 1985).

Instead of functional suites of characters as seem to link chimpanzee and gorilla, the argument to humans has assumed the form of a trait list, not different in principle from the spurious phylogenetic arguments brought forth earlier this century. The criteria for acceptance of such a trait-list are difficult to ascertain. Indeed, Kluge (1982) and Schwartz (1984) produce lengthier lists in defense of alternative heterodox phylogenies, with Kluge linking *Pan-Gorilla* to *Pongo,* and Schwartz linking *Pan-Gorilla* to *Pongo-Homo.*

The linkage of humans to the African ape clade may be supportable on anatomical grounds, but is established predominantly on the basis of genetic relationships. Here, as has been shown repeatedly over the last quarter-century, human, chimpanzee, and gorilla are virtually indistinguishable at the genetic level (Goodman, 1963; Bruce and Ayala, 1979; Goodman and Cronin, 1982). Thus, given that the coding regions of the α-globin genes of human and chimpanzee differ by only three base substitutions (Liebhaber and Begley, 1983), and those of human and orangutan differ by 21 (Marks and others, 1986a), it follows that for the orangutan to be a sister group of either human or chimpanzee, orangutan's α-globin genes would have to be evolving at least seven times as rapidly as human's or chimpanzee's. This is an unlikely circumstance, given the predominance of horotely in molecular evolution.

Repeated attempts to establish sister groups within the human-chimpanzee-gorilla clade on a genetic basis have been equivocal and contradictory (cf. Hixson and Brown, 1986; Koop and others, 1986). This is attributable to two causes: (1) the great similarity among their genomes; and (2) the profound superficiality of our understanding of genomic evolution. Ferris and others (1981) analyzed restriction cleavage sites in the mitochondrial DNA of the three genera, but were unable to establish any two as closer relatives on the basis of these data. Templeton (1983) re-analyzed the data, and statistically linked chimpanzee and gorilla. Hasegawa and others (1984) used a different algorithm, and linked chimpanzee and human. Brown and others (1982) were unable to resolve the problem with comparison of nearly a kilobase of mitochondrial DNA sequence. Sibley and Ahlquist (1984) used DNA hybridization to suggest that the unique-sequence DNA of the chimpanzee is slightly more similar to the human than to the gorilla, and therefore that human and chimpanzee are sister groups. However, their data on the human-chimpanzee DNA hybridization do not concord with the earlier studies of Benveniste and Todaro (1976) or Deininger and Schmid (1976). Further, performing a similar study, O'Brien and others (1985) obtained a tree which clustered gorilla and human (though they refrained from drawing the facile phylogenetic conclusion). Finally Templeton (1985) roundly criticized the Sibley-Ahlquist study for their statistical treatment.

What little we know about genomic processes may be instructive here as well. The chimpanzee has lost the highly repetitive DNA satellite II from its genome, although human, gorilla, and orangutan retain it (Jones, 1976). Ueda and others (1985) interpret the presence of a specific pseudogene to suggest either a genetic linkage between human and gorilla, or excision of the pseudogene from the chimpanzee.

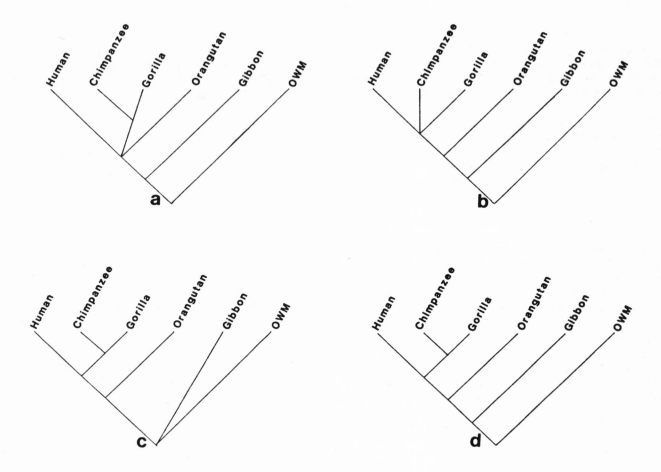

Figure 3. Relations of hominoid primates, with an out-group, the Old World monkeys (OWM). *a*, relations obtained from anatomy; *b*, relations obtained from genetics; *c*, relations obtained from chromosomes; and *d*, relations compatible with data from genome, chromosomes, and morphologies.

Judicious use of both anatomical and genetic data allow us to link chimpanzee and gorilla anatomically, but not genetically (and human to chimpanzee-gorilla genetically, but not anatomically).

The chromosomes tell a concordant story. While detection of inversions and translocations in primate chromosomes is not as reliable as might be hoped (due to difficulty in replicating observations), there are some telling phylogenetic characters. Specifically, we have the emergence of certain new cytological features linking the hominoids as follows: (1) multiplication of nucleolar organizer sites (NORs) in the ancestor of *Pongo, Pan, Gorilla,* and *Homo* (see Marks, 1983); (2) emergence of a chromosomal fraction which exhibits intense fluorescence under quinacrine staining in an ancestor of *Pan, Gorilla,* and *Homo* (see Pearson and others, 1971); and (3) emergence of terminal C-bands in the ancestor of *Pan* and *Gorilla* (see Stanyon and Chiarelli, 1982).

However, the chromosomes do not permit phylogenetic placement of the gibbons, insofar as their chromosomes are highly derived, and highly diverse within the genus as well (Bernstein and others, 1980; Marks, 1982).

Figure 3 summarizes phylogenetic data on the hominoids. In Figure 3*a*, anatomical data serve to group *Pan* with *Gorilla*, but are inconsistent in grouping this clade with either *Homo* or *Pongo*. Outside this ambiguity, the gibbons cluster with the great apes on the basis of molar tooth morphology, upper body complex, orthograde posture, and loss of tail. In Figure 3*b*, genetic data are ambiguous with respect to *Homo-Pan-Gorilla*, but clearly separate these from *Pongo*. Again the gibbons cluster with this clade, quite separate from the Old World monkeys. In Figure 3*c*, the karyotypic data group *Pan* and *Gorilla*, join this clade to *Homo*, join that clade to *Pongo*, but cannot cluster the gibbons relative to the out-group. Finally, in Figure 3*d*, consensus phylogeny is given, consistent with data obtained from all three evolutionary systems.

DISCUSSION

Interpretation of data drawn from the study of different evolutionary systems operating within the same taxa often involves a reliance upon the metaphor of a hierarchy (*e. g.*, Novikoff, 1945). This metaphor, however, has considerably greater utility above the organismal level than below it. For example, Vrba and Eldredge (1984) propose an increasing hierarchy of replicating entities as follows: "genomic constituents," organisms, demes, species, and monophyletic taxa. Eldredge and Salthe (1984) modify this by replacing "genomic constituents" with codons, then genes. In this hierarchy, the relation of each level to the one above it is one of composition: organisms compose demes, demes compose species, and so forth. Below the organism, however, the relation among levels becomes highly arbitrary. There is no real sense in which an organism can be considered to be composed of genes; the relation here is not part:whole, but cause:effect.

Further, an organism certainly is composed of cells, but these quintessential replicators are relegated to the hierarchy of interactors, rather than replicators. Chromosomes, which possess a structural integrity larger than molecules and smaller than cells, and which physically replicate, are accorded no place in either hierarchy (Eldredge and Salthe, 1984). It is at least semantically possible to construct a hierarchy of replicators below the level of the organism (which would be in greater accordance with biological realities), as follows: macromolecules, organelles, cells, and organisms. This still, however, obscures rather than clarifies the functional and formal relations among these levels; for example, only certain macromolecules are capable of reproduction. Further, even this hierarchy tells us nothing about either evolutionary process or phylogeny.

Additionally, the hierarchy metaphor has three drawbacks. First, the notion of hierarchy incorporates a conception of rank; some levels (usually anatomical) are "higher" than others (usually genetic), and are considered to result from "emergent properties" of the lower levels. While this is certainly valid in two senses, functional and physical, it does not seem to work in an evolutionary sense. It would be quite unfounded to assert that the uniquely divergent gibbon chromosomes are an emergent property of the minor genetic changes which have accompanied them, or that the minor anatomical changes are emergents of the vast chromosomal changes.

A second drawback with the notion of hierarchy is that it is often strongly tainted with ideology (Gould, 1980; Levins and Lewontin, 1985).

Third, the metaphor of a hierarchy obscures the dynamic component of evolution. Phylogeny is directional; it leads from past to present, but hierarchies are static, as are their levels.

The metaphor I suggested at the outset of this paper, the epicycle, has none of these drawbacks. It is ideologically neutral, as the Copernican controversy has not raised an eyebrow for centuries. It emphasizes independence of the evolutionary processes among the systems, as each cycle has its own orbit, and independence of rate, as each cycle revolves at a different speed from the others. Most significantly, however, it can convey the notion that there is a single main trajectory, a phylogenetic history, which each evolutionary system has followed — and further, that these systems are linked only in following this trajectory.

CONCLUSIONS

Evolutionary processes are now known to be vastly more complicated than any previous generation of biologists could have imagined. While we appreciate that the material basis of evolutionary change involves genetic alterations of some (or several) kinds, it is not at all clear how primarily non-adaptive genetic changes translate into primarily adaptive organismal changes.

I have attempted to set forth the primary modes of evolutionary change of genes, chromosomes, and organisms. These not only vary in their modes of change, but in their rates of change as well. Simply recognizing that these differences exist should make us cognizant that each has assets and liabilities when used as phylogenetic characters. Judicious use of data from all three systems enables a clarification of relationships which exist among extant hominoid primates.

We must not fall prey to the fallacy of granting hegemony to one evolutionary system over the rest. Study of evolution in the genetic system of a taxon, for example, has its advantages, but has its disadvantages as well. Discordant results have frequently arisen from simple interpretations of complex genetic data, and the temptation to re-write phylogenies and classifications is often difficult to resist (Hill and Buettner-Janusch, 1965). By first producing a conceptual model which emphasizes the differences among and difficulties within the data sets obtained from these evolutionary systems, we may ultimately better come to understand the overall system itself.

The epicycle theory in pre-Copernican astronomy was an unwieldy system which was superseded when the true relations among the planets were finally grasped. The relations among the evolutionary systems discussed here are not only not yet grasped, but are not likely to be grasped, if indeed such relations do exist, for some time to come. Until the next scientific revolution, the best we can do is to try to understand each of the systems in its own terms, and to use aspects of one system to clarify different areas of the larger evolutionary picture.

ACKNOWLEDGMENTS

I thank Nancy S. Jones for her illustrations; and Henry McHenry, G. Ledyard Stebbins, J. Bruce Walsh, David Dean, and Eric Delson for their helpful comments.

REFERENCES CITED

Arnason, U., 1972, The role of chromosomal rearrangement in mammalian speciation with a special reference to Cetacea and Pinnipedia: Hereditas, v. 70, p. 113-118.

Beadle, G. W., and Tatum, E. L., 1941, Genetic control of biochemical reactions in *Neurospora*: National Academy of Sciences, USA, Proceedings, v. 27, p. 499-506.

Benveniste, R. E., and Todaro, G. J., 1976, Evolution of type C viral genes: evidence for an Asian origin of man: Nature, v. 261, p. 101-107.

Bernstein, R., Pinto, M., Morcom, G., and Bielert, C., 1980, A reassessment of the karyotype of *Papio ursinus:* homoeology between human chromosome 15 and 22 and a characteristic submetacentric baboon chromosome: Cytogenetics and Cell Genetics, v. 28, p. 55-63.

Bonner, J. T., 1980, Evolution and development: New York, Springer-Verlag, 356 p.

Boveri, T., 1904, Ergebnisse uber die Konstitution der chromatischen Substanz des Zellkerns: Jena, Gustav Fischer, 446 p.

Britten, R. J., 1986, Rates of DNA sequence evolution differ between taxonomic groups: Science, v. 231, p. 1393-1398.

Britten, R. J., and Davidson, E. H., 1971, Repetitive and non-repetitive DNA sequences and a speculation on the origins of evolutionary novelty: Quarterly Review of Biology, v. 46, p. 111-138.

Brown, W. M., Prager, E. M., Wang, A., and Wilson, A. C., 1982, Mitochondrial DNA sequences of primates: tempo and mode of evolution: Journal of Molecular Evolution, v. 18, p. 225-239.

Bruce, E. J., and Ayala, F. J., 1979, Phylogenetic relationships between man and the apes: electrophoretic evidence: Evolution, v. 33, p. 1040-1056.

Chetverikov, S. S., 1926, On certain aspects of the evolutionary process from the standpoint of modern genetics: Zhurnal Eksperimental'noi Biologii, v. A2, p. 3-54 (reprinted, 1961, American Philosophical Society, Proceedings, v. 105, p. 167-195).

Chiarelli, B., 1966, Caryology and taxonomy of the catarrhine monkeys: American Journal of Physical Anthropology, v. 24, p. 155-170.

Ciochon, R. L., 1983, Hominoid cladistics and the ancestry of modern apes and humans, *in* Ciochon, R. L, and Corruccini, R. S., eds., New interpretations of ape and human ancestry: New York, Plenum, p. 783-843.

Deininger, P. L., and Schmid, C. W., 1976, Thermal stability of human DNA and chimpanzee DNA heteroduplexes: Science, v. 194, p. 846-848.

Dobzhansky, T., 1937, genetics and the origin of species: New York, Columbia University Press, 364 p.

_____ 1941, Genetics and the origin of species, 2nd ed.: New York, Columbia University Press, 446 p.

Doolittle, R. F., 1981, Similar amino acid sequences: chance or common ancestry?: Science, v. 214, p. 149-159.

Dover, G., 1982, Molecular drive: a cohesive model of species evolution: Nature, v. 299, p. 111-117.

Dutrillaux, B., 1977, New chromosome techniques, *in* Yunis, J., ed., Molecular structure of human chromosomes: New York, Academic, p. 255-263.

Eldredge, N., and Salthe, S. N., 1984, Hierarchy and evolution, *in* Dawkins, R., and Ridley, M., eds., Oxford surveys in evolutionary biology, v. 1: London, Oxford University Press, p. 184-208.

Elliot Smith, G., 1924, Essays on the evolution of man: London, Oxford University Press, 159 p.

Ferris, S. D., Wilson, A. C., and Brown, W. M., 1981, Evolutionary tree for apes and humans based on cleavage maps of mitochondrial DNA: National Academy of Sciences, USA, Proceedings, v. 78, p. 2432-2436.

Fisher, R., 1930, The genetical theory of natural selection: London, Oxford University Press, 291 p.

Frazzetta, T. H., 1970, From hopeful monsters to bolyerine snakes: American Naturalist, v. 104 p. 55-72.

Giles, E., 1956, Cranial allometry in the great apes: Human Biology, v. 28, p. 43-58.

Goldschmidt, R. B., 1937, Spontaneous chromatin rearrangements and the theory of the gene: National Academy of Sciences, USA, Proceedings, v. 23, p. 621-623.

_____ 1940, The material basis of evolution: New Haven, Yale University Press, 436 p.

_____ 1952, Evolution, as viewed by one geneticist: American Scientist, v. 40, p. 84-96, 135.

Goodman, M., 1963, Serological analysis of the systematics of recent hominoids: Human Biology, v. 35, p. 377-436.

Goodman, M., Braunitzer, G., Stangl, A., and Schrank, B., 1983, Evidence on human origins from haemoglobins of African apes: Nature, v. 303, p. 546-548.

Goodman, M., and Cronin, J. E., 1982, Molecular anthropology: its development and future directions, *in* Spencer, F., ed., A history of American physical anthropology, 1930-1980: New York, Academic, p. 105-146.

Goodman, M., Moore, G. W., and Matsuda, G., 1975, Darwinian evolution in the genealogy of haemoglobin: Nature, v. 253, p. 603-608.

Goodman, M., Weiss, M. L., and Czelusniak, J., 1982, Molecular evolution above the species level: branching pattern, rates, and mechanisms: Systematic Zoology, v. 31, p. 376-399.

Gould, S. J., 1977, Ontogeny and phylogeny: Cambridge, Belknap Press of Harvard University Press, ix + 501 p.

_____ 1980, Is a new and general theory of evolution emerging?: Paleobiology, v. 6, p. 119-130.

Gould, S. J., and Lewontin, R. C., 1979, The spandrels of San Marco and the Panglossian paradigm: a critique of the adaptationist programme: Royal Society of London, Proceedings, Series B, v. 205, p. 581-598.

Haldane, J. B. S., 1932, The causes of evolution: London, Longmans, Green, 235 p.

Hamilton, A., and Buettner-Janusch, J., 1977, Chromosomes of Lemuriformes. III. The genus *Lemur:* karyotypes of species, subspecies, and hybrids: New York Academy of Sciences, Annals, v. 293, p. 125-159.

Hasegawa, M., Yano, T., and Kishino, H., 1984, A new molecular clock of mitochondrial DNA and the evolution of the hominoids: Japan Academy, Proceedings, Series B, v. 60, p. 95-98.

Hill, R., and Buettner-Janusch, J., 1965, Molecules and monkeys: Science, v. 147, p. 836-842.

Hixson, J. E., and Brown, W. M., 1986, A comparison of the small ribosomal RNA genes from the mitochondrial DNA of the great apes and humans: sequence, structure, evolution, and phylogenetic implications: Molecular Biology and Evolution, v. 3, p. 1-18.

Ho, M. W., and Saunders, P. T., 1979, Beyond neo-Darwinism — an epigenetic approach: Journal of Theoretical Biology, v. 78, p. 573-591.

Jones, K. W., 1976, Comparative aspects of DNA in higher primates, in Goodman, M., Tashian, R., and Tashian, J., eds., Molecular anthropology: New York, Plenum, p. 357-368.

Jungers, W. L., and Susman, R. L., 1984, Body size and skeletal allometry in African apes, in Susman, R. L., ed., The pygmy chimpanzee: evolution, biology, and behavior: New York, Plenum, p. 131-177.

Keith, A., 1915, The antiquity of man: London, Williams and Norgate, 524 p.

King, M.-C., and Wilson, A. C., 1975, Evolution at two levels in humans and chimpanzees: Science, v. 188, p. 107-116.

Kluge, A., 1982, Cladistics and the classification of the great apes, in Ciochon, R. L., and Corruccini, R. S., eds., New interpretations of ape and human ancestry: New York, Plenum, p. 151-177.

Koop, B. F., Goodman, M., Xu, P., Chen, K., and Slightom, J. L., 1986, Primate eta-globin DNA sequence and man's place among the great apes: Nature, v. 319, p. 234-237.

Lauer, J., Shen, C.-K. J., and Maniatis, T., 1980, The chromosomal arrangement of human alpha-like globin genes: sequence homology and alpha-globin gene deletions: Cell, v. 20, p. 119-130.

Le Gros Clark, W. E., 1934, Early forerunners of man: London, Baillière, Tindall & Cox, 296 p.

Levins, R., and Lewontin, R., 1985, The dialectical biologist: Cambridge, Belknap, 303 p.

Lewontin, R. C., 1978, Adaptation: Scientific American, v. 239, p. 212-230 (November).

Liebhaber, S. A., and Begley, K. A., 1983, Structural and evolutionary analysis of the two chimpanzee alpha-globin mRNAs: Nucleic Acids Research, v. 11, p. 8915-8928.

Marks, J., 1982, Evolutionary tempo and phylogenetic inference based on primate karyotypes: Cytogenetics and Cell Genetics, v. 34, p. 261-264.

―――― 1983, Hominoid cytogenetics and evolution: Yearbook of Physical Anthropology, v. 26, p. 131-159.

―――― in press, Social and ecological aspects of primate cytogenetics, in Kinzey, W. G., ed., Primate models for the evolution of human behavior: Albany, State University of New York Press.

Marks, J., Shaw, J.-P., and Shen, C.-K. J., 1986a, The orangutan adult alpha-globin gene locus: duplicated structural genes and a new member of the primate alpha-globin gene family: National Academy of Sciences, USA, Proceedings, v. 83, p. 1413-1417.

―――― 1986b, Sequence organization and genomic complexity of the primate θ 1 globin gene, a novel αglobin-like gene: Nature, v. 322.

Martin, L., 1985, Significance of enamel thickness in hominoid evolution: ibid., v. 314, p. 260-263.

Mayr, E., 1942, Systematics and the origin of species: New York, Columbia University Press, 334 p.

―――― 1983, How to carry out the adaptationist program?: American Naturalist, v. 121, p. 324-334.

Miller, D. A., 1977, Evolution of primate chromosomes: Science, v. 198, p. 1116-1124.

Montagna, W., 1985, The evolution of human skin (?): Journal of Human Evolution, v. 14, p. 3-22.

Morton, D. J., 1927, Human origin: correlation of previous studies of primate feet and posture with other morphologic evidence: American Journal of Physical Anthropology, v. 10, p. 173-203.

Myers, R. H., and Shafer, D. A., 1978, Hybrid ape offspring of a mating of gibbon and siamang: Science, v. 205, p. 308-310.

Nagel, U., 1973, A comparison of anubis baboons, hamadryas baboons, and their hybrids at a species border in Ethiopia: Folia Primatologica, v. 19, p. 104-165.

Nei, M., and Koehn, R., 1982, Evolution of genes and proteins: Sunderland, Massachusetts, Sinauer, 331 p.

Nevers, P., and Saedler, H., 1977, Transposable genetic elements as agents of gene instability and chromosome rearrangements: Nature, v. 268, p. 109-115.

Novikoff, A., 1945, The concept of integrative levels in biology: Science, v. 101, p. 209-215.

O'Brien, S. J., Nash, W. G., Wildt, D. E., Bush, M. E., and Benveniste, R., 1985, A molecular solution to the riddle of the giant panda's phylogeny: Nature, v. 317, p. 140-144.

Pearson, P. L., Bobrow, M., Vosa, C. G., and Barlow, P. W., 1971, Quinacrine fluorescence in mammalian chromosomes: ibid., v. 231, p. 326-329.

Pilbeam, D., 1982, New hominoid skull from the Miocene of Pakistan: ibid., v. 295, p. 232-235.

Sarich, V. M., and Wilson, A. C., 1967, Immunological time scale for hominid evolution: Science, v. 158, p. 1200-1202.

Schmid, C. W., and Jelinek, W.R., 1982, The Alu family of dispersed repetitive sequences: ibid., v. 216, p. 1065-1070.

Schultz, A. H., 1930, The skeleton of the trunk and limbs of higher primates: Human Biology, v. 2, p. 303-438.

Schwartz, J., 1984, The evolutionary relationships of man and orangutans: Nature, v. 308, p. 501-505.

Selander, R. K., 1982, Phylogeny, in Milkman, R., ed., Perspectives in evolution: Sunderland, Massachusetts, Sinauer, p. 32-59.

Shafit-Zagardo, B., Maio, J. J., and Brown, F. L., 1982, Kpn families of long, interspersed repetitive DNAs in human and other primate genomes: Nucleic Acids Research, v. 10, p. 3175-3193.

Shea, B. T., 1981, Relative growth of the limbs and trunk in the African apes: American Journal of Physical Anthropology, v. 56, p. 179-202.

─── 1983, Allometry and heterochrony in the African apes: *ibid.,* v. 62, p. 275-289.

─── 1985, Ontogenetic allometry and scaling: a discussion based on the growth and form of the skull in African apes, *in* Jungers, W. J., ed., Size and scaling in primate biology: New York, Plenum, p. 175-205.

Sibley, C., and Ahlquist, J., 1984, The phylogeny of the hominoid primates, as indicated by DNA-DNA hybridization: Journal of Molecular Evolution, v. 20, p. 2-15.

Simpson, G. G., 1944, Tempo and mode in evolution: New York, Columbia University Press, 237 p.

─── 1945, The principles of classification and a classification of the mammals: American Museum of Natural History, Bulletin, v. 85, p. 1-350.

─── 1963, The meaning of taxonomic statements, *in* Washburn, S. L., ed., Classification and human evolution: Chicago, Aldine, p. 1-31.

─── 1964, Organisms and molecules in evolution: Science, v. 146, p. 1535-1538.

Sonntag, C. F., 1924, The morphology and evolution of the apes and man: London, John Bale, 364 p.

Stanyon, R., and Chiarelli, B., 1982, Phylogeny of the Hominoidea: the chromosome evidence: Journal of Human Evolution, v. 11, p. 493-504.

─── 1983, Mode and tempo in primate chromosome evolution: implications for hylobatid phylogeny: *ibid.,* v. 12, p. 305-315.

Stebbins, G. L., 1950, Variation and evolution in plants: New York, Columbia University Press, 643 p.

Stebbins, G. L., and Ayala, F. J., 1982, Is a new evolutionary synthesis necessary?: Science, v. 213, p. 967-971.

Stern, J. T., Jr., 1975, Before bipedality: Yearbook of Physical Anthropology, v. 19, p. 59-68.

Straus, W. L., Jr., 1949, The riddle of man's ancestry: Quarterly Review of Biology, v. 24, p. 200-223.

Sturtevant, A. H., 1913, The linear arrangement of six sex-linked factors in *Drosophila,* as shown by their mode of association: Journal of Experimental Zoology, v. 14, p. 43-59.

Sturtevant, A. H., and Dobzhansky, T., 1936, Inversions in the third chromosome of wild races of *Drosophila pseudobscura* and their use in the study of the history of the species: National Academy of Sciences, USA, Proceedings, v. 22, p. 448-450.

Sutton, W. S., 1902, The chromosomes in heredity: Biological Bulletin, v. 4, p. 231-248.

Templeton, A. R., 1982, Why read Goldschmidt?: Paleobiology, v. 8, p. 474-481.

─── 1983, Phylogenetic inference from restriction endonuclease cleavage site maps with particular reference to the evolution of humans and the apes: Evolution, v. 37, p. 221-244.

─── 1985, The phylogeny of the hominoid primates: a statistical analysis of the DNA-DNA hybridization data: Molecular Biology and Evolution, v. 2, p. 2-15.

Thompson, J., and Woodruff, R., 1978, Mutator genes — pacemakers of evolution: Nature, v. 274, p. 317-321.

Tuinen, P. van, and Ledbetter, D., 1983, Cytogenetic comparison and phylogeny of three species of Hylobatidae: American Journal of Physical Anthropology, v. 61, p. 453-466.

Tuttle, R. H., 1969, Knuckle-walking and the problem of human origins: Science, v. 166, p. 953-961.

Ueda, S., Takenaka, O., and Honjo, T., 1985, A truncated immunoglobulin epsilon pseudogene is found in gorilla and man but not in chimpanzee: National Academy of Sciences, USA, Proceedings, v. 82, p. 3712-3715.

Valentine, J., and Campbell, C., 1975, Genetic regulation and the fossil record: American Scientist, v. 63, p. 673-680.

Waddington, C. H., 1942, Canalization of development and the inheritance of acquired characters: Nature, v. 150, p. 563-565.

─── 1953, Genetic assimilation of an acquired character: Evolution, v. 7, p. 118-126.

─── 1957, The strategy of the genes: London, George Allen and Unwin, 262 p.

Weinert, H., 1932, Ursprung der Menscheit: Stuttgart, Ferdinand Enke Verlag, 606 p.

White, M. J. D., 1945, Animal cytology and evolution: London, Cambridge University Press, 375 p.

─── 1973, Animal cytology and evolution, 3rd ed.: London, Cambridge University Press, 961 p.

─── 1978, Chain processes in chromosomal speciation: Systematic Zoology, v. 27, p. 285-298.

Wilson, A. C., Carlson, S. S., and White, T. J., 1977, Biochemical evolution: Annual Review of Biochemistry, v. 46, p. 573-639.

Wilson, A. C., Sarich, V. M., and Maxson, L. R., 1974, The importance of gene rearrangement in evolution: evidence from studies on rates of chromosomal, protein, and anatomical evolution: National Academy of Sciences, USA, Proceedings, v. 71, p. 3028-3030.

Wilson, E. B., 1902, The cell in development and inheritance, 2nd ed.: New York, Macmillan, 483 p.

Wright, S., 1931, Evolution in mendelian populations: Genetics, v. 16, p. 97-159.

─── 1941, The material basis of evolution: Scientific Monthly, v. 53, p. 165-170.

─── 1968, Evolution and the genetics of populations. Volume 1: Genetic and biometric foundations: Chicago, University of Chicago Press, 469 p.

Zimmer, E. A., 1980, Evolution of primate globin genes [Ph.D. thesis]: Berkeley, California, University of California, 366 p.

MANUSCRIPT RECEIVED DECEMBER 18, 1985
REVISED MANUSCRIPT RECEIVED JUNE 2, 1986
MANUSCRIPT ACCEPTED JUNE 18, 1986

The evolutionary synthesis today: an essay on paleontology and molecular biology

EVERETT C. OLSON
CLIFFORD F. BRUNK

Department of Biology, University of California, Los Angeles, California 90024

ABSTRACT

From the introduction of the synthetic theory of evolution in the late 1930s and 1940s until the 1960s, the study of evolution was cast largely within its tenets. Starting in the 1960s changes, primarily introduced by paleontologists, began to have an impact. Although gradualism and phenotypic selection are still prominent, ideas of stasis, punctuation, and heirarchical selection have suggested that revisions of the synthetic theory may be necessary. Molecular biology and molecular evolution have developed in parallel with studies of phenotypic evolution during the last forty years. For the most part these two lines of investigation have remained discrete, although an increasing number of investigators is applying data and concepts from both fields. This convergence is particularly apparent in studies of rates of molecular clocks and the relationship between changes in the genome and changes in the phenotype. As a better understanding of the relationship between the genome and the phenotype emerges, an enlargement or possible abandonment of some aspects of the synthetic theory of evolution may prove necessary. Thus, the synthetic theory is presently being reshaped by data and concepts originating in areas as diverse as paleontology and molecular biology.

INTRODUCTION

"Haldane, . . . has called attention to the fact that one of the features of importance in human evolution is man's polymorphism. A striking illustration of this (though by no means mainly on the genetic level) is the coexistance of those two peculiar types known as paleontologists and geneticists. Fortunately there is a related feature even more important—plasticity. This not only allows for the change of any type, but the fusion of types, leaving newer and higher types formed by their synthesis, and this is a process which we are witnessing now. We may, of course, expect the synthetic types in due course to split again, along different lines of cleavage than before, as the evolutionary process continues."

H. J. Muller, 1949, p. 421

In the quotation from Muller (1949) lies a hint of the past, a statement of the synthesis in the 1940s, and a prediction for the future. Each was valid. The "Princeton Symposium" of January 1947, from which the quotation issued, saw a remarkable meeting of bodies and minds of two but recently disparate groups of evolutionists. Such a meeting was first conceived by a small group at Columbia University in 1942, inspired by Walter Bucher's suggestion of the need for a synthesis of genetics and paleontology. The impetus and theme of the meeting at Princeton were embodied in George Simpson's 1944 book titled "Tempo and Mode in Evolution." Started in the mid 1930s and interrupted by World War II, this work almost single-handedly accomplished a synthesis in these disparate fields and began an intimate and persistent relationship between paleontology and the general synthetic theory of evolution.

The seeds had been planted in the 1930s by the crossfertilization of new concepts of genetics, especially in the works of J. B. S. Haldane, S. Wright, and R. Fisher, and recognition of the significance of this work to their problems by a small handful of paleontologists among whom Simpson was the leading light. Demonstration of the compatibility of evolutionary information from paleontology and from genetics was the harbinger of the full-fledged entry of paleontology into the developing synthetic theory of evolution. There was, however, much more to "Tempo and Mode" than a mere demonstration of consistency between paleontology and genetics, important as this was. Most crucial was the elucidation of the relationships of rates and modes of change, and in particular the recognition that modes, based on biological data, and rates, inferred from paleontological findings, were inexorably interrelated and must be treated together in a common context. Biology and the temporal aspects of the fossil record became linked by demonstrated necessity.

The last decades have opened still newer trends of evolutionary thought, with revision or even rejection of the concepts of Neodarwinian evolution and with the burgeoning field of molecular biology. Will the increase in biological knowledge at the molecular level have a significant impact on evolutionary theory? In particular, will its interactions with current concepts of organismic evolution lead to significant advances in the understanding of evolution comparable to those signalled by "Tempo and Mode" and the development of synthetic theory?

Whatever answers are forthcoming must be cast in the contemporary understanding of the data and theoretical constructs of evolution, both of which are very different from what pertained in the 1940s. Data and theorizing must be divorced from vacuous, unconfirmed and unconfirmable speculations and based on hard facts. So it is appropriate to our purposes to outline briefly the major parallel developments in the two fields, organismic biology and molecular biology, that have brought us to the present.

ORGANISMIC EVOLUTION AFTER 1944

"Genetics, Paleontology and Evolution" edited by Jepson, Mayr, and Simpson (1949) comprises published results of the 1947 "Princeton Symposium." It reveals an essentially unanimous acceptance of the validity of interpretation of evolution in the gradualistic mode that traces its origin back to Darwin. Although there were many bricks in the foundation of the synthetic theory, the crucial time dimension was supplied mainly by "Tempo and Mode." As noted by LaPorte (1983), paleontologists at first seemed to pay little attention to this book, for it was reviewed primarily by neozoologists. By 1947 there had been a shift, but even then many paleontologists were not particularly interested in mechanisms of evolution beyond a general acceptance of the Neodarwinian framework, as emphasized by Romer (1949, p. 104).

> "Currently most paleontologists seriously interested in evolution subscribe in common with their neozoological fellows to a Neodarwinian doctrine. In part, however, they tend to worship at other shrines—those of Lamarckianism, orthogenesis and a variety of other mysterious, teleological cults . . . ."

By the time of the 1947 symposium at Princeton, an excitement had grown among a relatively small number of paleontologists. The stage was set for a very active expansion of the evolutionary branches of paleontology. Once initiated this development has continued, with some lulls, until the present.

General adherence to the basic tenets of the modern synthesis (Huxley, 1942) was characteristic of the late 1940s and 1950s. Under its banner studies on the rates and modes of evolution, on the applications of mathematics and statistics to paleontology, and on development of models of paleoecology all flourished. Field studies and collections of data from fossils were basic to these studies. Increased accuracy in the determination of clock-like dates in the geological calendar and a wide range of efforts to better interpret measurements of morphological change greatly aided the interpretation of rates of change.

The evolutionary synthesis was the dominant mode of evolutionary interpretations through this time, as is documented in "Evolution after Darwin, vol. I: The Evolution of Life" (Tax, 1960) which issued from the Darwinian Centennial at The University of Chicago in 1959. Granting some bias in those invited, one cannot but be impressed by the solid acceptance of the Neodarwinian framework of the participants as a general guide to evolutionary thought at the time. Only slightly questioning dissent was presented in Olson (1960), in which the adequacy of the synthesis, not its correctness, was questioned. Mayr and Provine (1980) in "The Evolutionary Synthesis" highlight the development and success of the synthetic point of view during this period, as does Gould (1980a) in his contribution to the volume. In Gould's paper the effects of new trends which developed during the 1960s and crystallized in the 1970s, also were discussed.

The late 1940s and 1950s were not a sterile period, but with some notable exceptions most paleontologists were more concerned with their proximate problems than with the more theoretical aspects of evolution. A search through paleontological journals led to the conclusion (Olson, 1971) that construction of phylogenetic trees was a primary goal of most paleontological research. The preoccupation with phylogenetic trees was grounded primarily in concerns for ancestral-descendant relationships, a position strongly rejected by some students later.

Stirrings of change were in the air even in the 1950s and early 1960s. They could be seen in the seemingly endless arguments on what species are and how they are formed, in the beginnings of numerical taxonomy (Sokal and Sneath, 1963), in the use of simple and multivariant statistics in paleoecological models (Simpson and Roe, 1939; Olson, 1951, 1952, 1957; Olson and Miller 1951a-b, 1958), and in the biogeographical analysis (Simpson, 1953). Things to come were initiated by reformulations of systematics (Hennig, 1950), by early ideas concerning the relationships of molecular biology to evolution, by new concerns with gaps in the fossil record, and by revival of concepts of continental drift. Through all of this period, and intensifying in subsequent decades, investigations of rates of change held center stage (e. g., Van Valen, 1973, 1974; Gingerich, 1983b). Of particular interest were ways of measuring and expressing rates of change and the relationships between modes and rates of evolution. Simpson's influence was pervasive in many of these areas, both among those who supported his concepts and those who denied or modified them.

THE PERIOD 1960-1985

During the 1960s, but difficult to pin-point by specific landmarks, a shift in concepts emerged in the thinking of some but by no means the majority of evolutionists. Selection, the natural selection of Darwin, remained the core of evolutionary thought, but a partitioning of selection into various hierarchical levels of development began to creep in. A hallmark of this change in the organismic field and in paleontology was the growing tendency to seek evolutionary information directly from the fossil record rather than use the record merely as an indicator of the historical pattern of change. Process and pattern began to be contrasted partners, adaptation was redefined, and its roles questioned. Dialectical thinking became a tool in evolutionary biology (see e. g., Olson, 1968, and Levin and Lewontin, 1980). Such trends became definitive and greatly elaborated during the last 25 years.

An indirect influence leading to a growing dichotomy of the "classic" and the "new" was the translation into English and publication of Hennig's "Phylogenetic Systematics" (Hennig, 1950) by Davis and Zangerl (Hennig, 1966). This book presented a method of phylogenetic and systematic investigation which was to a degree at odds with some of the basic tenets of Neodarwinian gradualism. It was only after the translation that an appreciable impact on New World evolutionists was felt. The methodologies of phylogenetic systematics were based upon the concept of origin of clades by bifurcation. This relegated phyletic speciation, which was important in some gradualistic hypotheses, to insignificance as a major evolutionary mechanism. Extensive studies of models for construction of phylogenetic trees, coalescing in what has come to be known as "cladistics," carried over into vicariant biogeography (Croizat, 1962; Nelson and Platnick, 1981). The concept of testable hypothesis was infused into phylogenetic analyses, and supported in particular by the philosophical scientific tenets of Popper.

A landmark in the trend to "trust the fossil record" was the publication of the model of punctuated equilibria (Eldredge and Gould, 1972), elaborating the concept developed by Eldredge (1971) and, before him, Simpson (1944). Initially, this approach involved a plea to view the fossil record objectively in the context of allopatric speciation rather than as a flawed exposition of "Darwinian gradualism." Abrupt appearance of species after long periods of little change, or stasis, lies at the heart of the concept. The fossil record does, in most instances, fail to provide evidence of species transitions as envisaged by Darwin. Those few recorded instances of species transitions may be considered examples of gradual phyletic change without major evolutionary significance.

The impact of the idea of punctuated equilibria, whether considered as a new concept forcing reconsideration of synthetic theory, at one extreme, or as scientifically trivial, at the other, has been immense in paleontological studies. Often with rather blind acceptance, punctuated equilibria has invaded neobiological interpretations of evolution as well. As a stimulant of research and rethinking, it cannot be dismissed as unimportant under any interpretation.

Reconsiderations of gradualism and gradation, tempo and mode, meanings of the gaps in the fossil record, revival of ideas of the "hopeful monster" of Goldschmidt (1940; e. g., Gould, 1972 and Telford, 1984), denials and reaffirmation of Darwin and Darwinism, and conformity with or replacement of the synthetic theory have formed a significant fraction of the evolutionary literature during the last decade and a half. The pages of such journals as Systematic Zoology, Evolution, Nature, Science, and most of all Paleobiology (which was spawned by the new look at paleontology) are carriers of the arguments. Several symposia, such as the Macroevolution Conference (Chicago 1979) revealed both the hopes and confusions symptomatic of intensive new studies of concepts that not long ago rested comfortably within the synthetic framework.

The jury is not in and resolution to the problems, which themselves are changing as time passes, is not in sight. Most likely the "furor" will subside into a calm integration of the good and the bad of it all with a generally accepted array of evolutionary concepts; and, as always, there will be some dissenters. The wedding, if such there be, will likely be followed by a bifurcation or fragmentation in Muller's sense.

We cannot predict where it will go, but a good sense of the current state of affairs in one broad area of evolutionary concerns can be had in the commentaries and replies by Rhodes (1983) and Gingerich (1983a) and in the article of Gould (1985) titled "The paradox of the first tier: an agenda for paleobiology." In the former, Rhodes and Gingerich agree that punctuated equilibria is not, in fact, a radical departure from Darwin and Darwinism. Rhodes, however, considers the concept of great importance as a catalyzer of evolutionary thought and research, whereas Gingerich considers it merely an interesting, but untestable hypothesis. Gould, within the framework of the currently popular hierarchical construct of evolution, expands upon the decoupling of microevolution and macroevolution and, in this sense, downgrades to insignificance Darwinian gradualism in macroevolutionary processes. Each of these views has had a long and tangled history, but in these statements the essences of the current status are enunciated.

The matter is far from settled, and now has become of wide concern among neobiologists as well as paleobiologists. Maynard Smith (1984) in a commentary on an evolutionary conference welcomed, if somewhat belatedly, paleontology into the halls of evolutionary biology in his paper "Paleontology at the High Table." The crucial question, he implies, revolves around whether the Darwinian concepts as reformulated in Neodarwinism and synthetic theory are supported or refuted by the concept of punctuated equilibria and all that has emerged from it. Attempts to answer this question have run the gamut from Stebbins and Ayala (1981) and Mayr (1982), who do not find a revision of synthesis necessary, to Gould (1985) and Stanley (1979, 1982), who feel the need for modification and, particularly, in decoupling of the hierarchies, negation of the basic tenets of the synthetic theory as pertinent to macroevolution, and the level at which significant evolutionary changes occur.

During this time of ferment in our understanding of organismic evolution, the main interest in biological research has shifted to areas of cell biology, microbiology, molecular biology, and biochemistry. It is perhaps significant that in his review of landmarks in the development of biology, Glass (1979) makes no mention of paleobiology. During the last two decades, efforts at a synthesis of molecular and organismic biology, while relatively small in volume, have been by no means insignificant. In spite of these efforts, the reductive studies of molecular biology have remained apart from the concepts of evolution that we have been discussing. This has left a gap that needs to be bridged to more deductive studies.

Attempts to bridge this gap have been made. Among them are integrations by Wilson and others (1977), Sarich (1971), Fitch and Margoliash (1967), Sibley and Ahlquist (1983), and by Runnegar (1982, 1986). As attempts at a synthesis of the broad areas of molecular and organismic evolution go on, the central question remains: are extensive modifications and perhaps an abandonment of Darwinian evolution necessary? If either occurs, would the change be as significant and controversial as that engendered by the punctuated views of evolution? Is a reexamination of the earlier synthesis of paleontology and genetics as developed by Simpson in 1944 required? Will, perhaps, a new synthesis follow? Before addressing these questions it is appropriate to undertake a brief review of the development and impact of molecular biology.

MOLECULAR BIOLOGY

A perspective arising from studies in the field of molecular biology has been sufficiently well developed in recent years that it must either be integrated into the current version of synthetic evolution theory or else a new evolutionary theory must be formulated. "Evolution at the Crossroads" (Depew and Weber, 1985) presents a spectrum of current thought on this matter in a primarily philosophical context. Gould (1980a, 1980b, 1985), Stebbins and Ayala (1981), Stebbins (1982), Mayr (1982), and Vrba and Eldredge (1984) are some, among many, who have given thought to the impact of this new perspective. For the most part, theory and philosophy dominate; relatively few experimental or field investigations have crossed the "iron curtain" between reductive and integrative perspectives. Few investigators have been in a position to bridge this barrier, which is crucial to a full grasp of the disparate information. The situation is in many ways similar to that which preceeded the integration of genetics and paleontology by Simpson (1944) in his "Tempo and Mode in Evolution."

CURRENT CONCEPTS IN MOLECULAR BIOLOGY

Current knowledge suggests that for eucaryotes: (1) the genome (DNA sequences) undergoes rapid and continous alteration, which we might characterize as the "restless genome"; and (2) only a tiny fraction of the genome codes for proteins. Thus, it is possible that a substantial portion of the genome has little or no effect on the phenotype of the organism. The function of the majority of the eucaryotic genome remains unclear, or at best the subject for educated speculation. These insights make previous ideas of the genome as a passive carrier of evolutionary messages inappropriate.

The eucaryotic genome is composed of a large ensemble of DNA sequences in constant flux. The changes in the DNA sequences and the arrangement of these sequences are governed by processes of replication, recombination, and repair operating at the molecular level. Many of these changes, however, have no detectable effect on the organismic phenotype, and thus do not appear to be subject to natural selection at the organismic level. Selection at other levels in the developmental hierarchy does occur, but this is of a different nature and currently the subject of much debate.

The shift in perspective regarding the genome emerges from a comparison of the following statements:

". . . (natural selection) is the composer of the genetic message, and DNA, RNA, enzymes, and other molecules in the system are successively its messengers."
Simpson, 1964, p. 1538.

"We can not agree with Simpson that DNA is a passive carrier of the evolutionary message. Evolutionary change is not imposed upon DNA from without; it arises from within. Natural selection is the editor, rather than the composer, of the genetic message. One thing the editor does not do is remove changes which it is unable to perceive."
King and Jukes, 1969, p. 788.

This latter perspective has had a profound effect on our view of evolution theory. Before probing the incorporation of these concepts and data into the existing evolutionary synthesis, a short review of the history of molecular biology is necessary to set the stage.

THE DEVELOPMENT OF MOLECULAR BIOLOGY

Molecular biology, or more specifically molecular genetics, has come into being since the early 1940s, beginning at about the same time as publication of "Tempo and Mode in Evolution" (Simpson, 1944). The first experimental evidence that DNA was the physical substance responsible for transmission of inherited characteristics was published by Avery and others (1944). General acceptance of these data came only with the experimental work of Hershey and Chase (1952).

The linkage of genes, or more precisely mutations in genes, to specific enzymes in a biochemical pathway was articulated in the famous "one gene one enzyme (or polypeptide chain)" of Beadle and Tatum (1941). This work changed the gene from an abstract concept to a solid biochemical entity. Identification of DNA as the genetic material, and protein as the phenotypic expression of genes, also shifted the view of genes from a nebulous concept to a tangible sequence of nucleotides in DNA.

Accelerated development of concepts and details of the molecular basis of life has characterized the past three and one half decades. A model for the structure of DNA was proposed by Watson and Crick (1953). A mechanism for the replication of the DNA molecule and for the transfer of genetic information to other macromolecules was implicit in the structure. Details of DNA replication were experimentally confirmed by Meselson and Stahl (1958), and the general flow of genetic information from DNA to RNA to protein (affectionately dubbed the "Central Dogma") was proposed (Crick, 1970). Details of the "Central Dogma" were quickly supported by extensive experimental evidence, particularly the role of transfer RNA molecules as the essential element in the translation of the information contained in the nucleotide sequence of messenger RNA into the amino acid sequence of protein. With the elements of the "Central Dogma" experimentally confirmed, the basis of genetic transmission and expression at the molecular level became well understood.

Deciphering of the genetic code, its triplet nature, and the assignment of the various codons (nucleotide triplets) to specific amino acids was again accomplished in record time. By the mid-1960s, through the separate efforts of Nirenberg and others (1966) and Khorana and others (1966), the complete genetic code became known. Further investigation indicated that the genetic code was virtually universal, identical in all organisms examined, which suggested that the pattern of redundance in the codons was fixed at an early stage in evolution. As so often happens to generalizations, some interesting exceptions to the universality of the genetic code recently have been demonstrated in the mitochondria (Anderson and others, 1981; Heckman and others, 1980; Kochel and others, 1981), ciliates (Horowitz and Gorovsky, 1985; Preer and others, 1985), and some bacteria (Yamao and others, 1985). These findings are prompting reevaluation of our ideas about stability of the genetic code and the hypothesis that the code was fixed very early in evolution.

The basic principles of control for gene expression were worked out in procaryotic cells, including the feedback processes governing expression of inducible enzymes. The complicated orchestration of gene expression occurring during bacteriophage infection also is known in remarkable detail. The even more intricate developmental processes in eucaryotic cells have not been established to the same degree; however, intense effort in this area promises a clearer understanding in the near future.

A great expansion of understanding concerning protein structure and function also occurred during this period. Examinations of a wide variety of proteins have yielded a detailed and consistent picture of the interactions determining protein structure and functions. The primary sequence of amino acids, dictated by information in the DNA, is completely responsible for the structure of protein. Thus the structure and function of proteins and, in turn, virtually all biochemical processes within the cell are traced back to the nucleotide sequence in DNA.

By the late 1960s, evolutionary implications of the flow of genetic information were clear. Stochastic changes in DNA sequences, by molecular processes, are the raw material of genetic variation. Expression of this genetic variation impinges to varying degrees on the phenotype of the organism, and affects the survival of lineages of organisms primarily through natural selection at the organismic level. Thus, mutations in DNA are only loosely coupled to the phenotype of the organism. There is no feedback from the phenotype to DNA, save a decrease in numbers of members in the lineage. In short, individual mutations can not be "line item vetoed" by natural selection; the whole organism must be passed or rejected. The processes of selection are recognized as acting at all developmental levels, and this recognition has lead to concepts of hierarchical subdivisions of selection effects (e.g., Van Valen, 1983). No matter how selection is defined, the molecular mechanisms clearly indicate that many of the mutations in the DNA are essentially neutral relative to the effect of natural selection at the organismic levels.

Two new technological advances in the last decade are providing an even more rapid accumulation of information about the molecular basis of living systems and the molecular basis of evolution. It is hard to overestimate the impact that recombinant DNA technology and techniques for determining DNA sequences have had on molecular biology.

Recombinant DNA technology makes possible the isolation of virtually any DNA sequence, and it also makes possible the production of large quantities of that sequence with relative ease. Purification of a specific DNA sequence in an experimentally significant quantity was virtually impossible prior to this technology. Along with recombinant DNA technology, the ability to determine the order of nucleotides within the DNA molecules has revolutionized molecular biology and its role in evolutionary biology. It is now feasible to read the genetic information directly from the DNA molecules. This not only has profound potential for elucidating molecular biological processes, it has equally profound implications for reading the evolutionary history of organisms as written in the DNA.

MOLECULAR EVOLUTION

As the synthetic theory was being questioned by organismic biologists, especially paleontologists, during the late 1960s and 1970s, molecular biology was more or less going its own way, with only occasional crossings of the two pathways. The development of molecular biology initially had little effect on evolutionary theory. This was due in part to the great speed with which basic principles were established, and in part to the technical nature of the supporting data. By the mid 1970s, however, the general concepts of molecular biology were firmly established, and supporting data were in most cases simple and direct. At this point the concepts of molecular biology began to impinge more directly upon evolutionary theory.

The bulk of this interaction can be seen in the emergence of two major concepts of molecular evolution: (1) the "neutral mutation" theory; and (2) the hypothesis of a "molecular clock." The "neutral mutation" theory states that a significant portion of the mutations in a DNA sequence produce such a slight change in the phenotype of the organism that they become fixed by chance, independent of natural selection at the organismic level. These mutations are selectively neutral. The concept of a "molecular clock" is simply that changes in the DNA sequences accumulate as a function of time. This implies that the differences in DNA sequence between two lineages of organisms are proportional to the time that has elapsed since they shared a common ancestor. The extent of neutral mutations and the dynamics of "molecular clocks" are open to question, but the eventual resolution of these questions will have great effect upon evolutionary theory.

NEUTRALITY

The final step in the flow of genetic information, the translation of a sequence of codons in the messenger RNA into a sequence of amino acids in the protein, is degenerate. The degeneracy arises because in most cases several different codons, referred to as synonymous codons, can specify a single amino acid. Changes in the DNA sequence that produce synonymous codons do not change the amino acid sequence of the protein encoded. Kimura (1968) proposed a neutral theory of mutation, based on this mechanism. The theory asserts that the great majority of evolutionary changes at the molecular level result, not from Darwinian selection, but by nonselective fixation of selectively neutral or nearly neutral mutants. This proposal was quickly followed by a paper by King and Jukes (1969) with the provocative title of "Non-Darwinian Evolution." This paper also suggested that a significant portion of the genetic variation in DNA sequences arises by nonselective processes, rather than reflecting selection at the organismic level. The fact that the molecular processes make nonselective fixation a probability is undeniable. The proportion of the genetic complement which is determined in a nonselective manner still remains open, and is the subject of intense investigation.

The effect of substituting one nucleotide for another in the DNA sequence is hard to assess. The function of the nucleotide, if it has one, is critical. If the nucleotide is part of a sequence coding for a protein or a region controlling gene expression, then it may affect the phenotype of the organism. Even among these nucleotides, some changes will be dramatic while others will have little if any consequence. The effects on the phenotype of changes in regions of the DNA that do not code for protein or control the expression of proteins, are usually negligible. The idea that every nucleotide, or even most nucleotides, in the DNA are the result of positive selection is patently untenable.

The types of changes in the DNA sequence considered to this point are single nucleotide substitutions. In recent years, the dynamic nature of DNA sequences on a much larger scale has become apparent. Substantial regions of DNA are capable of duplication and translocation to new sites in the genome. Translocation radically changes the context of the translocated sequence, often disrupting the expression of the sequences at the site of translocation. McClintock (1952) described the genetic consequences of this type of movement several decades ago, but it is only recently that the physical characterization of transposons has made it apparent how widespread these sequences are.

Along these lines, Doolittle and Sapienza (1980) and Orgel and Crick (1980) proposed the concept of "selfish DNA." Like the publication of "Nondarwinian Evolution," the title as well as the idea stirred a good deal of controversy. The principle is straightforward. They proposed that sequences of DNA within the genome, capable of promoting their own replication and accumulating heritable mutations, perpetuate themselves independent of their contribution to the phenotype of the organism. The main feature of this proposal is that if molecular processes allow proliferation of this type of sequence, then no positive contribution to the phenotype of the organism is required to rationalize existence of such sequences. Certain DNA sequences, particularly transposons, do seem to meet these criteria. Whole regions of the genome are duplicated and translocated within the genome by molecular processes that do not necessarily confer a positive advantage on the organism.

Errors in DNA replication that produce tandem duplication of DNA regions or the transposition of sequences within the genome, particularly eucaryotic genomes, result in a large number of highly related DNA sequences often referred to as gene families. The members of a gene family may be sequences that code for specific proteins like the histone or globin genes. They may be a family of transposons, or they may be a large group of highly repeated sequences such as the Alu family in the human genome, which has about 300,000 members. Occasionally, individual members of a gene family (which code for a protein) become altered, and no longer produce a functional protein. These non-functional DNA sequences are referred to as pseudogenes. Such a pseudogene ceases to directly influence the phenotype of the organism.

In general, DNA sequences of members of a gene family would be expected to diverge from one another rapidly. In many cases, however, the sequences remain virtually identical, a process which is referred to as concerted evolution. A combination of gene conversion, unequal crossing over, gene duplication, gene deletion, and transposition has been suggested as the molecular mechanism responsible for concerted evolution of families of repeated sequences. In such cases, the degree of homogeneity (among members of a repeated sequence family within a species) is much higher than otherwise would be expected. At the same time, the homologous gene family when found in another species shows the expected divergence between the species. Dover (1982 a,b) dubbed this process "molecular drive," and claimed that it acts largely independent of natural selection operative at the organismic level. He also suggested that this process may be responsible for biological incompatibility leading to speciation.

In summary, the picture that emerges from detailed molecular studies of the genome is one of a highly dynamic and rapidly changing entity. Sequences are duplicated and rearranged in addition to the substitution of nucleotides within sequences. A large fraction of this activity is only loosely coupled to the phenotype of the organism, because only a small fraction of the genome is involved in coding for proteins or controlling the expression of genes that do code for proteins. The vast majority of the genome is uncoupled from the phenotype, and therefore has little effect on selection operating at the organismic level. This concept is a far cry from the simple idea that earlier seemed to promise a direct, interpretable relationship between genomic divergences and phenotypic evolution by the feedback mechanisms of natural selection.

MOLECULAR CLOCKS

The second major concept of molecular evolution is that of a "molecular clock." An initial study of the evolutionary aspects of protein sequences by Zuckerkandl and Pauling (1965) demonstrated that the amino acid sequences of proteins apparently evolve at a constant rate. They referred to this steady amino acid substitution process as a "molecular evolutionary clock."

A large body of evidence is consistent with the concept that the accumulation of substitutions in homologous proteins in different lineages is proportional to the time of separation of the lineages. Times of separation, in some instances, may be provided by paleontological evidence. Among early students of this phenomenon, Fitch and Margoliash (1967) constructed extensive phylogenies based upon amino acid sequences utilizing cytochrome c from many species. Methods of phylogenetic reconstruction that maximize parsimony were employed. Work on albumins in primates by Sarich and Wilson (1967) began a revolution in interpretations of the times of divergence of simians and hominids. These were followed by many additional studies, among them those of Doolittle and others (1971) and Goodman (1976).

The concept that amino acid substitution in proteins occurs in a roughly clock-like fashion allows an estimate of the relative evolutionary distance among any extant organisms by mere comparison of the amino acid sequences of any one of a number of homologous proteins. This relative evolutionary distance can then be converted to an estimate of the time since these organisms shared a common ancestor by relating the number of amino acid substitutions observed to the substitution rate for the specific protein in question. This, in turn, is determined from the observed substitution rate measured between species for which independent paleontological determination of the divergence time is available. The concept of a steady, clock-like rate of substitution has been challenged and reevaluated repeatedly (*e. g.*, Fitch, 1976; Langley and Fitch, 1974; Wilson and others, 1977).

The regularity and relative rate of accumulation of amino acid substitutions appears to vary from protein to protein and lineage to lineage. Lee and others (1985) demonstrate a wide variation in accumulation of amino acid substitutions in different lineages for superoxide dismutase. Wu and Li (1985) show similar differences in accumulation of nucleotide substitutions in several mammalian lineages. Accumulation of amino acid substitution in the histone IV genes of yeast and *Tetrahymena* clearly is faster than accumulation rate in higher organisms. If the rate of amino acid substitution determined for histone IV, derived from higher organisms, is applied directly to *Tetrahymena,* it suggests that the ciliates last shared a common ancestor with higher organisms over 6 billion years ago. This clearly is unreasonable, and indicates different rates in different lineages.

The rate of evolution is some proteins and in some lineages is far from constant, and this clearly poses problems in the application of the general concept of a molecular clock. Furthermore, a strong correlation exists between the function of the protein and its rate of evolution. The supposition is that a highly conserved protein, like histone IV, is so functionally essential that few if any amino acid substitutions can be tolerated. Conversely, the fibrinopeptides, which essentially are spacers that are excised from fibrin during blood clotting, can accommodate extensive substitution.

Kimura and Ohta (1973) and Kimura (1982) have noted a similar variation in the rate of amino acid substitution within individual proteins. In these cases, some of the amino acids in a protein remain essentially invariant, ostensibly because they play a critical role in the function of the protein. Other amino acids in the same protein have a higher rate of substitution because they have a less essential role in the protein's function.

The ability to sequence DNA allows nucleotide substitutions to be used as "molecular clocks" as well as amino acid substitutions. Nucleotide substitutions may be an even more effective molecule clock than amino acid substitution in many cases. The nucleotide sequence is less coupled to the phenotype, and more directly reflective of the molecular processes repsonsible for generating genetic variation. Kreitman (1983), for example, found a great deal of nucleotide polymorphism in the alcohol dehydrogenase locus of *Drosophila*. In general, nucleotides that altered the protein sequence were more highly conserved while those outside of the coding region showed more polymorphism. However, some DNA sequences outside of the regions coding for proteins also show remarkable conservation.

DNA-DNA hybridization allows a general comparison of the sequence homology between two genomes. Sibley and Ahlquist (1983) have produced a detailed avian phylogeny based upon this process. This work is of particular interest because it deals with the majority of the genome as a basis for comparison, not merely regions involved in coding for proteins. These studies have provided new insights into avian phylogenetic relations. Such studies, involving large portions of the genome, indicate that regions that do not code for proteins still display an evolutionary record.

CURRENT STATUS OF EVOLUTIONARY CONCEPTS

The genome emerges as a "restless" system. The pathways of organismic evolution appear to be somewhat orderly. However, the route from the genome to the organismic phenotype appears to be an unpaved trail through epigenesis and ontogeny. The term "crossroads" used by Depew and Weber (1985) seems appropriate, but the junction is of many roads. Most organismic biologists are secure in the broad framework of the Synthetic Theory, a few are restless and are asking questions, others feel a need to scrap the whole thing, or at least make drastic alterations. The following quotes collected by Bakker (1985) are from evolutionists bred in the paleontological framework, who are trying to understand evolution and to reconcile the old with the new:

> "The message from the fossils is that most species don't evolve as fast as they should."
>
> R. Bakker

> "We need an expanded version of Darwinian theory to explain how natural selection operates, not only on organisms but on a hierarchy of units under selection—genes, organisms, local populations and species."
>
> S. J. Gould

> "Most biologists are still transfixed on Natural Selection acting on mechanisms of evolutionary change. But natural selection may well shrink in importance and we could find that mutations are anything but random."
>
> E. Vrba

Are these prescriptions for the future? Do the molecular evolutionists swim in a similar sea of doubt? Do the few who cross between the parallel courses of study see an emerging basis of stability? Of course, these questions are all too simplistic, but perhaps in this vein we can see where we now stand.

There is a long way to go if the impacts of molecular evolution, as briefly outlined here, are to be understood and integrated within current evolutionary thought, not in a framework of idle speculation and pious hopes, but in one based on concrete data. As the molecular data are integrated with morphological data, it is important to reemphasize the essential difference in the character of information from the two sources. DNA sequences of the genome ultimately are responsible for virtually the entire phenotype of the organism. Morphological characters, used to establish phylogenetic relationships, are first of all difficult to decipher and describe; moreover, they are almost always the results of a large number of genes, most of which have pleiotropic effects. Herein lies the gulf that we must cross to reach a formulation of a complete evolutionary theory. This can be accomplished not on the pathways of hopeful statements, but on the firm ground linking the vast plexus of data ranging from the molecular to beyond the species level.

At present, only the most rudimentary understanding of developmental processes exists, and current knowledge is insufficient for a full description of the transmission of genomic information through the network of ontogeny. Grand holistic and epigenetic schemes (e. g., Koestler and Smythies, 1969; Becht, 1974; Riedl, 1977) point up needs and possible directions, but lack molecular developmental data. Wilson and others (1977) have summarized and integrated the molecular data in a most coherent fashion. Many of the more recent studies seem to make the bridge between inductive and deductive aspects of evolutionary thought more difficult, rather than easier.

Current knowledge of molecular mechanisms and understanding of developmental pathways are consistent with rapid and dramatic changes in the phenoytpe. Differential effects on ontogenetic processes can result in radical shifts in evolutionary direction. Ontogenetic processes may act as a filter on molecular evolution, passing only specific collections of mutations through to the phenotype. Such filtering could result in very slow, bradytelic, evolution at the organismic level in spite of a rapid "molecular clock." Evolutionary stasis at the morphological level is supported strongly by paleontological evidence. Alternatively, ontogenetic pathways could possibly pass a whole series of interacting mutations into the phenotype at once. In extreme cases, this might result in the equivalent of Goldschmidt's "hopeful monster." Thus, the rapid formation of species suggested in punctuated equilibria might be accommodated. Punctuation, while an attractive way of viewing the rapid origins of species and "gaps" in the fossil record, is not actually deducible from the fossil record itself.

It can hardly be questioned that Neodarwinian concepts and the general synthesis provide valid explanations of many observed evolutionary phenomena. The question is, how significant is this essentially microevolutionary mode in the total agenda of evolution? This issue can be resolved into two questions: (1) is natural selection operative at levels other than that of the phenotype of the organisms (i. e., are there other definable hierarchies of selective action)?; and (2) is there significant autonomy in evolutionary processes (such as "molecular drive," random fixation of neutral mutations, or random drift at the population level) to constitute an independent mode of evolution?

Relative to the first question, Gould (1985) indicates that he believes selection mechanisms operating at populations, species, and higher taxonomic levels are dominant in macroevolution. He also asserts that these selective mechanisms are uncoupled from molecular processes responsible for microevolution. At an even higher level, catastrophic events, which are not predictable from microevolutionary processes, may be dominant. Clearly, at various hierarchical levels, different types of selective processes do change the course of evolutionary development. It is possible to treat such selective processes at different levels as discrete, but does such treatment provide a more effective paradigm for the total dynamics of evolution?

In the second instance, it is becoming clear that autonomy is a more pervasive phenomenon than is conceived in the usual formulation of the synthetic theory. Autonomy at various levels, from molecular through populations and beyond, will continue to be an important question as time passes. Autonomy at the molecular level relative to that at the organismic level is particularly difficult to reconcile at present. In the case of simple organisms, where a close relationship between gene and character can be established, the difficulties are less acute (*e. g.*, Ambrose and Horovitz, 1984). In the great majority of cases, however, the complexity of genetic relationships is so great that current understanding is inadequate to sort out details of the interactions (*e. g.*, Levin and Lewontin, 1980).

We now stand at a new crossroad; the dominant scheme of the Synthetic Theory, while of great correlative value, will require expansion and modification. It is still an open question whether our developing knowledge can be accommodated by modifications that do not disrupt some of the basic tenets of the Synthetic Theory. Although our situations are more complex because of data from a wider range of disciplines, in some ways our position resembles that of evolutionists in the 1930s. Today our crossroad is not merely the intersection of two or three paths, but many. Simpson's "Tempo and Mode in Evolution" resolved many of the problems current in the 1930s with solutions that appear straight-forward today. However, to the evolutionist of the 1930s, those problems seemed as complex as ours are today. Our present situation is well described by a non-molecular biologist and non-paleontologist Abraham Lincoln: "The dogmas of our quiet past are inadequate to the stormy present."

At this juncture, an incorporation of data and mechanisms into the modern synthesis seems more an appropriate course than an attempt to forge a totally new synthesis. Simpson's "Tempo and Mode in Evolution" and the works of his contemporaries present an apt paradigm for combining the best of the old with the best of the new. The development of a completely satisfying synthesis may require both the elucidation of the interactions between molecular and developmental processes and the emergence of a new Darwin, Simpson, Huxley, or biological Einstein who can grasp the whole and merge it into an ultimate evolutionary synthesis.

ACKNOWLEDGMENTS

The general content of this paper was stimulated by talks and seminars conducted by scientists from a wide array of disciplines at Wednesday Evening Evolution Group (WEEG) meetings at UCLA. Dr. J. William Schopf who organized and conducts the WEEG discussions and Dr. John Campbell of the Department of Anatomy at UCLA have read the paper critically and have made suggestions which have been incorporated into the text. To these two and to our many colleagues in WEEG, we express our sincere appreciation.

REFERENCES CITED

Ambrose, V., and Horvitz, H. R., 1984, Heterochronic mutants of the nematode, *Caenorhahtitis elegans*: Science, v. 226, p. 409-416.

Anderson, S. H., Bankier, A. T., Barrell, B. G., de Bruijn, M. H. L., Coulson, A. R., Drouin, J., Eperon, I. C., Nierlich, D. P., Roe, B. A., Sanger, F., Schreier, P. H., Smith, A. J. H., Staden, R., and Young, I. G., 1981, Sequence and organization of the mammalian genome: Nature, v. 290, p. 457-465.

Avery, O. T., McLeod, C. M., and McCarty, M., 1944, Studies on the chemical nature of the substance inducing transformation of pneumococcal types: Journal of Experimental Medicine, v. 79, p. 137-158.

Bakker, R. T., 1985, Evolution by revolution: Science 85, v. 6, no. 9, p. 72-80.

Beadle, G. W., and Tatum, E. L., 1941, Genetic control of biochemical reactions in *Neurospora*: National Academy of Science, USA, Proceedings, v. 27, p. 499-506.

Becht, G., 1974, Systems theory, the key to holism and reductionism: BioScience, v. 24, p. 569-597.

Crick, F. H. C., 1970, Central dogma of molecular biology: Nature, v. 227, p. 561-562.

Croizat, L., 1962, Space, time and form: the biological synthesis: Caracas, publication by author, 881 p.

Depew, O. J., and Weber, B. H., 1985, Evolution at the crossroads: Cambridge, Massachusetts, M.I.T. Press, 320 p.

Doolittle, R. F., Wooding, G. L., Lin, Y., and Riley, M., 1971, Hominoid evolution as judged by fibrinopeptide structure: Journal of Molecular Evolution, v. 1, p. 74-83.

Doolittle, W. F., and Sapienza, C., 1980, Selfish genes, the phenotype paradigm and genome evolution: Nature, v. 284, p. 601-603.

Dover, G., 1982a, A molecular drive through evolution: BioScience, v. 32, p. 526-533.

―――― 1982b, Molecular drive: a cohesive mode of species evolution: Nature, v. 299, p. 111-117.

Eldredge, N., 1971, The allopatric model and phylogeny in Paleozoic invertebrates: Evolution, v. 25, p. 156-167.

Eldredge, N., and Gould, S. J., 1972, Punctuated equilibria: an alternative to phyletic gradualism, *in* Schopf, T. J. M., ed., Models in paleobiology: San Francisco, Freeman, Cooper and Company, p. 82-115.

Fitch, W. M., 1976, Molecular evolutionary clocks, *in* Ayala, F., ed., Molecular evolution: Sunderland, Massachusetts, Sinauer, p. 160-178.

Fitch, W. M., and Margoliash, E., 1967, Construction of phylogenetic trees: Science, v. 155, p. 279-284.

Gingerich, P., 1983a, Darwin's gradualism and empiricism: Nature, v. 309, p. 116.

―――― 1983b, Rates of evolution: effects of time and temporal scaling: Science, v. 222, p. 159-161.

Glass, B., 1979, Milestones and rates of growth in the development of biology: Quarterly Review of Biology, v. 54, p. 31-53.

Goldschmidt, R., 1940, The material basis of evolution: New Haven, Yale University Press, 431 p.

Goodman, M., 1976, Protein sequences in phylogeny, *in* Ayala, F., ed., Molecular evolution: Sunderland, Massachusetts, Sinauer, p. 141-159.

Gould, S. J., 1972, The return of the hopeful monster: Natural History, v. 86, p. 22-30.

―――― 1980a, Is a new and general theory of evolution emerging?: Paleobiology, v. 6, p. 114-130.

―――― 1980b, G. G. Simpson, paleontology and the modern synthesis, *in* Mayr, E., and Provine, W. B., eds., The evolutionary synthesis: Cambridge, Massachusetts, Harvard University Press, p. 153-172.

―――― 1985, The paradox of the first tier: an agenda for paleobiology: Paleobiology, v. 11, p. 2-12.

Heckman, J. E., Sarnoff, J., Alzner-DeWeerd, B., Yin, S., and RajBhandary, U. L., 1980, Novel features in the genetic code and codons reading patterns in *Neurospora crassa* mitochondria based on sequences of six mitochondrial tRNAs: National Academy of Science, USA, Proceedings, v. 77, p. 3159-3163.

Hennig, W., 1950, Grundzuge einer Theorie der phylogenetichen Systematik: Berlin, Deutscher Zentralverlag, 370 p.

―――― 1966, Phylogentic systematics, Davis, D., and Zangerl, R., translators: Urbana, University of Illinois Press, 263 p.

Hershey, A. D., and Chase, M., 1952, Independent functions of viral protein and nucleic acid in growth of bacteriophage: Journal of General Physiology, v. 26, p. 36-56.

Horowitz, S., and Gorovsky, M. A., 1985, An unusual genetic code in nuclear genes of *Tetrahymena*: National Academy of Science, USA, Proceedings, v. 82, p. 2452-2455.

Huxley, J., 1942, Evolution, the modern synthesis: Oxford, Clarendon Press, 645 p.

Jepsen, G. L., Mayr, E., and Simpson, G. G., 1949, Genetics, paleontology and evolution: Princeton, New Jersey, Princeton University Press, 474 p.

Khorana, H. G., Bucchi, H., Ghosh, H., Gupta, N., Jacob, T. M., Kossel, H., Morgan, R., Narang, S. A., Ohtsuka, E., and Wells, R. D., 1966, Polynucleotide synthesis and the genetic code: Cold Spring Harbor Symposium on Quantitative Biology, v. 31, p. 39-49.

Kimura, M., 1968, Evolutionary rate at the molecular level: Nature, v. 217, p. 624-626.

―――― 1982, The neutral theory as a basis for understanding the mechanism of evolution and variation at the molecular level, *in* Kimura, M., ed., Molecular evolution, protein polymorphism and the neutral theory: Tokyo, Japan Scientific Press, p. 3-65.

Kimura, M., and Ohta, T., 1973, Mutation and evolution at the molecular level: Genetics, v. 73 (Suppl.), p. 19-35.

King, J. L., and Jukes, T. H., 1969, Non-Darwinian evolution: random fixation of selectively neutral mutations: Science, v. 164, p. 788-798.

Kochel, H. G., Lazarus, C. M., Basak, N., and Kuntzel, H., 1981, Mitochondrial tRNA gene clusters in *Aspergillus nidulans*: organization and nucleotide sequence: Cell, v. 23, p. 625-633.

Koestler, A., and Smythies, T. R., 1969, Beyond reductionism: Boston, Beacon Press, 438 p.

Kreitman, M., 1983, Nucleotide polymorphism at the alcohol dehydrogenase locus of *Drosophila melanogaster:* Nature, v. 304, p. 412-417.

Langley, C. H., and Fitch, W. M., 1974, An examination of the constancy of the rate of molecular evolution: Journal of Molecular Evolution, v. 3, p. 161-177.

Laporte, L. F., 1983, Simpson's Tempo and Mode in Evolution revisited: American Philosophical Society, Proceedings, v. 127, p. 365-417.

Lee, M. Y., Friedman, D. J., and Ayala, F. J., 1985, Superoxide dismutase: an evolutionary puzzle: National Academy of Science, USA, Proceedings, v. 82, p. 824-828.

Levin, R., and Lewontin, R., 1980, The dialectical biologist: Cambridge, Massachusetts, Harvard University Press, 303 p.

McClintock, B., 1952, Chromosome organization and gene expression: Cold Spring Harbor Symposium on Quantitative Biology, v. 16, p. 13-47.

Mayr, E., 1982, Speciation and macroevolution: Evolution, v. 36, p. 1119-1132.

Mayr, E., and Provine, W. B., 1980, The evolutionary synthesis: perspectives in the unification of biology: Cambridge, Massachusetts, Harvard University Press, 487 p.

Meselson, M., and Stahl, F., 1958, The replication of DNA in *E. coli*: National Academy of Science, USA, Proceedings, v. 44, p. 671-682.

Muller, H. J., 1949, Reintegration of the symposium on genetics, paleontology and evolution, *in* Jepsen, G. L., Mayr, E., and Simpson, G. G., eds., Genetics, paleontology and evolution: Princeton, New Jersey, Princeton University Press, p. 421-445.

Nelson, G., and Platnick, N., 1981, Systematics and biogeography: cladistics and vicariance: New York, Columbia University Press, 576 p.

Nirenberg, M., Caskey, C. T., Marshall, R., Brimacombe, R., Kelley, D., Doctor, B., Hatfield, D., Levin, J., Rottman, F., Pestka, S., Wilcox, F., and Anderson, W. F., 1966, The RNA code and protein synthesis: Cold Spring Harbor Symposium on Quantitative Biology, v. 31, p. 11-24.

Olson, E. C., 1951, *Diplocaulus,* a study in growth and variation: Fieldiana: Geology, v. 11, p. 57-154.

―――― 1952, The evolution of a Permian vertebrate chronofauna: Evolution, v. 6, p. 181-196.

―――― 1957, Size-frequency distribution in samples of extinct organisms: Journal of Geology, v. 65, p. 309-333.

―――― 1960, Morphology, paleontology and evolution, *in* Tax, S., ed., Evolution after Darwin, v. I, The evolution of life: Chicago, University of Chicago Press, p. 523-546.

―――― 1968, Dialectics in evolution studies: Evolution, v. 22, p. 426-436.

―――― 1971, Vertebrate paleontology: New York, Wiley and Sons, Interscience, 839 p.

Olson, E. C., and Miller, R. L., 1951a, Relative growth in paleontolgical studies: Journal of Paleontology, v. 25, p. 213-223.

―――― 1951b, A mathematical model applied to study of the evolution of species: Evolution, v. 5, p. 325-338.

―――― 1958, Morphological integration: Chicago, University of Chicago Press, 317 p.

Orgel, L. E., and Crick, F. H. C., 1980, Selfish DNA: the ultimate parasite: Nature, v. 284, p. 604-607.

Preer, J. R., Preer, L. B., Rudman, B. M., and Barnett, A. J., 1985, Deviation from the universal code shown by the gene for surface protein 51A in *Paramecium*: *ibid.*, v. 314, p. 188-190.

Rhodes, F. H. T., 1983, Gradualism, punctuated equilibrium and the origin of species: *ibid.*, v. 305, p. 209-272.

Riedl, R., 1977, A system-analysis approach to macroevolutionary phenomena: Quarterly Review of Biology, v. 52, p. 351-370.

Romer, A. S., 1949, Time series and trends in animal evolution, *in* Jepsen, G. L., Mayr, E., and Simpson, G. G., eds., Genetics, paleontology and evolution: Princeton, New Jersey, Princeton University Press, p. 103-120.

Runnegar, B., 1982, A molecular-clock date for the origin of the animal phyla: Lethaia, v. 15, p. 199-205.

―――― 1986, Molecular paleontology: Paleontology, v. 29, p. 1-24.

Sarich, V. M., 1971, A molecular approach to the question of human origins, *in* Dolhinow, P. L., and Sarich, V. M., eds., Background for man: readings in physical anthropology: Boston, Little Brown Company, p. 60-81.

Sarich, V. M., and Wilson, A. C., 1967, Immunological time scale for hominoid evolution: Science, v. 158, p. 1200-1203.

Sibley, C. G., and Ahlquist, J. E., 1983, Phylogeny and classification of birds based on data of DNA-DNA hybridization, *in* Johnston, R. F., ed., Current ornithology, v. 1, p. 245-292.

Simpson, G. G., 1944, Tempo and mode in evolution: New York, Columbia University Press, 237 p.

―――― 1953, Evolution and geography: Condon Lecture, Oregon State System of Higher Education, 64 p.

―――― 1964, Organisms and molecules in evolution: Science, v. 146, p. 1535-1538.

Simpson, G. G., and Roe, A., 1939, Quantitative zoology: New York, McGraw-Hill, 406 p.

Smith, M. J., 1984, Paleontology at the High Table: Nature, v. 309, p. 401-402.

Sokal, R., and Sneath, P. H. A., 1963, Principles of numerical taxonomy: San Francisco, W. H. Freeman and Company, 359 p.

Stanley, S. M., 1979, Macroevolution: pattern and process: San Francisco, W. H. Freeman and Company, 332 p.

―――― 1982, Macroevolution and the fossil record: Evolution, v. 36, p. 460-473.

Stebbins, G. L., 1982, Perspectives in evolutionary theory: *ibid.*, v. 36, p. 1109-1118.

Stebbins, G. L., and Ayala, F., 1981, Is a new evolutionary synthesis necessary?: Science, v. 213, p. 967-971.

Tax, S., 1960, Evolution after Darwin, v. I, The evolution of life: Chicago, University of Chicago Press, 629 p.

Telford, S. R., 1984, Goldschmidt reaffirmed: a positive step toward a new synthesis: Evolutionary Theory, v. 7, p. 52.

Van Valen, L., 1973, A new evolutionary law: *ibid.*, v. 1, p. 1-30.

―――― 1974, Molecular evolution as predicted by natural selection: Journal of Molecular Evolution, v. 3, p. 89-101.

―――― 1983, Molecular selection: Evolutionary Theory, v. 6, p. 297-298.

Vrba, E. S., and Eldredge, N., 1984, Individuals, hierarchies and processes: toward a more complete evolutionary theory: Paleobiology, v. 10, p. 146-171.

Watson, J. D., and Crick, F. H. C., 1953, Molecular structure of nucleic acids: Nature, v. 171, p. 737-738.

Wilson, A. C., Carlson, S. S., and White, T. J., 1977, Biochemical evolution: Annual Reviews of Biochemistry, v. 46, p. 573-639.

Wu, C.-I., and Li, W.-H., 1985, Evidence for higher rates of nucleotide substitution in rodents than in man: National Academy of Science, USA, Proceedings, v. 82, p. 1741-1745.

Yamao, F., Muto, A., Kawauchi, Y., Iwami, M., Iwagami, S., Azumi, Y., and Osawa, S., 1985, UGA is read as tryptophan in *Mycoplasma capricolum*: National Academy of Science, USA, Proceedings, v. 82, p. 2306-2309.

Zuckerkandl, E., and Pauling, L., 1965, Evolutionary divergence and convergence in proteins, *in* Bryson, V., and Vogel, H. J., eds., Evolving genes and protein: New York, Academic Press, p. 97-166.

MANUSCRIPT RECEIVED JANUARY 3, 1986
MANUSCRIPT ACCEPTED JUNE 16, 1986